Enhanced Mechanical Properties of Metals by Surface Treatments

Enhanced Mechanical Properties of Metals by Surface Treatments

Guest Editors

Pinghu Chen
Yun Zhang
Ruiqing Li

MDPI

Basel • Beijing • Wuhan • Barcelona • Belgrade • Novi Sad • Cluj • Manchester

Guest Editors

Pinghu Chen
College of Mechanical Engineering
University of South China
Hengyang
China

Yun Zhang
School of Mechanical Engineering
Hunan University of Science and Technology
Xiangtan
China

Ruiqing Li
Light Alloy Research Institute
Central South China
Changsha
China

Editorial Office
MDPI AG
Grosspeteranlage 5
4052 Basel, Switzerland

This is a reprint of the Special Issue, published open access by the journal *Coatings* (ISSN 2079-6412), freely accessible at: https://www.mdpi.com/journal/coatings/special_issues/Metal_Surface.

For citation purposes, cite each article independently as indicated on the article page online and as indicated below:

Lastname, A.A.; Lastname, B.B. Article Title. *Journal Name* **Year**, *Volume Number*, Page Range.

ISBN 978-3-7258-3497-6 (Hbk)
ISBN 978-3-7258-3498-3 (PDF)
https://doi.org/10.3390/books978-3-7258-3498-3

Contents

coatings

Review

Overview of Surface Modification Techniques for Titanium Alloys in Modern Material Science: A Comprehensive Analysis

Kang Gao [1], Yun Zhang [1], Junhao Yi [1], Fang Dong [2] and Pinghu Chen [3,*]

[1] School of Mechanical Engineering, Hunan University of Science and Technology, Xiangtan 411201, China; 18073993487@163.com (K.G.); yun_zhang66@163.com (Y.Z.); 18207338390@163.com (J.Y.)
[2] State Key Laboratory of Precision Manufacturing for Extreme Service Performance, Light Alloy Research Institute, Central South University, Changsha 410083, China; dongfang@csu.edu.cn
[3] College of Mechanical Engineering, Key Laboratory of Hunan Province of Equipment Safety Service Technology under Extreme Environment, University of South China, Hengyang 421001, China
* Correspondence: chenpinghu1986@163.com; Tel.: +86-137-8616-1244

Abstract: Titanium alloys are acclaimed for their remarkable biocompatibility, high specific strength, excellent corrosion resistance, and stable performance in high and low temperatures. These characteristics render them invaluable in a multitude of sectors, including biomedicine, shipbuilding, aerospace, and daily life. According to the different phases, the alloys can be broadly categorized into α-titanium and β-titanium, and these alloys demonstrate unique properties shaped by their respective phases. The hexagonal close-packed structure of α-titanium alloys is notably associated with superior high-temperature creep resistance but limited plasticity. Conversely, the body-centered cubic structure of β-titanium alloys contributes to enhanced slip and greater plasticity. To optimize these alloys for specific industrial applications, alloy strengthening is often necessary to meet diverse environmental and operational demands. The impact of various processing techniques on the microstructure and metal characteristics of titanium alloys is reviewed and discussed in this research. This article systematically analyzes the effects of machining, shot peening, and surface heat treatment methods, including surface quenching, carburizing, and nitriding, on the structure and characteristics of titanium alloys. This research is arranged and categorized into three categories based on the methods of processing and treatment: general heat treatment, thermochemical treatment, and machining. The results of a large number of studies show that surface treatment can significantly improve the hardness and friction mechanical properties of titanium alloys. At present, a single treatment method is often insufficient. Therefore, composite treatment methods combining multiple treatment techniques are expected to be more widely used in the future. The authors provide an overview of titanium alloy modification methods in recent years with the aim of assisting and promoting further research in the very important and promising direction of multi-technology composite treatment.

Keywords: titanium and titanium alloys; surface treatment; classification of titanium alloys

Citation: Gao, K.; Zhang, Y.; Yi, J.; Dong, F.; Chen, P. Overview of Surface Modification Techniques for Titanium Alloys in Modern Material Science: A Comprehensive Analysis. *Coatings* **2024**, *14*, 148. https://doi.org/10.3390/coatings14010148

Academic Editors: Hector R. Siller and Emmanuel Segura-Cárdenas

Received: 19 December 2023
Revised: 15 January 2024
Accepted: 19 January 2024
Published: 21 January 2024

1. Introduction

A titanium alloy is recognized as a high-strength lightweight structural material with excellent biocompatibility, high specific strength, robust corrosion resistance, and good performance under high–low-temperature conditions. Titanium alloys are widely used in various fields, for instance, biomedicine, shipbuilding, aerospace, and in daily life scenarios [1–5]. Generally, based on their microstructure, Ti alloys are typically categorized into α, α+compound, near-α, α+β, metastable β, and β alloys [6–10]. A near-α titanium alloy, Ti6242S, is notably employed in the hot zones of jet engines, effectively operating up to 550 °C for its remarkable high-temperature mechanical strength [11]. As a typical dual-phase alloy, Ti6Al4V is widely used in the aerospace, automotive manufacturing, and biomedical fields for its excellent comprehensive performance [12]. Additionally, the β titanium alloy Ti-13V-13Cr-3Al (wt. %) is extensively utilized in the SR-71 Blackbird aircraft because of its high specific strength and good high-temperature stability [6].

However, considering the need for high wear resistance, high quality, and long service life of titanium alloys in the industrial field, traditional manufacturing processes are increasingly inadequate to fulfill these stringent requirements. Accordingly, surface strengthening technology is proposed to improve the mechanical properties of titanium alloys and minimize their surface roughness in order to achieve excellent service characteristics. Huang et al. [13] demonstrated that the elongation of the TA15 alloy was increased to 46.7% after 1073 K heat treatment. The observed fracture morphology showed that the size and depth of dimples increased with heat treatment time, improving the toughness of TA15. Yan et al. [14] found that the fatigue strength limit increased by 37.6% after shot peening, which was attributed to the formation of ~260 μm thick gradient-hardened layers on the surface. Li et al. [15] performed ion implantation on titanium alloy surfaces and found that nitrogen ion implantation can notably improve the surface hardness of the titanium alloy.

Surface properties exhibit high sensitivity to treatment, and distinct surface treatment techniques yield titanium alloys with varying microstructural and mechanical properties, which, in turn, influence their appropriateness for particular industrial uses. Numerous advanced surface treatment technologies have been used to modify the surface of titanium alloys in tandem with the dramatic rise in demand for titanium. The majority of current research on titanium alloy surface treatment involves either a single treatment or a composite treatment including two treatment techniques. Current research summaries on different novel therapeutic approaches are rare. There is a relative lack of summarized research on various new treatment methods. This work covers the research development on alloy composition and surface treatment from the viewpoints of the preparation method, microstructure, mechanical characteristics, and strengthening mechanism of titanium alloys in an effort to improve the findings of previous studies in this area.

2. Titanium and Titanium Alloys

2.1. Basic Properties of Titanium

Titanium, ranking as the 10th most abundant element in the Earth's crust, predominantly exists in the form of titanium-rich ores: ilmenite and rutile. Renowned for its lightweight nature, formidable strength, ease of processing, and exceptional specific strength, titanium and its alloys have become staples in various industrial applications. A comparison of the density and specific strength of titanium with other metals is shown in Table 1.

Table 1. Comparison of density and specific strength of titanium and other metals.

Cooking Oil	Titanium (Alloy)	Aluminum (Alloy)	Magnesium (Alloy)	High-Strength Steel
Density/ (g/cm^3)	4.5	2.7	1.74	7.8
Specific strength	29	21	16	23

2.2. The Main Structure of Titanium Alloy and Its Properties

As an allotropic element, crystallographically titanium exists in two structures: hexagonal close-packed structure (hcp) α-phase at room temperature; and body-centered cubic (bcc) structure β-phase, prevalent under high-temperature conditions. The transition temperature between these two phases is approximately 882 °C. The crystal structure of titanium alloy is shown in Figure 1 [16]. In addition to elevated temperatures to force the phase transition, the addition of different alloys to change the alloy phase transition temperature and component content is the main method of obtaining different organizations of titanium alloys.

The alloying of titanium alloys mainly affects the properties of the titanium matrix by changing the lattice structure of titanium atoms, adjusting the lattice constants and electronic structure, etc. These changes give rise to a variety of alloy types and microstructures. Alloying serves as the principal means to improve the organization and enhance the mechanical properties of titanium alloys, enabling them to maintain the stability of the original crystal structure. With to different alloys, the stability preferences can be divided into three categories of titanium alloying elements: (1) α-stabilizers, mainly replacement elements Al and interstitial elements O, N, C, etc. (2) β-stabilizers, containing eutectic elements Mo, V, Nb, Ta, and the eutectic elements Fe, W, Cr, Si, Co, Mn, H, etc. (3) Neutral elements, notably Zr and Sn. After alloying, depending on the conditions of room temperature, the microstructure is broadly divided into: α-phase alloys, two-phase α+β alloys, and β-phase alloys.

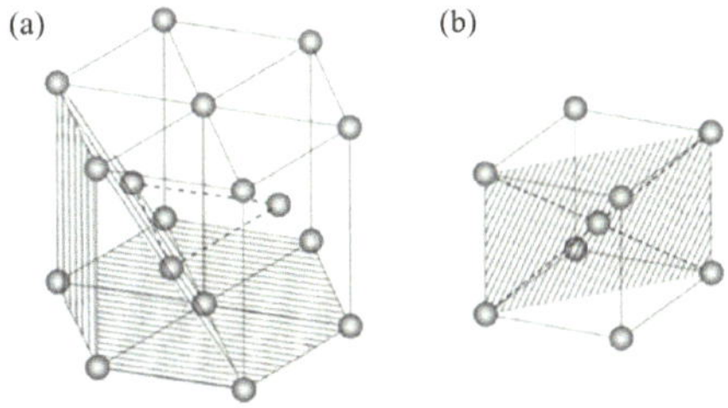

Figure 1. Crystal structure of titanium alloy: (**a**) α-phase crystal structure, (**b**) β-phase crystal structure [16].

According to the stabilizer content, α-type titanium alloys can be divided into α alloys with a single α-phase and near-α alloys with 1%–2% β-stabilizers [7]. The microstructure characteristics of α-type titanium alloys are shown in Figure 2. The low-symmetric hexagonal close-packed row (HCP) crystal structure makes α-type titanium alloys show obvious plastic anisotropy at both room and working temperatures. Additionally, the close-packed row tends to be transformed into quasi-disintegration surfaces when subjected to tension, which is prone to premature fracture initiation [17]. This tendency significantly reduces the fatigue resistance of the alloy and adversely affects its properties.

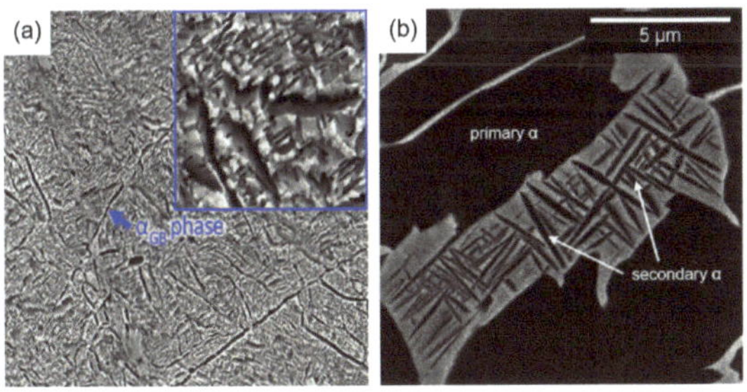

Figure 2. Microstructure of α-type titanium alloy: (**a**) α-phase boundary [18]; (**b**) primary α-phase and secondary α-phase [10].

Unlike at room temperature, α-type titanium alloys excel in high and low temperatures. For example, the α alloy Ti-5Al-2.5-Sn ELI shows good strength and toughness under 20 K low-temperature conditions due to its low interstitial element content [19]. So, it is an ideal choice for applications like low-temperature containers and impellers in rocket engine turbine oil pumps. In addition, the densely arranged hexagonal crystal structure gives it good thermal stability and high-temperature creep properties, which makes it have good material properties under high-temperature conditions. Taking the near-α alloy Ti-6Al-2Sn-4Zr-2Mo (Ti-6242) as an example, its mechanical properties are better than those of the two-phase alloy Ti-6Al-4V (TC4) under high-temperature conditions, positioning it as an optimal material for aerospace field [9].

β-type titanium alloys generally have the advantages of high strength, low modulus of elasticity and good toughness [16,20]. Their body-centered cubic structure gives them more slip systems, thereby granting β-type titanium alloys a more pronounced plastic deformation capacity compared to α-type titanium alloys. The lower densification and higher self-diffusivity under high-temperature conditions result in them being relatively poorly thermally stable, which limits their suitability for high-temperature applications. According to the different concentrations of β-phase stabilizing elements, β-type titanium alloys can be further subdivided into nearly β alloys, sub-stable β alloys, and stable β alloys.

Near-β alloys and substable β alloys are representative high-strength β alloys with a relatively low content of β-phase stabilizing elements and good mechanical properties, which are widely used in large load-bearing components in aerospace [21]. Among them, near-β alloys, which contain the least amount of β-phase stabilizing elements, boast exceptional qualities such as high specific strength and corrosion resistance. This makes them highly suitable for critical structural parts in aerospace and other industries, such as high-strength new experimental alloys Ti-7Mo-3Nb-3Cr-3Al, Ti-4Al-7Mo-3Cr-3Vl, etc. [6]. The microstructure of near-β alloys is shown in Figure 3 [20]. Near-β alloys have strict requirements for thermo-mechanical processing, and improper treatment processes are prone to form abnormally coarse grains, affecting the quality of the structural components [22].

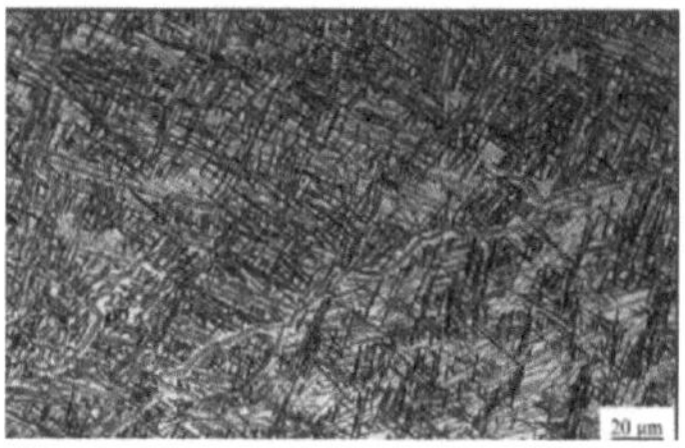

Figure 3. Microstructure of near-β alloy Ti-17 (Ti-5Al-2Sn-2Zr-4Mo-4Cr) [20].

Substable β alloys with a relatively high content of β-phase stabilizing elements are widely used in the aerospace industry and biomedical fields because of their excellent overall properties compared to those of near-β alloys. For example, the substable β alloy Ti-35Nb-7Zr-6Ta-2F-0.5Si alloy developed by Kopova et al. [23] has good biocompatibility, and its higher strength and lower elastic modulus than those of Ti-6Al-4V make it ideal for biomedical applications. The high-moldability and high-strength properties make substable β alloys such asTi-6Cr-5Mo-5V-4Al (Ti-6554) show good potential for application in making large parts such as landing gears of airplanes [24]. However, they are more expensive than other alloys; so, their development in some fields is constrained. The microstructure transformation of substable β alloys is shown in Figure 4 [25].

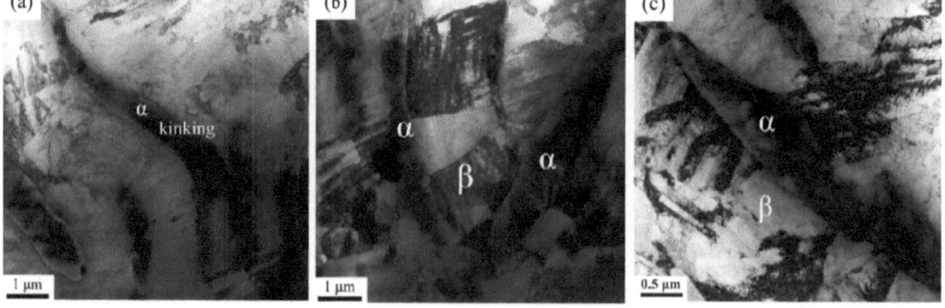

Figure 4. Microstructural transformation of substable β alloys at 785 °C. (**a**) Phase kinking; (**b**) interlaced lamellar α-phase restricting the growth of β subcrystals; (**c**) phase resistance to dislocation motion [25].

Stabilized β alloys represent the apex of β-type titanium alloys, enriched with the highest concentration of β-phase stabilizing elements. This unique composition endows them with distinct performance characteristics, diverging significantly from other β-type titanium alloys. Xin et al. [26] found that various heat treatment techniques applied to stabilized β-type titanium alloys can profoundly influence their properties. Notably, the presence of grain boundaries has been identified as a key factor in enhancing the high-temperature creep resistance of these alloys. This enhancement renders stabilized β-type titanium alloys particularly suitable for high-temperature applications. The current typical heat-strengthened β titanium alloys are Ti-35V-15Cr-0.3Si-0.1C, Ti-40, etc. The organization image of stabilized β titanium alloy is shown in Figure 5 [27].

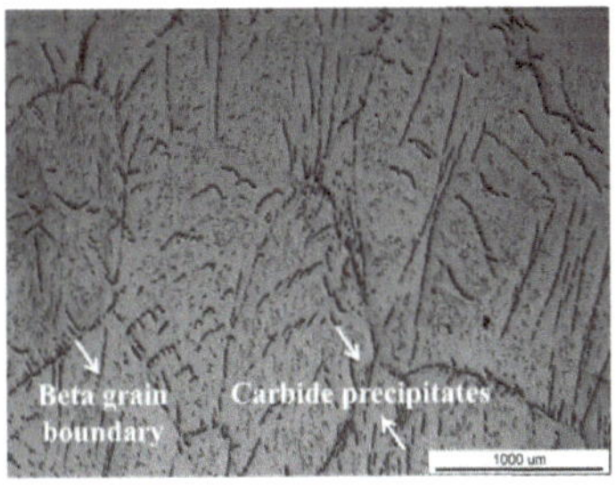

Figure 5. Microstructures of as-received Ti3515 alloy [27].

In addition to the two phases mentioned above, the transition phase occurring during the transformation from α-phase to β-phase is generally referred to as the α+β-phase. An alloy exhibiting this transitional phase is referred to as a α+β-type titanium alloy. As the most commonly used type of alloy in the industrial field, two-phase alloys have excellent comprehensive performance and are widely used in aerospace, medical, marine and other fields. For example, Ti-6Al-4V is employed in manufacturing engine compressor blades [28]. The excellent performance of two-phase alloys mainly depends on the specificity of their internal organization, such as α-phase, β-phase two-phase organization, so that it combines the advantages of both α-type titanium alloys and β-type titanium alloys and exhibits higher strength. A typical two-phase organization is shown in Figure 6 [29].

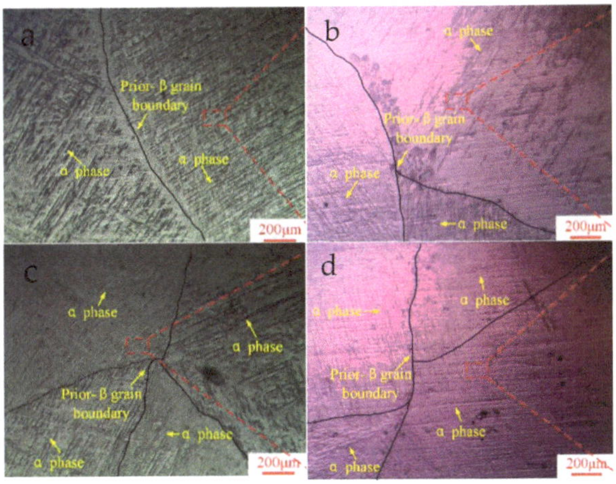

Figure 6. Typical two-phase organization chart [29]. (**a–d**): Non-IPM, IPM200, IPM300, IPM400 metallurgical microstructure of the top.

Diverse heat treatment methodologies can induce significant alterations in the microstructure of titanium alloys, consequently affecting their material properties. Alessandro et al. [10] obtained the following conclusions by comparing the changes in the organization and mechanical properties of Ti-6Al-2Sn-4Zr-6Mo alloys after heat treatments at different temperatures: At 600 °C, the α' phase undergoes an a + b transformation, resulting in a marked increase in material strength to 1383 MPa. At 875 °C, the appearance of the bilayer organization improves the ductility of the material, but the extensively roughened microstructure results in a significant decrease in the hardness (339 HV). At 950 °C, the strength (1012 MPa) and ductility (17%) decrease slightly, accompanied by a minor increase in hardness (352 HV). At the same time, the complete recrystallization of the microstructure of the isotropy was improved. The microscopic images of the material at different temperatures are depicted in Figure 7 [10] and Figure 8 [30].

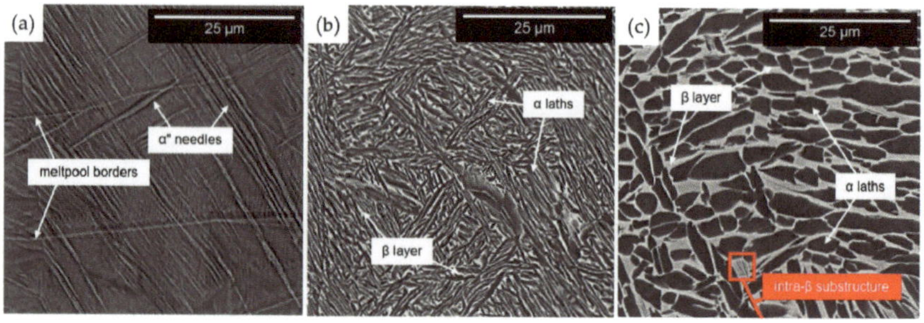

Figure 7. Microstructure of duplex alloys after treatment at different temperatures (**a**) 600 °C; (**b**) 875 °C; (**c**) 950 °C [10].

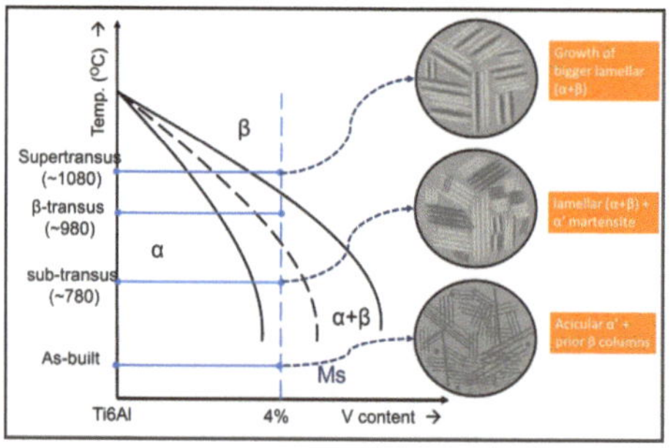

Figure 8. Summary of microstructure growth with heat treatment temperature [30].

2.3. *Classification of Titanium Alloys*

According to the different uses of titanium alloys, they can be categorized into corrosion-resistant alloys, heat-resistant alloys, low-temperature alloys and high-strength alloys, as well as alloys with special functions.

2.3.1. Corrosion-Resistant Alloys

Compared to other metallic materials, titanium exhibits higher activity at room temperature and reacts very easily in atmospheric and aqueous solutions to produce dense oxides. This formation of TiO_2 oxide film is the cornerstone of the exceptional corrosion resistance observed in titanium alloys [31]. Protected by the surface layer of oxide, titanium alloy products are well-suited for use in highly corrosive environments, such as marine engineering and oil extraction, maintaining their integrity under challenging work conditions. Since their inception, corrosion-resistant titanium alloys have been developed continuously. Currently, there are four relatively mature corrosion-resistant alloy families, namely titanium–molybdenum alloys, titanium–palladium alloys, titanium–nickel alloys, and titanium–tantalum alloys.

Accordingly, it is easy to see that the incorporation of different alloying elements is one of the main methods to improve the corrosion resistance of alloys. The results of Masahiko Morinaga's research have shown that the introduction of elements with high bonding sequence values (Bo) (e.g., Ta, Nb, etc.) effectively strengthens the chemical bonding between the matrix and the alloying elements. This enhancement leads to a reduction in the critical anodic current density on the polarization curve, consequently elevating the corrosion resistance properties of the alloys [32]. Bosung et al. [33] found that the introduction of elements such as Cr and Mo into titanium alloys can induce the passivation film to form Ti-Mo and Ti-Cr double hydroxides. This process accelerates the spontaneous passivation of the alloys, making the oxide film more stable and thus showing high corrosion resistance in highly concentrated reducing acids. Currently, titanium alloys with Pt or Pd are recognized as the most corrosion-resistant alloys. However, their widespread application is substantially limited by cost constraints and other practical factors.

The variability of the microstructure also has an important effect on the corrosion resistance of materials. Dong et al. [34] have shown that at a constant strain rate, the lamellar organization has higher stress corrosion sensitivity than the equiaxial organization due to a higher content of the α-phase. In addition, it has been shown that the lamellar organization can enhance the stress corrosion sensitivity of the material by changing the crack extension pathway. The effects of grain size and the number of grain boundaries are also critical factors in determining the corrosion resistance. When the alloy's grain size is reduced to microns or smaller by processing means, the material properties are significantly optimized with increased strength, hardness, and densification, coupled with a rise in the number of grain boundaries [35,36]. It is beneficial to increase the contact area of the alloy surface with the external environment, form a dense and uniform oxide layer, and improve corrosion resistance. Compared with alloying, such means (e.g., violent plastic deformation) can effectively reduce costs, and eliminate the original defects within the alloy material, thereby improving the mechanical properties of the material.

Surface treatment is also an effective means of improving the corrosion resistance of alloys, with the coating method being often considered to be the preferred method for this purpose [37]. In addition, processes such as heat treatment can significantly affect the corrosion resistance of materials by changing the properties of surface alloy materials. Taking nitriding treatment as an example, plasma nitriding (PN) is known to improve the surface hardness and corrosion resistance through the formation of a titanium nitride layer [38–41]. Li et al. [31] compared the corrosion current densities of the electron-beam melting (EBM) Ti-6Al-4V alloy in seawater before and after PN treatment. They found that the corrosion current density of the alloy after PN (PN-EMM) was significantly lower than that of the original sample, which indicated that the corrosion resistance of the treated material was effectively improved. The kinetic potential polarization curves of EBM and PN-EBM samples in natural seawater are illustrated in Figure 9 [31].

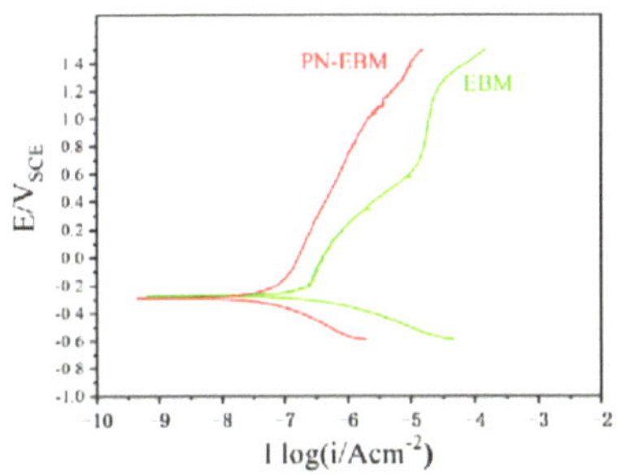

Figure 9. Dynamic potential polarization curves of EBM and PN-EBM samples [31].

2.3.2. Heat-Resistant Alloys

High-temperature alloys are widely used in the aerospace field because of their exceptional specific strength and good high-temperature mechanical properties. Currently, the service temperature of high-temperature alloys is mostly in the range of 550–650 °C. Predominantly, these alloys are near-α-type titanium alloys, such as Ti60, Ti-1100, etc. The continuous development of aviation technology has put forward higher requirements for the operating temperature and other critical properties of these high-temperature alloys. However, the increase in service temperature will result in a consequent decrease in the antioxidant properties of the alloys, which poses a significant limitation on their application in high-temperature components [42–44].

The addition of alloying elements and the use of ordered phase hardening are effective methods to improve the oxidation resistance and high-temperature performance of alloys. Xu et al. [45] found that the addition of high W content can increase the oxidation activation energy of the alloy, so that the alloy generates a more homogeneous and denser oxidation layer, which significantly improves the oxidation resistance of the alloy. The addition of W also plays an important role in refining the α-sheet layer and ordered α2-Ti3Al phase. This results in improved room-temperature yield strength and high-temperature yield strength, and better plasticity after extended thermal exposure. SEM images of the W-free alloy and the alloy with 4.0 W (wt.%) addition (W40) after 100 h exposure at 650 °C are shown in Figure 10 [45].

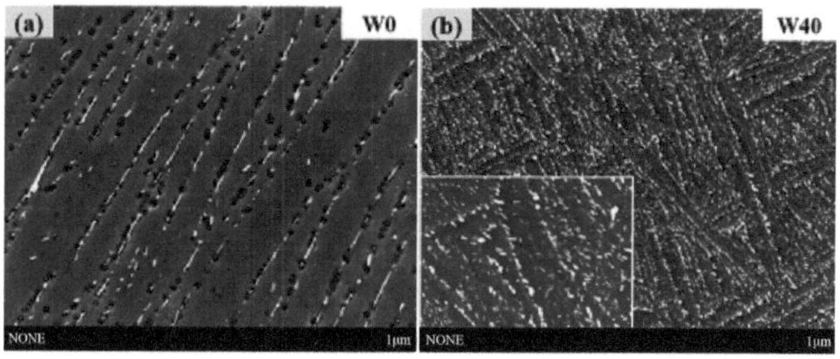

Figure 10. SEM images of the (**a**) W0 and (**b**) W40 alloys after thermal exposure [45].

Most high-temperature alloys frequently incorporate aluminum to reinforce the α-phase. However, aluminum-containing titanium alloys tend to experience coarsening due to the precipitation of the ordered α2-Ti-3Al phase after prolonged exposure to high temperatures. This would be detrimental to the alloy plasticity and make the alloy more susceptible to cold residence fatigue [46,47]. Increasing the Al solubility in the alloy can inhibit the growth of the ordered α2-Ti3Al phase, thereby enhancing the thermal strength and thermal stability of the material. Cao et al. [48] compared the interactions of Al with other alloy atoms in α-Ti, and found that the traps of Al atoms, such as Mo, Ru, etc., can increase the solubility of Al and inhibit the growth of the intermetallic compound of Ti3Al in α-Ti. This improves the thermodynamic properties of aluminum-containing high-temperature alloys. Rare earth elements can affect the thermomechanical properties of alloys by changing the metallographic organization of the alloys. Li et al. [49] found that the introduction of Re increased the β-phase content of the alloys significantly, and the tensile strength of the alloys was improved. This element also played a role in refining the grain size and the thickness of the sheet layer and effectively inhibited the Al, Ta, and other elements' diffusion. While these alloys exhibit higher thermal strength post-Re addition, their plasticity is somewhat compromised. The comparison of the organization images of Ti-Al-Ta-Nb alloy before and after the addition of Re is shown in Figure 11 [49].

Compared with single-element alloying, multi-element co-alloying has proven to be more effective in improving the high-temperature creep performance of alloys. For example, Wen et al. [50] found that when Al-Ta coexisted on the surface of the alloy, the superposition of hindering oxygen adsorption and oxygen diffusion effects could be realized. Thus, the antioxidant performance of the material could be optimized. Table 2 shows typical high-temperature alloys used in various countries [4].

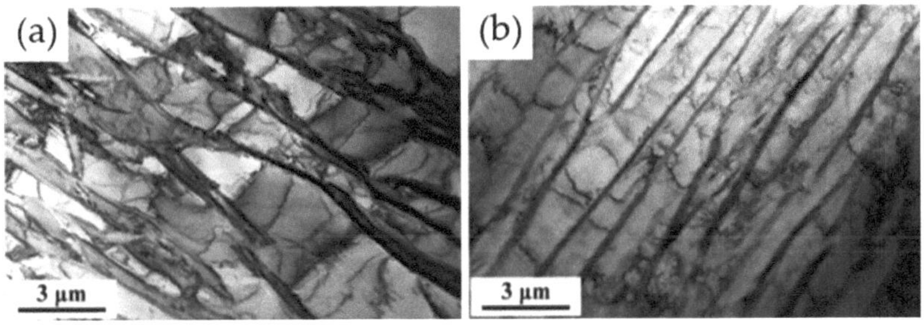

Figure 11. Ti-Al-Ta-Nb TEM image. (**a**) Original alloy. (**b**) Alloy after introduction of Re [49].

Table 2. Typical high-temperature alloys in various countries [4].

Service Temperature/°C	Grade	National	Chemical Composition/wt.%						
			Al	Sn	Zr	Mo	Nb	Si	Other
400	IMI550	UK	6	2		4		0.5	
500–550	BT9	RU	6.5	2		3.5		0.3	
500–550	BT18	RU	8		8	0.6	1	0.22	0.15Fe
500–550	BT25	RU	6.8	2	1.7	2		0.2	0.7W
550	Ti6242s	USA	6	2	4	2		0.1	
550	Ti55	CN	5	4	2	1		0.25	1Nd
580	IMI829	UK	5.5	3.5	3	0.3	1	0.3	
600	IMI834	UK	5.5	4	4	0.31	1	0.5	0.06C
600	Ti-1100	USA	6	2.7	4	0.4		0.45	
600	BT36	RU	6.2	2	3.6	0.7		0.35	5W
600	Ti60	CN	5.8	4.8	2	1		0.35	0.86Nd
600	TG6	CN	5.8	4	4		0.7	0.4	1.5Ta–0.06C

Currently, the physical mechanism of the effect of alloying elements on the thermal strength and thermal stability of high-temperature alloys remains an area of ongoing investigation [48]. It is believed that this will be one of the breakthrough directions to improve the performance of high-temperature alloys in the future.

2.3.3. Low-Temperature Alloys

Compared with other metal materials such as aluminum alloys and stainless steel, titanium alloys show excellent corrosion resistance, toughness, and high specific strength in cryogenic environments. They are commonly used in the manufacturing of cryogenic equipment, such as hydrogen storage tanks, hydrogen pump impellers, and structural parts for the aerospace industry. Currently, low-temperature alloys used both domestically and internationally are dominated by α-type titanium alloys, such as TA7 and TC4 ELI.

Normally, the alloy's hardness increases with decreasing temperature, while its elongation and fracture toughness tend to decrease. This makes the alloys exhibit significant brittleness at low temperatures. In addition to temperature, the low-temperature toughness and plasticity of alloys are also affected by the gap element content, alloy composition, and microstructure. Researchers both domestically and internationally have managed to enhance these properties by reducing the content of interstitial elements like carbon (C), hydrogen (H), and oxygen (O) in the alloy. For instance, TA7 ELI alloy can even be used at −253 °C. Compared with low-Al-content alloys, high-Al-content alloys are prone to chemical interactions with surrounding atoms due to the presence of Al, which hinders the dislocation movement and thereby reduces the alloy's plasticity [51].

Microstructure is another important factor affecting the low-temperature plasticity of the material. The lamellar organization under low-temperature conditions shows better low-temperature plasticity than the equiaxial organization because of the higher number of serrations on the deformation curve and the greater susceptibility to twinning [51].

Currently, the prevalent cryogenic titanium alloys are mostly near-α alloys and two-phase alloys with a minor β-phase component. Generally, these alloys exhibit poor plasticity. Moreover, α titanium alloys cannot be strengthened through heat treatment; so, the use of cryogenic alloys has been greatly restricted, relegating their use to low-stress components [19]. With the rapid development of science and technology, the plasticity and strength of alloys at low temperatures will certainly put forward higher requirements. Therefore, the development of low-cost, high-strength, and high-plasticity alloys will likely become a principal focus in the field of high-temperature alloys in the foreseeable future. Research into the low-temperature performance of β-phase alloys may emerge as a key direction in the evolution of low-temperature alloy research.

2.3.4. High-Strength Alloys

High-strength titanium alloys, characterized by a tensile strength greater than 1100 MPa at room temperature, primarily consist of near-β titanium alloys and sub-stable β titanium alloys, such as Ti5Si3 and TB17. High-strength titanium alloys usually have a high specific strength, good toughness, excellent corrosion resistance, and other characteristics. Consequently, they find extensive applications in the aerospace and defense industries.

The strength and toughness of titanium alloys are usually affected by grain size, metallographic structure, and alloying elements. Lu et al. [52] found that α lamellae within the grain can strengthen the β matrix and prolong the crack extension path, thus optimizing the alloy's strength and toughness. Mantri et al. [53] successfully optimized the strength and toughness of titanium alloys by aging the β-21S titanium alloy (Ti-15Mo-3Nb-2.7Al-0.2Si, wt.%) by adopting an aging treatment to precipitate lamellar αs phase, which effectively improved the strength of this alloy.

Heat treatment techniques, such as solid solution aging, are effective means to increase the strength of β alloys. Comparative observations of Figure 12 [54] show that the width of the α-phase grows rapidly as the heat treatment temperature is increased, and the number of strength-enhancing α′ phases precipitated in the alloy increases significantly. This phase not only becomes more refined but also contributes to an increase in the alloy's strength as the process continues. The incorporation of alloying elements is the most commonly used method to improve the strength of alloys. For instance, the introduction of Al elements can encourage the precipitation of the α-phase, contributing to solid solution strengthening and amplifying the solid solution aging strengthening effect. High-strength alloys are widely used in a variety of applications. The chemical composition of common high-strength titanium alloys is shown in Table 3 [55].

Table 3. Chemical composition of ordinary high-strength titanium alloys [55].

Alloy Code	Chemical Ingredients	Number of Alloying Elements	Alloying Element Content (wt.%)
Ti-1300	Ti-5Al-4Mo-4V-4Cr-3Zr	5	20%
Ti-7333	Ti-7Mo-3Nb-3Cr-3Al	4	16%
Ti-6554	Ti-6Cr-5Mo-5V-4Al	4	20%
Ti-5553	Ti-5Al-5V-5Mo-3Cr	4	18%
Ti-55531	Ti-5Al-5Mo-5V-3Cr-1Zr	5	19%
TB17	Ti-6.5Mo-2.5Cr-2V-2Nb-1Sn-1Zr-4Al	7	19%
BT22	Ti-5Al-5Mo-5V-1Cr-1Fe	5	17%
β-21S	Ti-11.5Mo-6Zr-4.5Sn	3	22%
BT32	Ti-8 V-8Mo-1.5Cr-1.6Fe-3Al	5	22.1%
Ti-153	Ti-15 V-3Al-3Cr-3Sn	4	24%
RT-1400	Ti-4.8Al-5.6Mo-3V-1.5Cr-1.6Fe	5	16.5%

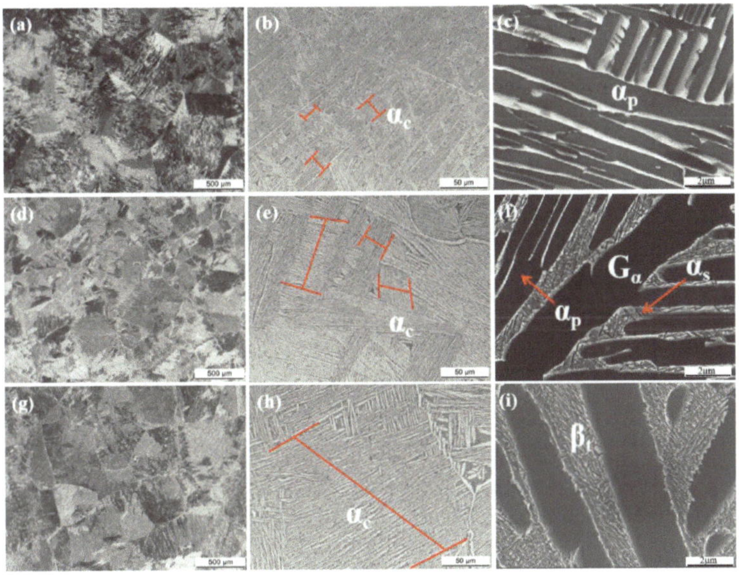

Figure 12. Microstructures of the alloy annealed at different temperatures: (**a–c**) 760 °C, (**d–f**) 800 °C, (**g–i**) 840 °C [54].

2.3.5. Special Function Alloys

Beyond the aforementioned high-performing alloys, titanium can be alloyed with various other metals to create materials with specialized functionalities. For example, the nickel–titanium alloy generated by the combination of titanium and nickel has a shape memory function. Ti-Nb alloy is a superconducting material commonly used in the industrial and medical fields. Additionally, Ti-Cr-based alloy is used as a solid-state hydrogen storage material, demonstrating the versatility of titanium when alloyed with different elements.

Shape memory function refers to the ability of a severely deformed alloy to recover its intended shape when heated above its transformation temperature. According to the different alloy compositions, they can be categorized into three types: copper-based, iron-based, and nickel–titanium-based alloys. Among them, Ni-Ti-based alloys are widely used as implants and surgical tools for medical applications due to their bidirectional shape memory, superplastic effect, and good biocompatibility [56]. In addition, Ni-Ti alloys are also commonly used in the industrial field. And in the early stage, the United States pioneered the use of NiTiFe alloy pipe joints in the industrial domain, notably on the F14 fighter jet. In recent years, the application of NiTiFe alloys has been further expanded as the research on the basic properties of shape memory alloys has been deepened in various countries.

The β-type Ti-Nb alloy's capacity to withstand stress during magnetic field establishment renders it a viable superconducting material for industrial applications. Niobium–titanium alloys offer several significant advantages over other materials, including high strength, good plasticity, superior mechanical properties, and lower cost. These benefits have led to their widespread use across various fields, making them the most extensively utilized material among low-temperature superconducting materials [57]. They can be seen in particle gas pedals, nuclear magnetic resonance, military minesweeping, superconducting power transmission and energy storage, and magnetic levitation trains. In recent years, with the continuous research on superconducting materials in various countries, the demand for niobium–titanium materials has been rising, and the annual consumption of niobium–titanium superconducting wires for only one item of nuclear magnetic resonance in medical research has reached more than 1000 tons. In the future, the field of superconducting materials is poised for rapid advancement. As an important component of superconducting materials, the niobium–titanium alloy is expected to find even broader and deeper applications.

In addition, researchers have found that certain titanium alloys can inhale hydrogen under specific conditions, and release it following a designated process, demonstrating that titanium alloys have the function of hydrogen storage. In recent years, the research and application of new energy sources such as hydrogen has become the main way to solve the current energy crisis, which puts forward higher requirements on the capacity, safety and cost of hydrogen storage materials. Ti-Cr-based alloys have been emphasized by researchers because of their low price and large hydrogen storage capacity. A study by Lv et al. [58] explored the impact of Mn on the hydrogen storage performance of Ti-Cr-based alloys. They discovered that introducing Mn not only effectively reduces the oxidation of Ti and Cr but also minimizes the loss of the hydrogen-absorbing C14-type phase in TiCr2 alloy. This modification results in superior hydrogen storage properties under low-temperature conditions.

In addition, the addition of multiple alloys, using the lattice distortion caused by differences in atomic size, enables hydrogen atoms to occupy more interstitial positions in the alloy, thereby effectively improving its hydrogen storage capacity. Multi-major element alloys are relatively inexpensive to produce, exhibit a large hydrogen storage capacity, and absorb hydrogen quickly, which is one of the main directions for the development of hydrogen storage materials in the future.

3. Titanium Alloy Surface Treatment

Titanium and its alloys are celebrated for their excellent material properties, contributing to their extreme versatility. However, in most industrial cases, they cannot directly meet the requirements of the working environment. Therefore, it is usually necessary to improve the mechanical, force, and chemical properties of the material to suit varying operational conditions. Metal properties are usually directly related to the friction and corrosion resistance of the surface material and many other physical and chemical surface properties [59]. The performance of titanium alloys depends greatly on the quality of the surface material, which is affected by the wear and corrosion of the surface metal. Zhang et al. [60] demonstrated that, according to the variations in the working environment, selecting the processing technique and parameters wisely is essential to achieving a good functional performance of the surface. The qualities of the material are positively impacted by good surface morphology. To increase a part's fatigue resistance, surface strengthening is often applied as the final step in the manufacturing process [61].

All established metal surface treatment methods have been adapted for titanium alloys, which can be categorized into mechanical, thermal, and chemical treatments depending on the nature of the process. Essentially, these treatments aim to either improve the surface properties of the alloy or form a protective layer on it, mitigating the effects of hydrogen, oxygen, or corrosive agents on the substrate. Selecting an appropriate treatment method according to the specific working environment can significantly enhance the alloy's performance while preserving its inherent material and mechanical properties. This section will delve into the mechanical treatment methods for titanium and its alloys, as well as discuss various heat treatment techniques.

3.1. Mechanical Processing

Mechanical processing, including techniques such as mechanical shot peening and mechanical grinding treatment, is a common means of mechanical treatment. The underlying principle shared by these techniques is the application of kinetic energy to a solid medium, which then interacts with the surface of the workpiece. This interaction utilizes the imparted force to modify the surface contour of the workpiece. The primary objectives of these methods are to strengthen the material's surface or to achieve desired surface morphology characteristics, such as specific patterns or surface roughness levels.

3.1.1. Mechanical Processes

Mechanical processing is one of the common surface treatment technologies for titanium alloys. However, given their status as materials typically challenging to machine, enhancing the surface quality and reducing tool wear during the machining of titanium alloys present considerable challenges [62]. Du et al. [62] propose a novel hybrid machining process to improve the machinability and mechanical properties of alloys. This process is shown in Figure 13 [62]. In this hybrid machining process, laser-assisted diamond turning is used to improve machinability by softening material surfaces. Additionally, slow-tool-servo cutting is employed to create various microstructure arrays.

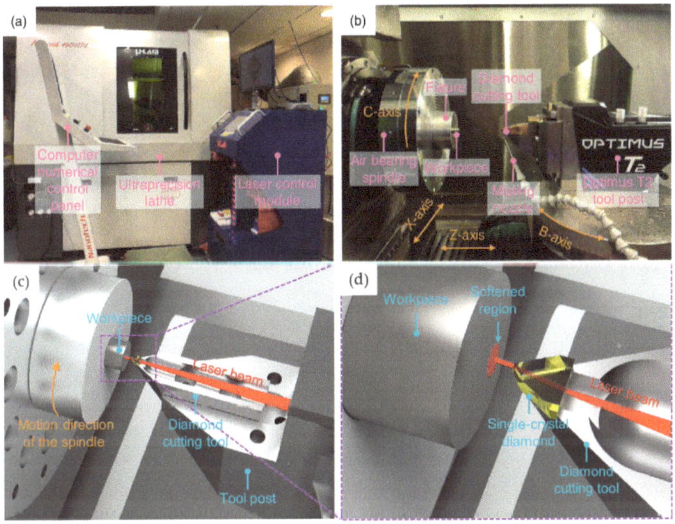

Figure 13. (**a**,**b**) Experiment setup of machining high-quality microstructure arrays; (**c**,**d**) working principle of the in situ laser-assisted diamond turning [62].

In the machining process, the high temperature caused by friction can increase the titanium's chemical activity, so that it reacts with the oxygen in the air to generate high-hardness oxides. These reactions often produce high-hardness oxides, complicating the subsequent processing steps. Moreover, the elevated chemical activity can cause portions of the titanium to cold weld onto the cutting tool. Therefore, titanium machining, due to different hardness requirements, is not easy to process with the titanium in the bonding diffusion of the tool material. Ahmed et al. [63] investigated the machinability of titanium alloy (Ti-6Al-4V) through electric discharge machining and identified the most appropriate tool material for better machining performance. This shows that superhard materials are the first choice for turning tools. Research [64] further supports that superhard cutting tool materials can effectively mitigate bonding issues, facilitating the finishing and high-speed machining of titanium alloys. For example, polycrystalline cubic boron nitride tools can effectively cut titanium alloy materials under the conditions of high cutting speed, low feed, and low back draft, maintaining a stable cutting force and low machining surface roughness. Additionally, polycrystalline diamond tools can still maintain good tool life and machining surface quality when machining a titanium alloy at a speed of more than 200 m/min.

Furtherly, Cui et al. [65] evaluated the grindability of the titanium alloy by considering the grinding temperature, grinding force, surface roughness, and defect ratio. The experimental results indicated that the values of surface roughness and grinding force obviously decreased under the assistance of the cryogenic nano-lubricant minimum quantity lubrication. Moreover, they observed a substantial 84.5% reduction in the defect ratio of the workpiece surface compared to cryogenic air cooling. Fábio et al. [66] compared the machinability of the casting Ti-6Al-4V with the selective laser-melted Ti-6Al-4V in terms of cutting forces, surface roughness, burr formation analysis, and microchips morphology, and found that the selective laser-melted Ti-6Al-4V had higher machinability with lower cutting forces, lower surface roughness, and less burr formation.

Apart from grinding and milling, turning is another machining process to remove the workpiece material and create the desired shape. Dandekar et al. [67] combined laser-assisted machining with the cryogenic cooling of the cutting tool to improve the machinability of a titanium alloy (Ti-6Al-4V) during the turning process. But the nanometric surface roughness was not achieved due to the limitations of the lathe and cutting tool. Addressing a similar challenge, Li et al. [68] utilized the femtosecond laser to texture the submillimeter-scaled structures on the rake faces of the uncoated cemented carbide cutting tool. Their findings showed that the textured cutting tool enhanced the machinability of Ti-6Al-4V because the cutting forces were reduced. Ni et al. [69] investigated the influences of anisotropic mechanical properties and microstructure features of additively manufactured Ti-6Al-4V alloys on machining performance in terms of cutting force and surface roughness. Yip et al. [70] studied the single-point diamond turning of titanium alloy Ti-6Al-4V under the assistance of the magnetic field. They found that the surface quality had been improved and marks on the workpiece surface had been obviously reduced. Considering that the mechanical properties of titanium alloys were dramatically influenced by their microstructures, Zhao et al. [71] applied a commercial electropulsing treatment to Ti-6Al-4V. The results showed that the single-point diamond turning improved the machinability by adding the magnetic field, electropulsing treatment, or using the hybrid cooling strategy.

In addition to traditional tool machining, Electrical Discharge Machining (EDM), as a new precision machining method, is commonly used for the treatment of difficult-to-machine materials such as titanium alloys. This method achieves material removal by applying a regulated electrical pulse between the tool electrode and the workpiece specimen [72]. The EDM setup is shown in Figure 14 [73].

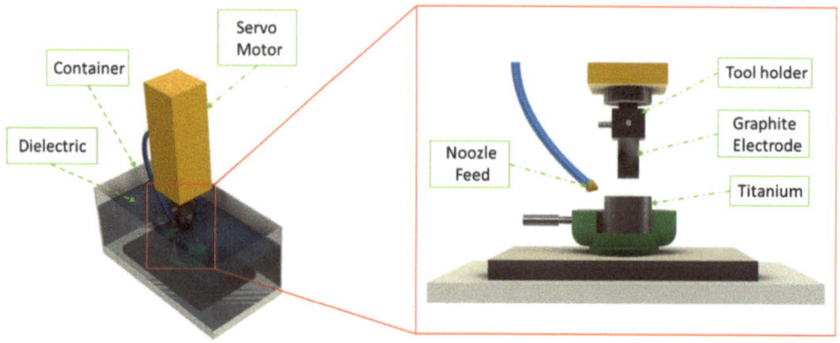

Figure 14. Graphical representation of EDM setup [73].

The surface quality and machining efficiency of EDM are mainly affected by the electrode material and pulse on conductive current. Panagiotis et al. [73] found that the effect of pulse on conductive current (IP) and pulse on time (Ton) on the material removal rate is significant by comparing the effect of pulse on conductive current (IP) and pulse on time (Ton) on the material removal rate. Ahmed et al. [74] found the effect of pulse on conductive current (IP) and pulse on time (Ton) on the material removal rate of α-β titanium alloys by exploring the effect of tool electrode on the machinability of EDM. The evaluation was based on material removal rate and surface morphology. It was finally determined that the copper electrode enhances the material removal mechanism, and the tungsten carbide electrode produces small craters after machining due to its high melting point. And the surface roughness after processing is superior to the other electrode materials. Brass electrodes have a lower melting point, produce larger craters on the surface after machining, and have higher surface roughness and electrode wear.

EDM can be used to fabricate complex shapes and geometries with high dimensional accuracy compared to conventional manufacturing processes [73]. And EDM is relatively inexpensive compared to other mechanical treatments. However, EDM can produce machined surfaces with high roughness, high white layer thickness and large heat-affected zones. Therefore, the selection of a suitable pulse generator is critical to improve the quality of EDM machining.

3.1.2. Machine Shot Peening

Shot peening (SP) is a robust technique for surface layer strengthening, effectively enhancing the surface life and durability of the components while preserving their original internal chemical structure. Its operating principle involves utilizing compressed air, pressurized water, ultrasonic energy, or centrifugal force to propel spherical particles at high velocities against the material's surface. The impact of the projectile produces compressive residual stress on the surface of the component [75]. The introduction of compressive stresses effectively prevents crack initiation and extension and improves the fatigue resistance of the material. In addition, shot peening can also effectively refine the grain structure, thus playing a role in improving the fatigue strength and resistance to stress corrosion cracking [76]. Figure 15 shows a schematic diagram of the SP process and the plastic deformation introduced [77].

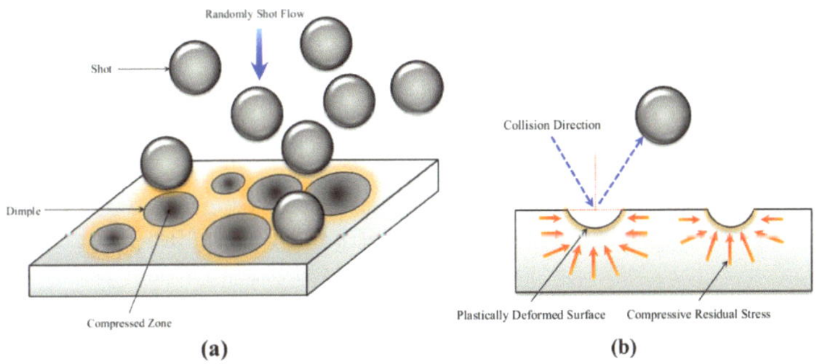

Figure 15. Schematic illustration of (**a**) the SP process and (**b**) the plastic deformation made by SP [77].

The plastic deformation capacity of metallic materials depends on the dislocation mobility and plastic strain rate. Constraining the movement of dislocations can lead to an increase in the material's strength [78]. The degree of plastic deformation during SP is directly related to the amount of total impact energy. This suggests three major factors that determine the effectiveness of SP: blast size, blast velocity, and number of blasts. Each of these factors plays a pivotal role in determining the level of impact energy and, consequently, the degree of plastic deformation achieved during the SP process.

A comparative study of conventional SP (CSP) and micro-SP (MSP) by Su et al. [79] found that micro-shot peening can better inhibit crack initiation than the impact of high-speed, large-size particles. This efficacy results in surfaces exhibiting a superior finish and enhanced fatigue properties. However, the smaller size used in MSP also equates to lower kinetic energy, leading to a shallower impact layer depth. Consequently, MSP is generally not preferred for alloys that require stringent corrosion and wear resistance. Wang et al. [80] investigated the effect of different shot sizes on the microstructure via coupled intrinsic modeling and found that larger shot sizes increased the depth of the dislocation density layer. The alloy was made to possess higher compressive residual stresses and exhibit greater corrosion resistance.

Research indicates that employing a dual shot peening approach can enhance the fatigue strength of an alloy more effectively than a single shot peening method. Shi et al. [81] determined the fatigue strength of the TC17 titanium alloy using a combination of classical shot peening and vibratory shot peening. Their findings demonstrated that this combined technique resulted in a greater increase in fatigue strength compared to the application of classical shot peening alone.

Shot peening has the advantage of being a simple process without heat exposure, and adaptable to complex and intrinsic conditions [82]. However, the determination of its process parameters remains a challenging task. The use of the small-size shot peen may not affect the deeper alloys, thus limiting the improvements in corrosion resistance. Conversely, the bombardment of large-size particles tends to increase the surface roughness and friction coefficients, which leads to pitting corrosion and the creation of stress concentration defects. Therefore, optimizing the coefficient of the shot peening process and developing better means for the shot peening process are an effective way to improve the shot peening technology.

3.1.3. Surface Mechanical Grinding

Surface mechanical grinding (SMAT) is a new surface nanosizing technology based on the high-frequency impact of balls. It operates on the principle of high-frequency ball impacts, inducing intense local plastic deformation on the surface. This process achieves grain size refinement within the impacted layer, alongside the introduction of dislocation density and residual stresses. Notably, SMAT enhances the material properties without altering the material's chemical composition [83].

The surface generated by the SMAT technique is mainly affected by the vibration frequency and the treatment time. Yao et al. [84] found that the thickness of the surface nanocrystalline layer increased with increasing treatment time by comparing the thickness of the crystalline layer after STMA treatment of different durations. Additionally, Aoudia et al. [83] investigated the impact of SMAT on the mechanical properties of coatings and their corrosion resistance. They discovered that the SMAT treatment effectively increased the hardness of the affected zone and improved the wear rate of the coating. SEM images of the cross-section microstructure of titanium alloy at different processing speeds are shown in Figure 16 [85]. Furthermore, they noted a positive correlation between the wear rate of the material and the amplitude of ultrasonic electrode vibration. Figure 17 [85] shows the 3D morphology of the wear trajectories of the specimens after different processing treatments. Chamgordani et al. [86] demonstrated that an ultrafine grain layer treated via a surface mechanical attrition treatment was generated on the surface of commercial pure titanium, resulting in a 60% reduction in the friction coefficient.

SMAT treatment technology is inexpensive and flexible, significantly improving the surface properties of materials. It has substantial potential value addition, particularly when integrated with other surface treatment technologies such as coatings. The working principle is similar to that of mechanical shot peening, but there are still some differences. For instance, SMAT involves random directional impact that facilitates grain refinement. Additionally, the shots used in SMAT are typically larger and smoother. In addition, compared with ordinary machining and mechanical shot peening, SMAT can achieve high-speed shear deformation and large strain gradients on large-scale materials without size constraints, and achieve the purpose of surface auto-nanosizing [87].

The methods discussed above represent some of the more commonly employed mechanical treatments for titanium alloys. These techniques are characterized by their simplicity in operation and the ability to be applied to large-sized metals. They enhance the overall chemical and mechanical properties of the material. Moreover, certain methods among these can even boost the material's strength and further improve its comprehensive mechanical properties. This enhancement is achieved through surface nanonization, which is accomplished while preserving the original toughness of the material.

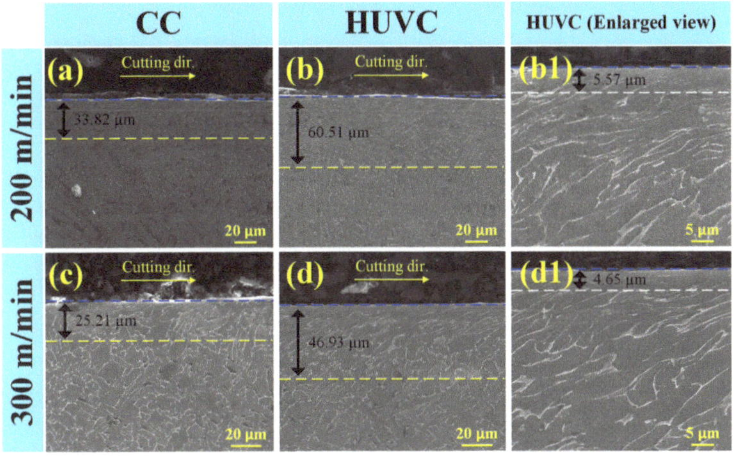

Figure 16. SEM images of cross-sectional microstructure of (**a**,**c**) CC and (**b**,**b1**,**d**,**d1**) HUVC samples with cutting speeds of 200 m/min (**a**,**b**,**b1**) and 300 m/min (**c**,**d**,**d1**) [85].

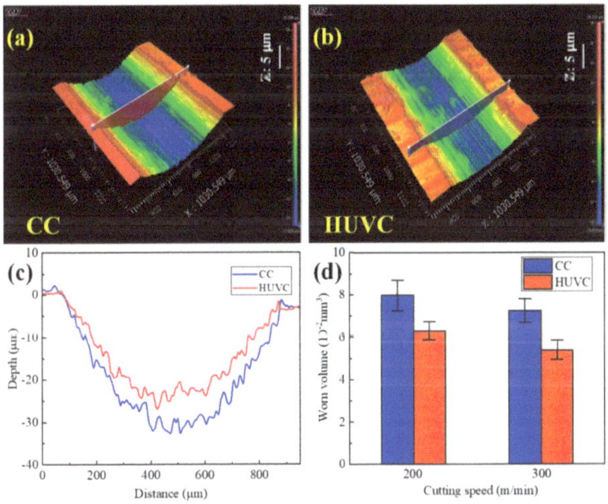

Figure 17. Three-dimensional topography of wear tracks of (**a**) CC- and (**b**) HUVC-machined samples; (**c**) cross-sectional profile curves at cutting speed of 200 m/min; (**d**) summary of worn volume [85].

3.2. Surface Heat Treatment

As a traditional treatment process, heat treatment is widely utilized to prevent the formation of cracks and voids, as well as to overcome the generation of unstable phases and internal stresses [30]. By heating and cooling the material, the metal undergoes phase transformation or recrystallization to achieve changes in the internal organization and structure so that the alloy achieves the specific desired material properties. The properties of titanium alloys are strongly related to the properties of their surface metals; so, it is necessary to perform targeted surface heat treatments on titanium alloys according to the demands of the working environment.

According to researchers, titanium alloys can be broadly classified into two categories based on the introduction of chemical elements: general heat treatment and thermochemical treatment [30]. The first method can take advantage of the hot and cold deformation of the material to change the metal organization structure, optimize the stress distribution within the material, and obtain better wear resistance and fatigue resistance. Thermochemical treatment, on the other hand, is used to achieve surface modification by injecting different alloying elements, resulting in the formation of alloys with higher hardness. At present, the prevalent thermochemical methods include boron infiltration, carburization, nitriding, and carbonitriding.

3.2.1. Normal Surface Heat Treatment

Surface heat treatment of titanium alloys facilitates alterations in the microstructure and organization of the surface material, consequently influencing the material's properties. Surface heat treatment operations have also been proven to be an effective way to extend the service life of titanium alloy materials and improve the wear resistance of titanium alloy components. This section focuses on surface quenching and annealing processes as applied to titanium alloys.

- Surface hardening

Quenching is known to significantly enhance material properties through the alteration of phase composition and redistribution of elements within the alloy. It has been shown that quenching above and below the solid-phase temperature results in a uniformly distributed single-phase organization. This process effectively eliminates defects associated with the segregation of the constituent elements and mitigates selective phase corrosion. As a result, the wear resistance of the material is substantially improved [88].

Dang et al. [89] developed a new forming process of fast gas forming with in-die quenching for titanium alloys. The formed component could be significantly strengthened through the formation of abundant fine martensitic microstructure during in-die quenching. For example, the yield strength and tensile strength of the Ti-3Al-2.5V alloy component increased by 26.5% and 15.2%, respectively, while the elongation still reached up to 16.2%. Traditional quenching methods, which typically involve treating the entire workpiece, can lead to reduced plasticity and potential brittle fracture, thus adversely affecting the mechanical toughness of the substrate [90]. This challenge led to the development of surface quenching, which targets only the surface layer of the material, forming a supersaturated layer with uniform element distribution and enhanced corrosion resistance. Since surface quenching primarily affects the surface, the internal microstructure and properties of the metal matrix remain largely unchanged, preserving the original mechanical characteristics of the interior metal. At present, laser quenching and induction quenching are the two main ways of titanium alloy quenching.

Laser quenching is currently the most prevalent method for quenching titanium alloys. This process involves directing a high-density energy beam onto the metal surface, where the photothermal effect instantaneously melts and then rapidly solidifies the surface metal. This results in the formation of ultrafine, acicular martensite with a uniform distribution and well-developed vertical grain boundaries, significantly enhancing the material's wear and corrosion resistance. Laser quenching has the advantages of high processing efficiency, and controllable quenching depth. It also minimizes stress introduction and is less likely to induce porosity defects. Additionally, the rapid cooling effect associated with laser treatment ensures minimal changes in surface roughness and contour.

As a rapid thermal treatment technology, high-frequency induction quenching has been widely used in the industrial field due to its advantage of process controllability, fast heating rate, cleanliness, low energy consumption, and environmental friendliness. A temperature gradient can be formed via fast quenching. The surface temperature of the material reduces rapidly, in contrast to the more gradual temperature decline within the substrate. This differential cooling rate facilitates the formation of distinct microstructures across the material's depth during high-frequency induction quenching, thereby engendering heterogeneous mechanical properties at the surface and the core. As reported by Jian et al. [91], a gradient microstructure was successfully fabricated through a high-frequency induction quenching treatment to improve the mechanical behavior of Ti-6Al-4V alloy. The results showed that the gradient microstructure of the alloy varied from a fine, as lamellae decomposed from a'-martensite at the surface layer, to a bimodal microstructure at the center.

Moreover, the alloy with a gradient microstructure presents an optimal strength–ductility synergy. However, it is worth noting that the application of high-frequency induction quenching for surface enhancement in titanium alloys remains underexplored, signifying a potential avenue for further research in material science.

During a quenching treatment, the thermal gradients induce internal stresses and plastic strains. The mechanical behavior depends on the local temperature and microstructure. In an effort to demystify this phenomenon, Teixeira et al. [92] conducted a comprehensive numerical simulation, meticulously examining the Ti17 alloy. Their study was pioneering in considering the synergistic effects of thermal, mechanical, and microstructural evolutions during the quenching process. Simultaneously, emphasis was put on the influence of the β and α+β phases transformation on the internal stress evolutions during quenching. Furthermore, for the quantitative assessment of the parameters related to the morphological structure generated during laser treatment, Lytvynenko et al. [93] introduced a method of mathematical analysis. This method, coupled with the consideration of the stochastic and cyclic nature inherent in laser treatment, is employed to analyze the ordered undulations that manifest on the surface of titanium nanoparticles following laser shock wave treatment.

- Surface annealing

Annealing stands as a pivotal process in the realm of material fabrication, primarily for its efficacy in mitigating or entirely rectifying the non-uniform microstructures typically induced by uneven deformation during processing, as outlined in [94]. This thermal treatment is instrumental in alleviating residual stresses inherent within the material, thereby enhancing both the stability of the alloy's microstructure and its overall toughness. Furthermore, annealing contributes significantly to the homogenization of the material's microstructure, ensuring a more consistent and predictable response to subsequent processing or operational stresses. To illustrate this process, Figure 18, as referenced in [95], presents a detailed schematic that delineates both the preparation of specimens for annealing and their subsequent characterization.

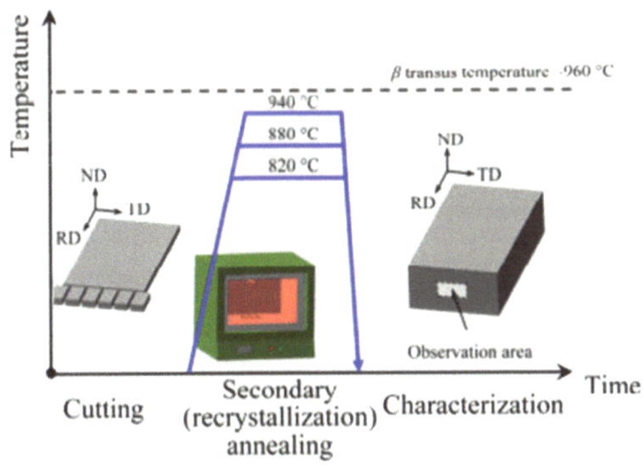

Figure 18. Schematic diagram of specimen preparation and characterization of annealed specimens [95].

Annealing temperature and annealing time are the main factors affecting the usable properties of the treated material. In a seminal study conducted by He et al. [94], the impact of annealing on the microstructure of the TB8 alloy post cold rolling was meticulously investigated. Their findings underscored that the annealing temperature positively correlates with the grain growth time index. Furthermore, as the holding time extends, the activation energy required for grain growth exhibits an increasing trend. This phenomenon is vividly illustrated in Figure 19 [96], which compares the microstructural transformation of TA10 alloys under different heat treatment regimes. The micrographs show that the recrystallization drive increases with annealing time. On the other hand, a high annealing time will lead to the consumption of small grains by large grains. The β-phase dispersed between α-phases gradually decreases, while the volume fraction of the α-phase shows an increasing tendency [96]. This transition leads to a morphological change from an equiaxed to a Weissian organization. Such microstructural evolution results in a material that displays enhanced plasticity, albeit at a slight compromise in strength.

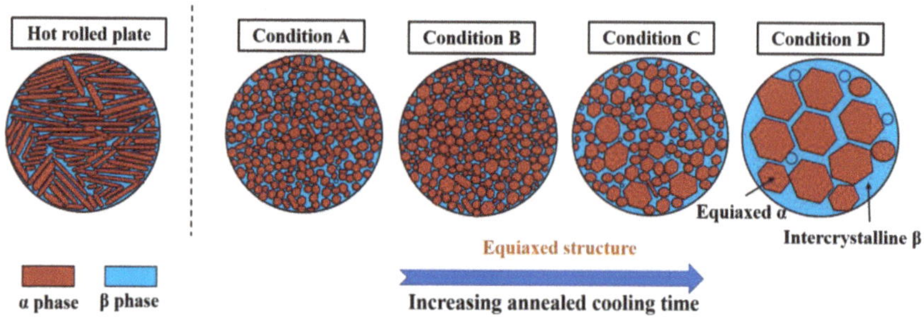

Figure 19. Microstructural evolution of the TA10 sheet at different annealing times [96].

Annealing is also commonly used to improve the organization and mechanical properties of welded joints and so on. In this context, it serves as a crucial technique for improving both the microstructural organization and mechanical properties of the joints. A notable investigation by Kumar [97] delved into the effects of annealing on titanium alloy welded joints under various thermal conditions. This study illuminated the multifaceted impact of annealing on the microstructure of these joints.

Surface heat treatment is one of the pivotal techniques to regulate the microstructure of the alloy and change the mechanical properties. Compared with deformation processing, it has the advantages of energy saving and high efficiency, short cycle time and convenient operation. Therefore, this tailored approach to heat treatment not only conserves energy and reduces processing time but also unlocks the potential for customizing the mechanical profile of alloys to meet specific application requirements [98].

3.2.2. Thermochemical Treatment

Thermochemical treatments, as one of the most common surface methods, include carburizing, nitriding, and other elemental carburizing surface treatments. These methods fundamentally enhance the material properties of surface alloys by infusing elements such as carbon and nitrogen into the alloy matrix. This infusion results in the formation of a hardened layer, which is a composite of ceramic particles and a solid solution, as detailed in [99]. The quality of the hardened layer formed via this method is influenced by a number of factors such as the oxide layer, gas atmosphere, treatment time and temperature. The oxide layer is a significant impediment to the penetration of elements such as carbon and nitrogen. So, it is usually necessary to remove the surface oxide layer by grinding and cleaning before carburizing. Increasing the treatment temperature and time can also lead to the generation of a thicker saturated penetration layer. This enhanced penetration contributes to a marked improvement in the treated material's wear resistance and hardness, thereby extending the material's operational life and broadening its application spectrum in demanding environments.

- Carburization

Carbon is one of the elements with the greatest influence on the surface hardness of titanium alloys. And the injection of carbon into the surface layer of the alloy can effectively improve the hardness and wear-resistant properties of the surface metal. It is usually believed that the main strengthening effect on the surface of the material is the hard TiC particles in the carburized layer. Take the carburization treatment of Ti-6Al-4V as an example: a hardened layer composed of carbide particles is formed on the surface of the material after the carbonization treatment. Compared with the pre-treatment material, the hardness of the treated material was increased by about 128%, and the fatigue strength and wear resistance of the material were improved [100]. The thickness of the penetration layer generated by the carburizing treatment is thickened with the increase in the treatment temperature. And there is no obvious sharpness between the penetration layer and the subsurface. These methods include solid carburizing, where the alloy is exposed to a carbon-rich solid medium; liquid carburizing, involving immersion in a carbonaceous liquid; gas carburizing, which employs a carbon-rich gas atmosphere; and plasma carburizing, which utilizes ionized gases in a plasma state for the process.

1. Solid carburization

Solid carburization, a prominent method in hydrogen-free carburization, stands out as an efficient technique to enhance the surface properties of titanium alloys, notably with minimal alteration to the matrix properties of the material. Prior to the carburizing process, it is imperative to remove any oxide layer present on the sample's surface. Subsequently, the sample is enveloped in a carburizing agent and placed within a sealed chamber. The solid vacuum carburizing unit is shown in Figure 20 [99]. After solid carburizing, TiC reinforcing phase and Ti-C solid solution are introduced into the α-Ti phase of the alloy. The carbon content of α-Ti is increased and the hardness of the surface layer is improved. With the increase in sample depth, the volume fraction of α-Ti decreases gradually, and the carbon content of the inner metal layer decreases. So, the solid-nitriding-treated samples usually show a decrease in metal hardness with the increase in the samples' layer depth. This phenomenon is corroborated by the experimental findings of Duan et al. [99] on the Ti-6Al-4V alloy, which demonstrate a significant decrease in the volume fraction of the α-Ti phase in the diffusion layer as the sample depth increases. In addition, the hardness test results showed that the microhardness of both carburized surfaces significantly increased by about 100% compared to the untreated material.

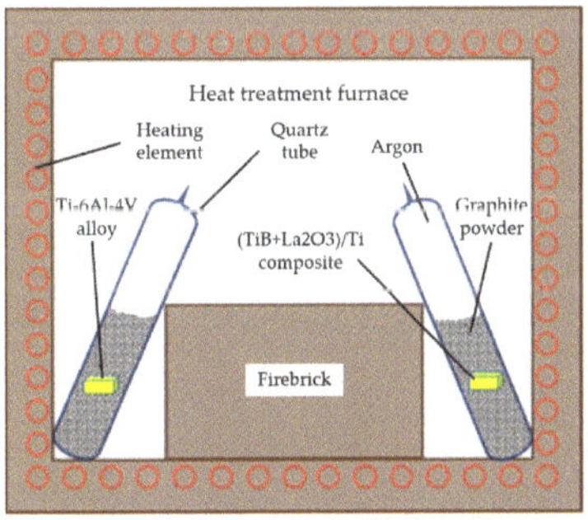

Figure 20. Schematic diagram of solid vacuum carburization process [99].

The diffusion coefficient of carbon in titanium is very low, and the titanium carbide layer generated on the surface will hinder the diffusion of carbon. Therefore, more efficient carburizing methods and better carburizing processes have become the main direction for the development of solid carburizing. Zhao et al. [101] proposed a contact solid carburization method to fabricate TiC coatings on titanium alloy using a "carbon sponge" cast iron. The schematic illustration of the contact solid carburization is shown in Figure 21 [101].

Solid titanium alloy (depicted in blue) and "carbon sponge" (illustrated in yellow) contact in atomic scale at high temperatures to enable the directional diffusion of interstitial carbon atoms (represented in black) from "carbon sponge" to the substrate. Specifically, when titanium alloy (Ti6Al4V) and cast iron contact in atomic scales at high temperatures below the melting point of the cast iron (e.g., 1100 °C), the interstitial carbon atoms in the cast iron diffuse into the titanium alloy, forming a TiC layer. This process is marked by the negligible interdiffusion of metallic atoms, which notably facilitates the subsequent removal of iron, leaving behind a titanium alloy coated with TiC. The resultant coating is characterized by its composition of equiaxed TiC grains and its complete density, signifying a noteworthy advancement in the field of surface-coating technologies for titanium alloys.

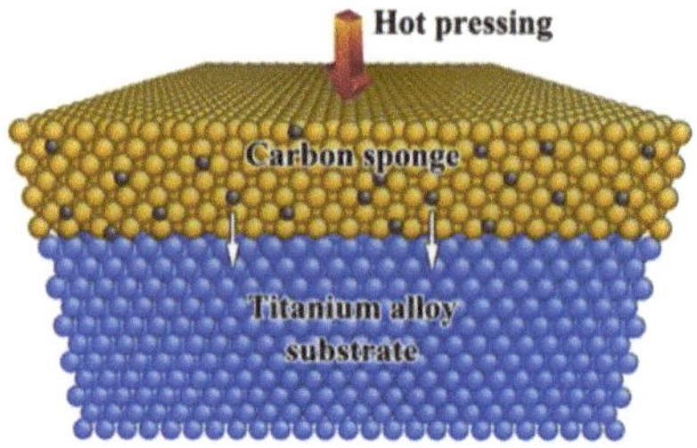

Figure 21. Schematic illustration of the contact solid carburization [101].

Solid carburizing, recognized for its operational simplicity and cost-effectiveness, plays a pivotal role in augmenting the hardness and wear resistance of alloys. However, this technique is currently limited by its relatively low carburizing efficiency. The resulting penetration layer often exhibits limitations such as being thin, porous, and prone to uncontrollable oxidation reactions. Moreover, the quality of this layer is significantly and adversely influenced by the process temperature, highlighting intrinsic drawbacks in its application under certain conditions.

Given these constraints, we posit that the development of innovative process technologies and treatments aimed at enhancing the quality of the solid carburized layer is imperative. Such advancements will not only address the existing limitations but also significantly broaden the application spectrum of solid carburization. Therefore, it stands to reason that refining and innovating in this domain will constitute a primary direction for future development in solid carburizing techniques.

2. Liquid carburization

Liquid carburizing, also known as salt bath carburizing, is a method of carburizing in a liquid medium. Molten salts can provide a liquid reaction media at high temperature for surface treatment of metals. Comparing to an aqueous solution, they have excellent thermal conductivity, rapid ion migration and diffusion at elevated temperature, resulting in a faster reaction rate at the solid/liquid (S/L) and metal/carbide interfaces [102]. In addition, inorganic carbon sources such as CO_2 [103] and carbonates [104–106] can be employed as raw materials for the preparation of carbides. These sources can be converted into carbon via electrochemical reduction or oxidation [107,108] to prepare carbon films or metal carbides. Furthermore, due to the good wettability of molten salt to the metal substrate, it also offers possibility for preparing coatings on shaped structural parts. Zhao et al. [109] demonstrated a general salt-thermo-carburizing method to prepare TiC coatings, as shown in Figure 22 [109]. In molten $CaCl_2$-CaC_2, TiC coatings were efficiently prepared on the Ti substrate at a temperature below 900 °C. The spontaneous carbonization of Ti is attributed to the negative formation of Gibbs free energies [110]. The growth of TiC is determined by the diffusion of C in the TiC coating, which is driven by the high carbon potential at the molten salt/TiC interface. The growing kinetics of the coating follow a parabolic law, demonstrating a diffusion-controlled process [111].

This simplicity and effectiveness of the salt-thermo-carburizing method suggests a general way to prepare metallic carbide coatings on the relative metal substrate at a medium temperature. However, due to the limitations of the container, it is difficult to carburize larger sizes and slender parts via salt bath carburizing, rendering this method more suitable for small batch production rather than large-scale manufacturing.

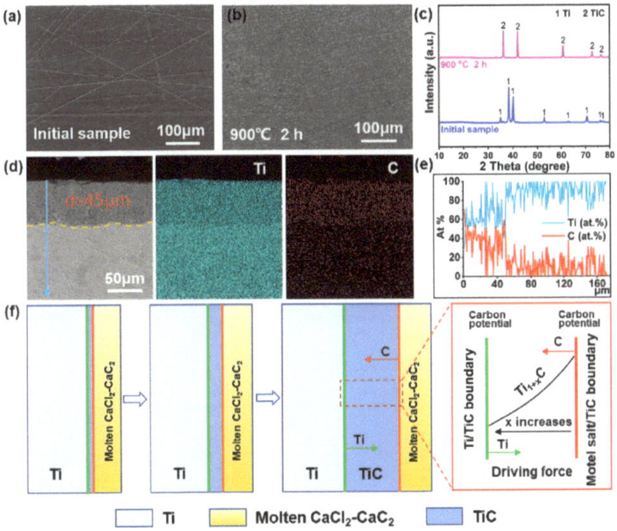

Figure 22. (**a**–**c**) The SEM images and structural composition of the initial sample and the sample after carburizing for 2 h at 900 °C, respectively. (**d**) The SEM-mappings, (**e**) linear scan of the elements, and (**f**) the carburizing mechanism process of the TiC coating [109].

3. Gas carburization

Gas carburization stands as a prevalent method for the carburization of titanium alloys, predominantly utilizing high-carbon-content gases such as methane and carbon monoxide (CO) as the carbon sources. This process typically involves constant-temperature heating. Concurrently, inert gases like nitrogen (N_2) are introduced as protective agents to circumvent the potential oxidation of the material or to prevent hydrogen embrittlement, a common concern in high-temperature treatments. The elevated furnace temperatures facilitate the disassociation of carbon from the gas source, subsequently enabling the activated carbon atoms to react with the surface titanium atoms, leading to the formation of TiC (titanium carbide). A depiction of a typical gas nitriding setup is illustrated in Figure 23.

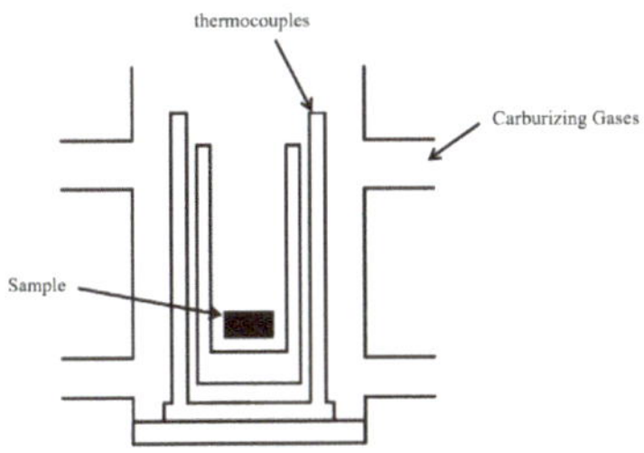

Figure 23. Schematic diagram of gas carburization device.

This process not only ensures a controlled introduction of carbon into the titanium alloy but also maintains the integrity of the material by preventing adverse reactions that could compromise its mechanical properties. The resultant TiC layer is known for its enhanced hardness and wear resistance, making gas carburizing a vital technique in the surface treatment of titanium alloys.

The carburizing temperature is the factor that has the greatest influence on the quality of the layer. A lower temperature will lead to lower carbon atom activity, resulting in a thinner thickness of the generated layer. With increasing carburization temperature, there is a decrease in reaction activity but an increase in the availability and diffusion capability of active carbon atoms. This change facilitates deeper penetration of carbon into the titanium matrix, leading to the formation of a thicker carburized layer. Such temperature-dependent dynamics are crucial for tailoring the depth and quality of the carburized layer in titanium alloys. Wu et al. [112] compared the XRD physical and metallographic organization of TA2 alloy under different carburizing temperatures. It was also concluded that the thickness of the carburized layer increased with the increase in the carburizing temperature. And when the treatment temperature exceeds 882.5 °C, the organization of the titanium alloy changes, while the abnormal coarsening of martensitic organization provides more possibilities for carbon atoms to enter the interstices. The XRD physical phase diagrams and metallographic histograms of TA2 alloy carburized at different temperatures are shown in Figure 24 [112]. And the related microstructural characteristics are shown in Figure 25 [112].

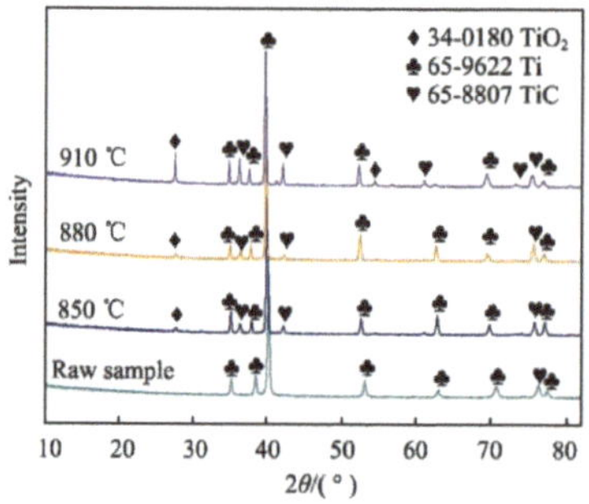

Figure 24. Phase analysis of TA2 titanium alloy via XRD at different temperature [112].

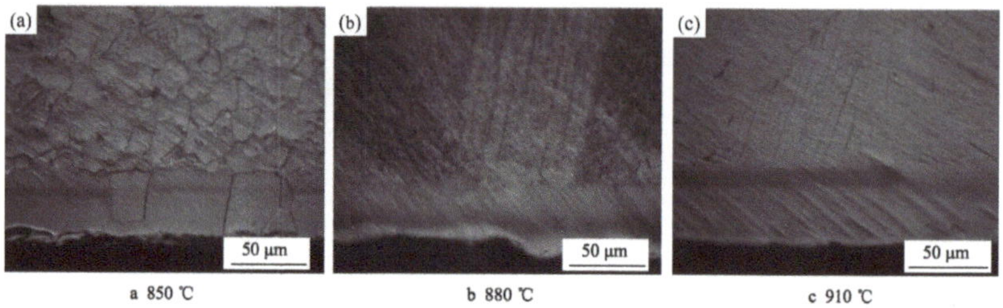

Figure 25. Microstructure of TA2 titanium alloy carburized at different temperature, (**a**) 850 °C, (**b**) 880 °C, (**c**) 910 °C [112].

The second important factor affecting the quality of the penetrating layer is the treatment time. It is generally believed that the longer the treatment time, the greater the thickness of the carburized layer, and the more carbon atoms there are in the carburized layer. The hardness distribution of the carburized layer is positively correlated with the carbon concentration distribution [113]. Therefore, longer treatment times typically yield a carburized layer of superior quality. Amar et al. [114] observed that the carburized layer's thickness and the tribological properties of TC4 alloys improve proportionally with increased treatment durations.

Upon examining the comparative graph of the tribological properties in Figure 26 [114], a distinct trend is evident. Alloys subjected to prolonged carburization times demonstrate significantly lower friction coefficients. This reduction in friction is directly linked to an increased resistance to mass loss over distance, indicating enhanced wear properties.

Gas carburizing can effectively improve the material hardness, coefficient of friction and wear resistance. Characterized by high processing temperatures, this method offers rapid carburization, with cycles approximately half the duration of those in solid carburizing. Its operational simplicity, coupled with minimal pollution, makes gas carburizing a preferable choice for continuous and large-scale production. Furthermore, the ability to control the thickness of the carburized layer by adjusting the carbon potential is a significant advantage of this technique, allowing for precise tailoring of surface properties to meet specific application requirements. But too high carburizing temperature will make the martensite grain abnormally coarse, resulting in a decline in the mechanical properties of the material.

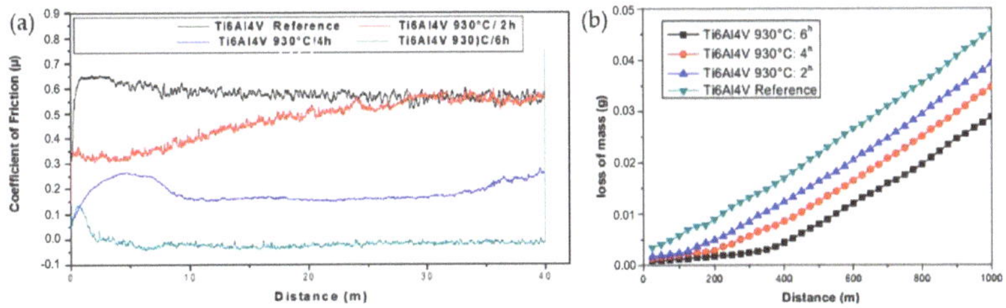

Figure 26. Comparison chart of friction performance. (**a**) Coefficient of friction (untreated and carburizing 930 °C: 2 h, 4 h, 6 h). (**b**) Variation in mass loss (untreated state (reference) and carburizing 930 °C: 2 h, 4 h, 6 h) [114].

Therefore, the authors believe that optimizing gas carburizing to ensure material properties while lowering the carburizing temperature represents a key direction for future advancements in this field. This approach not only maintains the desired surface characteristics but also aims to enhance the overall efficiency and sustainability of the carburizing process. Research has shown that intense plastic deformation and surface nano-means such as active ions can provide diffusion channels, effectively improve the diffusion of active ions, and reduce the temperature required for thermochemical treatment.

Integrating gas carburizing with additional surface strengthening techniques, such as shot peening, could potentially reduce the required carburizing temperature while simultaneously enhancing the mechanical properties of the carburized alloy and the quality of the carburized layer. This combined approach may offer a synergistic effect, optimizing both the process efficiency and the performance of the treated material. In addition, constant-temperature, low-pressure intermittent carburizing can repeat the saturation and diffusion cycle by repeatedly pumping in and out the carburizing gas. It effectively avoids hydrogen embrittlement and maintains a high carbon ion concentration to generate a high-quality carburized layer [115].

4. Ion carburizing

Ion carburizing is a process of surface carburization using the glow discharge effect between the cathode and anode in a low carburizing atmosphere, which is commonly used to improve the hardness, corrosion resistance and tribological properties of alloys. Compared with the traditional means of carburizing, ionic carburizing can obtain a higher quality and greater thickness of the layer in 3–6 h. This represents a substantial improvement in both processing efficiency and layer quality. Ionic carburizing is derived from a variety of technologies, including plasma electrolytic carburizing and double-layer glow plasma carburizing, broadening its application and efficacy in surface treatment processes.

The quality of ionic carburizing layers depends mainly on the carburizing temperature. Numerous studies have shown that the carburizing temperature plays a decisive role in the diffusion process of atoms. Under the same conditions, high temperatures are more likely to form a composite layer on the metal surface, and the composite layer is likely to hinder the penetration of carbon atoms into deeper specimens. Zhang et al. [116] compared the effects of different temperatures on the quality of TC4 carburized layers, and found that the carburized layers obtained at 950 °C were of higher quality compared with those obtained at 1000 °C and 900 °C, highlighting the significance of optimizing the temperature to achieve the best carburizing results.

The SEM cross-section of SLM-C-TC4 is shown in Figure 27 [117]. A black TiC layer is formed on the surface, followed by a gray transition layer. In contrast, the titanium substrate contains α-Ti and β-Ti, while the transition layer contains only β-Ti. This suggests that the diffusion of carbon in the titanium matrix promotes the transition of titanium from the α-phase to the β-phase. On the other hand, a thicker transition layer indicates a good bonding between the carburized layer and the titanium substrate [117].

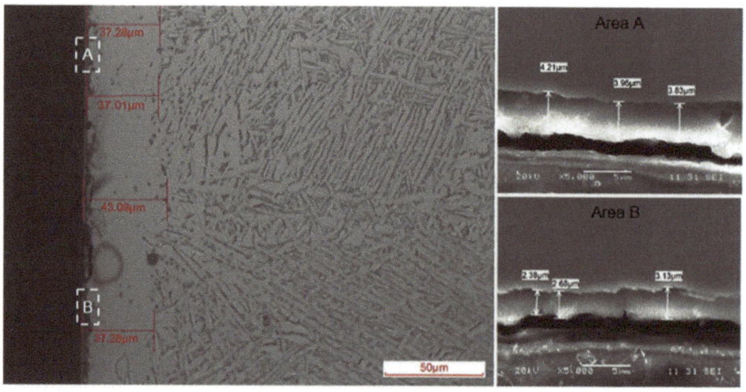

Figure 27. Typical microstructures of SLM-C-TC4 specimen (**left**) and the enlarged views of areas A and B (**right**) [117].

Dong et al. [118] investigated the effect of carburizing on the bond strength of titanium alloy DLC coatings and found that the carburizing treatment can effectively reduce the grain size of DLC coatings and thus improve the densification of the coatings. The carburizing layer also inhibits grain boundary migration in the bulk of the DLC coatings, preventing grain coarsening and imparting higher thermal stability [118]. SEM micrographs and 3D images of the carburized DLC coatings are shown in Figure 28 [118]. The cross-sectional SEM micrographs and EDS line scan analysis of the carburized DLC coatings are shown in Figure 29 [118].

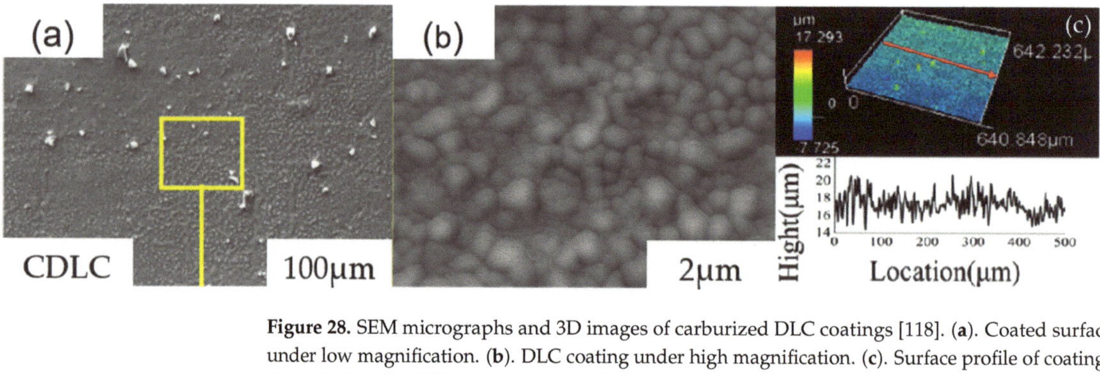

Figure 28. SEM micrographs and 3D images of carburized DLC coatings [118]. (**a**). Coated surface under low magnification. (**b**). DLC coating under high magnification. (**c**). Surface profile of coatings deposited on the different pretreated substrates.

Compared with glow plasma carburization, dual-glow plasma carburization technology further improves the hardness and service life of the carburized alloy by forming two different layers of glow ion carburization. Zheng et al. [119] demonstrated that dual-glow plasma carburization can introduce supersaturated vacancies to form a gradient structure layer, which accelerates the transport of metallic elements to the outside. This process accelerates the outward transport of metallic elements, leading to a more rapid development of the gradient structure layer. Post carburization, there is a significant improvement in the surface layer hardness due to the vacancy concentration gradient.

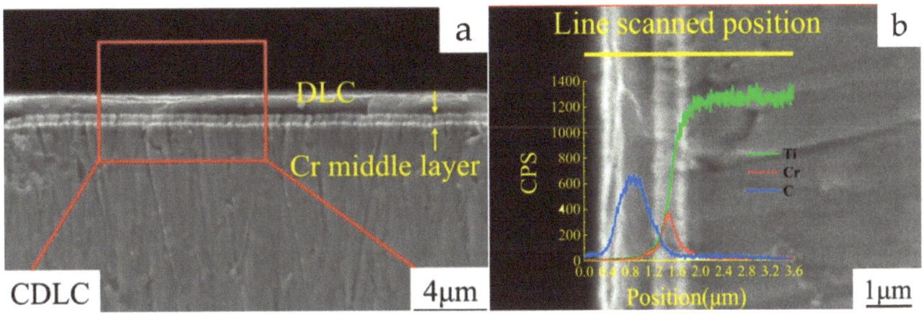

Figure 29. Cross-sectional SEM micrographs and EDS line scan analysis of carburization-treated DLC layers [118]. (**a**). SEM micrograph of DLC coated sample after carburizing treatment. (**b**). EDS line scan analysis of DLC coated sample after carburizing treatment.

Ion carburizing, known for its brief heat treatment duration, produces a high-quality penetration layer and is versatile enough to treat workpieces of varied shapes and complexities in a clean manner [120]. It can improve the hardness and wear resistance of the material surface without damaging the original properties of titanium. However, its drawbacks such as an unavoidable hollow cathode effect and difficulty in scale production remain significant obstacles to its broader adoption. It is worth mentioning that the emergence of new treatment methods such as physical vapor deposition and chemical vapor deposition provides a new choice and development direction for the ion carburization of titanium alloys.

Carburizing is an effective means to improve the surface properties of titanium alloy. Compared with nitrogen atoms, carbon has a weaker solubility in titanium, which makes carburizing titanium alloys more difficult than nitriding [121]. In addition, the nitrogen element has a greater effect on the surface hardness of titanium alloys than carbon. Consequently, both domestically and internationally, there is a notable prevalence of research focusing on the nitriding of titanium alloys.

- Nitriding

Nitrogen significantly influences the surface hardness of titanium alloys, with nitriding treatments notably enhancing their surface hardness and friction corrosion resistance. Currently, the more common nitriding treatments include solid nitriding, gas nitriding and plasma nitriding. These treatments typically require several hours to effectuate the desired surface modification. However, solid and gas nitriding often result in a loosely structured and porous layer, which may lead to hydrogen embrittlement and thus fail to meet quality standards for the workpieces. Another research report shows that plasma nitriding can complete the surface alloy nitriding within one hour, offering a marked improvement in both nitriding efficiency and layer quality compared to the other two methods [122]. In this section, three different nitriding methods for titanium alloys are presented.

1. Salt bath nitridation

Salt bath nitridation is a surface modification technique that involves the decomposition of cyanide into cyanate in a molten salt medium at temperatures ranging from 500 to 600 °C. This process facilitates the diffusion of nitrogen from the salt into the surface of the workpiece, with nitrogen atoms infiltrating the surface in the form of an interstitial solid solution. Because of a relatively low working temperature, salt bath nitriding is categorized as a low-temperature surface treatment. The primary nitrogen source in salt bath nitriding originates from the cyanate ion in the nitriding salt. However, the high temperatures involved in the nitriding process can lead to the gradual decomposition of cyanate and the formation of cyanogen, which poses health risks. Consequently, the development of salt bath nitriding techniques has been oriented towards low-cyanogen, cyanide-free solutions, and faster processing rates to mitigate these concerns.

Typically, salt bath nitriding involves extended treatment durations. However, prolonged exposure to high temperatures during nitriding can lead to porosity in the nitride layer, adversely affecting the hardness and wear resistance of the alloy [123]. In order to improve the nitriding efficiency of salt bath nitriding, researchers have developed many directions, including the addition of rare earth elements to change the metallographic structure of the alloy. For example, Zhu et al. [124] compared the metallographic structures of nitride titanium alloys before and after the addition of rare earth elements. They discovered that rare earth atoms, due to their strong surface adsorption capabilities, can attract a significant amount of reactive nitrogen atoms. These atoms are adsorbed onto the workpiece's surface at low energy, significantly increasing the nitrogen concentration in the substrate's surface layer.

Salt bath nitriding has the advantages of low equipment cost, reduced nitriding temperatures, and minimal deformation of the workpiece. This method effectively enhances the roughness, hardness, and wear resistance of samples, thereby broadening their application spectrum. The hardness of the workpiece quenched in water after salt bath nitriding will be greatly improved, and the workpieces treated by this means can usually be used for making fasteners and automotive parts (e.g., nuts and bolts), etc. [125]. However, the highly toxic cyanogen produced by decomposition in the salt bath process is still an urgent problem to be solved.

2. Gas nitriding

Gas nitriding, a method for surface modification of titanium alloys, involves exposing these alloys to an atmosphere of nitrogen or a nitrogen–hydrogen mixture and heating them to form a hardened layer of Ti2N or TiN on the surface. The nitrogen source mainly comes from nitrogen or nitrogen–hydrogen and other gaseous media. According to the different heating methods, gas nitriding can be divided into vacuum nitriding and laser nitriding.

Vacuum nitriding is an innovative technique developed in recent years, which involves a series of steps to enhance the surface properties of titanium alloys. The process begins with placing the titanium alloy into the furnace, followed by purging impurity gases using an inert gas. The furnace is then evacuated, and this vacuum–gas washing cycle is repeated 2–3 times. The alloy is heated to 800 °C and maintained at this temperature for approximately an hour. This step is crucial for decomposing the titanium surface oxidation layer, thereby reducing the hindrance it poses to nitrogen atom infiltration into the substrate. Such a temperature setting also facilitates the formation of a fine lamellar structure on the alloy's surface, aiding in grain refinement and reducing lamellar organization. Finally, high-purity nitrogen is introduced into the nitriding treatment for 30 min, so that nitrogen is fully diffused into the substrate, and the above operation is repeated until the nitriding treatment time is reached [126], which can generate a layer of dense nitriding layer. Intermittent vacuum nitriding treatment is particularly effective in breaking down residual oxide layers on the titanium alloy's surface, enhancing the number of reactive nitrogen atoms, and improving both nitrogen adsorption rates and reaction speeds on the surface. A higher nitrogen potential further facilitates nitrogen's inward diffusion, resulting in a thicker nitriding layer.

Vacuum nitriding, characterized by its simplicity, cost effectiveness, and flexibility in accommodating various workpiece geometries, has certain limitations. Typically, the treatment duration is lengthy, and as the nitriding time extends, the thickness of the nitride layer increases. However, this newly formed nitriding layer can hinder further nitrogen diffusion, leading to issues such as brittleness, thinness, and poor adhesion strength with the substrate. Managing the duration of the nitriding process is crucial; excessively long treatment times can cause the nitride layer to peel off due to prolonged high temperatures, while insufficient treatment time may not satisfy the material's wear resistance requirements [127]. To overcome these shortcomings, laser nitriding has been gradually used for surface modification of titanium alloys. This technique, developed through continuous exploration, offers a solution to the challenges posed by traditional vacuum nitriding.

High-purity nitrogen then reacts with this molten pool to form a nitride layer, significantly improving the wear and abrasion corrosion resistance of the treated surface. A key advantage of this technique is its surface-specific treatment, which leaves the mechanical properties of the titanium alloy matrix largely unaffected. By restricting the process parameters during the treatment, it is possible to control the thickness and hardness of the titanium alloy nitriding layer. The working schematic of laser nitriding is shown in Figure 30 [128]. Figure 31 shows a schematic diagram of the wear mechanism for samples with different laser nitriding energies [128].

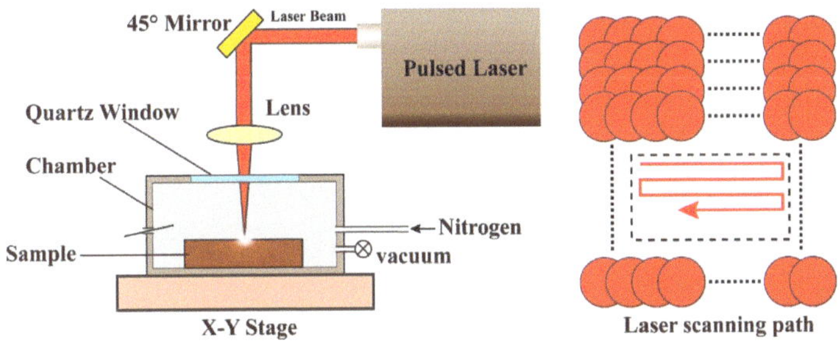

Figure 30. Pulsed laser nitriding process flow diagram [128].

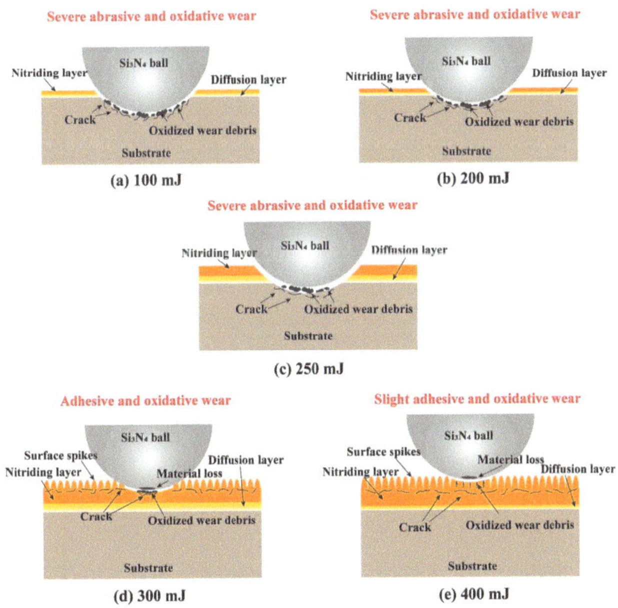

Figure 31. Schematic of the wear mechanism of different laser nitriding energy samples [128].

With the change in depth, the nitrogen concentration shows a gradient from the surface of the percolated layer to the bottom of the molten pool. Meanwhile, the formed nitrides also show different metallographic structures. On the near surface of the percolated layer, TiN exists in the form of dendrites, and the secondary dendrites grow perpendicular to the primary dendrites, as shown in Figure 32a [129]. Originating from the bias effect of TiN dendrites on nitrogen, the near-surface region has the largest nitride concentration and the highest hardness. This inference is consistent with the results exhibited in Figure 32d. In the intermediate region, often referred to as the heat-affected zone, the microstructure primarily consists of parallel rows of martensite, exhibiting a predominantly needle-like organization, as illustrated in Figure 32b [129]. The TiN content in the heat-affected region is relatively lower than that in the near-surface region. Its hardness is relatively low compared to the nitride region, which is generally twice the hardness of the matrix. With increasing depth, nitrides precipitate in the interphase region in the form of particles (Figure 32c) [129]. In this region, the heat-affected zone is less pronounced, and the overall structure of the alloy remains largely unchanged.

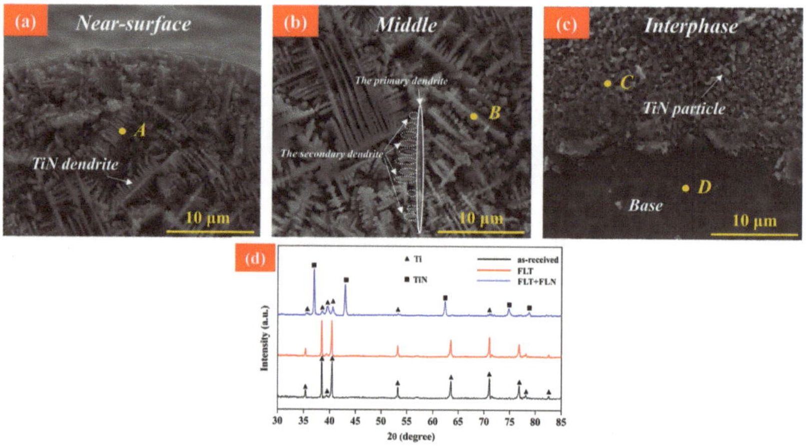

Figure 32. (**a**–**c**) SEM images of different nitride zones. (**d**) The microhardness and nitrogen contents measured at points A, B, C, and D [129].

Compared with vacuum nitriding, the processing time of laser nitriding is greatly shortened. The thickness of the nitride layer can be increased via repeated laser scanning. Among the nitriding treatments discussed in this paper, laser nitriding achieves the highest surface hardness post treatment. However, defects, such as surface cracks, triggered by the heat buildup effect are still a challenge that needs to be solved urgently. Therefore, maintaining high hardness and limiting the generation of cracks are still important issues in optimizing the parameters of the laser nitriding process. It is worth mentioning that in recent years, scholars have proposed a new surface-strengthening strategy combining surface nitriding and surface weaving. This approach has been shown to produce a uniform, crack-free nitride layer of 40–60 µm on the weave's surface [129]. Kang et al. [130] also reported that the composite treatment can significantly reduce the material wear rate and friction coefficient. And it can effectively control the thermal deformation and inhibit the generation of cracks, addressing one of the primary drawbacks of traditional laser nitriding.

Gas nitriding, as one of the most convenient and effective nitriding methods, has an extremely wide range of applications in the field of industry and scientific research. But prolonged high-temperature treatment will cause a large amount of energy loss, and may lead to the growth of secondary α-phase or the decomposition of the substable phase, which will affect the internal organization and structure [131]. Therefore, reducing the nitriding temperature and improving the nitriding efficiency is a major critical issue in the development of gas nitriding. Recent developments in low-temperature nitriding have led researchers to explore innovative strategies for improving nitrogen diffusion. These strategies include creating diffusion channels and utilizing surface metal phase transitions to provide auxiliary driving forces for nitrogen diffusion [131]. For example, Liu et al. [132] reduced the nitriding temperature of the Ti-6Al-4V alloy to 650 °C via surface nanosizing. Additionally, Jiang et al. [131] increased the effective thickness of the nitride layer by 15 µm from 500 °C nitriding treatment through post-laser nitriding, demonstrating the potential for significant improvements in gas nitriding efficiency.

3. Plasma nitridation

Plasma nitriding (PN) is an important surface modification technique that significantly increases surface hardness and improves the anti-corrosion ability by forming a titanium nitride layer [38–41]. When the nitrogen ions hit the surface of the workpiece at high speed, the kinetic energy is converted into thermal energy, which increases the temperature of the surface of the titanium alloy. In this process, part of the nitrogen ions seizes electrons to become nitrogen atoms and penetrate into the alloy surface [133]. Another part causes cathodic sputtering upon impact with the alloy surface. The sputtered titanium atoms react with nitrogen atoms to form nitrides, which subsequently deposit on the surface of the workpiece [134]. Under the effect of sputtering and diffusion of ions, nitrogen continuously diffuses to the inside of titanium and forms a new nitride layer. The schematic diagram of the plasma nitriding device is shown in Figure 33 [135].

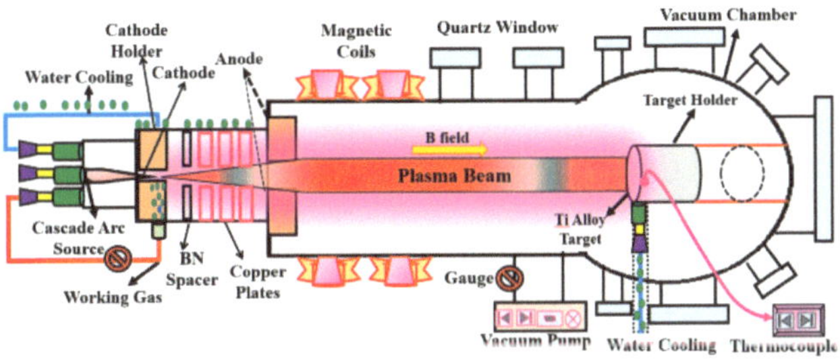

Figure 33. A schematic representation of the plasma nitriding setup [135].

From the working principle of ion nitriding, the rate of ion nitriding has a greater relationship with the nitrogen content in the gas medium and the voltage or temperature of the glow discharge. A higher nitrogen content in the gas medium accelerates the nitriding rate, but it can also lead to rapid formation of surface nitrides, which may hinder the inward diffusion of nitrogen and the growth of the nitriding layer, thereby impacting the wear resistance of the material. It has been found that gas nitriding of Ti6Al4V alloy is conducted at temperatures near or exceeding 1000 °C, whereas plasma nitriding is optimally performed within the 700 to 900 °C range. This enables the formation of a protective surface layer of TiN compound while preserving the mechanical properties of the workpiece [136]. The nitride layer is composed of a compound layer (TiN and Ti2N) on the surface and a subsequent nitrogen diffusion zone (α-Ti(N)) [41,137,138]. In previous work, researchers have demonstrated that a protective Ti–N layer could be formed on Ti6Al4V surface via hollow cathode plasma source nitriding [139]. And a special compound layer structure with nanocrystalline/amorphous TiN formed during this process.

Ionic nitriding can significantly enhance the surface hardness and wear resistance of titanium alloys, and it can form a fine grain layer by changing the microstructure of the alloy, thus enabling it to exhibit better mechanical properties [140]. However, prolonged high-temperature nitriding is prone to lead to the growth and precipitation of undesirable grains, which is not conducive to the enhancement of the material properties of titanium alloys [141]. The study of She et al. [142] also showed that the hardness and brittleness of the nitride layer increase with the increase in the temperature, which has a significant effect on the abrasion resistance. The low-temperature environment is not conducive to the growth of nitriding layer. Therefore, the nitriding layer may not meet the requirements of surface modification. So, the selection of the appropriate treatment temperature and treatment time is an important part to ensure the quality of the nitriding layer.

Ion nitriding's characteristics are its high energy efficiency, fast speed, small deformation of workpiece, low pollution levels, and suitability for a complex surface workpiece. However, it has a high cost, energy consumption and other shortcomings, which limit the application of this technology. And traditional DC plasma nitriding often results in surface damage, edge effect and hollow cathode effect, which can negatively impact the quality of the nitride layer [143]. In order to overcome these defects, researchers have made many efforts. Active screen plasma nitriding technology is one of the active screen plasma nitriding technologies. This innovative approach utilizes a metal mesh cage to transfer the cathodic potential, shifting the formation of plasma from the component's surface to the cage. Consequently, it effectively mitigates many of the inherent problems associated with DC plasma nitriding. The experimental setup of this technology is shown in Figure 34 [144].

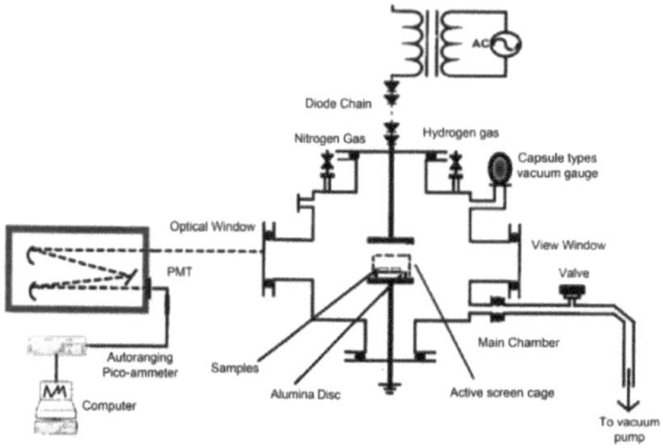

Figure 34. Experimental diagram of active screen plasma nitriding device [144].

To address these issues, significant efforts have been made, leading to the development of active screen plasma nitriding technology. This innovative approach utilizes a metal mesh cage to transfer the cathodic potential, shifting the formation of plasma from the component's surface to the cage. Consequently, it effectively mitigates many of the inherent problems associated with DC plasma nitriding. The experimental setup of this technology is illustrated in Figure 34 [144], demonstrating its unique approach to improving the nitriding process.

Although this method of active screen plasma nitriding solves many of the inherent shortcomings of traditional plasma nitriding, it also introduces a new problem: the process's high nitriding temperature of around 800 °C. When applied to titanium alloys, this elevated temperature may lead to the growth of large grains, potentially compromising the mechanical properties of the titanium alloy matrix. Therefore, when employing this method, additional measures are often necessary to mitigate the required temperature of titanium nitriding. Some studies have shown that shot peening can obtain a nanocrystalline structure with a surface gradient. The high-energy grain boundaries present in the surface nanocrystalline layer can provide channels for the diffusion of nitrogen, which is conducive to saving the nitriding time and lowering the nitriding temperature [143].

For example, Yao et al. [145] carried out nitriding treatment of pure titanium after shot peening and found that the nitriding temperature was reduced to 550 °C. Similarly, Wen et al. [146] reduced the optimum plasma nitriding temperature to 500 °C for TC4 titanium alloy via ASPN after SP. Zhang et al. [143] further confirmed the effectiveness of this approach by comparing nitride TA17 samples, both pre-treated with shot peening and without pre-treatment. They found that the surface nanocrystals generated by shot peening significantly enhanced the nitriding kinetics and promoted the formation of nitrides. Observing the surface morphology and SEM cross-sectional images of the original TA17 and the SP-treated TA17 samples in Figure 35 [143], it can be found that the nitriding samples treated with shot peening have more titanium nitride particles generated in the surface layer, forming a reticulated structure. Additionally, the composite layer generated is also thicker. Combined with the previous research, it can be found that SP-ASPN nitriding technology not only effectively reduces the nitriding temperature but also yields a nitriding layer of superior quality and thickness.

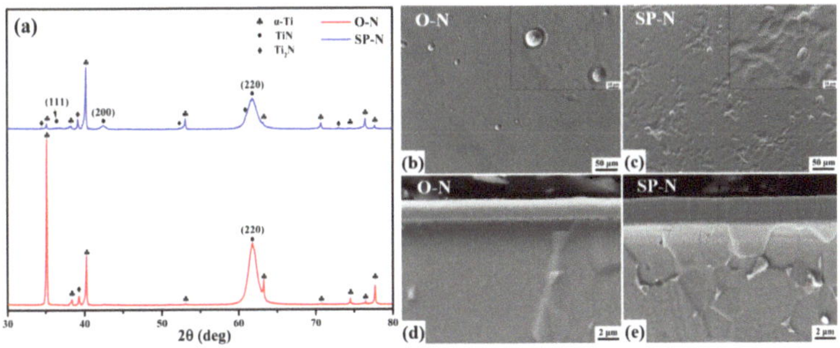

Figure 35. The O-N and SP-N samples nitrided at 500 °C for 20 h via ASPN: (**a**) XRD patterns, (**b**,**c**) surface morphologies, (**d**,**e**) cross-sectional SEM images [143].

In summary, the combination of shot peening and ASPN nitriding technology has demonstrated significant advantages in the surface treatment of titanium alloys. This approach not only effectively reduces the required temperature for nitriding but also enhances the quality and thickness of the nitride layer.

- Oxidized coating

Oxygen is another important element that significantly affects the quality of titanium alloys, and this paper focuses on plasma electrolytic oxidation (PEO) technology. Compared with the nitriding and carbonizing treatments, this coating method is a more cost-effective and environmentally friendly surface treatment technique. Taking flame-retardant alloys as an example, the preparation of flame-retardant coatings on the surface of parts can effectively maintain the original mechanical properties of the metal and improve the flame-retardant properties of the parts, avoiding the introduction of low-strength and high-cost alloys through alloying [147].

Plasma electrolytic oxidation (PEO), as a promising high-pressure plasma-assisted electrochemical method, can produce oxide ceramic-like coatings on metal substrates submerged in electrolyte [148]. Their coatings are grown in situ and consist of a two-layer structure: the inner layer is dense, while the outer layer is porous [149–152]. Currently, PEO has been shown to be an effective surface treatment for additive manufacturing of Ti6Al4V alloys to obtain coatings with high hardness and high abrasion and corrosion resistance [153]. However, inherent defects such as porosity and cracks limit the development of this coating in the medical field [154].

Composite surface modification technology is an effective measure to compensate for the defects of a single type of technology. Arash et al. [148] showed that pre-treatment or post-treatment methods, such as laser surface melting (LSM) or laser surface texturing (LST), can effectively improve the densification of PEO coatings, which leads to a reduction in the porosity and number of the pores and an improvement of the coating quality.

With ongoing advancements in technology and metal surface treatment, the application of titanium alloys in biomedical, shipbuilding, and aerospace industries is expected to experience further expansion [155–157]. Simultaneously, these advanced technologies are likely to be adapted for the fabrication of a broader spectrum of alloys, such as aluminum alloys and high-temperature-resistant alloys [158,159]. The advancements discussed herein not only highlight the evolution of surface treatment techniques but also pave the way for future research in optimizing the mechanical and chemical properties of metal alloys. Our comprehensive review underscores the importance of continued innovation and interdisciplinary research in material science to develop more efficient, cost-effective, and environmentally friendly surface treatment methods.

4. Conclusions

In conclusion, this review comprehensively examines the characteristics of titanium alloys with diverse microstructures and investigates the impact of various surface processing methods, including machining, conventional heat treatment, and thermochemical treatment, on the surface quality and material properties of Ti and its alloys at the microstructural level. The findings underscore the superiority of composite surface modification technologies in enhancing the material properties of metals compared to single treatment methods.

Addressing the challenges posed by the difficult machinability of titanium alloys, Electrical Discharge Machining (EDM) emerges as an effective technique for Ti alloy cutting. Additionally, shot peening and surface mechanical grinding play crucial roles in enhancing surface metal densification through nanosizing, thereby contributing to improved material properties. Innovative approaches such as dual shot peening and the combination of vibratory and conventional shot peening exhibit superior efficacy in increasing metal fatigue strength.

Common heat treatment methods like quenching and annealing are identified as effective means to alter the metallurgical structure of alloys, thereby enhancing the performance of the metal's surface layer. Thermochemical treatment, involving the introduction of titanium-sensitive elements, leads to the formation of a protective layer on titanium alloy surfaces, mitigating external influences and contact with the alloy matrix. Carburizing elements, when introduced before heat treatment, nanostructure the surface, facilitating element diffusion and reducing the carburization temperature requirements, ultimately improving heat treatment efficiency and forming higher quality carburized alloys.

The authors assert that composite surface modification technology serves as a potent strategy for elevating the surface hardness and friction corrosion resistance of titanium alloys. This review aims to aid researchers seeking to enhance the surface properties of titanium and its alloys by delving into the mechanisms through which diverse processing methods bolster the surface characteristics of these alloys. Given the widespread use of titanium, continuous development and refinement of surface modification techniques remain imperative, leaving room for further exploration and summarization of various technical approaches.

Author Contributions: Y.Z.: Conceptualization, Investigation, Resources, Supervision, Project Administration, Writing—Review and Editing; K.G.: Investigation, Methodology, Validation, Data curation, Writing—Original Draft Preparation; P.C.: Methodology, Validation, Supervision; J.Y. and F.D.: Investigation, Data Curation. All authors have read and agreed to the published version of the manuscript.

Funding: This work was supported by the Research Project Founded by the Education Department of Hunan Province (23B0499), and the National Natural Science Foundation of China (No. 52005517).

Institutional Review Board Statement: Not applicable.

Informed Consent Statement: Not applicable.

Data Availability Statement: Data are contained within the article.

Conflicts of Interest: The authors declare no conflict of interest.

References

1. Yang, C.; Li, M.; Liu, Y. Severe plastic deformation induced precipitation of the ordered α_2-Ti_3Al phase in Ti-5Al-2Sn-2Zr-4Mo-4Cr. *J. Alloys Compd.* **2021**, *854*, 157277. [CrossRef]
2. Ma, R.; Zhang, X.F. Improvement of mechanical properties and microstructural refining of cast titanium alloys by coupling of electropulsing and temporary alloying element hydrogen. *Mater. Sci. Eng. A* **2022**, *858*, 144176. [CrossRef]
3. Gao, Y.; Yang, L.; Fan, Q.B.; Lei, W.; Chen, K.; Zhu, X.J.; Mu, X.N.; Yao, J.H. Effect of Fe content on microstructure and hardness of Ti-4.5Mo–5Al-1.8Zr-2.5Cr-1.1Sn titanium alloy based on high-throughput diffusion couple. *Mater. Sci. Eng. A* **2022**, *842*, 143089. [CrossRef]

4. Zhao, E.T.; Sun, S.C.; Zhang, Y. Recent advances in silicon containing high temperature titanium alloys. *J. Mater. Res. Technol.* **2021**, *14*, 3029–3042. [CrossRef]
5. Tan, Y.M.; Chen, R.R.; Fang, H.Z.; Liu, Y.L.; Cui, H.Z.; Su, Y.Q.; Guo, J.J.; Fu, H.Z. Enhanced strength and ductility in Ti46Al4Nb1Mo alloys via boron addition. *J. Mater. Sci. Technol.* **2022**, *102*, 16–23. [CrossRef]
6. Kolli, P.R.; Devaraj, A. A Review of Metastable Beta Titanium Alloys. *Metals* **2018**, *8*, 506. [CrossRef]
7. Singh, S.S.; Harpreet, S.; Hady, A.M.G. A review on alloy design, biological response, and strengthening of β-titanium alloys as biomaterials. *Mater. Sci. Eng. C* **2020**, *121*, 111661.
8. Masoud, S.; Erfan, G.R.; Saeid, A.; Seeram, R.; Nazatul, S.L. A state-of-the-art review of the fabrication and characteristics of titanium and its alloys for biomedical applications. *Bio-Des. Manuf.* **2021**, *5*, 371–395.
9. Fan, H.; Yang, S. Effects of direct aging on near-alpha Ti–6Al–2Sn–4Zr–2Mo (Ti-6242) titanium alloy fabricated by selective laser melting (SLM). *Mater. Sci. Eng. A* **2020**, *788*, 139533. [CrossRef]
10. Alessandro, C.; Alberta, A.; Paolo, F.; Mariangela, L. Towards customized heat treatments and mechanical properties in the LPBF-processed Ti-6Al-2Sn-4Zr-6Mo alloy. *Mater. Des.* **2022**, *215*, 110512.
11. Optasanu, V.; Berger, P.; Vincent, B.; Lucas, D.M.C.M.; Herbst, F.; Montesin, T. Strong correlation between high temperature oxidation resistance and nitrogen mass gain during near alpha titanium alloys exposure in air. *Corros. Sci.* **2023**, *224*, 111547. [CrossRef]
12. Si, Y.H.; Li, M.S.; Liu, H.Y.; Jiang, X.Z.; Yu, H.Y.; Sun, D.B. Evaluation of tribocorrosion performance of Ti6Al4V alloy in simulated inflammatory and hyperglycemic microenvironments. *Wear* **2023**, *532–533*, 205077. [CrossRef]
13. Huang, S.; Sun, B.B.; Guo, S.Q. Microstructure and property evaluation of TA15 titanium alloy fabricated by selective laser melting after heat treatment. *Opt. Laser Technol.* **2021**, *144*, 107422. [CrossRef]
14. Yan, Z.; Liang, Y.L.; Zhang, Z.J.; Xu, J.; Yi, Y.L. Effect of shot peening on high-frequency fatigue properties of TC11 titanium alloy. *Chin. J. Rare Met.* **2014**, *38*, 554–560. (In Chinese)
15. Li, Z.Y.; Cai, Z.B.; Wu, Y.Q.; Zhu, M.H. Effect of nitrogen ion implantation dose on torsional fretting wear behavior of titanium and its alloy. *Trans. Nonferr. Met. Soc.* **2017**, *27*, 324–355. [CrossRef]
16. Nihal, Y.; Kubilay, A. A review on heat treatment efficiency in metastable β titanium alloys: The role of treatment process and parameters. *J. Mater. Res. Technol.* **2020**, *9*, 15360–15380.
17. Xu, S.; Zhang, H.M.; Xiao, N.M.; Qiu, R.S.; Cui, Z.S.; Fu, M.W. Mechanisms of macrozone elimination and grain refinement of near α Ti alloy via the spheroidization of the Widmannstätten structure. *Acta Mater.* **2023**, *260*, 119339. [CrossRef]
18. Sun, S.C.; Fang, H.Z.; Li, Y.L.; Chen, R.R.; Zhu, B.H.; Guo, J.J. Formation mechanism of titanium solid solution and its influence on equiaxed behavior of α phase of Ti–5Al–5Mo–5Cr–2Zr-xNb alloys. *J. Mater. Res. Technol.* **2023**, *26*, 434–444. [CrossRef]
19. Xu, A.J.; Wan, H.F.; Liang, C.Z.; Niu, Y.T.; Tao, Q.; Tang, Z.J. Current status and development trend of low-temperature titanium alloy materials. *J. Netshape Form. Eng.* **2020**, *12*, 145–156. (In Chinese)
20. Huang, L.; Li, C.M.; Li, C.L.; Hui, S.X.; Yu, Y.; Zhao, M.J.; Guo, S.Q.; Li, J.J. Research progress on microstructure evolution and hot processing maps of high strength β titanium alloys during hot deformation. *Trans. Nonferr. Met. Soc.* **2022**, *32*, 3835–3859. [CrossRef]
21. Shi, S.X.; Liu, X.S.; Zhang, X.V.; Zhou, K.C. Comparison of flow behaviors of near beta Ti-55511 alloy during hot compression based on SCA and BPANN models. *Trans. Nonferr. Met. Soc.* **2021**, *31*, 1665–1679. [CrossRef]
22. Zhang, H.; Zhang, S.; Zhang, S.; Liu, X.J.; Wu, X.X.; Zhang, S.Q.; Zhou, G. High Temperature Deformation Behavior of Near-β Titanium Alloy Ti-3Al-6Cr-5V-5Mo at α+β and β Phase Fields. *Crystals* **2023**, *13*, 371. [CrossRef]
23. Kopova, I.; Stráský, J.; Harcuba, P.; Landa, M.; Janeček, M.; Bačákova, L. Newly developed Ti–Nb–Zr–Ta–Si–Fe biomedical beta titanium alloys with increased strength and enhanced biocompatibility. *Mater. Sci. Eng. C* **2016**, *60*, 230–238. [CrossRef]
24. Li, C.M.; Huang, L.; Zhao, M.J.; Guo, S.Q.; Li, J.J. Hot deformation behavior and mechanism of a new metastable β titanium alloy Ti–6Cr–5Mo–5V–4Al in single phase region. *Mater. Sci. Eng. A* **2021**, *814*, 141231. [CrossRef]
25. Li, C.H.; Cui, Y.M.; Zheng, W.W.; Song, L.; Wu, Y.L.; Wu, Y.D.; Yu, C.Y.; Hui, X.D. Unveiling the dynamic softening mechanism via micromechanical behavior for a near-β titanium alloy deformed at a high strain rate. *J. Mater. Res. Technol.* **2023**, *26*, 9392–9405. [CrossRef]
26. Xin, S.W.; Zhao, Y.Q.; Lu, Y.F.; Li, Q.; Yang, H.Y. Role of grain boundaries in the high-temperature performance of a highly stabilized beta titanium alloy II: Creep behavior. *Mater. Sci. Eng. A* **2013**, *559*, 7–13. [CrossRef]
27. Zheng, Y.P.; Zeng, W.D.; Wang, Y.B.; Zhou, D.D.; Gao, X.X. High strain rate compression behavior of a heavily stabilized beta titanium alloy: Kink deformation and adiabatic shearing. *J. Alloys Compd.* **2017**, *708*, 84–92. [CrossRef]
28. Xu, Q.B.; Liu, S.Y. Titanium and titanium alloy grades and applications in foreign aerospace industry. *Lead. Edge Technol.* **2022**, *16*, 96–99. (In Chinese)
29. Chen, C.; Feng, T.T.; Zhang, Y.W.; Ren, B.Q.; Hao, W.; Zhao, X.H. Improvement of microstructure and mechanical properties of TC4 titanium alloy GTAW based wire arc additive manufacturing by using interpass milling. *J. Mater. Res. Technol.* **2023**, *27*, 1428–1445. [CrossRef]
30. Iliana, J.F.; Mohd, F.F.; Ahmad, B.M.; Bakar, S.A.; Afiqah, M.R.N.; Norhamidi, M.; Fadhlina, M.I.; Hani, J.N.; Seah, T.K. Influence of heat treatment parameters on microstructure and mechanical performance of titanium alloy in LPBF: A brief review. *J. Mater. Res. Technol.* **2023**, *24*, 4091–4110.

31. Li, Y.; Zhou, Z.L.; Yi, X.N.; Yan, J.W.; Xiu, J.J.; Fang, D.Z.; Shao, M.H.; Ren, P.; He, Y.Y.; Qiu, J.X. Improved seawater corrosion resistance of electron beam melting Ti6Al4V titanium alloy by plasma nitriding. *Vacuum* **2023**, *216*, 112463. [CrossRef]
32. Morinaga, M. The Molecular Orbital Approach to Titanium Alloy Design. *Key Eng. Mater.* **2018**, *770*, 217–223. [CrossRef]
33. Bosung, S.; Hyung-Ki, P.; Chang-Soo, P.; Kwangsuk, P. Effect of alloying elements on corrosion properties of high corrosion resistant titanium alloys in high concentrated sulfuric acid. *Mater. Today Commun.* **2023**, *34*, 105113.
34. Dong, Y.C.; Huang, S.; Wang, Y.Y.; Zhang, B.; Alexandrov, I.V.; Chang, H.; Dan, Z.H.; Ma, L.; Zhou, L. Stress corrosion cracking of TC4 ELI alloy with different microstructure in 3.5% NaCl solution. *Mater. Charact.* **2022**, *194*, 112357. [CrossRef]
35. Miyamoto, H.; Yuasa, M.; Rifai, M.; Fujiwara, H. Corrosion Behavior of Severely Deformed Pure and Single-Phase Materials. *Mater. Trans.* **2019**, *60*, 1243–1255. [CrossRef]
36. Chuvil'deev, V.; Kopylov, V.; Nokhrin, A.; Tryaev, P.; Kozlova, N.; Tabachkova, N.; Lopatin, Y.; Ershova, A.; Mikhaylov, A.; Gryaznov, M.; et al. Study of mechanical properties and corrosive resistance of ultrafine-grained α-titanium alloy Ti-5Al-2V. *J. Alloys Compd.* **2017**, *723*, 354–367. [CrossRef]
37. Abhinay, T.; Savaş, K.; Ashish, K. Recent Trends in the Characterization and Application Progress of Nano-Modified Coatings in Corrosion Mitigation of Metals and Alloys. *Appl. Sci.* **2023**, *13*, 730–763.
38. Zhang, F.; Yan, S.; Yan, M. In-Situ fabrication of novel (Ti, Cr)-N/aluminide multilayer coatings by plasma nitriding Ti-Cr coated Al alloy. *Ceram. Int.* **2018**, *44*, 7259–7266. [CrossRef]
39. Naeem, M.; Shabana, A.; Shafiq, M.; Raza, H.A.; Javed, I.; Díaz-Guillén, J.; Sousa, R.; Jelani, M.; Abrar, M. Wear and corrosion studies of duplex surface-treated AISI-304 steel by a combination of cathodic cage plasma nitriding and PVD-TiN coating. *Ceram. Int.* **2022**, *48*, 21473–21482. [CrossRef]
40. Keisuke, F.; Masataka, I.; Yoichi, I.; Shoichi, K. Rapid Nitriding of Titanium Alloy with Fine Grains at Room Temperature. *Adv. Mater.* **2021**, *33*, 2008098.
41. Ahmadi, M.; Hosseini, S.; Hadavi, S. Comparison of auxiliary cathode and conventional plasma nitriding of gamma-TiAl alloy. *Vacuum* **2016**, *131*, 89–96. [CrossRef]
42. Dai, J.J.; Zhu, J.Y.; Chen, C.Z.; Weng, F. High temperature oxidation behavior and research status of modifications on improving high temperature oxidation resistance of titanium alloys and titanium aluminides: A review. *J. Alloys Compd.* **2016**, *685*, 784–798. [CrossRef]
43. Jiang, B.B.; Wen, D.H.; Wang, Q.; Che, J.D.; Dong, C.; Peter, K.L.; Xu, F.; Sun, L.X. Design of near-α Ti alloys via a cluster formula approach and their high-temperature oxidation resistance. *J. Mater. Sci. Technol.* **2019**, *35*, 1008–1016. [CrossRef]
44. Ebach-Stahl, A.; Eilers, C.; Laska, N.; Braun, R. Cyclic oxidation behaviour of the titanium alloys Ti-6242 and Ti-17 with Ti–Al–Cr–Y coatings at 600 and 700 °C in air. *Surf. Coat. Technol.* **2013**, *223*, 24–31. [CrossRef]
45. Xu, Y.Q.; Fu, Y.; Li, J.; Xiao, W.L.; Zhao, X.Q.; Ma, C.L. Effects of tungsten addition on the microstructural stability and properties of Ti-6.5Al-2Sn-4Hf-2Nb-based high temperature titanium alloys. *J. Mater. Sci. Technol.* **2021**, *93*, 147–156. [CrossRef]
46. Ozturk, D.; Pilchak, A.; Ghosh, S. Experimentally validated dwell and cyclic fatigue crack nucleation model for α–titanium alloys. *Scr. Mater.* **2016**, *127*, 15–18. [CrossRef]
47. Hémery, S.; Villechaise, P. On the influence of ageing on the onset of plastic slip in Ti-6Al-4V at room temperature: Insight on dwell fatigue behavior. *Scr. Mater.* **2016**, *130*, 157–160. [CrossRef]
48. Cao, S.; Zhang, S.Z.; Liu, J.R.; Li, S.J.; Sun, T.; Li, J.P.; Gao, Y.; Yang, R.; Hu, Q.M. Interaction between Al and other alloying atoms in α-Ti for designing high temperature titanium alloy. *Comp. Mater. Sci.* **2021**, *197*, 110620. [CrossRef]
49. Li, J.; Xu, Y.Q.; Xiao, W.L.; Ma, C.L.; Huang, X. Development of Ti-Al-Ta-Nb-(Re) near-α high temperature titanium alloy: Microstructure, thermal stability and mechanical properties. *J. Mater. Sci. Technol.* **2022**, *109*, 1–11. [CrossRef]
50. Wen, P.C.; Yuan, L.J.; Tao, R.; Li, J.; Li, D. First-principles investigation of interaction between surface oxygen and other alloy atoms in α-Ti (0001) for designing high-temperature titanium alloy. *Appl. Surf. Sci.* **2022**, *604*, 154535. [CrossRef]
51. Huang, C.W.; Ge, P.; Zhao, Y.Q.; Xin, S.W.; Zhou, W.; Li, Q.; Zeng, W.D. Research progress of low-temperature titanium alloy. *Rare Met. Mat. Eng.* **2016**, *45*, 254–260.
52. Lu, J.W.; Zhao, Y.Q.; Ge, P.; Niu, H.Z.; Zhang, Y.S.; Zhang, W.; Zhang, P.X. Microstructure and mechanical properties of new high strength beta-titanium alloy Ti-1300. *Mater. Sci. Eng. A* **2015**, *621*, 182–189. [CrossRef]
53. Mantri, S.; Choudhuri, D.; Alam, T.; Viswanathan, G.; Sosa, J.; Fraser, H.; Banerjee, R. Tuning the scale of α precipitates in β-titanium alloys for achieving high strength. *Scr. Mater.* **2018**, *154*, 139–144. [CrossRef]
54. Wang, J.X.; Ye, X.W.; Li, Y.H.; Wan, M.P.; Huang, C.W.; Huang, F.; Lei, M.; Liu, D.; Ma, R.; Ren, X.L. Effect of annealing temperature on mechanical properties of TC21 titanium alloy with multilevel lamellar microstructure. *Mater. Sci. Eng. A* **2023**, *869*, 144788. [CrossRef]
55. Li, C.C.; Xin, C.; Wang, Q.; Ren, J.Q.; Zhao, B.; Wu, J.P.; Pan, X.L.; Lu, X.F. A novel low-cost high-strength β titanium alloy: Microstructure evolution and mechanical behavior. *J. Alloys Compd.* **2023**, *959*, 170497. [CrossRef]
56. Karolina, D.; Tomasz, G.; Mateusz, D.; Bronisław, P.; Agnieszka, S.; Zdzisław, L. Functionalization of the Implant Surface Made of NiTi Shape Memory Alloy. *Materials* **2023**, *16*, 1609–1620.
57. Wang, S.Y. Investigation of chemical analysis methods of niobium-titanium alloys for superconductivity. *Ind. Technol. Innov.* **2022**, *4*, 33–35. (In Chinese)
58. Lv, P.; Zhong, C.L.; Huang, D.F.; Zhou, X.S.; Liu, Z.C.; Huang, D.J. Effect of introducing manganese as additive on microstructure, hydrogen storage properties and rate limiting step of Ti–Cr alloy. *Int. J. Hydrogen Energy* **2022**, *47*, 459–469. [CrossRef]

59. Ilias, G.S.; Papazoglou, E.L.; Panagiotis, K.O.; Karkalos, N.E.; Markopoulos, A.P. Surface antibacterial properties enhanced through engineered textures and surface roughness: A review. *Colloid. Surf. B* **2023**, *231*, 113584.
60. Zhang, Y.B.; Bai, Q.S.; Wang, P. 3D surface topography analysis and functionality-related performance of the machined surface in slot micro-milling titanium alloy Ti6Al4V. *Int. J. Adv. Manuf. Technol.* **2023**, *127*, 1609–1629. [CrossRef]
61. Yu, W.W.; Wu, J.; Li, Y.G.; An, Q.L.; Ming, W.W.; Chen, D.; Wang, H.W.; Chen, M. Investigations on surface modification of nickel-based superalloy subjected to ultrasonic surface rolling process. *Int. J. Adv. Manuf. Technol.* **2023**, *129*, 1473–1488. [CrossRef]
62. Du, H.H.; Chen, H.W.; Zhu, Z.W.; Wang, Z.K.; Suet, T. Novel hybrid machining process of titanium alloy for texturing high-quality microstructure array surfaces. *Surf. Coat. Technol.* **2023**, *462*, 129494. [CrossRef]
63. Ahmed, N.; Ishfaq, K.; Moiduddin, K.; Ali, R.; Al-Shammary, N. Machinability of titanium alloy through electric discharge machining. *Mater. Manuf. Process.* **2019**, *34*, 93–102. [CrossRef]
64. Rebecka, L.; Filip, L.; Henrik, P.; Rachid, M.; Eric, S.; Volodymyr, B. Performance and wear mechanisms of PCD and pcBN cutting tools during machining titanium alloy Ti6Al4V. *Wear* **2020**, *454–455*, 203329.
65. Cui, X.; Li, C.; Zhang, Y.; Said, Z.; Debnath, S.; Sharma, S.; Ali, H.M.; Yang, M.; Gao, T.; Li, R.Z. Grindability of titanium alloy using cryogenic nanolubricant minimum quantity lubrication. *J. Manuf. Process.* **2022**, *80*, 273–286. [CrossRef]
66. Fábio, C.O.D.; Carla, A.A.; Luiz, A.M.J.; Gopal, S.K. The influence of additive manufacturing on the micromilling machinability of Ti6Al4V: A comparison of SLM and commercial workpieces. *J. Manuf. Process.* **2020**, *60*, 299–307.
67. Dandekar, R.C.; Shin, C.Y.; Barnes, J. Machinability improvement of titanium alloy (Ti–6Al–4V) via LAM and hybrid machining. *Int. J. Mach. Tool Manuf.* **2010**, *50*, 174–182. [CrossRef]
68. Li, N.; Chen, Y.J.; Kong, D.D.; Tan, S.L. Experimental investigation with respect to the performance of deep submillimeter-scaled textured tools in dry turning titanium alloy Ti-6Al-4V. *Appl. Surf. Sci.* **2017**, *403*, 187–199. [CrossRef]
69. Ni, C.B.; Zhu, L.D.; Zheng, Z.P.; Zhang, J.Y.; Yang, Y.; Yang, J.; Bai, Y.C.; Weng, C.; Lu, W.F.; Wang, H. Effect of material anisotropy on ultra-precision machining of Ti-6Al-4V alloy fabricated by selective laser melting. *J. Alloys Compd.* **2020**, *848*, 156457. [CrossRef]
70. Yip, S.; To, S. Reduction of tool tip vibration in single-point diamond turning using an eddy current damping effect. *Int. J. Adv. Manuf. Technol.* **2019**, *103*, 1799–1809. [CrossRef]
71. Zhao, Z.J.; To, S.; Sun, Z.W.; Ji, R.J.; Yu, K.M. Microstructural effects of Ti6Al4V alloys modified by electropulsing treatment on ultraprecision diamond turning. *J. Manuf. Process.* **2019**, *39*, 58–68. [CrossRef]
72. Vijaykumar, S.J. Multi-characteristics optimization in EDM of NiTi alloy, NiCu alloy and and BeCu alloy using Taguchi's approach and utility concept. *Alex. Eng. J.* **2017**, *57*, 2807–2817.
73. Panagiotis, K.O.; Papazoglou, E.L.; Beata, L.; Krzysztof, Z.; Markopoulos, A.P. Surface and Subsurface Quality of Titanium Grade 23 Machined by Electro Discharge Machining. *Materials* **2021**, *15*, 164–185.
74. Khoshaim, A.B.; Muthuramalingam, T.; Moustafa, E.B.; Elsheikh, A. Influences of tool electrodes on machinability of titanium α-β alloy with ISO energy pulse generator in EDM process. *Alex. Eng. J.* **2023**, *63*, 465–474. [CrossRef]
75. Unal, O.; Maleki, E.; Karademir, I.; Husem, F.; Efe, Y.; Das, T. Effects of conventional shot peening, severe shot peening, re-shot peening and precised grinding operations on fatigue performance of AISI 1050 railway axle steel. *Int. J. Fatigue* **2022**, *155*, 106613. [CrossRef]
76. Dunja, R.; Roman, Š.; Sebastjan, Ž. Effect of Shot Peening on the Strength and Corrosion Properties of 6082-T651 Aluminium Alloy. *Materials* **2023**, *16*, 4976–4990.
77. Maleki, E.; Unal, O.; Amanov, A. Novel experimental methods for the determination of the boundaries between conventional, severe and over shot peening processes. *Surf. Interfaces* **2018**, *13*, 233–254. [CrossRef]
78. Zheng, Z.B.; Balint, S.D.; Dunne, P.F. Investigation of slip transfer across HCP grain boundaries with application to cold dwell facet fatigue. *Acta Mater.* **2017**, *127*, 43–53. [CrossRef]
79. Su, K.X.; Zhang, J.W.; Li, H.; Ji, D.D.; Hu, L.K. Anti-fatigue strengthening mechanism of conventional shot peening and micro-shot peening on bare and micro-arc oxidation coated 6082-T6 aluminum alloy. *Mater. Lett.* **2023**, *331*, 133442. [CrossRef]
80. Wang, Z.Y.; Xie, L.L.; Zhang, Q.; Abd, A.R.; Chen, W.L.; Zhou, L.J. Surface layer strengthening mechanism of 2060 aluminum–lithium alloy after shot-peening. *J. Mater. Res. Technol.* **2023**, *23*, 4615–4633. [CrossRef]
81. Shi, H.L.; Liu, D.X.; Pan, Y.F.; Zhao, W.D.; Zhang, X.H.; Ma, A.M.; Liu, B.; Hu, Y.H.; Wang, W. Effect of shot peening and vibration finishing on the fatigue behavior of TC17 titanium alloy at room and high temperature. *Int. J. Fatigue* **2021**, *151*, 106391. [CrossRef]
82. Efe, Y.; Karademir, I.; Husem, F.; Maleki, E.; Unal, O. Surface Severe Plastically Deformed Nanostructured AA7075 Alloy: Assessment on Tribological and Axial Fatigue Behaviors. *J. Mater. Eng. Perform.* **2020**, *29*, 1774–1783. [CrossRef]
83. Aoudia, K.; Retraint, D.; Verdy, C.; Langlade, C.; Creus, J.; Sanchette, F. Enhancement of Mechanical Properties and Corrosion Resistance of HVOF-Sprayed NiCrBSi Coatings Through Mechanical Attrition Treatment (SMAT). *J. Therm. Spray Technol.* **2020**, *29*, 2065–2079. [CrossRef]
84. Yao, Q.T.; Tong, W.P.; Li, M.Y.; Zhang, G.L. Neutral Molten Salt-Bath Carburizing of Ti6Al4V Alloy with Nanocrystalline Surface Layer at Low Temperature Assisted by Surface Mechanical Attrition Treatment. *Key Eng. Mater.* **2017**, *727*, 1001–1008. [CrossRef]
85. Peng, Z.L.; Zhang, X.Y.; Liu, L.B.; Xu, G.T.; Wang, G.; Zhao, M.H. Effect of high-speed ultrasonic vibration cutting on the microstructure, surface integrity, and wear behavior of titanium alloy. *J. Mater. Res. Technol.* **2023**, *24*, 3870–3888. [CrossRef]
86. Chamgordani, A.S.; Miresmaeili, R.; Aliofkhazraei, M. Improvement in tribological behavior of commercial pure titanium (CP-Ti) by surface mechanical attrition treatment (SMAT). *Tribol. Int.* **2018**, *119*, 744–752. [CrossRef]

87. Li, C.; Cui, W.F.; Zhang, Y.S. Surface self-nanocrystallization of α+β titanium alloy by surface mechanical grinding treatment. *Met. Mater. Int.* **2017**, *23*, 512–518. [CrossRef]
88. Qin, Z.B.; Zhang, Q.; Luo, Q.; Zhong, W.; Shen, B.; Liu, L.; Hu, W.B. Microstructure design to improve the corrosion and cavitation corrosion resistance of a nickel-aluminum bronze. *Corros. Sci.* **2018**, *139*, 255–266. [CrossRef]
89. Dang, K.X.; Wang, K.H.; Chen, W.T.; Liu, G. Study on fast gas forming with in-die quenching for titanium alloys and the strengthening mechanisms of the components. *J. Mater. Res. Technol.* **2022**, *18*, 3916–3932. [CrossRef]
90. Qin, Z.B.; Xia, D.H.; Zhang, Y.W.; Wu, Z.; Liu, L.; Lv, Y.T.; Liu, Y.C.; Hu, W.B. Microstructure modification and improving corrosion resistance of laser surface quenched nickel–aluminum bronze alloy. *Corros. Sci.* **2020**, *174*, 108744. [CrossRef]
91. Jian, S.C.; Wang, J.X.; Xu, D.; Ma, R.; Huang, C.W.; Lei, M.; Liu, D.; Wan, M.P. Gradient microstructure and mechanical properties of Ti-6Al-4V titanium alloy fabricated by high-frequency induction quenching treatment. *Mater. Des.* **2022**, *222*, 111031. [CrossRef]
92. Teixeira, J.; Denand, B.; Aeby-Gautier, E.; Denis, S. Simulation of coupled temperature, microstructure and internal stresses evolutions during quenching of a β-metastable titanium alloy. *Mater. Sci. Eng. A* **2016**, *651*, 615–625. [CrossRef]
93. Lytvynenko, I.V.; Maruschak, P.O. Analysis of the state of the modified nanotitanium surface with the use of the mathematical model of a cyclic random process. *Optoelectron. Instrum.* **2015**, *51*, 254–263. [CrossRef]
94. He, Q.W.; Du, Z.X.; Wang, X.P.; Cui, X.M.; Liu, F.; Chen, Y.F.; Cheng, J.; Zhao, X.P.; Chen, Y.Y. Effect of cold rolling reduction on grain growth kinetics of TB8 titanium alloy during annealing heat treatment. *Mater. Res. Express* **2019**, *6*, 116586. [CrossRef]
95. Wang, K.; Wu, M.Y.; Ren, Z.; Zhang, Y.; Xin, R.L.; Liu, Q. Static globularization and grain morphology evolution of α and β phases during annealing of hot-rolled TC21 titanium alloy. *Trans. Nonferr. Met. Soc.* **2021**, *31*, 2664–2676. [CrossRef]
96. Huang, Z.R.; Xiao, H.; Yu, J.X.; Zhang, H.Y.; Huang, H.G.; Yu, K.; Zhou, R.F. Effects of different annealing cooling methods on the microstructure and properties of TA10 titanium alloys. *J. Mater. Res. Technol.* **2022**, *18*, 4859–4870. [CrossRef]
97. Kumar, U.; Chattopadhyaya, S.; Das, K.A.; Seikh, A.; Sharma, S.; Dwivedi, P.S.; Nagai, K.; Kumar, A.; Agrawal, A.; Singh, S. Effect of Pulsation in Microstructure and Mechanical Properties of Titanium Alloy-Annealed Welded Joints at Different Temperatures. *Photonics* **2023**, *10*, 372. [CrossRef]
98. Zhang, M.Y.; Yun, X.B.; Fu, H.W. Effect of annealing process on the organization and impact properties of TC10 titanium alloy. *Rare Met. Mat. Eng.* **2023**, *52*, 3106–3115.
99. Duan, H.Q.; Han, Y.F.; Lu, W.J.; Mao, J.W.; Wang, L.Q.; Zhang, D. Effect of solid carburization on surface microstructure and hardness of Ti-6Al-4V alloy and (TiB+La_2O_3)/Ti-6Al-4V composite. *Trans. Nonferr. Met. Soc.* **2016**, *26*, 1871–1877. [CrossRef]
100. Jacek, G.; Damian, B.; Witold, K.; Bartosz, P.; Bartłomiej, J.; Barbara, B.; Małgorzata, C.; Marcin, M.; Piotr, N. Comparison of Different Thermo-Chemical Treatments Methods of Ti-6Al-4V Alloy in Terms of Tribological and Corrosion Properties. *Materials* **2020**, *13*, 5192–5207.
101. Zhao, Z.Y.; Hui, P.F.; Wang, T.; Wang, X.; Xu, Y.H.; Zhong, L.S.; Zhao, M.X. New strategy to grow TiC coatings on titanium alloy: Contact solid carburization by cast iron. *J. Alloys Compd.* **2018**, *745*, 637–643. [CrossRef]
102. Liu, W.; Wang, X.T.; Wang, F.; Du, K.F.; Zhang, Z.F.; Guo, Y.Z.; Yin, H.Y.; Wang, D.H. A durable and pH-universal self-standing MoC–Mo_2C heterojunction electrode for efficient hydrogen evolution reaction. *Nat. Commun.* **2021**, *12*, 6776–6785. [CrossRef] [PubMed]
103. Yu, R.; Deng, B.W.; Du, K.F.; Chen, D.; Gao, M.X.; Wang, D.H. Modulating carbon growth kinetics enables electrosynthesis of graphite derived from CO_2 via a liquid–solid–solid process. *Carbon* **2021**, *184*, 426–436. [CrossRef]
104. Song, Q.S.; Xu, Q.; Xu, L.; Ning, Z.Q.; Lou, T.P.; Xie, H.W.; Qi, Y.; Yu, K. Synthesis of Ni-TiC composite powder electrochemically in molten chlorides. *J. Alloys Compd.* **2017**, *690*, 116–122. [CrossRef]
105. Mao, Y.; Xie, H.W.; Chen, X.; Zhao, Y.; Qu, J.K.; Song, Q.S.; Ning, Z.Q.; Xing, P.F.; Yin, H.Y. A combined leaching and electrochemical activation approach to converting coal to capacitive carbon in molten carbonates. *J. Clean. Prod.* **2020**, *248*, 119218. [CrossRef]
106. Jiang, R.; Gao, M.X.; Mao, X.H.; Wang, D.H. Advancements and potentials of molten salt CO_2 capture and electrochemical transformation (MSCC-ET) process. *Curr. Opin. Electrochem.* **2019**, *17*, 38–46. [CrossRef]
107. Liu, J.C.; Dolan, P.K.; Liu, J.X.; Wang, J.Q.; Long, D.W. Dense carbon film coated 316L via In-Situ synthesized CaC_2 in FLiNaK molten salts and its high performance of anti-corrosion property. *Electrochim. Acta* **2019**, *317*, 232–239. [CrossRef]
108. Zhao, M.Y.; Du, P.; Liu, W.; Du, K.F.; Ma, Y.S.; Yin, H.Y.; Wang, D.H. Anodic carbidation of tantalum in molten $CaCl_2$-CaC_2. *J. Solid State Electrochem.* **2022**, *26*, 791–798. [CrossRef]
109. Zhao, M.Y.; Ma, Y.S.; Zhang, Y.; Liu, X.L.; Sun, H.O.; Liang, R.H.; Yin, H.Y.; Wang, D.H. An efficient salt-thermo-carburizing method to prepare titanium carbide coating. *Surf. Coat. Technol.* **2023**, *465*, 129546. [CrossRef]
110. Sveidy, V.; Oscar, P.; Rodríguez, A.G.; Laura, N.S.; Naser, Q.; Laura, O. Photolithographically-patterned C-MEMS graphene by carbon diffusion through nickel. *Nanotechnology* **2021**, *32*, 265302.
111. Dariel, M.; Klein, O.; Frage, N. Enhanced Mass Transport In Titanium Carbide At Large Departures From Stoichiometry. *Powder Metall. Met. C.* **2003**, *42*, 460–467. [CrossRef]
112. Wu, X.; Guan, J.; Liu, J.; Li, K.M.; Yang, F.; Dai, Y. Corrosion behavior of TA2 titanium alloy vacuum induction carburized layer in fluorinated mixed acid. *Surf. Technol.* **2019**, *48*, 304–311. (In Chinese)
113. Jiang, Y.; Sun, N.; Peng, Y.W.; Gong, J.M. Stability of low-temperature-gaseous-carburization layer in AISI316L stainless steel at high temperature. *Surf. Interfaces* **2021**, *23*, 100898. [CrossRef]

114. Amar, T.; Zine, M.T.; Kamel, F. Effect of gaseous carburizing thermochemical treatment on tribological behavior of Ti-6Al-4V alloy. *Frat. Integrita. Strut.* **2021**, *58*, 179–190.
115. Xue, Y.J.; Yan, Y.M.; Yu, W.C.; Liu, K.; Shi, J.; Wang, M.Q. Variation of hydrogen content in gear steel after carburizing and heat treatment and its effect on fatigue properties. *Int. J. Fatigue* **2023**, *177*, 107967. [CrossRef]
116. Zhang, Y.; Wei, Q.L.; Xing, Y.Z.; Jiang, C.P.; Li, X.H.; Zhao, Z.Y. Evaluation of microstructure and wear properties of Ti-6Al-4V alloy plasma carbonized at different temperatures. *J. Wuhan Univ. Technol.* **2015**, *30*, 631–638. [CrossRef]
117. Yang, W.L.; He, X.J.; Li, H.P.; Dong, J.; Chen, W.; Xin, H.; Jin, Z.M. A tribological investigation of SLM fabricated TC4 titanium alloy with carburization pre-treatment. *Ceram. Int.* **2020**, *46*, 3043–3050. [CrossRef]
118. Dong, B.Z.; Guo, X.H.; Zhang, K.D.; Zhang, Y.P.; Li, Z.H.; Wang, W.S.; Cai, C. Combined effect of laser texturing and carburizing on the bonding strength of DLC coatings deposited on medical titanium alloy. *Surf. Coat. Technol.* **2022**, *429*, 127951. [CrossRef]
119. Zheng, Y.; Zhong, J.; Lv, X.P.; Zhao, Y.J.; Zhou, W.; Zhang, Y.X. Microstructure and performance of functionally graded Ti (C, N)-based cermets prepared by double-glow plasma carburization. *Int. J. Refract. Met. Hard* **2014**, *44*, 109–112. [CrossRef]
120. Semboshi, S.; Iwase, A.; Takasugi, T. Surface hardening of age-hardenable Cu–Ti alloy by plasma carburization. *Surf. Coat. Technol.* **2015**, *283*, 262–267. [CrossRef]
121. Oliveira, D.A.C.M.V.; Silva, D.L.C.M.; Pinto, G.C.; Suzuki, A.P.; Machado, B.P.J.; Chad, M.V.; Barboza, R.J.M. Short-term creep properties of Ti-6Al-4V alloy subjected to surface plasma carburizing process. *J. Mater. Res. Technol.* **2015**, *4*, 359–366. [CrossRef]
122. Zhou, Y.; Jing, Z.; Jun, H. Study on surface properties of nanosecond laser textured plasma nitrided titanium alloy. *Mater. Today Commun.* **2022**, *31*, 103746.
123. Peng, T.T.; Dai, M.Y.; Cai, W.; Wei, W.; Wei, K.X.; Hu, J. The enhancement effect of salt bath preoxidation on salt bath nitriding for AISI 1045 steel. *Appl. Surf. Sci.* **2019**, *484*, 610–615. [CrossRef]
124. Zhu, Y.; Lu, W.; Zuo, D.; Cao, D. A novel rare earth-salt bath nitriding of TC21-DT titanium alloy. *Surf. Eng.* **2018**, *34*, 128–131. [CrossRef]
125. Deepak, J.; Raja, B.V.; Kumar, A.K.; Radhakrishnan, V.; Thomas, S.S.H. Salt Bath Nitriding of CP Titanium Grade-2 and TI-6AL-4V Grade-5. *IOP Conf. Ser. Mater. Sci. Eng.* **2017**, *197*, 012066. [CrossRef]
126. Wang, L.; Yang, G.; Yang, L.; Duan, Z.G.; Yang, Y.X. Study on corrosion performance of TC4 titanium alloy treated by low pressure vacuum nitriding. *Foundry Technol.* **2014**, *35*, 503–505. (In Chinese)
127. Ye, M.; Song, X.L.; Liu, Q.; Liu, X.F. Research on surface heat treatment technology of titanium alloy. *Met. Process.* **2010**, *19*, 31–32+34. (In Chinese)
128. Chen, Z.L.; Wang, Z.G.; Liu, J.Q.; Ye, Z.P.; Hu, Y.; Wu, J.H.; Liu, K.Z.; Cai, Z.B. Enhancing the tribological properties of TA1 pure titanium by modulating the energy of pulsed laser nitriding. *Opt. Laser Technol.* **2024**, *169*, 110118. [CrossRef]
129. Xin, Z.D.; Ren, N.F.; Ren, Y.P.; Yue, X.L.; Han, Q.; Zhou, W.F.; Tao, Y.F.; Ye, Y.X. In-Situ nitriding on the textured titanium alloy using femtosecond laser. *J. Mater. Res. Technol.* **2022**, *19*, 466–471. [CrossRef]
130. Kang, J.J.; Wang, M.Z.; Yue, W.; Fu, Z.Q.; Zhu, L.N.; She, D.S.; Wang, C.B. Tribological Behavior of Titanium Alloy Treated by Nitriding and Surface Texturing Composite Technology. *Materials* **2019**, *12*, 301. [CrossRef]
131. Jiang, X.J.; Wang, S.Z.; Feng, Z.H.; Qi, H.B.; Fu, H.; Liu, R.P. Improving vacuum gas nitriding of a Ti-based alloy via surface solid phase transformation. *Vacuum* **2021**, *197*, 110860. [CrossRef]
132. Liu, J.; Suslov, S.; Vellore, A.; Ren, Z.C.; Amanov, A.; Pyun, Y.; Martini, A.; Dong, Y.L.; Ye, C. Surface Nanocrystallization by Ultrasonic Nanocrystal Surface Modification and its Effect on Gas Nitriding of Ti6Al4V Alloy. *Mater. Sci. Eng. A* **2018**, *736*, 335–343. [CrossRef]
133. Li, Q.X.; Zhang, X.F.; Chen, M.; Li, W.; Luo, X.F. Development of gaseous element diffusion wear-resistant treatment technology for titanium surfaces. *Iron Steel Vanadium Titan.* **2021**, *42*, 28–35. (In Chinese)
134. Wang, M.Z. Effects of Ion Nitriding/Laser Surface Weaving Composite Treatment on the Organizational Structure of Titanium Alloys and Their Vacuum Tribological Properties. Master's Thesis, China University of Geosciences (Beijing), Beijing, China, 2018.
135. Harse, S.; Guan, F.Y.; Syedul, B.H.; Muhammad, I.; Muzamil, I.; Muhammad, B.A.; Guo, L.B.; Luo, W. Investigating the impact of plasma nitriding on Ti6Al4V surface, structural, and mechanical properties and their simultaneous evaluation via laser opto-ultrasonic dual detection (LOUD) approach. *Appl. Surf. Sci.* **2024**, *642*, 158539.
136. Morgiel, J.; Szymkiewicz, K.; Maj, Ł.; Tarnowski, M.; Wierzchoń, T. TEM studies of low temperature cathode-plasma nitrided Ti6Al7Nb alloy. *Surf. Coat. Technol.* **2019**, *359*, 183–189. [CrossRef]
137. Szymkiewicz, K.; Morgiel, J.; Maj, Ł.; Pomorska, M.; Tarnowski, M.; Tkachuk, O.; Pohrelyuk, I.; Wierzchoń, T. Effect of nitriding conditions of Ti6Al7Nb on microstructure of TiN surface layer. *J. Alloys Compd.* **2020**, *845*, 156320. [CrossRef]
138. Takesue, S.; Kikuchi, S.; Misaka, Y.; Morita, T.; Komotori, J. Rapid nitriding mechanism of titanium alloy by gas blow induction heating. *Surf. Coat. Technol.* **2020**, *399*, 126160. [CrossRef]
139. Li, Y.; Wang, Z.W.; Shao, M.H.; Zhang, Z.H.; Wang, C.X.; Yan, J.W.; Lu, J.P.; Zhang, L.; Xie, B.; He, Y.Y.; et al. Characterization and electrochemical behavior of a multilayer-structured Ti–N layer produced by plasma nitriding of electron beam melting TC4 alloy in Hank's solution. *Vacuum* **2023**, *208*, 111737. [CrossRef]
140. Farokhzadeh, K.; Edrisy, A. Plasma Nitriding of Titanium Alloys. In *Plasma Science and Technology*; Institute of Plasma Physis: Hefei, China, 2016; pp. 67–105.
141. Zhang, H.; Qin, H.F.; Ren, Z.C.; Zhao, J.Y.; Hou, X.N.; Doll, G.; Dong, Y.L.; Ye, C. Low-temperature nitriding of nanocrystalline Inconel 718 alloy. *Surf. Coat. Technol.* **2017**, *330*, 10–16. [CrossRef]

142. She, D.S.; Yue, W.; Fu, Z.Q.; Wang, C.B.; Yang, X.K.; Liu, J.J. Effects of nitriding temperature on microstructures and vacuum tribological properties of plasma-nitrided titanium. *Surf. Coat. Technol.* **2015**, *264*, 32–40. [CrossRef]
143. Zhang, C.W.; Wen, K.; Gao, Y. Columnar and nanocrystalline combined microstructure of the nitrided layer by active screen plasma nitriding on surface-nanocrystalline titanium alloy. *Appl. Surf. Sci.* **2023**, *617*, 156614. [CrossRef]
144. Saeed, A.; Khan, A.; Jan, F.; Abrar, M.; Khalid, M.; Zakaullah, M. Validity of "sputtering and re-condensation" model in active screen cage plasma nitriding process. *Appl. Surf. Sci.* **2013**, *273*, 173–178. [CrossRef]
145. Yao, Q.T.; Sun, J.; Shen, D.P.; Tong, W.P.; Zuo, L. Large-Scale Synthesis of Nanostructured Nitride Layer on Ti Plate Using Mechanical Shot Peening and Low-Temperature Nitriding. *Adv. Eng. Mater.* **2017**, *19*, 1700157. [CrossRef]
146. Wen, K.; Zhang, C.W.; Gao, Y. Influence of gas pressure on the low-temperature plasma nitriding of surface-nanocrystallined TC4 titanium alloy. *Surf. Coat. Technol.* **2022**, *436*, 128327. [CrossRef]
147. Zhang, F.Y.; Qiu, Y.; Hu, T.T.; Clare, A.T.; Li, Y.; Zhang, L.C. Microstructures and mechanical behavior of beta-type Ti-25V-15Cr-0.2Si titanium alloy coating by laser cladding. *Mater. Sci. Eng. A* **2020**, *796*, 140063. [CrossRef]
148. Arash, F.A.; Maryam, M. A review of functionalizing plasma electrolytic oxidation (PEO) coatings on titanium substrates with laser surface treatments. *Appl. Surf. Sci. Adv.* **2023**, *18*, 100506.
149. Wu, G.L.; Wang, Y.; Sun, M.; Zhang, Q.L.; Yao, J.H. Influence of microstructure of TC4 substrate on the MAO coating. *Surf. Eng.* **2019**, *36*, 827–836. [CrossRef]
150. Wang, H.B.; Zhai, D.J.; Feng, K.Q. Effect of the Microstructure of a Titanium Alloy Fabricated Using Selective Laser Melting on Microarc Oxidation Film. *Metall. Mater. Trans. A* **2021**, *52*, 4691–4702. [CrossRef]
151. Yang, X.; Wang, F.H.; Wang, W.L.; Liu, S.F.; Chen, Y.Q.; Tang, H.P. Comparison of two-step surface treatment on surface roughness and corrosion resistance of TC4 alloy parts prepared by SLM and SEBM. *J. Alloys Compd.* **2022**, *921*, 165929. [CrossRef]
152. Van Hengel, I.A.J.; Laçin, M.; Minneboo, M.; Fratila-Apachitei, L.E.; Apachitei, I.; Zadpoor, A.A. The effects of plasma electrolytically oxidized layers containing Sr and Ca on the osteogenic behavior of selective laser melted Ti6Al4V porous implants. *Mater. Sci. Eng. C* **2021**, *124*, 112074. [CrossRef]
153. Vargas, C.A.; Zuleta, A.A.; Botero, C.A.; Baena, L.M.; Castaño, J.G.; Gómez, M.A.; Tamayo, J.A. Morphological analysis of plasma electrolytic oxidation coatings formed on Ti6Al4V alloys manufactured by electron beam powder bed fusion. *Heliyon* **2023**, *9*, e19289. [CrossRef] [PubMed]
154. Sun, D.J.; Wu, X.Q.; Xie, F.Q.; He, P.; Li, Z.; He, J.Y. Fretting wear properties and microstructure evolution in micro-arc oxidation bioceramic coating pretreated using laser remelting. *Ceram. Int.* **2023**, *49*, 4979–4986. [CrossRef]
155. Yu, F.L.; Zhang, Y.; Kong, C.L.; Yu, H.L. Microstructure and mechanical properties of Ti-6Al-4V alloy sheets via room-temperature rolling and cryorolling. *Mater. Sci. Eng. A* **2022**, *834*, 142600. [CrossRef]
156. Yu, F.L.; Zhang, Y.; Kong, C.L.; Yu, H.L. High strength and toughness of Ti-6Al-4V sheets via cryorolling and short-period annealing. *Mater. Sci. Eng. A* **2022**, *854*, 143766. [CrossRef]
157. Chen, P.H.; Li, B.C.; Liu, Z.; Zhou, Y.H.; Li, R.Q.; Zhang, Y. Understanding high-temperature oxidation behaviors and mechanical properties of TiAlCrMn HEAs during heat treatment. *Trans. Nonferr. Metals Soc.* 2023, *in press*.
158. Zhang, Y.; Lei, G.; Luo, K.G.; Chen, P.H.; Kong, C.L.; Yu, H.L. Tribological behavior of high-entropy alloy particle reinforced aluminum matrix composites and their key impacting factors. *Tribol. Int.* **2022**, *175*, 107868. [CrossRef]
159. Zhang, Y.; Luo, K.G.; Lei, G.; Yu, H.L. Interfacial characteristics and enhanced mechanical properties of Al0.5CoCrFeNi high-entropy alloy particles reinforced Al matrix composites. *Metall. Mater. Trans. A* **2022**, *53*, 4161–4167. [CrossRef]

coatings

MDPI

Article

A Comparison Study on the Microstructure, Mechanical Features, and Tribological Characteristics of TiN Coatings on Ti6Al4V Using Different Deposition Techniques

Zhen Liu [1], Shuai Ren [1], Tijun Li [2,*], Pinghu Chen [3,*], Liangbin Hu [3], Wenxing Wu [3], Sheng Li [3], Hao Liu [4], Ruiqing Li [5] and Yun Zhang [6]

1 College of Mechatronics and Control Engineering, Shenzhen University, Shenzhen 518060, China; liuzhen598352368@126.com (Z.L.); shuai.ren@szu.edu.cn (S.R.)
2 State-Owned Sida Machinery Manufacturing Company, Xianyang 610400, China
3 Key Laboratory of Hunan Province of Equipment Safety Service Technology under Extreme Environment, College of Mechanical Engineering, College of Nuclear Science and Technology, School of Resource Environment and Safety Engineering, University of South China, Hengyang 421001, China; huliangbin83@sina.com (L.H.); wuzhiaifrank@hotmail.com (W.W.); lisheng325@126.com (S.L.)
4 College of Mechanical Engineering, Hunan Institute of Engineering, Xiangtan 411104, China; liuhao10212510@163.com
5 State Key Laboratory of Precision Manufacturing for Extreme Service Performance, Light Alloys Research Institute, Central South University, Changsha 410083, China; liruiqing@csu.edu.cn
6 School of Mechanical Engineering, Hunan University of Science and Technology, Xiangtan 411201, China; yun_zhang66@163.com
* Correspondence: litj@pku.edu.cn (T.L.); chenpinghu1986@163.com (P.C.)

Abstract: Titanium alloys are considered lightweight alloys and are widely applied across various industries. However, their low hardness, poor wear resistance, and limited oxidation resistance restrict their prospects for wider application. In this paper, nitride coatings were prepared using three preparation processes, namely laser surface nitriding (LSN), physical vapor deposition (PVD), and plasma ion implantation (PII). Their microstructure, microhardness, tribological behavior, and high-temperature oxidation characteristics were compared. The experimental results revealed that nitrided coatings were successfully prepared using the three methods. However, a comparison of these data shows that the LSN coating exhibited superior comprehensive performance. It achieved the maximum thickness within the shortest preparation time: the thickness was about 280 μm and the deposition rate of the LSN method was 2250 and 90,000 times higher than those of the PVD and PII methods. Nitrides have high hardness, but the carrying capacity could be attributed to the thickness of the coatings: the PVD coating could withstand a force of 500 g, while the PII coating only withstood a force of less than 25 g. In addition, as hardness is the most important factor for excellent wear resistance, the average volumetric wear rate of the LSN and PVD coatings was about 9×10^{-6} $mm^3/m{\cdot}N$, and their relative wear resistance was 49.2 times that of Ti6Al4V. Meanwhile, the excellent bond between the LSN coating and the substrate was evidenced by a high-temperature oxidation test during a rapid heating–cooling cycle.

Keywords: Ti alloys; surface nitriding; microhardness; wear resistance; high-temperature oxidation resistance

Citation: Liu, Z.; Ren, S.; Li, T.; Chen, P.; Hu, L.; Wu, W.; Li, S.; Liu, H.; Li, R.; Zhang, Y. A Comparison Study on the Microstructure, Mechanical Features, and Tribological Characteristics of TiN Coatings on Ti6Al4V Using Different Deposition Techniques. *Coatings* **2024**, *14*, 156. https://doi.org/10.3390/coatings14020156

Academic Editor: Alessandro Patelli

Received: 7 December 2023
Revised: 19 January 2024
Accepted: 22 January 2024
Published: 24 January 2024

1. Introduction

Titanium and its alloys have the advantages of low density, high strength, good corrosion resistance, etc., and they are used in the automobile, aerospace, military, and biomedical industries [1–4]. However, low hardness and susceptibility to wear limit their wide application [5–7]. Especially in aerospace, a complex service environment (ocean or desert zone) can cause a deterioration of mechanical properties and early failure [8–10].

Therefore, the requirement for lightweight in key transmission pairs restricts the use of titanium alloys in aircraft [11]. To meet the needs of industrial applications, titanium alloys should be enhanced for excellent wear resistance through different methods, such as novel materials development, surface alloying, and coating techniques.

A novel Ti alloy with high hardness was designed by Chen, P.H. [12]. The refractory high entropy alloy $Ti_{30}Hf_{20}Nb_{20}Ta_{10}V_{10}Mo_7W_3$, with excellent wear resistance, was designed by Wei, Q. [13]. The Cu element was added to the Ti-10Mo alloy to improve microhardness and tribo-corrosion resistance [14]. Two medium-entropy alloy coatings were prepared on a CP-Ti sheet by pulsed laser cladding, achieving a hardness of 762 HV and a specific wear rate of 1.7×10^{-5} $mm^3 \cdot N^{-1} \cdot m^{-1}$ [15]. A NiTi coating was prepared by the tungsten inert gas (TIG) cladding process, resulting in a microhardness of 680 $HV_{0.05}$ to enhance the wear resistance of the Ti6Al4V substrate [16]. Pure boron powders were pre-placed on the surface of the Ti6Al4V substrate, and high-hardness TiB, AlTi, $AlTi_2$, and $AlTi_3$ were precipitated to improve wear resistance [17]. An oxygen-charging method was employed to fabricate β-Ti alloys with ultrahigh surface hardness and exceptional wear resistance [18]. In addition, laser shock peening was performed to enhance the impact wear behavior of the Ti6Al4V substrate [19]. The ball burnishing process was applied to minimize the friction coefficient and increase micro-hardness, achieving a specific wear rate decrease of 52% as compared to Ti6Al4V [20]. The duplex treatment of plasma Ti alloying and plasma nitriding was applied to prepare a TiN coating, thus resulting in better wear resistance on the surface of the C17200 alloy [21]. Rossi et al. reported that a hard surface was created on different titanium alloys by plasma nitriding treatment and compared their microhardness and wear resistance [22]. Compared with the aforementioned surface modification methods, surface nitriding treatment is one of the simplest methods to enhance wear resistance. However, it could introduce new issues, such as high costs or long manufacturing cycles.

In this paper, to obtain a coating with high hardness for wear resistance, three nitriding methods were employed, namely laser surface nitriding (LSN), physical vapor deposition (PVD), and plasma ion implantation (PII), to deposit coatings on the Ti6Al4V substrate. Scanning electron microscopy (SEM), X-ray diffraction (XRD), and energy dispersive spectroscopy (EDS) were applied to analyze the microstructure and phases. The relationship between the thickness and deposition time was discussed to assess the thickness and deposition efficiency of the coatings. Microhardness and tribological behaviors were measured using a Vickers hardness tester and a high-temperature friction-wear machine. Meanwhile, the bonding strength was explored by high-temperature oxidation testing. And the microstructure, mechanical properties, and tribological behaviors of the coatings deposited by different methods were compared in detail.

2. Materials and Methods

A Ti6Al4V alloy was used as the substrate in this work, cut into rectangular blocks measuring 20 mm × 20 mm × 5 mm. The surfaces of all samples were polished using metallographic sandpaper with 400, 800, 1200, 1500, 2000, and 5000 mesh and polishing suspension with 0.5 μm diamonds, followed by drying for use. Firstly, the HEMII-80 ion implanter (Plasma Technology Ltd., Hong Kong, China) was utilized to conduct N ion implantation, as depicted in the schematic diagram reported in Refs. [23,24]. Before implantation, the plates were sputter-cleaned using argon plasma ion bombardment. During implantation, N_2 was introduced into the vacuum chamber, maintaining N plasma at a power of 900 W, a working pressure of 8×10^{-2} Pa, and a gas flow of 25 mL/min. Ion implantation was conducted at an accelerating voltage of 2 kV for 2.5 h at a base pressure of 1×10^{-3} Pa, with an N ion implantation dose of ~4×10^{17} ions/cm^2. Secondly, a TN target was supplied by Beijing Hezong Technology Co., Ltd., and a TSU-650 multifunctional coating machine (PVD technique, as in a previously reported schematic diagram [25]) was used to prepare the TiN coating. This process included a substrate bias voltage of ~100 V, target current of 1.3 A, a deposition time of 300 min, a duty cycle of 50%, a target base distance of 170 mm, and a substrate temperature of 300 °C. Thirdly, the XL-500 Laser cladding system (Guangzhou Xinglai Laser Technology Co., Ltd., Guangzhou, China) was adopted, equipped with a 500 W laser with a pulse width of 10–50 ns. The laser parameters included a power of 36 W, a scanning speed of 1800 mm/min, and a lapping rate of 90%; the deposition time was about 3 min (a similar schematic diagram was reported in Ref. [26]). In addition, the reaction atmosphere of high-purity N_2 at 99.99% with a 10 L/min flow rate was maintained during laser surface nitriding. The treated samples are referred to as PII, PVD, and LSN coatings, with detailed deposition method information presented in Table 1.

Table 1. The working parameters of three deposition methods.

Method	Process Parameters						
LSN	Power 36 W	Scanning speed 1800 mm/min		Lapping rate 90%		Deposition time 3 min	N_2 Flow rate 10 L/min
PVD	Target TiN	Target distance 170 mm	Substrate preheat 300 °C	Target current 1.3 A	Bias voltage ~100 V	Deposition time 300 min	Duty cycle 50%
PII	Power 900 W	Work pressure 8×10^{-2} Pa	Voltage 2 kV	Base pressure 1×10^{-3} Pa	Ion dose ~4×10^{17} ions/cm^2	Deposition time 150 min	Gas flow 25 mL/min

Microhardness on the surface of different coatings was measured using a Vickers hardness tester (Micro Vickers HV1000Z, MEGA INSTRUMENTS, Shanghai, China). Various forces of 10 g, 25 g, 100 g, 200 g, and 500 g were applied to obtain the microhardness value and carrying capacity. For each coating, 5 points were measured, and their average value was taken as the final result. The high-temperature friction-wear machine (HT-1000, Zhongke Kaihua, Lanzhou, China) was employed to measure wear characteristics [27], with testing parameters set as follows: a friction force of 3 N, a friction speed of 500 r/min, a rotation radius of 4 mm, a friction time of 30 min, and a φ6 mm Si_3N_4 counterpart ball. And then the mechanical profiler was utilized to delineate the wear tracks, and the average wear volume was calculated based on data collected four times at quarter positions. The volume wear rate was expressed by the equation $w = V/(F \cdot L)$ [28], where w is the volume wear rate (mm^3/m·N), V is the wear volume (mm^3), F is the friction force (N), and L is the run length (m). To explore the bonding between the coatings and substrate, an individual oxidation test was performed in an electric resistance furnace, similar to the test reported in Ref. [29]. Firstly, the furnace temperature was heated and stabilized at 700 °C. Secondly, the samples were placed in the furnace for holding times of 5 and 10 h, and the surface characteristics after high-temperature oxidation were characterized using a digital camera.

A high-energy X-ray diffractometer (HE-XRD, D8 discover, Bruker, Karlsruhe, Germany), using Cu-Kα radiation (λ = 0.15418 nm), was employed to examine the phase constitutions of the coatings deposited by different methods. The device was operated at 40 kV and 80 mA over a 2θ range of 30–80°, with a scanning rate of 2°/min. The microstructure of the surface and cross-section was characterized by scanning electron microscopes (SEM, TESCAN MIRA three LMH/LMU, Brno, Czech Republic). In addition, energy-dispersive X-ray spectroscopy (EDS) was utilized to determine the elemental distribution in the cross-section.

3. Results and Discussion

Figure 1 shows the XRD patterns of the samples deposited by different methods. The phase components of the three coatings differ significantly owing to different preparation processes. High crystallinity occurs on the surface of the sample fabricated by laser surface nitriding. The black line in Figure 1 indicates that abundant titanium nitride (TiN), a few aluminum nitride (AlN), and hcp-Ti can exist on the coating. The PVD coating exhibits broad peaks indicating, on the one hand, that crystallinity is relatively low under the current parameter conditions. On the other hand, a single peak contains three peaks from different phases; for example, the (111) crystal plane (111) of Vanadium nitride (VN), (111) of TiN, and (100) of hcp-Ti appear at ~36°. However, abundant hcp-Ti is discovered on the surface of the PII sample, with only a small presence of TiN and AlN detected. When comparing the three samples, the peak intensity of hcp-Ti gradually increased from the LSN coating to the PII coating. This can be because a difference exists in the thickness and surface characteristics of the coatings.

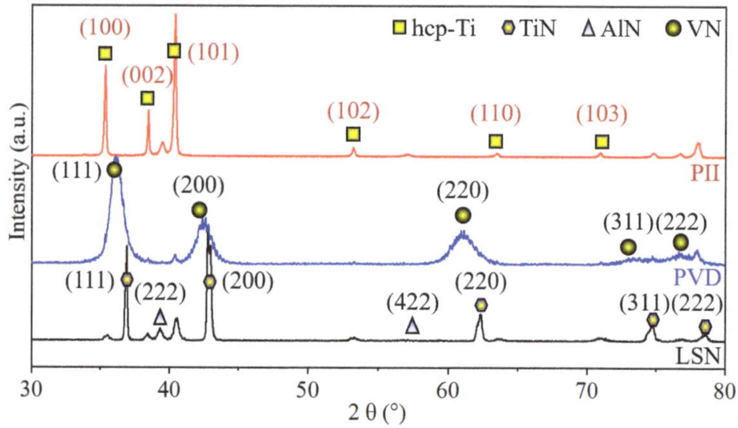

Figure 1. XRD patterns of the samples deposited by different methods.

Just as shown in Figure 2, the surface and cross-section microstructure of the samples with different methods used. High energy can contribute to the reaction region where nitrides are formed in situ owing to the embedment of supersaturated N_2. Obvious laser scanning traces are visible on the surface of the LSN sample, as shown in Figure 2a; the width between the two traces is about 100 μm, and holes and cracks appear. The surface roughness is also the maximum because of the violent fluctuation within the molten pool caused by high energy density. Meanwhile, the cross-section morphology shows obvious stratification in the coatings, consisting of a loose upper layer, a compact middle layer, and a transition layer, with a total thickness of about 280 μm. The deposition rate is determined by the relationship between the thickness and deposition time, with the values for LSN, PVD, and PII coatings being 70, 0.04, and 0.001 μm/min, respectively. This indicates that the deposition rate of the LSN coating is the highest, as shown in Table 2. Figure 2b reveals that some lateral and longitudinal cracks are present within the upper and middle layers. During X-ray diffraction testing, X-rays cannot penetrate the thick coating completely, resulting in a lower peak intensity of hcp-Ti in the sample fabricated by laser surface nitriding.

Table 2. The thickness of the coatings deposited by different methods.

Methods	Deposition Time (min)	Deposition Thickness (μm)	Deposition Rate (μm/min)
LSN	3	280	~70
PVD	300	11	~0.04
PII	150	0.19	~0.001

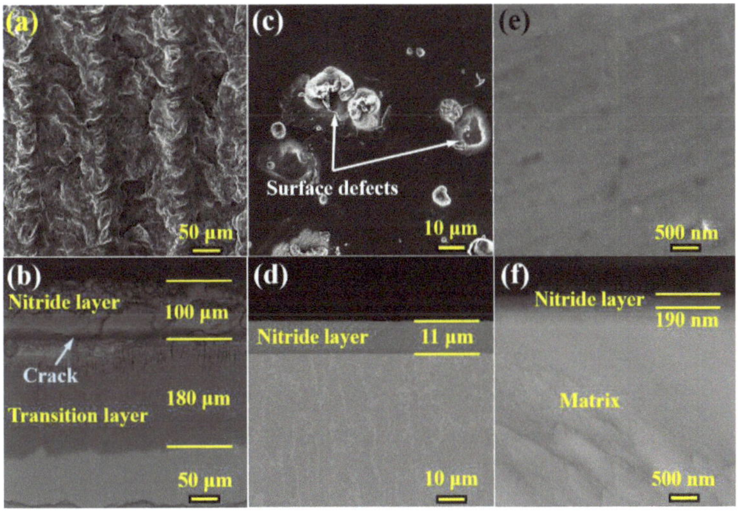

Figure 2. Surface and cross-section morphology of the samples deposited by different methods. (**a**,**b**) LSN; (**c**,**d**) PVD; (**e**,**f**) PII.

In the PVD sample, surface defects approximately ~20 μm in size are evident, as shown in Figure 2c. These defects can peel out during the preparation process. However, the cross-section morphology reveals that the ~11 μm coating is precipitated. It is incredibly dense, with a clearly defined interface between the coating and the Ti6Al4V substrate, as shown in Figure 2d. Just like the existence of the surface defects, a relatively high peak intensity of hcp-Ti is observed relative to the LSN sample. But fortunately, cracks were never discovered in the samples fabricated by PVD and plasma ion implantation. Due to the deposition rate of plasma ion implantation, only a 190 nm nitride layer is formed on the PII sample. Therefore, higher intensities of the hcp-Ti peaks are detected, indicated by a red line.

Figure 3 shows the elemental distributions in the line scanning of the cross-section for three samples. For the LSN sample, the upper layer includes AlN and TiN, but the TiN ratio can be higher in the transition layer, as shown in Figure 3a. More interestingly, an obvious AlN layer approximately 400 nm thick appears on the upper surface, and the AlN layer reappears at the interface between the coating and the Ti6Al4V substrate, as shown in Figure 3b. For the PII sample, VN can form in the 0.5–0.75 μm range from the top surface of the coating. These nitrides can contribute to microhardness and wear resistance. The microhardness of different samples under various forces is shown in Figure 4 and Table 3. The microhardness of Ti6Al4V is only 341.2 $\pm$ 8 $HV_{0.2}$; the microhardness values for PII, PVD, and LSN samples increase by 7%, 257.2%, and 274.1%, respectively. These values do not explain why the PII sample has low microhardness. We believe the coating of the PII sample is too thin to withstand the force. Therefore, a series of microhardness measurements were performed with different forces. The measured results reveal that a microhardness of 1260.8 $\pm$ 13 HV is obtained on the surface of the PII sample under the force of 10 g, with the average diagonal length on the top surface of the micro-indentation being about 3.88 μm according to the measured results (the facial angle of the indenter is 136°), yielding an indentation depth of about 554 nm. This depth is 2.9 times the thickness of the PII coating, yet the PII coating still has a preferable carrying capacity during the hardness test process. When the force is increased to 25 g, there is a 40.7% reduction in microhardness compared to the force of 10 g, with the depth of the indentation exceeding 1.1 μm, which indicates that the carrying capacity is weakened at 25 g. Table 3 reveals that the carrying capacity of the PVD sample can be up to 500 g, and that of the LSN sample can be higher than 500 g, but testing above 500 g was not carried out owing to the rough surface.

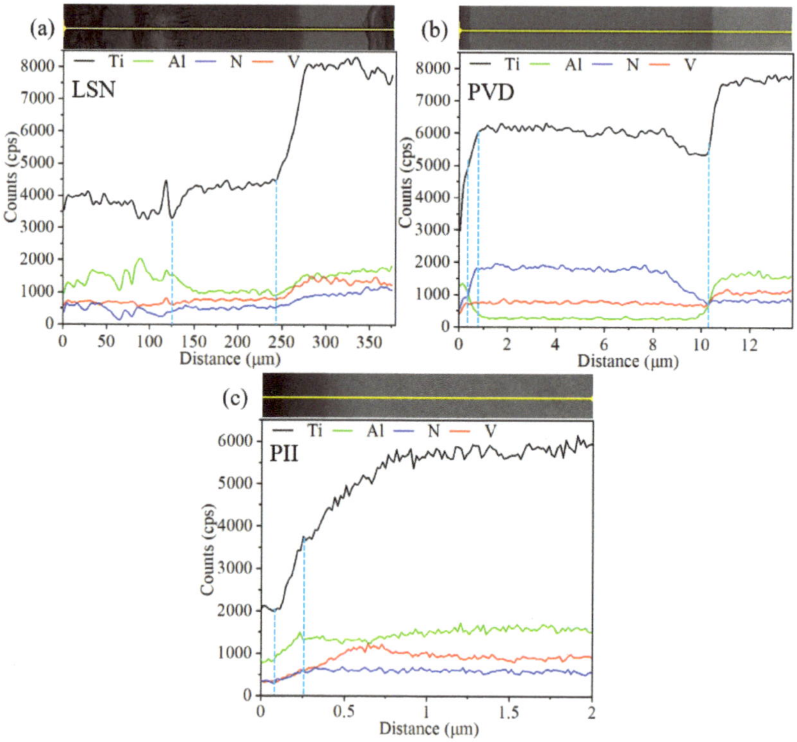

Figure 3. EDS elemental distribution in the cross-section of the samples deposited by different methods. (**a**) LSN; (**b**) PVD; (**c**) PII.

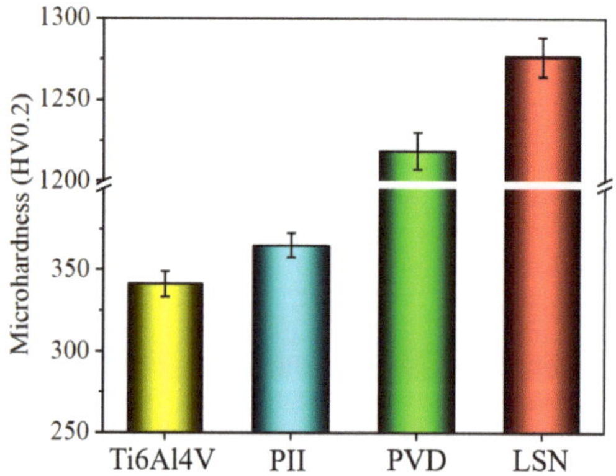

Figure 4. Microhardness of the samples deposited by different methods.

Table 3. Microhardness of the samples deposited by different loads.

Samples	Microhardness (HV)				
	10 g	25 g	100 g	200 g	500 g
Ti6Al4V	/	/	/	341.2 ± 8	/
PII	1260.8 ± 13	748 ± 10	377.5 ± 8	365 ± 8	351.7 ± 7
PVD	/	/	1597 ± 15	1218.6 ± 11	663 ± 10
LSN	/	/	1734.4 ± 21	1276.3 ± 12	/

It is well-known that hardness is one of the most important factors for wear resistance. Figure 5 shows the macro-scale morphology of the samples deposited by different methods. There is a huge difference in the characteristics of the wear tracks. A width of 1.16 mm is observed in the Ti6Al4V substrate, accompanied by a metallic color. A similar phenomenon occurs in the PII sample, with a width also about 1.16 mm and a metallic color, which indicates that the carrying capacity is low owing to the smaller thickness of the coating. The wear tracks of the PVD and LSN samples show a coated color. The width of the PVD and LSN samples decreased by 45.7% and 37.9%, respectively. This reveals that the coatings can withstand a wear force of 3 N given a certain thickness, and relatively excellent wear resistance is present in the PVD and LSN samples. Combined with the outline of the wear tracks in the different coatings, the most severe wear is seen on the Ti6Al4V substrate, where the depth of the wear track exceeds 30 μm. The nitride coatings can contribute to a relatively shallow wear track owing to their resistance to a 3 N wear force. However, in the PII coating, insufficient resistance can be caused by the ultrathin nitride coating, thus resulting in a depth of ~20 μm. Interestingly, the PVD and LSN coatings have excellent wear resistance with a wear track depth of ~1 μm, as shown in Figure 6. The volume wear rate was also obtained and is shown in Figure 7. A volume wear rate of 444×10^{-6} mm^3/m·N is observed in the Ti6Al4V, while the volume wear rates for the PII, PVD, and LSN coatings are 305×10^{-6} mm^3/m·N, 8.96×10^{-6} mm^3/m·N, and 9.02×10^{-6} mm^3/m·N, respectively. Their relative wear resistance is 1.45, 49.5, and 49.2 times higher than that of the Ti6Al4V substrate.

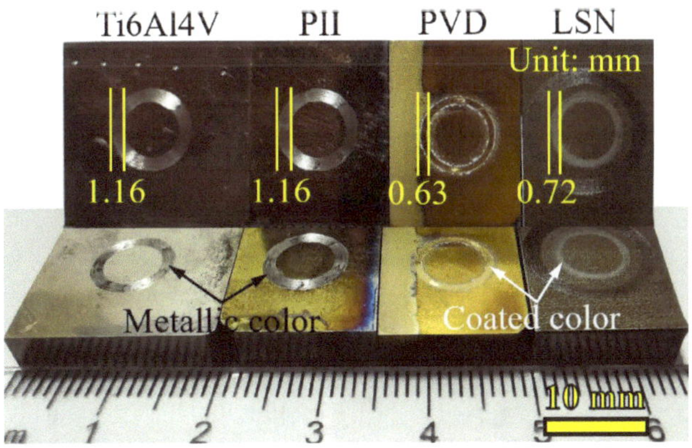

Figure 5. Macro-scale morphology of the samples deposited by different methods after friction-wear testing.

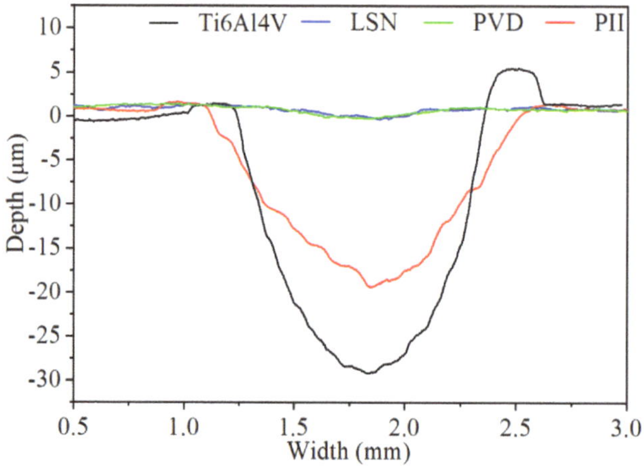

Figure 6. The outline of the wear tracks for the coatings deposited by different methods.

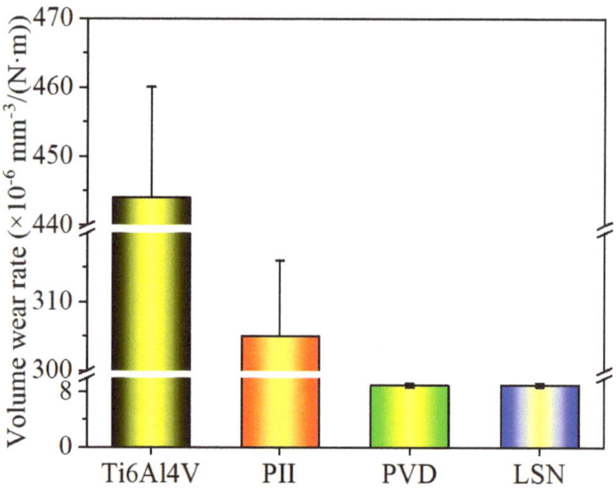

Figure 7. Average volume wear rate of the coatings deposited by different methods.

In addition to affecting hardness and wear resistance, the coating characteristics also contribute to high-temperature oxidation resistance. The affinity between titanium and oxygen is considerably high at high temperatures [30]. Ref. [31] reveals that a poor bond between the oxide layer and the base material causes serious spalling; this observation is consistent with our work, as shown in Figure 8a. For the LSN sample, obvious exfoliation is never observed in the LSN coating, but the oxidation phenomenon can occur in localized regions, as shown in Figure 8b. Vast areas of the nitride coating peel off after 5 h of oxidation, and a new oxide layer forms in the spalled areas, as shown in Figure 8(c2). This could be due to the preparation process; a clear interface between the PVD coating and Ti6Al4V substrate is noted, and poor bonding is an important factor causing spalling, owing to the hardness difference between the coating and substrate [32]. In addition, higher thermal stress can be another important factor for the failure of the coating during rapid heating–cooling cycles. The most important reason for spalling might be that some defects within the coating [25] act as channels for oxygen entering the substrate, primarily causing oxidation at the substrate, and the coating may crack and peel off owing to internal expansion caused by the formation of abundant oxides. A metallic bonding is evident between the PII coating and the substrate. Therefore, high-temperature oxidation resistance is relatively higher; spalling begins at the surface edges after 5 h oxidation and continues to spread to the entire surface, as shown in Figure 8(d2).

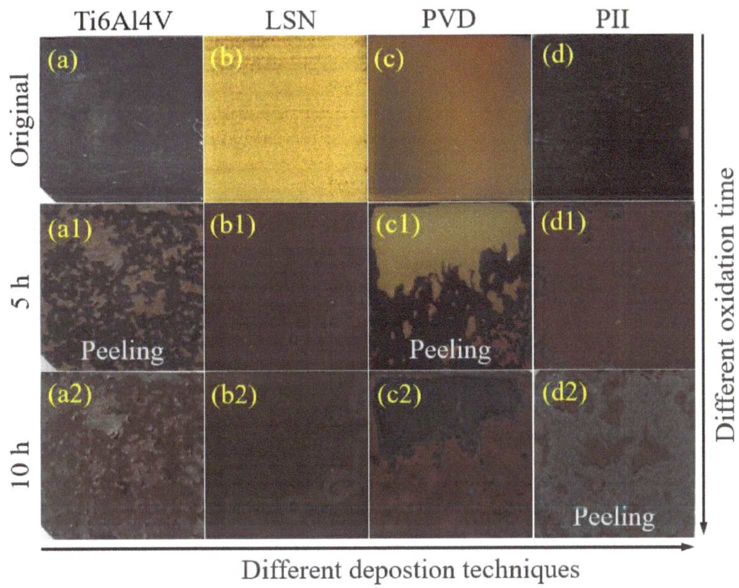

Figure 8. High-temperature oxidation of the samples deposited by different methods at 700 °C. (**a**) Ti6Al4V; (**b**) LSN; (**c**) PVD; (**d**) PII.

4. Conclusions

In this paper, three coatings were prepared using LSN, PVD, and PII on the Ti6Al4V. The microstructure, microhardness, wear behaviors, and high-temperature oxidation resistance were measured, and the main conclusions are as follows:

1. The nitriding coatings were successfully prepared. The LSN coating had the greatest roughness and thickness, with a thickness of about 280 μm, while the thicknesses of the PII and PVD coatings were 190 nm and 11 μm, respectively. The deposition rates for the LSN, PVD, and PII methods were about 90 μm/min, 0.04 μm/min, and 0.001 μm/min, respectively.
2. Owing to the thickness differences, the carrying capacities of the coatings varied significantly. The PII coating withstood forces of less than 100 g during microhardness testing. Relative to Ti6Al4V, the microhardness increased by 7%, 257.2%, and 274.1%, respectively.
3. Hardness and carrying capacity contribute to the wear resistance of the coatings. The volume wear rate of the Ti6Al4V substrate was only 4.44×10^{-4} $mm^3/m{\cdot}N$. The relative wear resistance of the PII, PVD, and LSN coatings was 1.45, 49.5, and 49.2 times higher than that of the Ti6Al4V substrate. The PVD and LSN coatings exhibited a wear rate of ~9.0×10^{-6} $mm^3/m{\cdot}N$.
4. Surface defects were present in the PVD coating; these could lead to the formation of oxides, which in turn could cause the coating to spall owing to differences in hardness, internal expansion, and thermal stress. Despite the presence of cracks in the LSN coating, it still had the best high-temperature oxidation resistance because a thick, dense layer impeded oxygen penetration. The excellent high-temperature oxidation resistance could be attributed to the metallurgical bonding.

Author Contributions: Conceptualization, Z.L. and T.L.; methodology, Z.L., L.H., S.L., H.L. and P.C.; software, Y.Z. and W.W.; validation, Z.L., R.L. and S.R.; formal analysis, W.W. and P.C.; investigation, Z.L., S.R. and T.L.; resources, T.L.; data curation, L.H., S.L., H.L. and P.C.; writing—original draft preparation, Z.L., P.C. and T.L.; writing—review and editing, Y.Z. and R.L.; visualization, Z.L.; supervision, T.L.; project administration, P.C.; funding acquisition, Y.Z., L.H. and R.L. All authors have read and agreed to the published version of the manuscript.

Funding: This work was funded by the National Natural Science Foundation of China (grant no. 52371032), a research project funded by the Education Department of Hunan Province (grant no. 23B0499), the Project of the State Key Laboratory of Precision Manufacturing for Extreme Service Performance, Central South University (grant no. ZZYJKT2021-01), and a Hengyang City research project of the Science and Technology Bureau (grant no. 202150063472).

Institutional Review Board Statement: Not applicable.

Informed Consent Statement: Not applicable.

Data Availability Statement: Data are contained within the article.

Conflicts of Interest: Author Tijun Li was employed by the State-Owned Sida Machinery Manufacturing Company. The remaining authors declare that the research was conducted in the absence of any commercial or financial relationships that could be construed as a potential conflict of interest.

References

1. Wu, Z.; Kou, H.; Chen, N.; Xi, Z.; Fan, J.; Tang, B.; Li, J. Recent developments in cold dwell fatigue of titanium alloys for aero-engine applications: A review. *J. Mater. Res. Technol.* **2022**, *20*, 469–484. [CrossRef]
2. Alipour, S.; Moridi, A.; Liou, F.; Emdadi, A. The Trajectory of Additively Manufactured Titanium Alloys with Superior Mechanical Properties and Engineered Microstructures. *Addit. Manuf.* **2022**, *60*, 103245. [CrossRef]
3. Zhao, Q.; Sun, Q.; Xin, S.; Chen, Y.; Wu, C.; Wang, H.; Xu, J.; Wan, M.; Zeng, W.; Zhao, Y. High-strength titanium alloys for aerospace engineering applications: A review on melting-forging process. *Mater. Sci. Eng. A* **2022**, *845*, 143260. [CrossRef]
4. Tshephe, T.S.; Akinwamide, S.O.; Olevsky, E.; Olubambi, P.A. Additive manufacturing of titanium-based alloys- A review of methods, properties, challenges, and prospects. *Heliyon* **2022**, *8*, e09041. [CrossRef] [PubMed]
5. Yang, C.; Cui, S.; Wu, Z.; Zhu, J.; Huang, J.; Ma, Z.; Fu, R.K.Y.; Tian, X.; Chu, P.K.; Wu, Z. High efficient co-doping in plasma electrolytic oxidation to obtain long-term self-lubrication on Ti6Al4V. *Tribol. Int.* **2021**, *160*, 107018. [CrossRef]
6. Shi, Y.; Gong, S.; Xu, H.; Wang, Z.; Yang, G.; Qi, B. Alloying strategies for additive manufacturing of Ti6Al4V based alloys, composites and functionally graded materials: Microstructure and phase evolution of intra and inter-layer. *J. Mater. Res. Technol.* **2023**, *27*, 6225–6263. [CrossRef]
7. Murmu, A.M.; Parida, S.K.; Das, A.K.; Kumar, S. Evaluation of laser cladding of Ti_6Al_4V-ZrO_2-CeO_2 composite coating on Ti6Al4V alloy substrate. *Surf. Coat. Technol.* **2023**, *473*, 129988. [CrossRef]
8. Suo, H.; Wei, Z.; Luo, B.; Wang, L.; Zhang, K.; Liang, B.; Deng, K.; Cheng, H. Interfacial wear damage mechanism between Ti-alloy and Al-alloy in interference-fit joint and influence of surface coatings: Experimental and numerical study. *Eng. Fail. Anal.* **2023**, *143*, 106931. [CrossRef]
9. Asl, H.G.; Sert, Y.; Küçükömeroğlu, T.; Bayrak, Ö. The comparison of wear performances of CP-Ti, Ti6Al4V, Ti45Nb alloys oxidized by anodic oxidation under ambient air and vacuum conditions. *Mater. Today Commun.* **2023**, *34*, 105466. [CrossRef]
10. Suo, H.; Wei, Z.; Zhang, K.; Deng, K.; Cheng, H.; Luo, B.; Li, H.; Wang, L.; Liang, B. Interfacial wear damage of CFRP/Ti-alloy single-lap bolted joint after long-term seawater aging. *Eng. Fail. Anal.* **2022**, *139*, 106464. [CrossRef]
11. Wu, F.; Zhang, N.; Sun, Y.; Li, X.; Liu, Z.; Yang, Y. Effect of applied load on tribological behaviour of Ti–48Al–2Cr–2Nb alloy under oil-lubricated condition. *Wear* **2024**, *538–539*, 205210. [CrossRef]
12. Chen, P.; Li, B.; Liu, Z.; Zhou, Y.; Li, R.; Zhang, Y. High-temperature oxidation behaviors and mechanical properties of TiAlCrMn HEAs during heat treatment. *Trans. Nonferrous Met. Soc. China* **2024**, *34*, 237–252.
13. Wei, Q.; Zhang, A.; Han, J.; Xin, B.; Su, B.; Wang, X.; Ma, Q.; Meng, J. Development of a $Ti_{30}Hf_{20}Nb_{20}Ta_{10}V_{10}Mo_7W_3$ refractory high entropy alloy with excellent mechanical properties and wear resistance. *J. Alloys Compd.* **2023**, *966*, 171571. [CrossRef]
14. Lu, X.; Zhang, D.; Xu, W.; Yu, A.; Zhang, J.; Tamaddon, M.; Zhang, J.; Qu, X.; Liu, C.; Su, B. The effect of Cu content on corrosion, wear and tribocorrosion resistance of Ti-Mo-Cu alloy for load-bearing bone implants. *Corros. Sci.* **2020**, *177*, 109007. [CrossRef]
15. Xiang, K.; Chai, L.; Zhang, C.; Guan, H.; Wang, Y.; Ma, Y.; Sun, Q.; Li, Y. Investigation of microstructure and wear resistance of laser-clad CoCrNiTi and CrFeNiTi medium-entropy alloy coatings on Ti sheet. *Opt. Laser Technol.* **2022**, *145*, 107518. [CrossRef]
16. Waghmare, D.T.; Kumar Padhee, C.; Prasad, R.; Masanta, M. NiTi coating on Ti-6Al-4V alloy by TIG cladding process for improvement of wear resistance: Microstructure evolution and mechanical performances. *J. Mater. Process. Technol.* **2018**, *262*, 551–561. [CrossRef]
17. Wu, Y.; Wang, A.H.; Zhang, Z.; Xia, H.B.; Wang, Y.N. Microstructure, wear resistance and cell proliferation ability of in situ synthesized Ti–B coating produced by laser alloying. *Opt. Laser Technol.* **2015**, *67*, 176–182. [CrossRef]

18. Wang, X.-Q.; Zhang, Y.-S.; Han, W.-Z. Design of high strength and wear-resistance β-Ti alloy via oxygen-charging. *Acta Mater.* **2022**, *227*, 117686. [CrossRef]
19. Yin, M.-g.; Cai, Z.-b.; Li, Z.-y.; Zhou, Z.-r.; Wang, W.-j.; He, W.-f. Improving impact wear resistance of Ti-6Al-4V alloy treated by laser shock peening. *Trans. Nonferrous Met. Soc. China* **2019**, *29*, 1439–1448. [CrossRef]
20. Revankar, G.D.; Shetty, R.; Rao, S.S.; Gaitonde, V.N. Wear resistance enhancement of titanium alloy (Ti–6Al–4V) by ball burnishing process. *J. Mater. Res. Technol.* **2017**, *6*, 13–32. [CrossRef]
21. Liu, L.; Shen, H.H.; Liu, X.Z.; Guo, Q.; Meng, T.X.; Wang, Z.X.; Yang, H.J.; Liu, X.P. Wear resistance of TiN(Ti2N)/Ti composite layer formed on C17200 alloy by plasma surface Ti-alloying and nitriding. *Appl. Surf. Sci.* **2016**, *388*, 103–108. [CrossRef]
22. Rossi, M.C.; Bazaglia Kuroda, P.A.; Solano de Almeida, L.; Rossino, L.S.; Moreira Afonso, C.R. A detailed analysis of the structural, morphological characteristics and micro-abrasive wear behavior of nitrided layer produced in α (CP–Ti), α+β (Ti–6Al–4V), and β (TNZ33) type Ti alloys. *J. Mater. Res. Technol.* **2023**, *27*, 2399–2412. [CrossRef]
23. Li, Z.-y.; Cai, Z.-b.; Wu, Y.-p.; Zhu, M.-h. Effect of nitrogen ion implantation dose on torsional fretting wear behavior of titanium and its alloy. *Trans. Nonferrous Met. Soc. China* **2017**, *27*, 324–335. [CrossRef]
24. Wu, H.; Xi, K.; Xiao, S.; Qasim, A.M.; Fu, R.K.Y.; Shi, K.; Ding, K.; Chen, G.; Wu, G.; Chu, P.K. Formation of self-layered hydrothermal coating on magnesium aided by titanium ion implantation: Synergistic control of corrosion resistance and cytocompatibility. *Surf. Coat. Technol.* **2020**, *401*, 126251. [CrossRef]
25. Naeem, M.; Awan, S.; Shafiq, M.; Raza, H.A.; Iqbal, J.; Díaz-Guillén, J.C.; Sousa, R.R.M.; Jelani, M.; Abrar, M. Wear and corrosion studies of duplex surface-treated AISI-304 steel by a combination of cathodic cage plasma nitriding and PVD-TiN coating. *Ceram. Int.* **2022**, *48*, 21473–21482. [CrossRef]
26. Guo, J.; Shi, Y.; Li, C.; Zhang, G. Investigation of nitrogen ionization state and its effect on the nitride layer during fiber laser gas nitriding of Ti-6Al-4V alloy. *Surf. Coat. Technol.* **2021**, *418*, 127254. [CrossRef]
27. Chen, P.; Wu, W.; Liu, H.; Li, H.; Qiu, C. Metastable FeCrMnCo HEAs with the reinforcement of tungsten carbide fabricated by laser melting: Microstructure, mechanical properties and tribological behaviors. *Mater. Charact.* **2023**, *206*, 113397. [CrossRef]
28. Chen, P.-h.; Zhang, Y.; Li, R.-q.; Liu, Y.-x.; Zeng, S.-s. Influence of carbon-partitioning treatment on the microstructure, mechanical properties and wear resistance of in situ VCp-reinforced Fe-matrix composite. *Int. J. Miner. Metall. Mater.* **2020**, *27*, 100–111. [CrossRef]
29. Chen, P.; Liu, Z.; Li, R.; Li, X. The effect of manganese additions on the high temperature oxidation behaviour of the high-vanadium cast iron. *J. Alloys Compd.* **2018**, *767*, 181–187. [CrossRef]
30. Pede, D.; Li, M.; Virovac, L.; Poleske, T.; Balle, F.; Müller, C.; Mozaffari-Jovein, H. Microstructure and corrosion resistance of novel β-type titanium alloys manufactured by selective laser melting. *J. Mater. Res. Technol.* **2022**, *19*, 4598–4612. [CrossRef]
31. Patel, P.; Zala, A.; Parekh, T.; Kahar, S.D.; Jamnapara, N.I. High temperature oxidation behavior of thermal and plasma processed aluminide coated Ti6Al4V alloys. *Surf. Coat. Technol.* **2022**, *447*, 128839. [CrossRef]
32. Sun, F.; Liu, X.-L.; Luo, S.-Q.; Xiang, D.-D.; Ba, D.-C.; Lin, Z.; Song, G.-Q. Duplex treatment of arc plasma nitriding and PVD TiN coating applied to dental implant screws. *Surf. Coat. Technol.* **2022**, *439*, 128449. [CrossRef]

coatings

Article

Effect of the Presence of a Silane Coupling Agent on Reaction Kinetics of Cationic Thermopolymerization of Epoxy Resin Adhesive

Jiangcong Chen [1,2], Bingxuan Li [1], Shujuan Zhang [3] and Hengfeng Li [1,*]

[1] School of Materials Science and Engineering, Central South University, Changsha 410083, China; zhongnancong@163.com (J.C.)
[2] GYGAIR Intelligent Technology Co., Ltd., Guangzhou 510000, China
[3] Ji'an MuLinSen Industrial Co., Ltd., Ji'an 360800, China
* Correspondence: lihf@csu.edu.cn

Abstract: The effect of the presence of a silane coupling agent containing different functional groups on the reaction kinetics and physical properties of epoxy resin generated via cationic thermopolymerization was investigated. The kinetics of cationic polymerization of epoxy resin was studied using a nonisothermal DSC method. The polymerization is propagated by activated chain end (ACE) and activated monomer (AM) processes. With an increase in silane content, the ACE apparent activation energy first decreased and then increased, while the AM apparent activation energy increased. Moreover, the shear strength, coefficient of linear thermal expansion, gelation time, environmental testing of damp heat, and steady state with varied contents of silane coupling agents were evaluated. The results indicated that the presence of a silane coupling agent has more adhesion and lower water absorption than pure EP. The coefficient of linear thermal expansion and gelation time increased with the increase in silane content, while the water absorption decreased with the increase in silane content. Comparing the overall performance of three silane coupling agents used in the research, the curing system obtained by blending with vinyltrimethoxysliane (VTS) is a much better option for practical application as an adhesive.

Keywords: epoxy resin; silane; cationic polymerization; cure kinetics; adhesive

Citation: Chen, J.; Li, B.; Zhang, S.; Li, H. Effect of the Presence of a Silane Coupling Agent on Reaction Kinetics of Cationic Thermopolymerization of Epoxy Resin Adhesive. *Coatings* **2023**, *13*, 1782. https://doi.org/10.3390/coatings13101782

Academic Editor: Csaba Balázsi

Received: 13 September 2023
Revised: 10 October 2023
Accepted: 12 October 2023
Published: 17 October 2023

1. Introduction

Epoxy resin (EP) refers to a class of substances generated from precursors containing two or more epoxy groups [1]. It is one of the most widely used matrix resins in polymer matrix composites, such as in aerospace and electronic materials, because of its excellent mechanical properties, low cost, and excellent weather resistance [2–5]. However, the uncured epoxy monomer has no application value and must be cured using an appropriate curing agent, such as amines, imidazoles, or anhydrides, to form a three-dimensional network structure [6]. After a suitable curing process, EP displays excellent chemical and mechanical properties. In practice, several hours are required for curing EP, which may be accompanied by significant energy consumption [7].

In recent decades, the fast curing of EP initiated by cationic initiators has attracted great concern; it is characterized by the formation of no byproducts, fast reaction rate, and absence of oxygen inhibition [8,9]. Detailed studies on the cationic polymerization of EP have been reported [10–13]. The cationic polymerization of EP consists of a series of reactions. Taking cationic initiator 4-hydroxyphenyl-2-methylbenzyl-methyl sulfonium hexafluorophosphate as an example, the cationic curing agent protonates and produces the strong Brønsted acid HPF6 during the pyrolysis process. This Brønsted acid triggers the ring-opening polymerization of EP in the subsequent process. After a rapid chemical equilibrium with inactive initiators, the active initiators react with epoxy groups, producing active propagating dimer species containing tertiary oxonium ions. Relevant reports suggest that the chain propagation process for EP cationic polymerization consists of activated chain end (ACE) and activated monomer (AM) processes [14,15]. In the early stage of the chain propagation process, polymer chains grow primarily by ACE processes with a fast curing rate. In this process, the chain end active species are propagated by the nucleophilic attack of the heteroatom on the epoxy groups, and the tertiary oxonium ions react with epoxy groups to form new tertiary oxonium ions. The AM process is dominant in late-stage curing and occurs at a low rate, while the hydroxyls react with the secondary oxonium ions to form new hydroxyls and protons, involving the nucleophilic attack of chain ends on the monomers activated by the cationic initiators. The processes for EP cationic polymerization are illustrated in Scheme 1.

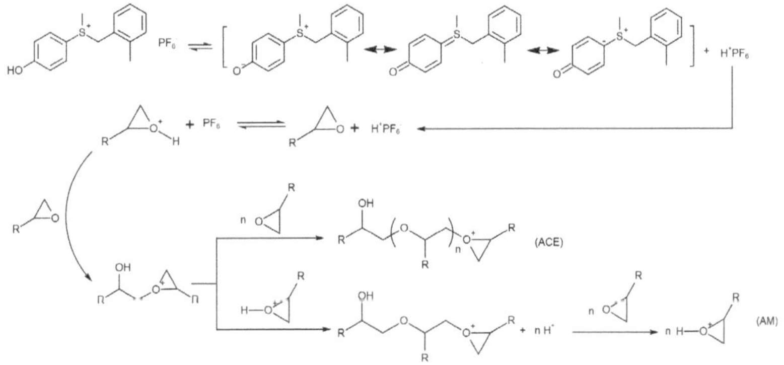

Scheme 1. The cationic polymerization of EP.

The highly crosslinked structure of EP leads to high brittleness and poor impact resistance that greatly limits its further application in some high-tech fields. Thus, it is of great importance to modify EP to improve its mechanical properties. Silane coupling agents are widely applied as additives in EP adhesives to improve this problem [16–18]. In some research, via the application of a coupling agent, the durability of the interfacial bond was significantly enhanced because the hydrogen bonding plays a large role in the degradation of the interfacial bond between epoxy and substrate [19,20]. Silanol groups can be produced both by siloxane hydrolysis and interacting with the Brønsted acid released by the cationic curing agent. The low-molecular-weight byproducts produced by silane hydrolysis function as plasticizers to reduce the high brittleness of EP. Furthermore, silanol groups can react with related groups on the surface of the bonded materials to form strong covalent bonds or form hydrogen bonds with polar groups to increase the bond strength of an EP adhesive [21–23].

Therefore, the effect of the presence of silane coupling agents containing different functional groups on the cationic thermopolymerization of EP was evaluated utilizing reaction kinetics. The physical properties of these systems were determined.

2. Materials and Method

Alicyclic EP (ERL-4221) was supplied by Daicel Co., Ltd. (Tokyo, Japan). 4-hydroxyphenyl-2-methylbenzyl-methyl sulfonium hexafluorophosphate, which was used as a cationic curing agent, was provided by Jiangsu Lingnuoxuan Co., Ltd. (Tongzhou, China). Propylene carbonate was purchased from Sigma-Aldrich (St. Louis, MO, USA). The silane coupling agents, including vinyltrimethoxysliane (VTS), 3-glycidoxypropyltrimethoxysilane (GPTS), and 3-methacryloxypropyltrimethoxysilane (MPTS), were purchased from Sigma-Aldrich. The chemical structures of all of these materials are illustrated in Scheme 2. The aluminum oxide chips (3.3 × 5.1 mil) and silver-plated brackets were provided by Ji'an MuLinSen Industrial Co., Ltd. (Ji'an, China).

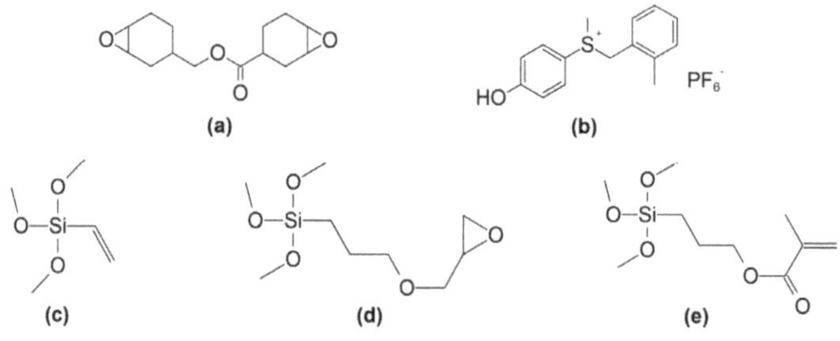

Scheme 2. The chemical structure of (**a**) ERL-4221; (**b**) cationic curing agent; (**c**) VTS; (**d**) GPTS; (**e**) MPTS.

The cationic initiator solution was prepared by dissolving the cationic curing agent in propylene carbonate (with a mass ratio of 1:1). The ingredients EP, cationic initiator solution (1 wt%), and one of the selected silane coupling agents with a certain proportion (0%, 10%, 20%, or 30%) were blended in given amounts depending on the experiment. Then, the mixtures were stirred for 10 min at a revolution of 1500 rpm and a rotation of 900 rpm using a double planetary centrifugal debubbling stirring machine. The curing process of this mixed resin system was 80 °C/0.5 h + 100 °C/0.5 h + 130 °C/0.5 h.

Fourier-transform infrared spectroscopy (Nicolet-iS20) was performed to identify the reaction between ingredients used in the experiment. The DSC studies of the curing behavior of EP were performed using a TA Instruments DSC 25 differential scanning calorimeter from 50 to 200 °C under a nitrogen atmosphere. All of the samples (7~10 mg) were contained within sealed aluminum DSC pans. The heating rate were 5, 10, 15, and 20 K min^{-1}, respectively. Shear strength tests were conducted using a bong tester (DAGE 4000, Nordson DAGE, Aylesbury, UK) with a shear height of 35μm and a shear speed of 200 μm s^{-1}. The samples were prepared by using an adhesive to paste the die on the silver-plated bracket. The samples were tested under each condition, and the average values were reported.

The samples for the coefficient of linear thermal expansion were cut into a dimension of 5 × 5 × 16 mm^3. Then, the samples were transferred in TMA 402 F3 Hyperion for testing at a heating rate of 5 K min^{-1} from 30 to 300 °C. The coefficient of linear thermal expansion from 40 to 110 °C was recorded. The generation time was measured using an indentation method. A glass slide was heated to a desired temperature (90, 110, 130, 150, 170, and 190 °C) for 5 min, and the mixed resin system was added to the glass slide. The surface of the resin mixture was pierced at a certain time, and the time was recorded when the pin did not pierce the surface.

The cured samples were cut into a dimension of 10 × 10 × 16 mm^3 and then transferred in a Jeasi S constant temperature and humidity testing chamber (set as 85 °C and 85% humidity) for a damp heat, steady-state experiment. The weight of each sample was recorded every 48 h.

3. Results and Discussion

3.1. Curing Process of the EP System

Infrared spectroscopy analysis is an important method to analyze the change of functional groups in EP during polymerization. The infrared spectrum of pure EP and EP blending with silane coupling agents containing different functional groups is shown in Figure 1 (The added content of each silane is 10%). The discussion below focuses on certain absorption peaks that suggest the processes of EP cationic polymerization.

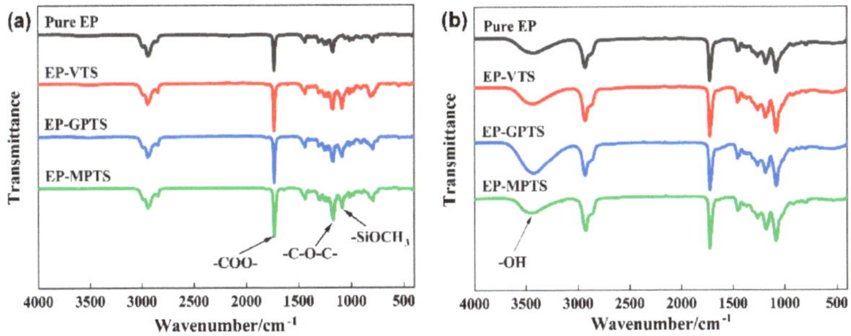

Figure 1. The infrared spectra of pure EP and EP blending with silane system. (**a**) Uncured; (**b**) after curing process.

Based on the previous research [24,25], in the infrared spectra of uncured EP adhesive, the absorption peaks at 1435 cm^{-1} and 789 cm^{-1} are related to the bending vibration of $-CH_2$ and –CH, respectively. At the same time, the characteristic bands at 2938 cm^{-1} correspond to the stretching vibration of alkyl. In addition, the absorption peaks at 1729 cm^{-1} and 1174 cm^{-1} are attributed to the stretching vibration of C=O in the ester groups and C–O–C in the epoxy groups on the EP of ERL-4221, respectively. The absorption peaks at 1081 cm^{-1} indicate the presence of $-SiOCH_3$ on the silane coupling agent. After the designed curing process, the peak area of absorption peak at 1081 cm^{-1} increases obviously due to the superposition of $-SiOCH_3$ and C–O–C in the ether bonds. It suggests the formation of ether bonds via the ring-opening reaction of epoxy groups, which corresponds to the ACE chain propagation processes. On the other hand, a broad shoulder at 3425 cm^{-1} due to the stretching of –OH indicates that the monomer was bonded to the chain end to form hydroxyl groups after the epoxy groups ring opening, and it is consistent with the AM chain propagation processes. It is worth noting that in the EP GPTS system, the peak area at 3425 cm^{-1} is obviously larger than the other three systems, showing that the epoxy groups in GPTS are also involved in the polymerization.

3.2. Kinetics of Epoxy Cationic Polymerization

A statistical modeling procedure is established for predicting the cure behavior of the cationic polymerization. Based on the general equation of dynamics (Equation (1)) [26,27], the reaction rate $\frac{d\alpha}{dt}$ is determined by the rate constant $k(T)$ and the reaction model function $f(\alpha)$. Additionally, the rate constant $k(T)$ can be further defined by Equation (2) according to the Arrhenius equation.

As mentioned above, cationic polymerization of EP is propagated by two processes with different rates in different stages. Thus, the reaction kinetics of this system should be divided into $(\frac{d\alpha}{dt})$ ACE and $(\frac{d\alpha}{dt})$ AM for further discussion. Kissinger's method [28] was used to evaluate the apparent activation energy of theoretically simulated complex processes that can be obtained by the slope of the linear fit between $\ln(\beta/T_i^2)$ and $1/T_i$ (Equation (3)).

$$\frac{d\alpha}{dt} = k(T)f(\alpha) \quad (1)$$

$$k(T) = Ae^{(-\frac{E_\alpha}{RT})} \quad (2)$$

$$\frac{d\left(ln\frac{\beta}{T_i^2}\right)}{d\left(\frac{1}{T_i}\right)} = -\frac{E_\alpha}{R} \quad (3)$$

where α is the conversion degree, T is the absolute temperature, E_α is the apparent activation energy, R is the universal gas constant, A is the frequency factor, β is the heating rate, and T_i is the exothermic peak temperature.

The pure EP system cured at temperatures ranging from 50 °C to 200 °C with a heating rate of 5 K min^{-1} is demonstrated in Figure 2a. It reveals that there are two exothermic peaks corresponding to the ACE and AM chain propagation processes mentioned above, respectively. To study the reaction kinetics of the EP system and monitor the change of apparent activation energy during the polymerization process, the DSC curves of each EP system at different heating rates were tested, and the corresponding peak temperatures were recorded. Each DSC curve has similar exothermic peak shapes and different peak temperatures. Thus, Kissinger's model is adopted for linear fitting to calculate apparent activation energy, as shown in Figure 2b,c. It is noticed that the $R^2 > 0.999$ in each fitting line indicates that the EP cationic polymerization system conforms to the Kissinger theory. The apparent activation energy E_α calculated by the slope of the fitting line $-\frac{E_\alpha}{R}$ is summarized in Table 1. In the systems of EP blending with 30 wt% VTS and blending with 30 wt% MPTS, the apparent activation energy corresponding to ACE propagation processes cannot be calculated due to the overlap of two exothermic peaks.

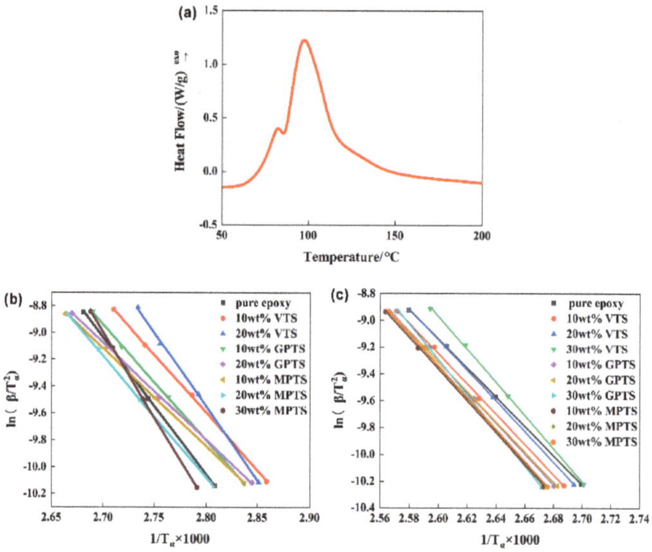

Figure 2. (**a**) Nonisothermal DSC curve of pure EP system with 5 K min^{-1}; (**b**) linear fit for calculation of reaction activation energy of ACE processes; and (**c**) linear fit for calculation of reaction activation energy of AM processes.

Table 1. Apparent activation energy of EP system blending with different silane coupling agents.

Content of Silane Coupling Agent/wt%	Activation Energy E_α/(kJ mol^{-1})					
	VTS		GPTS		MPTS	
	$E_{\alpha(ACE)}$	$E_{\alpha(AM)}$	$E_{\alpha(ACE)}$	$E_{\alpha(AM)}$	$E_{\alpha(ACE)}$	$E_{\alpha(AM)}$
0	84.81	90.38	84.81	90.38	84.81	90.38
10	71.27	93.7	71.61	93.14	60.43	98.51
20	91.35	94.87	60.05	95.3	75.07	98.57
30	/	102.5	/	110.15	98.57	99.01

At the beginning of the curing process, a number of active centers at a relatively low temperature are formed, and the polymer chain propagates primarily via ACE processes. The effect of the silane coupling agent on the activation energy of EP cationic polymerization is mainly reflected in two aspects: the small molecular products that are produced by the decomposition of siloxane groups have a plasticizing effect to improve molecular activity and reduce reaction activation energy. On the other hand, the introduction of inorganic chains blocks the movement of molecules and increases the reaction activation energy. Under the joint influence of these two processes, the activation energy of ACE processes first decreases and then increases with the increasing of silane coupling agent content. However, the activation energy of ACE processes decreases from 84.81 kJ mol^{-1} to 60.05 kJ mol^{-1} with the increasing of silane coupling agent content because the ACE processes in the system of EP-GPTS is propagated by the reaction between tertiary oxonium ion and epoxy groups, at the late stage of the curing process, the reaction is dominated by AM processes. The reaction rate changes from chemical concentration control to diffusion control. The activation energy is increased from 90.38 to 102.5 kJ mol^{-1}, 110.15 kJ mol^{-1}, and 99.01 kJ mol^{-1}, respectively, due to the concentration of reactive substance is decreased with the increasing of silane coupling agent content. The added value of activation energy of the system of EP-GPTS is obviously greater than that of the other two systems because the higher content of epoxy groups in this system makes a better crosslinking density and a more difficult molecular movement.

3.3. Adhesion Properties of the EP System

Figure 3 shows the shear strength of EP adhesive blending with different types of silane coupling agents evaluated. The slow curing group adopts the curing process of 60 °C/3 h + 90 °C/3 h + 130 °C/6 h for the pure EP system. At the same time, the fast-curing group and the groups containing silane coupling agents used the curing process of 80 °C/0.5 h + 100 °C/0.5 h + 130 °C/0.5 h. When the silane content is 10 wt%, the order of shear strength was found to be slow curing > VTS ≈ MPTS > GPTS > fast curing, and the values of shear strength are 117.4 g, 92.1 g, 91.7 g, 64.1 g, and 20.2 g, respectively. This is consistent with the results of relevant reports [21,28]. For each system, the shear strength decreased with the increase in silane content due to the reduction in EP concentration. There are two factors that affect the shear strength of EP adhesive: chemical properties of organic functional groups and shrinkage or deformation of adhesives during the curing process. The sequence of polarity of functional groups is –COO– > –C–O–C > –C=C–. In addition, the vinyl and methacrylic groups contained in silane can promote free radical polymerization, which also contributes to the shear strength [29,30]. The shrinkage and deformation of adhesives is regarded as the main factor affecting the shear strength. In the pure EP fast-curing group, the internal stress cannot be released in time due to the rapid reaction rate that results in cracking and greatly affects its shear strength. However, for pure EP slow curing group, although with high shear strength, it costs more time and energy. For EP blending with the silane system, it will be further discussed.

Expansion and shrinkage are two important factors that lead to the deformation of adhesives and affect the shear strength. The coefficient of linear thermal expansion (CTE) of cured EP adhesive is evaluated using the TMA method, and the average values between 40 and 110 °C were recorded in Figure 4. When the silane content is 30 wt%, the order of CTE was found to be MPTS > GPTS > VTS ≈ pure EP and the values are 180.3, 155.1, 95.3, and 90.0 ($\times 10^{-6}$ K^{-1}), respectively. For each system, the CTE increased with the increase in silane content.

Another factor that causes the deformation of adhesives is the cracks that are generated during the curing process due to the fast curing rate. Therefore, it is necessary to control the curing rate. Gelation refers to the phenomenon of loss of fluidity in a linear polymer system due to the infinite increase in viscosity during the crosslinking reaction process. It is associated with the curing rate of EP. The gelation time–temperature relation curves are shown in Figure 5. Compared to pure EP, the introduction of silane agents reduces the activity of polymerization. For each system, the gelation time decreased with the curing temperature increasing or the silane content decreasing. The order of gelation time is found to be EP-MPTS > EP-VTS > EP-GPTS > pure EP. It is worth noting that, compared to the other two silanes, the EP-GPTS system has a shorter gelation time due to its higher content of epoxy groups and lower activation energy that was described in the nonisothermal DSC test.

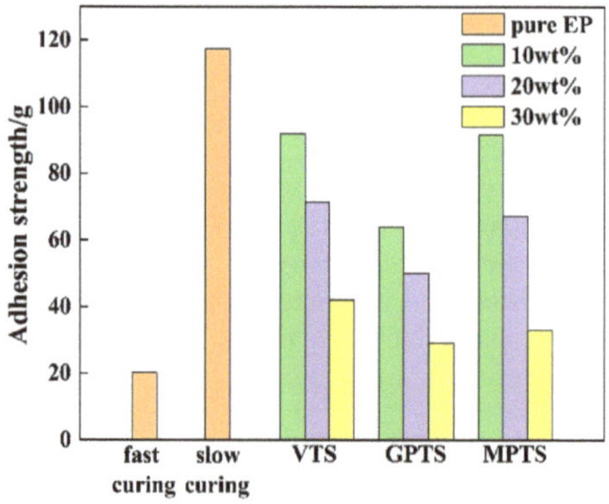

Figure 3. Shear strength of EP blending with different types of silane coupling agents.

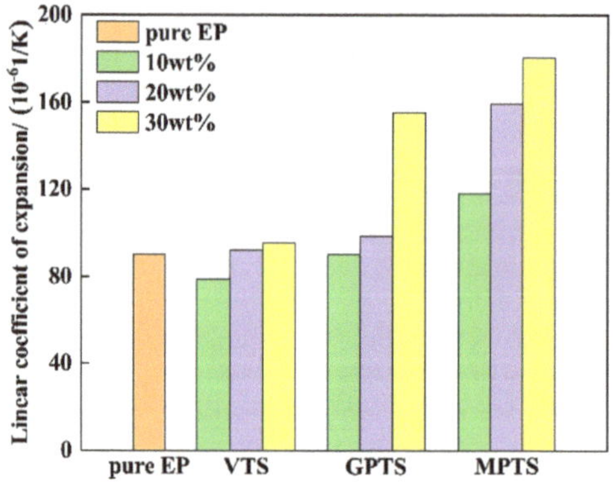

Figure 4. Linear expansion coefficient of EP blending with different types of silane coupling agents.

The EP-VTS system has a low linear expansion coefficient, and the EP-MPTS system has a longer gelation time that is conducive to the release of internal stress. Thus, the two systems both have excellent adhesion ability.

Ensuring excellent adhesion ability and great environmental stability are important aspects for the reliability of the EP curing system as an adhesive. The weight change of each system at the condition of 85 °C and 85% humidity was tested, and the weight change curves are shown in Figure 6. When the content of the silane coupling agent is 10 wt%, the EP-VTS, EP-GPTS, and EP-MPTS systems begin to lose weight on day 12, day 8, and day 4, respectively. For a certain silane coupling agent, the weight is lost earlier with the increase in silane content. That is because the content of the inorganic chain segment in the system increases, resulting in an increase in the degree of phase separation and instability.

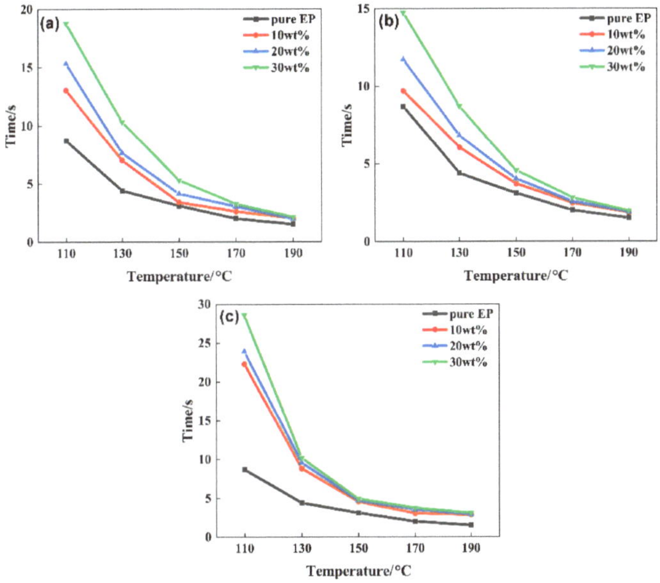

Figure 5. The gelation time–temperature diagram of EP blending with (**a**) VTS; (**b**) GPTS; and (**c**) MPTS.

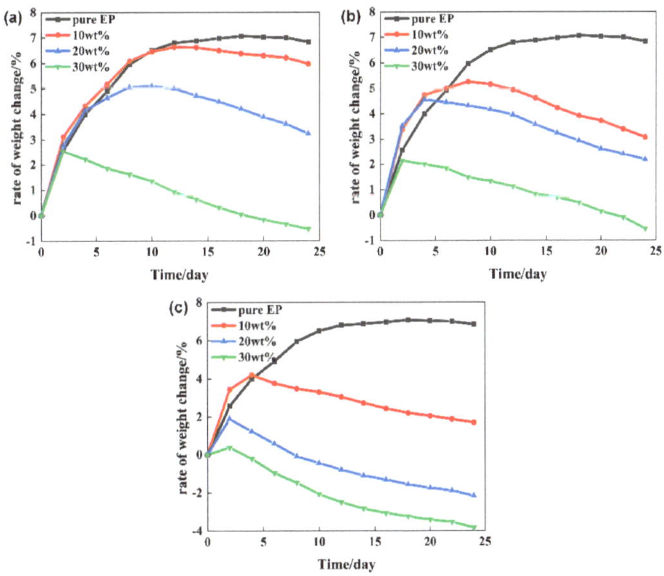

Figure 6. The rate of weight change of EP blending with (**a**) VTS; (**b**) GPTS; and (**c**) MPTS.

4. Conclusions

The effect of the presence of a silane coupling agent on the reaction kinetics and physical properties of epoxy resin made via cationic thermopolymerization was studied. For the ACE processes, with the increasing of silane content, the activation energy of the EP-VTS system decreases from 84.81 kJ mol^{-1} to 71.27 kJ mol^{-1}, and then increases to 91.35 kJ mol^{-1}. The activation energy of the EP-MPTS system decreases from 84.81 kJ mol^{-1} to 60.43 kJ mol^{-1} and then increases to 98.57 kJ mol^{-1}. At the same time, the activation energy of the EP-GPTS system decreased from 84.81 kJ mol^{-1} to 60.05 kJ mol^{-1} due to the increasing epoxy group content. For the AM chain growth processes, the apparent activation energy of the EP-VTS system, EP-GPTS system, and EP-MPTS system increased from 90.38 kJ mol^{-1} to 102.50 kJ mol^{-1}, 110.15 kJ mo^{l-1}, and 99.01 kJ mol^{-1} with the increase in silane content, respectively. For each system, the shear strength is 30.2 g, 92.1 g, 64.1 g, and 91.7 g, respectively, when the content of silane is 10 wt%. Taking the results of the thermal expansion coefficient and thermal stability into consideration, the EP-VTS system is suitable for practical application as an adhesive.

Author Contributions: Conceptualization, J.C. and H.L.; methodology, B.L. and S.Z.; software, J.C.; validation, J.C., B.L. and H.L.; formal analysis, J.C.; investigation, B.L.; resources, H.L.; data curation, J.C. and S.Z.; writing—original draft preparation, J.C.; writing—review and editing, B.L.; visualization, J.C.; supervision, H.L.; project administration, H.L.; funding acquisition, H.L. All authors have read and agreed to the published version of the manuscript.

Funding: This research received no external funding.

Institutional Review Board Statement: Not applicable.

Informed Consent Statement: Not applicable.

Data Availability Statement: The data that support the findings of this study are available from the corresponding author upon reasonable request.

Conflicts of Interest: The authors declare no conflict of interest.

References

1. Jin, F.L.; Li, X.; Park, S.J. Synthesis and application of epoxy resins: A review. *J. Ind. Eng. Chem.* **2015**, *29*, 1–11. [CrossRef]
2. Ham, Y.R.; Kim, S.H.; Shin, Y.J.; Lee, D.H.; Yang, M.; Min, J.H.; Shin, J.S. A comparison of some imidazoles in the curing of epoxy resin. *J. Ind. Eng. Chem.* **2010**, *16*, 556–559. [CrossRef]
3. Xu, Y.J.; Wang, J.; Tan, Y.; Qi, M.; Chen, L.; Wang, Y.Z. A novel and feasible approach for one-pack flame-retardant epoxy resin with long pot life and fast curing. *Chem. Eng. J.* **2018**, *337*, 30–39. [CrossRef]
4. Zhang, B.L.; Tang, G.L.; Shi, K.Y.; You, Y.C.; Du, Z.J.; Yang, J.F.; Huang, J.F. A study on properties of epoxy resin toughened by functionalized polymer containing rigid, rod-like moiety. *Eur. Polym. J.* **2020**, *36*, 205–213. [CrossRef]
5. Han, X.; Han, Y.; Jin, Y.; Tian, H.; Wang, Z.; Huang, J.; Men, S.; Kumar, R. Tailoring compatibility and toughness of microbial poly(3-hydroxybutyrate-co-3-hydroxyvalerate)/bio-based polyester elastomer blends by epoxy-terminated hyperbranched polyester. *Int. J. Biol. Macromol.* **2022**, *220*, 1163–1176. [CrossRef] [PubMed]
6. Ham, Y.R.; Lee, D.H.; Kim, S.H.; Shin, Y.J.; Yang, M.; Shin, J.S. Microencapsulation of imidazole curing agent for epoxy resin. *J. Ind. Eng. Chem.* **2010**, *16*, 728–733. [CrossRef]
7. Wang, Y.R.; Liu, W.S.; Qiu, Y.P.; Wei, Y. A One-Component, Fast-Cure, and Economical Epoxy Resin System Suitable for Liquid Molding of Automotive Composite Parts. *Materials* **2018**, *11*, 685. [CrossRef]
8. Unnikrishnan, K.P.; Thachil, E.T. Toughening of epoxy resins. *Des. Monomers Polym.* **2006**, *9*, 129–152. [CrossRef]
9. Sangermano, M.; Malucelli, G.; Amerio, E.; Priola, A.; Billi, E.; Rizza, G. Photopolymerization of epoxy coatings containing silica nanoparticles. *Prog. Org. Coat.* **2005**, *54*, 134–138. [CrossRef]
10. Sangermano, M.; Roppolo, I.; Chiappone, A. New Horizons in Cationic Photopolymerization. *Polymers* **2018**, *10*, 136. [CrossRef]
11. Penczek, S. Cationic ring-opening polymerization (CROP) major mechanistic phenomena. *J. Polym. Sci. Pol. Chem.* **2000**, *38*, 1919–1933. [CrossRef]
12. Penczek, S.; Kubisa, P.; Szymański, R. Activated monomer propagation in cationic polymerizations. *Macromol. Symp.* **2011**, *3*, 203–220. [CrossRef]
13. Gonzalez, L.; Ramis, X.; Salla, J.M.; Mantecon, A.; Serra, A. Kinetic analysis by DSC of the cationic curing of mixtures of DGEBA and 6,6-dimethyl (4,8-dioxaspiro[2.5]octane-5,7-dione). *Thermochim. Acta* **2007**, *464*, 35–41. [CrossRef]
14. Damien, D.; Blanca, M.V.; Didier, B.; Christophe, N. Ring-Opening Polymerization of Trimethylene Carbonate Catalyzed by Methanesulfonic Acid: Activated Monomer versus Active Chain End Mechanisms. *Macromolecules* **2010**, *43*, 8828–8835.
15. Anderson, B.J. Cationic cure kinetics of a polyoxometalate loaded epoxy nanocomposite. *J. Polym. Sci. Pol. Chem.* **2012**, *300*, 4507–4515. [CrossRef]
16. Nazir, T.; Afzal, A.; Siddiqi, H.M.; Ahmad, Z.; Dumon, M. Thermally and mechanically superior hybrid epoxy-silica polymer films via sol-gel method. *Prog. Org. Coat.* **2010**, *69*, 100–106. [CrossRef]
17. Mascia, L.; Prezzi, L.; Wilcox, G.D.; Lavorgna, M. Molybdate doping of networks in epoxy-silica hybrids: Domain structuring and corrosion inhibition. *Prog. Org. Coat.* **2006**, *56*, 13–22. [CrossRef]

18. Li, B.X.; Xia, W.X.; Shen, L.M.; Tan, W.M.; Luo, X.L. Preparation of cellulose nanoparticles/epoxy resin composites using the in-situ reaction method for strengthening and toughening epoxy resin film simultaneously. *Mater. Lett.* **2023**, *349*, 134790. [CrossRef]
19. Fiore, V.; Orlando, V.; Sanfilippo, C.; Badagliacco, D.; Valenza, A. Effect of Silane Coupling Treatment on the Adhesion between Polyamide and Epoxy Based Composites Reinforced with Carbon Fibers. *Fibers* **2020**, *8*, 48. [CrossRef]
20. Choi, S.; Maul, S.; Stewart, A.; Hamilton, H.R.; Douglas, E. Effect of silane coupling agent on the durability of epoxy adhesion for structural strengthening applications. *Polym. Eng. Sci.* **2013**, *53*, 283–294. [CrossRef]
21. Chiang, T.H.; Hsieh, T.E. The effect of organo-functional silanes on the adhesion of epoxide resins to ITO glass. *J. Adhes. Sci. Technol.* **2005**, *19*, 1–18. [CrossRef]
22. Kim, J.G.; Choi, I.; Lee, D.G.; Seo, I.S. Flame and silane treatments for improving the adhesive bonding characteristics of aramid/epoxy composites. *Compos. Struct.* **2011**, *93*, 2696–2705. [CrossRef]
23. Lee, C.Y.; Bae, J.H.; Kim, T.Y.; Chang, S.H.; Kim, S.Y. Using silane-functionalized graphene oxides for enhancing the interfacial bonding strength of carbon/epoxy composites. *Compos. Part. A-Appl. S* **2015**, *75*, 11–17. [CrossRef]
24. Canavate, J.; Colom, X.; Pages, P.; Carrasco, F. Study of the curing process of an epoxy resin by FTIR spectroscopy. *Polym-Plast. Technol.* **2000**, *39*, 937–943. [CrossRef]
25. Barkoula, N.M.; Karabela, M.; Zafeiropoulos, N.E.; Tsotra, P. Fast curing versus conventional resins—Degradation due to hygrothermal and UV exposure. *Express Polym. Lett.* **2020**, *14*, 401–415. [CrossRef]
26. Olmos, D.; Aznar, A.J.; Gonzalez-Benito, J. Kinetic study of the epoxy curing at the silica particles/epoxy interface using the fluorescence of pyrene label. *Polym. Test.* **2005**, *24*, 275–283. [CrossRef]
27. Sani, K.; Manisha, C.; Ashis, B. Solid-State Reaction of Ferrocene Controlled by Co-Precursor and Reaction Atmosphere Leading to Hematite and Cohenite Nanomaterials: A Reaction Kinetic Study. *J. Phys. Chem. C* **2023**, *127*, 18397–18408.
28. Peinado, C.; Alonso, A.; Catalina, F.; Schnabel, W. Using linear and branched polysilanes for the photoinitiated polymerization of a commercial silicone-acrylate resin. A real time FTIR study. *J. Photoch. Photobio. A Chem.* **2001**, *141*, 85–91. [CrossRef]
29. Choi, D.H.; Oh, S.J.; Cha, H.B.; Lee, J.Y. Photochemically bifunctional epoxy compound containing a chalcone moiety. *Eur. Polym. J.* **2001**, *37*, 1951–1959. [CrossRef]
30. Crivello, J.V. Redox Initiated Cationic Polymerization: Reduction of Triarylsulfonium Salts by Silanes. *Silicon* **2009**, *1*, 111–124. [CrossRef]

coatings

Article

The Relevance of Process Parameter Optimization and Geometric Figure for Direct Laser Deposition of Inconel 738 Alloy and Its Theoretical Modeling

Kun Qi [1], Wenxing Wu [2], Pinghu Chen [1], Hao Liu [1] and Changjun Qiu [1,*]

[1] Key Laboratory of Hunan Province of Equipment Safety Service Technology under Extreme Environment, College of Mechanical Engineering, University of South China, Hengyang 421001, China; qikun18955500214@163.com (K.Q.); chenpinghu1986@163.com (P.C.); liuhao10212510@163.com (H.L.)

[2] School of Nuclear Science and Technology, University of South China, Hengyang 421001, China; wuzhiaifrank@hotmail.com

* Correspondence: qcj@usc.edu.cn

Abstract: In order to minimize the gaps between the direct laser deposition channels and improve the quality and performance of the formed parts, the process of direct laser deposition is utilized in laser additive manufacturing to create sequential, single- and double-channel deposition layers on 304 stainless steel plates. Under the premise of keeping the layer rate and defocusing amount unchanged, this study investigates the effects of laser power, scanning speed, and powder feeding rate on the morphology and inclusions of single- and double-channel deposited layers. The aim is to determine the optimal process parameter values for direct laser deposition of single-layer, single-channel Inconel 738. The effects of the three process parameters on the response values were investigated using a multi-factor, multi-level experimental design. The evaluation indexes for the analysis included the deposited layer wetting angle and aspect ratio. The analysis involved one-way extreme analysis and ANOVA analysis. The optimal process parameters are a laser power of 550~750 W, a scanning speed of 7~13 mm/s, and the powder feeding rate was 2.1~4.33 g/min. At the same time, the relationship between surface tension and gravity was integrated with the spherical coronal model and Young's equation to develop a mathematical model of the direct laser deposition process at a theoretical level. The mathematical model of the direct laser deposition process was utilized to analyze the correlation between the geometric parameters of the cross-section of the deposited layer. This analysis provides a valuable data reference for future Inconel 738 direct laser deposition.

Keywords: laser additive manufacturing; process parameter optimization; geometrical shape; theoretical derivation

Citation: Qi, K.; Wu, W.; Chen, P.; Liu, H.; Qiu, C. The Relevance of Process Parameter Optimization and Geometric Figure for Direct Laser Deposition of Inconel 738 Alloy and Its Theoretical Modeling. *Coatings* **2023**, *13*, 1926. https://doi.org/10.3390/coatings13111926

Academic Editor: Roberto Teghil

Received: 18 October 2023
Revised: 8 November 2023
Accepted: 9 November 2023
Published: 10 November 2023

1. Introduction

Nowadays, it is common to investigate the effects of various process parameters on the properties of molded parts prepared using Inconel 738 laser direct deposition molding. However, the properties of the molded parts are more affected by changes in the morphology of individual passes and layers. The influence of inclusions between passes on the properties of the molded parts is even more significant under different combinations of process parameters. Currently, there are very few studies of this nature. Inconel 738 exhibits an excellent high-temperature strength, toughness, hardness, and resistance to heat corrosion, making it a popular choice for various applications, such as ships, industrial gas turbines, nuclear reactors, and petrochemicals. It is also used in high-end components, such as aeroengines and supersonic airplanes [1,2]. However, Inconel 738 faces problems with weldability and susceptibility to hot cracking, which significantly impact its machinability and the performance of its components. At present, the inclusion cracking issue of Inconel 738 alloy can be addressed through three main approaches: optimizing the powder composition, optimizing the process parameters of laser additive manufacturing, and incorporating physical field technology. It has attracted great attention in various fields, and related research work has been carried out [3–5]. Compared to other popular alloys, such as Inconel 718, Inconel 600, Monel 400, and Ti6Al4V, Inconel 738 holds a greater economic and industrial significance. Driven by the increasing demand in recent years, the market size of nickel-based alloys has grown significantly. It is projected that the market value in 2020 will reach approximately CNY 11.096 billion. In the field of aerospace engines, the use of nickel-based alloy materials comprises from 40% to 60% of the total, with a market value of CNY 6.203 billion. In the field of nuclear power, the market is worth CNY 1.979 billion. In recent years, China has experienced a steady growth in the production of nickel-based alloys. In 2020, the output of nickel-based alloys reached approximately 27,600 metric tons, with a sales volume of about 50,900 metric tons. Among them, the use of Inconel 738 has surpassed one-third of the total utilization of nickel-based alloys. With the continuous growth of the downstream nickel-based alloy industry, the industry scale is also projected to expand in the future. It is expected that the industry value will reach CNY 55.034 billion by 2027.

Laser additive manufacturing (LAM) can create a three-dimensional model of the metal parts by using professional software to slice and layer the data, resulting in a two-dimensional deposition path. Through the use of digital technology, a high-energy laser beam will be employed to achieve the direct formation of three-dimensional parts with a high degree of freedom. This will be accomplished by melting and depositing the material point by point, line by line, and surface by surface [6,7]. Direct laser deposition (DLD) is based on the principle of discrete/stacking by layering the 3D CAD model of the part, obtaining the 2D contour information of each layer's cross-section, and generating the machining path. In an inert gas-protected environment, high-energy-density lasers are used as the heat source to melt and stack powders or wires layer by layer along a predetermined machining path. This process enables the direct manufacturing and repair of metal parts. The most effective approach to address the issues of poor forming and cracking in the manufacturing of Inconel 738 using direct laser deposition technology is to optimize the parameters of the forming process. The change in process parameters will directly affect the deposition geometry, microstructure, residual stress, and mechanical properties. Therefore, it is necessary to conduct relevant studies on the process parameters in order to obtain dense and high-performance depositions [8]. The selection of process parameters, such as the laser power, scanning speed, and powder feeding rate, plays a crucial role in the quality of the mold [9]. The deposition gradually transitions from a single to multiple coats, and the quality of the single coat directly affects the overall quality [10,11]. Quality assessment requires a defect-free surface, without cracks, with appropriate values for the height, width, height-to-width ratio, and wetting angle size. In recent years, numerous scholars have developed models and empirical formulas to examine the impact of process parameters and their combinations on deposition morphology and properties. This has been achieved through the use of response surface analysis and linear regression analysis [12]. Perevoshchikova et al. [13] achieved a microcrack-free and macroscopic porosity of only 0.5% in molded samples by using a low-energy laser, setting a medium scanning speed, and employing a small scanning pitch. The interaction among scanning speed, laser power, and scanning spacing greatly influences the formation of microcracks and pores. Mondal et al. utilized the width and depth of the deposition as indicators to investigate the impact of each process parameter on deposition. This was achieved through the implementation of an orthogonal experimental design, an artificial neural network, and analysis of variance [14]. Barekat et al. investigated the relationship between the laser power, scanning speed, powder feed rate, and the geometrical characteristics of single-layer, single-pass depositions. They conducted a linear regression analysis and designed a guide for selecting the process parameters to achieve the desired deposition characteristics [15]. Ansari [16] proposed the use of a regression model to predict the deposition morphology.

This was achieved by analyzing the effect of laser additive manufacturing parameters on the deposition morphology using nickel-based alloy powder. T. Muthuramalingam [17] utilized the Taguchi–Grey correlation method to examine the impact of process parameters on surface property measurements while machining titanium alloys. It was found that the laser power (3 kW), nozzle distance (1.5 mm), focal length (−2 mm), and gas pressure (2 bar) were determined to be the optimal combinations of process parameters, resulting in less plasma energy and an accuracy of 2.2%. Ammar H. Elsheikh [18] conducted statistical analysis to examine the impact of various laser cutting parameters on the kerf characteristics. The results showed that increasing in cutting parameters leads to an increase in the top and bottom width of the saw kerf, while increasing in cutting speed or laser power leads to an increase in the taper of the kerf. Ahmed B. Khoshaim [19] utilized the Taguchi L18 hybrid design to investigate the impact of four process parameters (laser power, cutting speed, auxiliary gas pressure, and plate thickness) on five process responses: the kerf deviation, top heat-affected zone, bottom heat-affected zone, maximum surface roughness, and roughness area. The results show that as the laser power increases, the cutting speed decreases, the surface roughness worsens, and the heat-affected zone widens. Ammar H. Elsheikh [20] proposed a solution to address the margin problem in order to accurately calculate the transient temperature field during laser heating. The experiments were conducted using a continuous-wave CO_2 laser with a maximum power of 150 W to validate the findings. Ammar H. Elsheikh [21] utilized the Taguchi L18 orthogonal array design to examine the impact of four input variables—gas pressure, sheet thickness, laser power, and cutting speed—on five kerf quality metrics: the rough zone ratio, upper and lower heat-affected zone width, maximum surface roughness, and kerf taper angle. The results show that the RVFLEO model has a strong ability to accurately predict the laser-cutting characteristics of PMMA sheets.

In this paper, a multi-factorial and multi-level experimental design was adopted to investigate the effects of changes in the process parameters on the morphology of the deposited layer and the inclusions between channels. This was achieved through one-way analysis, extreme analysis, and analysis of variance (ANOVA) using three process parameters: the laser power, scanning speed, and powder feeding rate, as the influencing factors. The responses measured were the wetting angle, height, width, and their ratios. The goal was to determine the optimal values of the process parameters that would result in the lowest number of inclusions in the multi-channel deposited layer. The optimal value for each process parameter is obtained when there is a minimum of multi-channel sedimentary layer inclusions. At the same time, a mathematical model was developed to characterize the relationship between the geometrical parameters of the cross-section and the surface tension and gravity. This was conducted in conjunction with the theoretical derivation of the spherical crown model.

2. Experimental

2.1. Experimental Equipment and Materials

In this experiment, the coaxial powder feeding method is utilized for laser material deposition on the substrate surface. The laser material deposition platform equipment comprises an XL-1000 laser, an HW-05SF powder feeder, a CWFL-1000 refrigeration controller, a CW-5000 water-cooling system, and a 4-axis CNC6090 numerical control platform. The laser's rated power is 1000 W, and argon is used as both the protective gas and the carrier gas.

The substrate used for the experiment is 304 stainless steel, and the forming material is Inconel 738 alloy powder. The chemical composition (mass fraction) is shown in Table 1.

Table 1. Mass fractions of chemical elements in 304 stainless steel and Inconel 738 alloy powder.

Material	Chemical Composition (%)								
	C	Si	Mn	Cr	Co	S	P	N	Al
304	≤0.08	≤1.0	≤2.0	18.0~20.0	-	≤0.03	≤0.035	≤0.1	-
Inconel 738	0.095	-	-	15.88	8.66	-	-	-	3.51
	Mo	Nb	Ta	Ti	W	Zr	B	Fe	Ni
304	-	-	-	-	-	-	-	Bal	8.0~10.5
Inconel 738	1.76	0.86	1.76	3.49	2.65	0.022	0.009	0.09	Bal

The comparison of the properties of Inconel 738 with other popular alloys, namely Inconel 718, Inconel 600, Monel 400, and Ti6Al4V, is shown in Table 2 below.

Table 2. Comparison of the properties of Inconel 738 with other superalloys.

Materials \ Capability	Inconel 738	Inconel 718	Inconel 600	Monel 400	Ti6Al4V
Density	8.11 kg/dm^3	8.22 kg/dm^3	8.4 kg/dm^3	8.83 kg/dm^3	2.5 kg/dm^3
Hardness	41–43 HR	23–46 HR	65 HR	60–96 HR	25–40 HR
Strength	Yield strength: 951 MPa Tensile strength: 1096 MPa	Yield strength: 317.2–1300 MPa Tensile strength: 800–1950 MPa	Yield strength: ≥245 MPa Tensile strength: ≥550 MPa	Yield strength: 172–1104 MPa Tensile strength: 400–1100 MPa	Yield strength: 827–1034 MPa Tensile strength: ≥896 MPa
Ductility	5.5%	3%–42%	≥30%	2%–50%	≥3%
Fatigue life	124 Mpa	3950 MPa	283 MPa	279 MPa	1619 MPa
Thermal conductivity	13–84 W/(m·°C)	10.9–11.4 W/(m·°C)	12.4–14.4 W/(m·°C)	21.8 W/(m·°C)	7.04–7.23 W/(m·°C)
Thermal expansion coefficient	12.87–15.93 × 10^{-6}/°C	2.14–12.31 × 10^{-6}/°C	11.69–11.89 × 10^{-6}/°C	13.8 × 10^{-6}/°C	8.68–8.75 × 10^{-6}/°C

2.2. Experimental Methods

Before the experiment, the surface of the substrate was polished successively with a grinder and sandpaper. It was then cleaned with ethanol and acetone to remove any surface grease and stains. The substrate was subsequently dried in a drying oven, along with the powder for 2 h. After the experiment, the substrate was cooled to room temperature, and long-strip specimens were cut along the direction perpendicular to the fusion sediment. Then, each group of long strip specimens was cut into small samples, and the small samples were mounted and polished using 180–2000 grit sandpaper. This process allowed the creation of a cross-section of the deposited layer, as depicted in Figure 1. The height H and width W of the deposit layer were photographed and measured using an optical imager. The pictures were then imported into AutoCAD to measure the wetting angle. The cross-section of the deposit layer and the geometric dimensions are shown in Figure 2.

(a) (b)

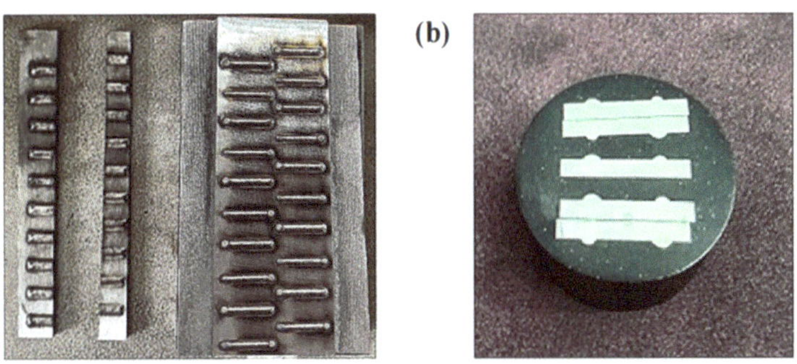

Figure 1. (**a**) Macrographs of the surfaces of single- and double-pass deposited layers using direct laser deposition, as well as a long-cut specimen. (**b**) Macroscopic view of the cross-section of a long specimen after wire cutting and mounting.

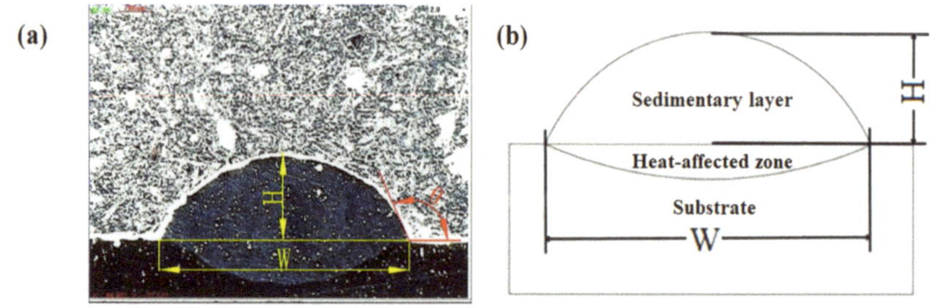

Figure 2. (**a**) Cross-sectional morphology of a single-layer, single-pass object formed through laser direct deposition. (**b**) Theoretical cross-sectional morphology of single-layer, single-pass laser direct deposition molding.

Figure 1a shows the laser deposition on the substrate, which is directly formed by a single layer of a single-channel, macro-photomorphological map and its wire-cut, long specimens. Figure 1b was obtained from Figure 1a based on the long specimens, which can be used for paneling. Figure 1b shows the macro-photomorphological map of the cross-section after paneling. Figure 2a depicts a microscopic cross-sectional geometrical profile of a single-channel deposition layer captured using an optical imager. In contrast, Figure 2b illustrates a theoretical cross-sectional profile of a single-channel deposition layer created using CAD.

3. Experimental Results and Analysis

Based on the multi-factorial and multi-level experimental design, five levels of laser power, four levels of scanning speed, and three levels of powder feeding rate were selected. The width of the deposited layer W, the height H, the aspect ratio H/W, and the wetting angle θ were used as the four response variables.

The results of the experimental design are shown in Table 3 and Figures 3 and 4.

Table 3. Multi-factor and multi-level experimental design scheme and results.

Depots	Laser Power P/W	Scanning Speed Vs/(mm/s)	Powder Feeding Rate Vf/(g/min)	Welding Width W/mm	Welding Height H/mm	Ratio of Deposition Height to Deposition Width H/W	Wetting Angle of Deposition Layer θ/°
1	550	7	6.67	1.495	0.996	0.666	81.50
2	600	7	6.67	1.561	0.827	0.530	90.00
3	650	7	6.67	1.705	0.884	0.518	85.06
4	700	7	6.67	1.825	0.971	0.532	94.61
5	750	7	6.67	1.803	1.065	0.591	72.51
6	550	9	6.67	1.478	0.636	0.430	105.38
7	600	9	6.67	1.582	0.759	0.480	103.53
8	650	9	6.67	1.666	0.728	0.437	107.93
9	700	9	6.67	1.700	0.763	0.449	77.29
10	750	9	6.67	1.763	0.785	0.445	101.22
11	550	11	6.67	1.395	0.532	0.381	85.13
12	600	11	6.67	1.495	0.609	0.407	104.77
13	650	11	6.67	1.589	0.664	0.418	110.05
14	700	11	6.67	1.665	0.725	0.435	96.97
15	750	11	6.67	1.750	0.698	0.399	118.13
16	550	13	6.67	1.502	0.617	0.411	116.94
17	600	13	6.67	1.481	0.564	0.381	116.32
18	650	13	6.67	1.522	0.503	0.330	118.26
19	700	13	6.67	1.666	0.557	0.334	119.47
20	750	13	6.67	1.726	0.606	0.351	120.56
21	550	7	4.33	1.666	0.616	0.370	100.96
22	600	7	4.33	1.766	0.652	0.369	114.87
23	650	7	4.33	1.809	0.611	0.338	101.80

Table 3. *Cont.*

Depots	Laser Power P/W	Scanning Speed Vs/(mm/s)	Powder Feeding Rate Vf/(g/min)	Welding Width W/mm	Welding Height H/mm	Ratio of Deposition Height to Deposition Width H/W	Wetting Angle of Deposition Layer θ/°
24	700	7	4.33	1.927	0.654	0.339	113.11
25	750	7	4.33	1.920	0.676	0.352	99.28
26	550	9	4.33	1.431	0.521	0.364	116.18
27	600	9	4.33	1.629	0.541	0.332	120.96
28	650	9	4.33	1.692	0.649	0.384	119.02
29	700	9	4.33	1.734	0.637	0.367	112.93
30	750	9	4.33	1.797	0.618	0.344	115.37
31	550	11	4.33	1.413	0.589	0.417	105.85
32	600	11	4.33	1.502	0.583	0.388	111.36
33	650	11	4.33	1.603	0.493	0.308	121.13
34	700	11	4.33	1.659	0.598	0.360	112.59
35	750	11	4.33	1.703	0.538	0.316	108.98
36	550	13	4.33	1.384	0.358	0.258	117.55
37	600	13	4.33	1.510	0.384	0.254	128.31
38	650	13	4.33	1.532	0.424	0.277	118.13
39	700	13	4.33	1.623	0.476	0.293	124.48
40	750	13	4.33	1.715	0.495	0.289	128.14
41	550	7	2.1	1.629	0.553	0.339	116.34
42	600	7	2.1	1.710	0.549	0.322	106.60
43	650	7	2.1	1.817	0.536	0.295	127.67
44	700	7	2.1	1.854	0.498	0.269	123.53
45	750	7	2.1	1.904	0.470	0.247	132.36
46	550	9	2.1	1.494	0.412	0.276	122.12
47	600	9	2.1	1.581	0.432	0.273	133.36
48	650	9	2.1	1.610	0.444	0.276	118.60
49	700	9	2.1	1.700	0.522	0.307	120.46
50	750	9	2.1	1.780	0.460	0.259	124.04
51	550	11	2.1	1.452	0.347	0.239	129.27
52	600	11	2.1	1.547	0.322	0.208	132.94
53	650	11	2.1	1.615	0.335	0.207	129.92
54	700	11	2.1	1.817	0.367	0.202	133.90
55	750	11	2.1	1.868	0.378	0.202	139.60
56	550	13	2.1	1.472	0.286	0.194	148.21
57	600	13	2.1	1.484	0.283	0.191	135.64
58	650	13	2.1	1.555	0.251	0.161	135.54
59	700	13	2.1	1.606	0.252	0.157	142.18
60	750	13	2.1	1.671	0.237	0.142	135.57

As shown in Figures 3 and 4, 60 sets of single-layer, single-pass, single-layer multi-pass, and single-layer, single-pass theoretical cross-sections are demonstrated at laser powers of 550 W, 600 W, 650 W, 700 W, and 750 W; scanning speeds of 7 mm/s, 9 mm/s, 11 mm/s, and 13 mm/s; and powder feeding rates of 2.1 g/min, 4.33 g/min, and 6.67 g/min. The single-layer, single-pass theoretical cross-sections were drawn using Shapr3D software (Version number: 3.46.1). Table 2 displays the 60 groups of single-layer, single-pass physical cross-sections, along with the corresponding height and width measurements obtained using the optical imager. By using Shapr3D software, the height and width of each physical cross-section were sequentially defined. This allows the creation of a theoretical single-layer, single-channel cross-section. In this case, each group of process parameter combination experiments was repeated three times, which ensures the sufficient reproducibility and high accuracy of the obtained data.

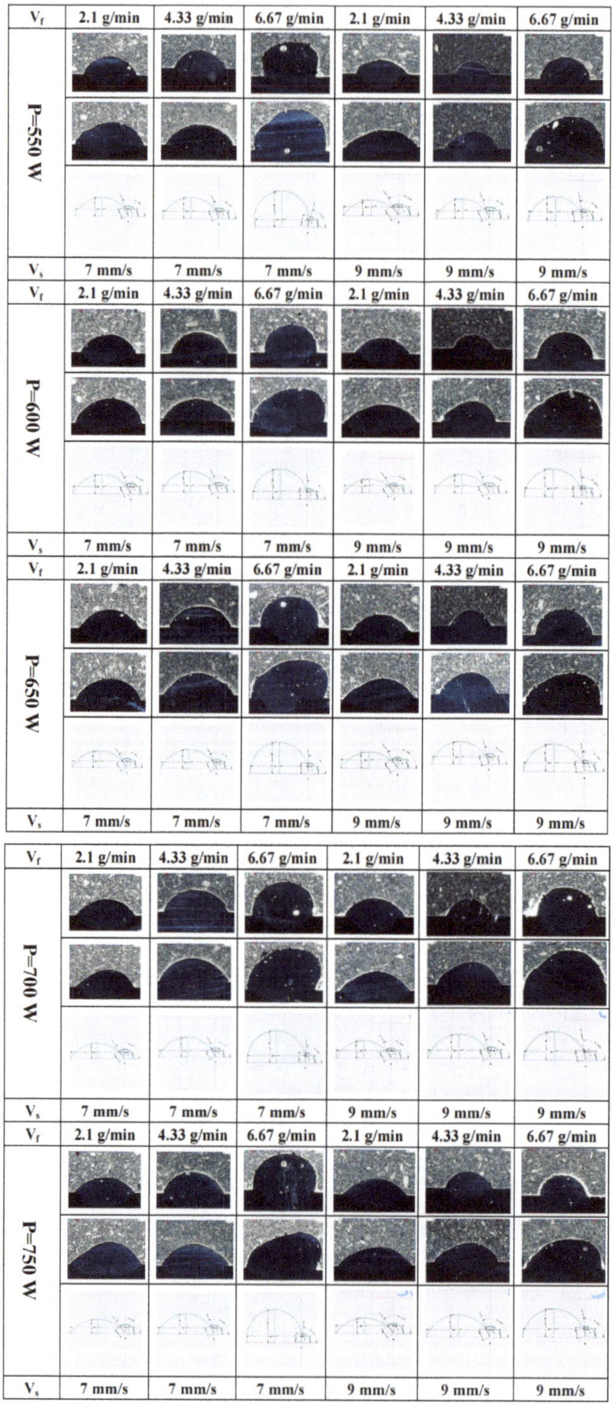

Figure 3. At scanning speeds of 7 mm/s and 9 mm/s, different process parameters are combined with single-channel, double-channel, and single-channel theoretical cross-section morphologies.

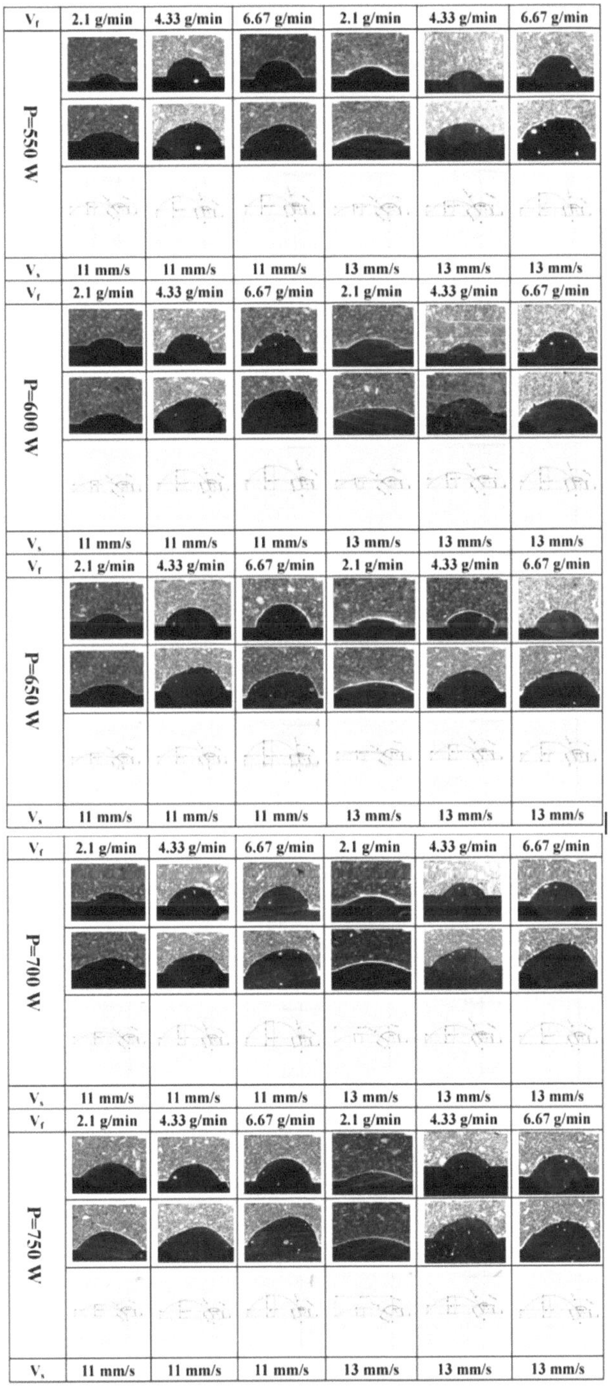

Figure 4. At scanning speeds of 11 mm/s and 13 mm/s, different process parameters are combined with single-channel, double-channel, and single-channel theoretical cross-section morphologies.

3.1. One-Way Analysis of Variance

(1) *The widths of the depositions*

We used Origin 2018 to plot the trend curve for the width of the layer deposited via laser power. As shown in Figure 5, the width of the deposited layer exhibits an increasing trend with the rise in laser power, assuming all the other conditions remain constant. This is because as the laser power increases, it primarily impacts two factors. Firstly, the energy input to the substrate's surface increases, leading to an expansion in the size of the molten pool. Consequently, the width of the molten pool also increases. Secondly, the quantity of laser-melted metal powder increases, resulting in a larger amount of material involved in the formation of the deposited layer. This, in turn, contributes to a wider width. The scanning speed has a negative effect on width, and increasing the scanning speed will, to some extent, cause the width to decrease. This is because a scanning speed that is too fast reduces the amount of linear energy input per unit length.

(2) *The heights of the depositions:*

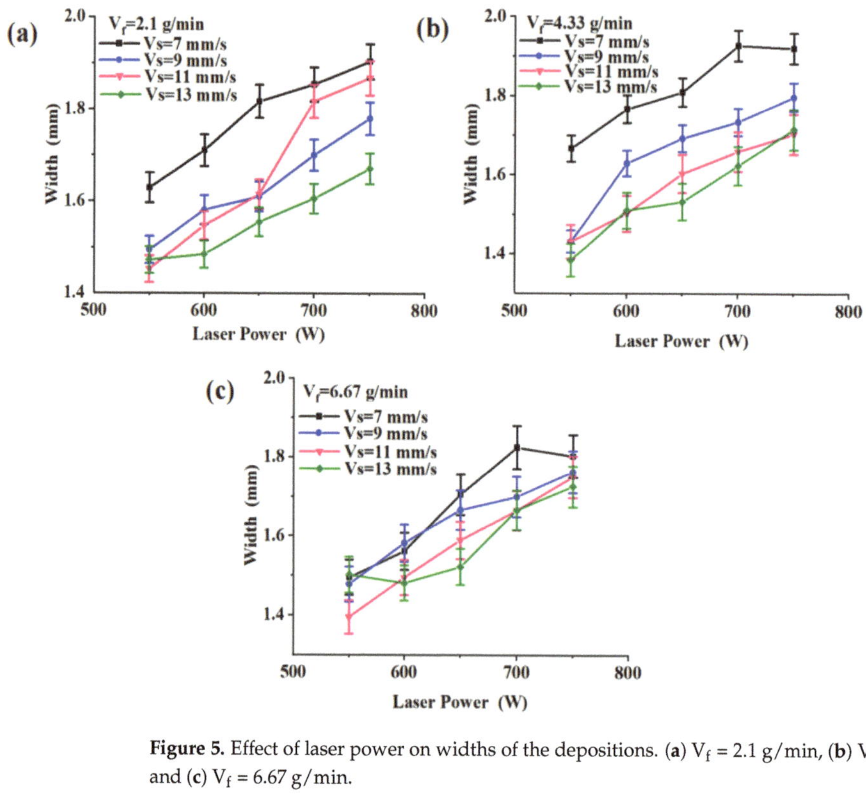

Figure 5. Effect of laser power on widths of the depositions. (**a**) V_f = 2.1 g/min, (**b**) V_f = 4.33 g/min, and (**c**) V_f = 6.67 g/min.

As can be seen in Figures 6 and 7, the height of the deposited layer decreases with an increase in the scanning speed and increases with an increase in the powder delivery rate, while keeping the other parameters constant. The reason for this phenomenon can be explained as follows: The increase in scanning speed causes the laser to spend less time at a specific position on the substrate. This leads to a decrease in the amount of energy applied per unit length, resulting in a reduction of melted metal powder and a smaller or shallower molten pool. Secondly, under the premise of maintaining a constant powder feeding rate, increasing the scanning speed will result in a decrease in the mass flow rate per unit length of the powder. The increase in the powder feed rate results in a higher amount of powder that can be melted by the laser per unit of time. This, in turn, leads to a greater amount of powder being involved in the formation of the deposition layer, resulting in an increase in height.

(3) *The wetting angles of the depositions:*

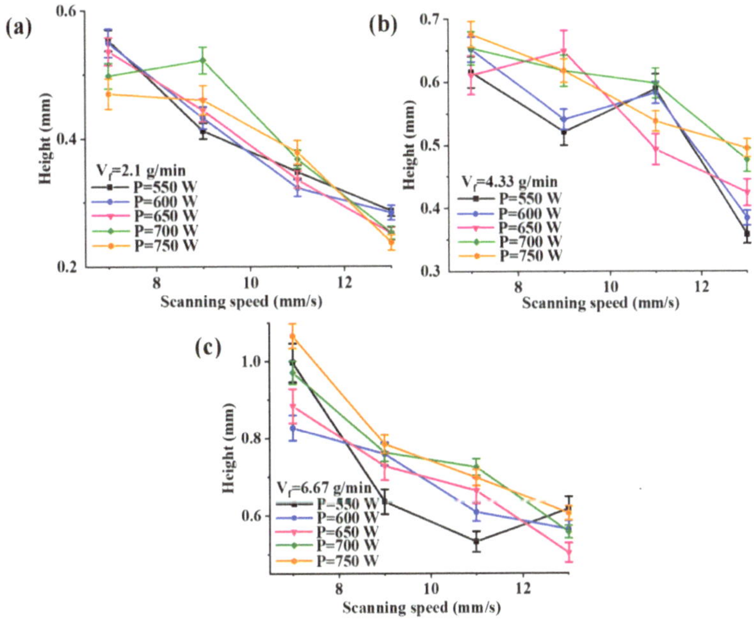

Figure 6. Effect of scanning speed on heights of the depositions. (**a**) V_f = 2.1 g/min, (**b**) V_f = 4.33 g/min, and (**c**) V_f = 6.67 g/min.

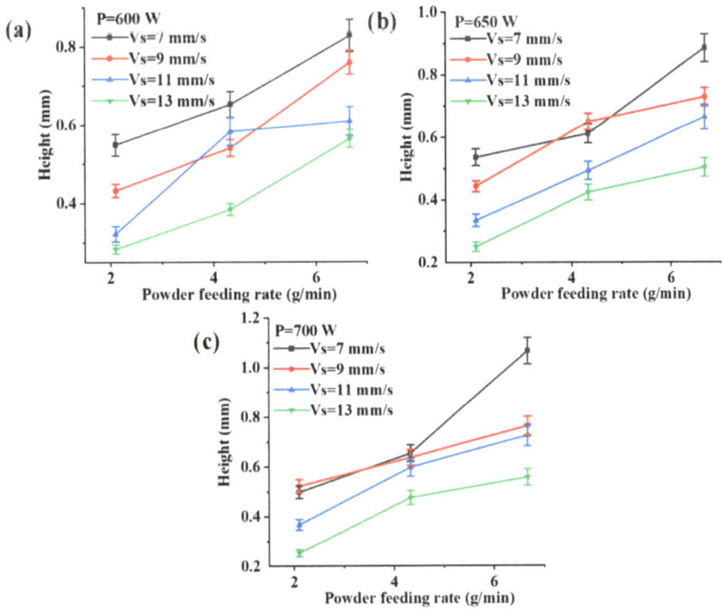

Figure 7. Effect of powder feeding rate on heights of the depositions. (**a**) V_f = 2.1 g/min, (**b**) V_f = 4.33 g/min, and (**c**) V_f = 6.67 g/min.

From Figure 8, it can be observed that the wetting angle gradually decreases as the powder feeding rate increases, while keeping the laser power and scanning speed constant. This can be attributed to the previous variations in height. From Figure 7, it can be observed that when other parameters are held constant, an increase in the powder feeding rate leads to a gradual increase in height. Consequently, the wetting angle becomes steeper and smaller. With the increase in scanning speed, the wetting angle gradually becomes larger. This is because the increase in scanning speed leads to a decrease in height, resulting in a larger wetting angle. As shown in the figure, when the laser power is increased to over 650 W, the wetting angle formed at a scanning speed of 7 mm/s is greater than the angle formed at 9 mm/s, while the powder feeding rate remains at 2.1 g/min. This indicates that, under this specific combination of parameters, the excessively high power level leads to a sudden alteration in the formation of the wetting angle.

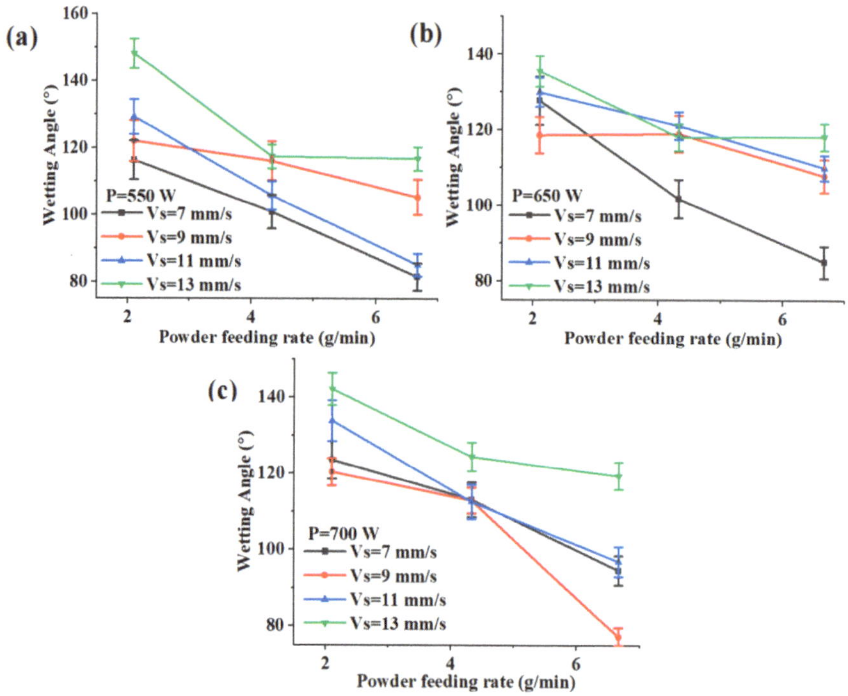

Figure 8. Effect of powder feeding rate on wetting angle of the depositions. (**a**) P = 550 W, (**b**) P = 650 W, and (**c**) P = 700 W.

3.2. Extreme Analysis

Based on the 60 groups of multi-factorial and multi-level experimental data in Table 4, single-factor analysis was conducted to determine the impact of each group of process parameter combinations on the geometrical shape of the fused deposition layer. Therefore, based on single-factor analysis, it is important to select representative combinations of process parameters for orthogonal experimental extreme analysis. Using the wetting angle as the reference index, the data results were obtained from 23 groups of parameter combinations with a single-pass wetting angle greater than 120°. These results were selected from a total of 60 groups of process parameter combinations. The laser power values of 600 W, 650 W, and 700 W; scanning speeds of 9 mm/s, 11 mm/s, and 13 mm/s; and powder feeding rates of 2.1 g/min, 4.33 g/min, and 6.67 g/min were analyzed using polar analysis and an orthogonal experimental design. The analysis aimed to determine the primary and secondary orders of influence for each factor. An orthogonal experimental design was used for polar analysis to determine the primary and secondary orders of influence of each factor. According to the calculation rule of the orthogonal experiment, the results of the orthogonal experiment with three factors and three levels were calculated and are listed in Table 4.

Table 4. Orthogonal experiment scheme and result analysis.

Test Pieces	Laser Power P/W	Scanning Speed Vs/(mm/s)	Powder Feeding Rate Vf/(g/min)	Welding Width W/mm	Welding Height H/mm	Wetting Angle of Deposition Layer θ/°	Ratio of Deposition Height to Deposition Width H/W
1	600	9	2.1	1.581	0.432	133.359	0.273
2	600	11	4.33	1.502	0.583	111.360	0.388
3	600	13	6.67	1.481	0.564	116.316	0.381
4	650	9	4.33	1.692	0.649	119.020	0.384
5	650	11	6.67	1.589	0.664	110.052	0.418
6	650	13	2.1	1.555	0.251	135.535	0.161
7	700	9	6.67	1.700	0.763	77.287	0.449
8	700	11	2.1	1.817	0.367	133.902	0.202
9	700	13	4.33	1.623	0.476	124.480	0.293
KW1	1.521	1.658	1.651	-	-	-	-
KW2	1.612	1.636	1.606	-	-	-	-
KW3	1.713	1.553	1.590	-	-	-	-
Range RW	0.192	0.105	0.061	-	-	-	-
KH1	0.526	0.615	0.350	-	-	-	-
KH2	0.521	0.538	0.569	-	-	-	-
KH3	0.535	0.430	0.664	-	-	-	-
Range RH	0.014	0.185	0.314	-			
Kθ1	120.345	109.889	134.265	-	-	-	-
Kθ2	121.536	118.438	118.287	-	-	-	-
Kθ3	111.890	125.444	101.218	-	-	-	-
Range Rθ	9.646	15.555	33.047	-	-	-	-

In Table 4, K1, K2, and K3 represent the average values of the three levels of the respective indicators in the columns where the factors are located. R represents the calculated range, and the subscripts W, H, and θ represent the values associated with the width, height, and wetting angle of the fused deposition layer, respectively.

From Table 4, it can be observed that RW (P) > RW (V_S) > RW (V_f), RH (V_f) > RH (V_S) > RH (P), and Rθ (V_f) > Rθ (V_S) > Rθ (P). This indicates that the deposition width is primarily influenced by the laser power, followed by the scanning speed, and lastly, the powder feeding rate. On the other hand, both the deposition height and the wetting angle are mainly affected by the powder feeding rate, followed by the scanning speed, and finally the laser power.

According to the results of orthogonal experiments, Mini Tab software (The version number: 21) was used to analyze the height of the deposition layer H, width W, aspect ratio H/W, and wetting angle θ as the response indices for data processing. The order of the influence of each factor on the response indices as well as the trend of change were determined using the mean value response table (in which the levels of A, B, and C represent the laser power, scanning speed, and powder feeding rate, respectively). The design is summarized in Table 5, and the mean response tables for the four responses are shown in Tables 6–9.

Table 5. Design summary.

Taguchi	L9 (3^3)
Factors	3
Number of experiments	9
Columns of the L9 (3^4) array	1 2 3

Table 6. The heights of the depositions mean response table.

Levels	A	B	C
1	0.5263	0.6147	0.3500
2	0.5213	0.5380	0.5693
3	0.5353	0.4303	0.3137
Delta	0.0140	0.1843	0.3137
Ordered list	3	2	1

Table 7. The widths of the depositions mean response table.

Levels	A	B	C
1	1.521	1.658	1.651
2	1.612	1.636	1.606
3	1.713	1.553	1.590
Delta	0.192	0.105	0.061
Ordered list	1	2	3

Table 8. Aspect ratio mean response table.

Levels	A	B	C
1	0.3473	0.3687	0.2120
2	0.3210	0.3360	0.3550
3	0.3147	0.2783	0.4160
Delta	0.0327	0.0903	0.2040
Ordered list	3	2	1

Table 9. Wetting angles mean response table.

Levels	A	B	C
1	120.3	109.9	134.3
2	121.5	118.4	118.3
3	111.9	125.4	101.2
Delta	9.6	15.6	33.0
Ordered list	3	2	1

As can be seen from the mean response tables in Tables 6–9, the results obtained from processing the data through Mini Tab are consistent with the experimental results calculated in Table 4. This indicates that the experimental calculations are reasonable.

From the orthogonal experiments, it can be seen that the deposition inclusions and morphology are determined jointly by various process parameters. Therefore, when selecting process parameters, it is important to consider different parameters for deposition inclusions and the morphology they cause due to their distinct effects. Based on this, ANOVA will be performed to investigate the effect of the interaction between the two factors on the deposition inclusions and morphology.

3.3. ANOVA (Analysis of Variance)

3.3.1. Influence of Process Parameters on Heights of the Depositions

It is important to control the height in order to obtain a homogeneous sediment layer. Therefore, it is important to study the effect of each factor on height and determine the percentage contributed by each factor. ANOVA was used to analyze the effect of the height of the sedimentary layer in relation to each factor, and the results are presented in Table 10.

Table 10. Results of variance analysis of heights of the depositions.

Source	DF	Adj SS	Adj MS	F	*p*
Mold	35	1.86943	0.053412	19.58	0.000
Biometrics	9	1.75167	0.194630	71.36	0.000
Laser power	4	0.02786	0.006964	2.55	0.065
Scanning speed	3	0.64929	0.216429	79.36	0.000
Powder feeding rate	2	1.07452	0.537262	196.99	0.000
2 Factor interactions	26	0.11776	0.004529	1.66	0.108
Laser power * Scanning speed	12	0.02250	0.001875	0.69	0.747
Laser power * Powder feeding rate	8	0.01915	0.002394	0.88	0.549
Scanning speed * Powder feeding rate	6	0.07610	0.012684	4.65	0.003
Error	24	0.06546	0.002727		
Total	59	1.93488			

S = 0.0522235; R-sq = 96.62%; R-sq (adj) = 91.68%; R-sq (pre) = 78.86%.

From Table 10, the F value of the model is 19.58, with a p-value < 0.0001, indicating that the model is statistically significant. The regression model explains 96.62% of the variation in the coating height, as shown by the regression coefficient R-sq in Table 10. The value of R-sq in this model is also very close to unity, which indicates that the adopted DOE model explains the relationship between height and the other factors well. The p value for the scanning speed, powder delivery rate, and their interaction is less than 0.05. Therefore, statistically, the scanning speed and powder delivery rate have the most significant impact on height. The p value for laser power was found to be 0.065, indicating that its impact on height was not statistically significant.

To determine the proportional effect of each factor, pie charts were created using the sum of the squares from the ANOVA table, as depicted in Figure 9a, and Pareto charts, as illustrated in Figure 9b.

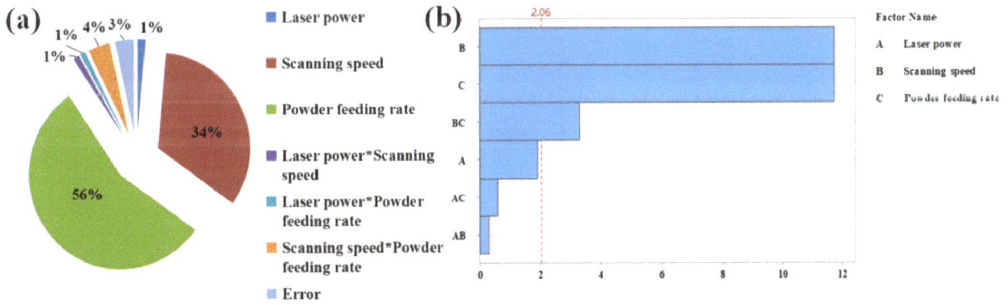

Figure 9. (**a**) Pie chart of practical significance factor of height of the depositions. (**b**) Pareto diagram in response to a height H/mm normalization effect.

As shown in Figure 9a, the powder feeding rate is responsible for 56% of the variation in height, making it the most significant parameter affecting the height. The scanning speed accounts for 34% of the variation, making it the second most important parameter. The laser power, on the other hand, only contributes to 1% of the variation in the height of the deposited layer. Figure 9b also indicates that the scanning speed, powder feeding rate, and their interaction have the most significant impact on the height, which aligns with the findings presented in Table 9 and Figure 9a.

The main effect plots and interaction plots of the main study parameters on height are shown in Figures 10 and 11, respectively.

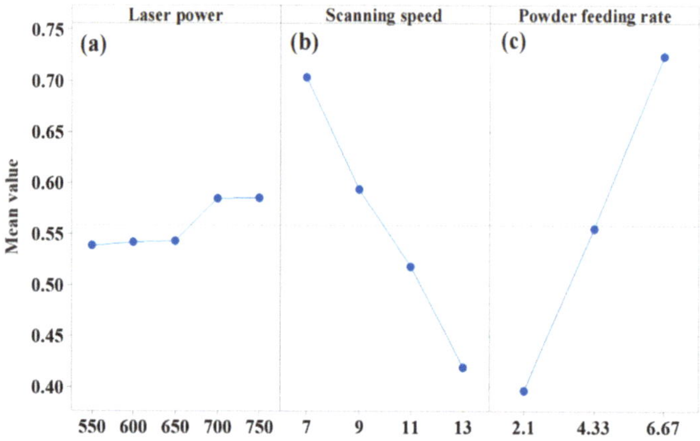

Figure 10. The response is a height H/mm main effect diagram. (**a**) Laser power. (**b**) Scanning speed. (**c**) Powder feeding rate.

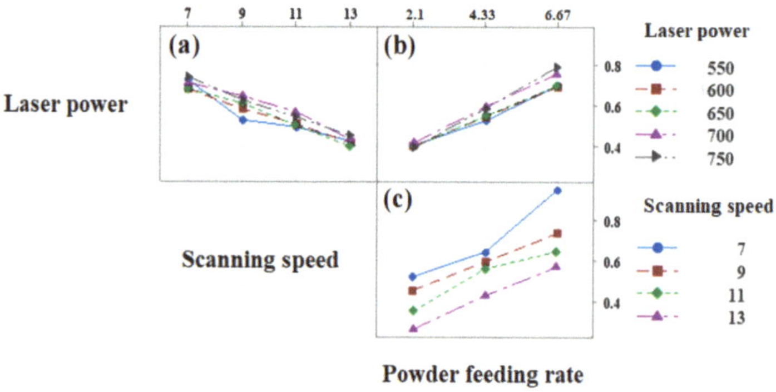

Figure 11. The response is a height H/mm interaction graph. (**a**) Laser power and scanning speed. (**b**) Laser power and powder feeding rate. (**c**) Scanning speed and powder feeding rate.

From the main effect plot, it can be observed that the maximum impact on the height of the deposited layer was achieved when the laser power was set at 750 W, the scanning speed was set at 7 mm/s, and the powder feeding rate was maintained at 6.67 g/min. And the effect on the height is much greater than the mean value level when the powder feeding rate is 6.67 g/min, followed by the scanning speed, and finally, the laser power. The change in laser power has no significant effect on the height. In fact, the heights remain almost the same at 550 W, 600 W, 650 W, 700 W, and 750 W. The heights of the layers at 550 W, 600 W, 650 W, 700 W, and 750 W are almost the same. This phenomenon can be explained by the fact that as the laser power increases, the amount of energy input also increases, resulting in a larger melt pool. However, the growth rate of the deposited layer will no longer increase significantly when the powder feed rate is maintained at a constant level. Therefore, it is also necessary to control the domain value of the laser power; otherwise, it will result in some degree of wastage. From the interaction plot, it can be seen that there is no significant interaction between the laser power and scanning speed. This is because the trend of height change is consistent with the increase in scanning speed at various laser powers. Similarly, there is no interaction between laser power and powder feeding rate.

3.3.2. Influence of Process Parameters on Widths of the Depositions

The width of the single-pass deposition layer is an important factor that affects its forming efficiency. As indicated by the regression coefficient R-Sq, the model accurately represents 97.51 percent of the width variance, demonstrating a strong fit. This is evident in Table 11.

Table 11. Results of variance analysis of widths of the depositions.

Source	DF	Adj SS	Adj MS	F	*p*
Mold	35	1.12756	0.032216	26.85	0.000
Biometrics	9	1.03346	0.114829	95.71	0.000
Laser power	4	0.69514	0.173786	144.86	0.000
Scanning speed	3	0.32040	0.106799	89.02	0.000
Powder feeding rate	2	0.01792	0.008961	7.47	0.003
2 Factor interactions	26	0.09410	0.003619	3.02	0.004
Laser power * Scanning speed	12	0.02595	0.002162	1.80	0.106
Laser power * Powder feeding rate	8	0.00688	0.000861	0.72	0.674
Scanning speed * Powder feeding rate	6	0.06126	0.010210	8.51	0.000
Error	24	0.02879	0.001200		
Total	59	1.15635			

S = 0.0346367; R-sq = 97.51%; R-sq (adj) = 93.88%; R-sq (pre) = 84.44%.

As shown in Table 11, the model has an F value of 26.85, with a p value < 0.0001, indicating its statistical significance. Additionally, the R-Sq value is close to unity, suggesting that the DOE model accurately represents the relationship between the widths and each parameter. The standard error of the model was only 0.03, indicating that the model is accurate. The analysis results of the p values indicate that the laser power, scanning speed, powder feeding rate, and the interaction between scanning speed and powder feeding rate all have statistically significant effects on width, with values less than 0.05. The sum of squares values of the individual factors in the ANOVA table were used to determine the ratio effect of the influencing factors on width. Pie charts illustrating the practical significant factors for width were constructed, as depicted in Figure 12a, and Pareto charts are shown in Figure 12b.

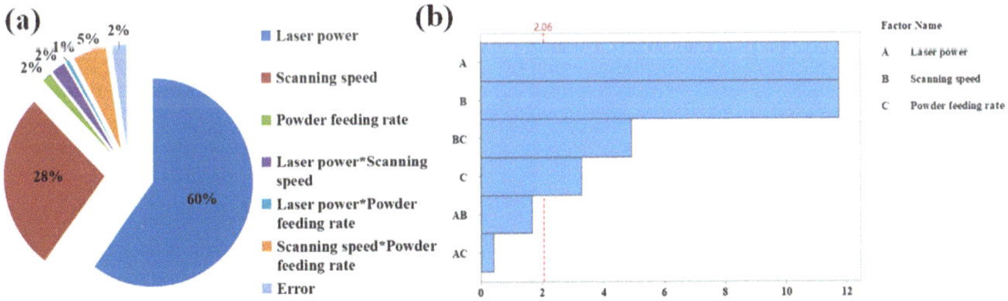

Figure 12. (**a**) Pie chart of practical significance factor of width of the depositions. (**b**) Pareto diagram in response to a width W/mm normalization effect.

As can be seen from Figure 12a, the laser power affects the width by 60% and is the most significant factor in determining the width. The scanning speed is the second most important factor, accounting for 28%. Among all the factors, only 2% of the variation is caused by the powder feeding rate. The Pareto chart also demonstrates that the laser power, scanning speed, and the interaction of the scanning speed and powder feed rate have the greatest influence on width.

The main effect plots and interaction plots of the main study parameters on width are shown in Figures 13 and 14, respectively.

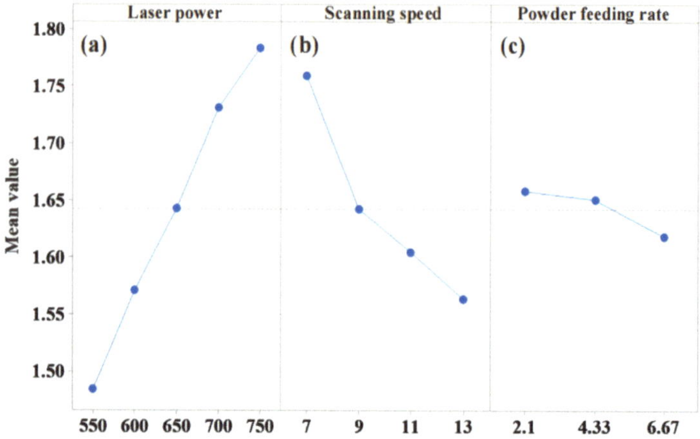

Figure 13. The response is a width W/mm main effect diagram. (**a**) Laser power. (**b**) Scanning speed. (**c**) Powder feeding rate.

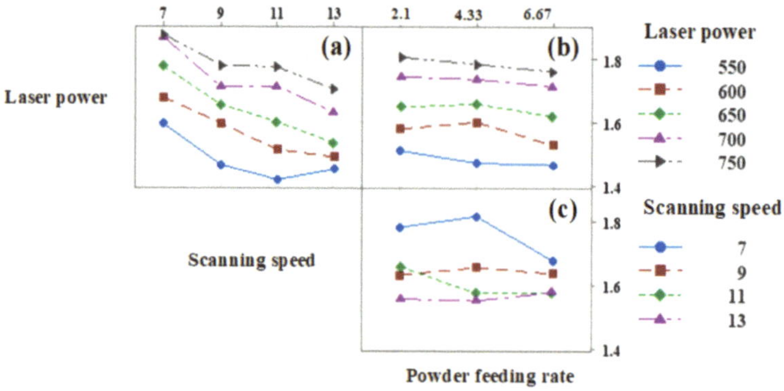

Figure 14. The response is a width W/mm interaction graph. (**a**) Laser power and scanning speed. (**b**) Laser power and powder feeding rate. (**c**) Scanning speed and powder feeding rate.

From Figure 13, it can be seen that the width of the deposited layer is most affected when the laser power is set to 750 W, the scanning speed is 7 mm/s, and the powder feeding rate is 2.1 g/min. The effect on the width is much greater than the mean value when the laser power is 750 W, followed by the scanning speed, and finally, the powder feeding rate. The variation in the powder feed rate does not have a significant impact on the mean width value. Additionally, there is minimal change in the width at powder feed rates of 2.1 g/min, 4.33 g/min, and 6.67 g/min. From the interaction plots, it can be seen that there is no interaction between the laser power and scanning speed, or between the laser power and powder feeding rate. On the contrary, when the scanning speed is 9 mm/s and the powder feeding rate is 11 mm/s, and the powder feeding rate is 2.1 g/min and 4.33 g/min, the response quantity changes differently at different levels of the other parameter. This indicates that there is a clear interaction between these two factors.

3.3.3. Influence of Process Parameters on Wetting Angles of the Depositions

The wetting angle is one of the most crucial parameters for assessing the uniformity of the deposited layer, and a wetting angle of less than 120° results in an inadequate formation.

From Table 12, the F value of the model is 6.99, with a p value < 0.0001, indicating that the model is statistically significant. The regression coefficients indicate that the laser power, scanning speed, powder feed rate, and their interaction collectively explain 91.06%

of the variation in the wetting angle. The comparison with the mean square of the other parameters yielded a low standard error, indicating that the model is satisfactory. From the p value results, it is clear that the p values of the scanning speed and powder feeding rate are both less than 0.05. Therefore, statistically, these two factors have a significant effect on the wetting angle results. The p value for the laser power and the interaction between two of the three factors is greater than 0.05. Therefore, none of their effects on the wetting angle are statistically significant. The sum of the squares values of the individual factors in the ANOVA table were used to determine the effect ratio of each influencing factor on the wetting angle. Pie charts depicting the practical significant factors were constructed, as shown in Figure 15a, and Pareto charts are shown in Figure 15b.

Table 12. Results of variance analysis of wetting angles of the depositions.

Source	DF	Adj SS	Adj MS	F	p
Mold	35	14,048.8	401.40	6.99	0.000
Biometrics	9	12,120.5	1346.73	23.44	0.000
Laser power	4	169.3	42.32	0.74	0.576
Scanning speed	3	4040.3	1346.77	23.44	0.000
Powder feeding rate	2	7911.0	3955.48	68.84	0.000
2 Factor interactions	26	1928.3	74.17	1.29	0.266
Laser power * Scanning speed	12	920.9	76.74	1.34	0.263
Laser power * Powder feeding rate	8	318.2	39.77	0.69	0.695
Scanning speed * Powder feeding rate	6	689.2	114.87	2.00	0.105
Error	24	1379.0	57.46		
Total	59	15,427.9			

S = 7.58023; R-sq = 91.06%; R-sq (Adj) = 78.03%; R-sq (Pre) = 44.13%.

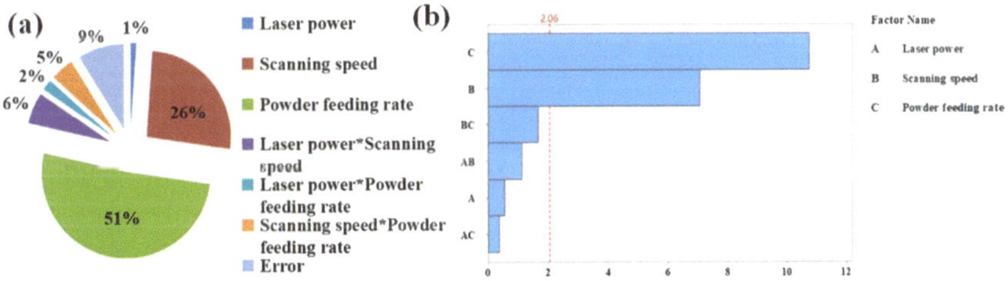

Figure 15. (**a**) Pie chart of practical significance factor of wetting angle of the depositions. (**b**) Pareto diagram in response to a wetting angle θ/° normalization effect.

From Figure 15a, it can be seen that the powder feeding rate has a significant effect on the wetting angle, accounting for 51% of the variation in the wetting angle. This indicates that the powder feeding rate is the most influential factor in determining the wetting angle. The scanning speed is the second most important factor, accounting for 26%. Of all the factors, only 1% of the variation is attributed to the laser power. The Pareto chart also demonstrates that the powder feeding rate and scanning speed are the most influential factors affecting the wetting angle.

The main effect plots and interaction plots of the studied parameters on the wetting angle are shown in Figures 16 and 17, respectively.

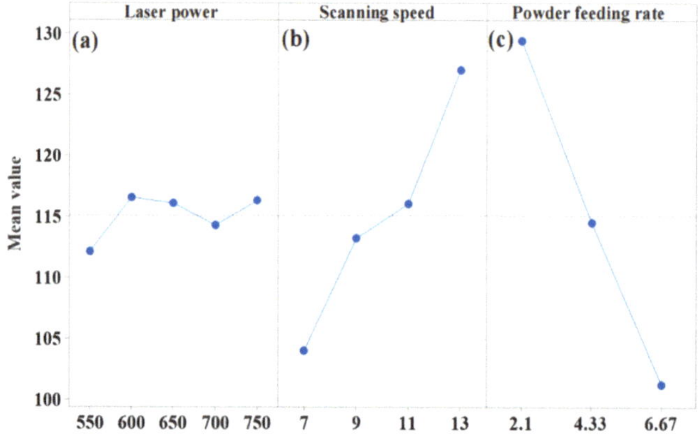

Figure 16. The response is a wetting angle θ/° main effect diagram. (**a**) Laser power. (**b**) Scanning speed. (**c**) Powder feeding rate.

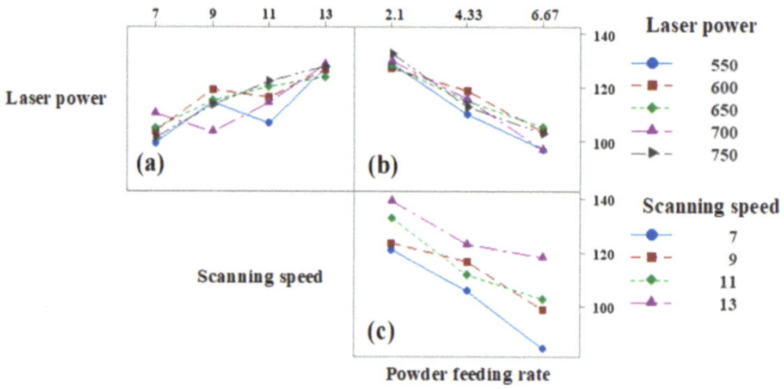

Figure 17. The response is a wetting angle θ/° interaction graph. (**a**) Laser power and scanning speed. (**b**) Laser power and powder feeding rate. (**c**) Scanning speed and powder feeding rate.

As shown in Figure 16, the wetting angle of the deposited layer is most affected when the laser power is 600 W, the scanning speed is 13 mm/s, and the powder feeding rate is 2.1 g/min. The effect on the wetting angle is significantly greater than the mean value when the powder feeding rate is 2.1 g/min, followed by the scanning speed, and finally, the laser power. The change in laser power does not have a significant effect on the mean value of the wetting angle, and there is almost no noticeable change in the wetting angle from 550 W to 750 W. However, the impact of laser power on the wetting angle is still considerable. From the interaction plot, it can be seen that there is no significant effect of two-by-two interactions among the three factors. Additionally, with different parameter values, the wetting angle shows a consistent change trend.

The results of the residual plots for the height, width, and wetting angle of the sedimentary layer are presented alongside the analysis of variance (ANOVA) results. Additionally, pie charts depicting the significant influences, Pareto charts, main effect plots, and interaction plots are included. These are given to ensure the validity of the results by strictly adhering to the assumptions of homogeneity of variance and normality of residuals. As depicted in Figure 18, the analysis of the height, width, and wetting angle of the 60 groups of sedimentary layers reveals a close approximation to a straight line. This suggests that the assumption of normality is satisfied, indicating a strong fit for the model.

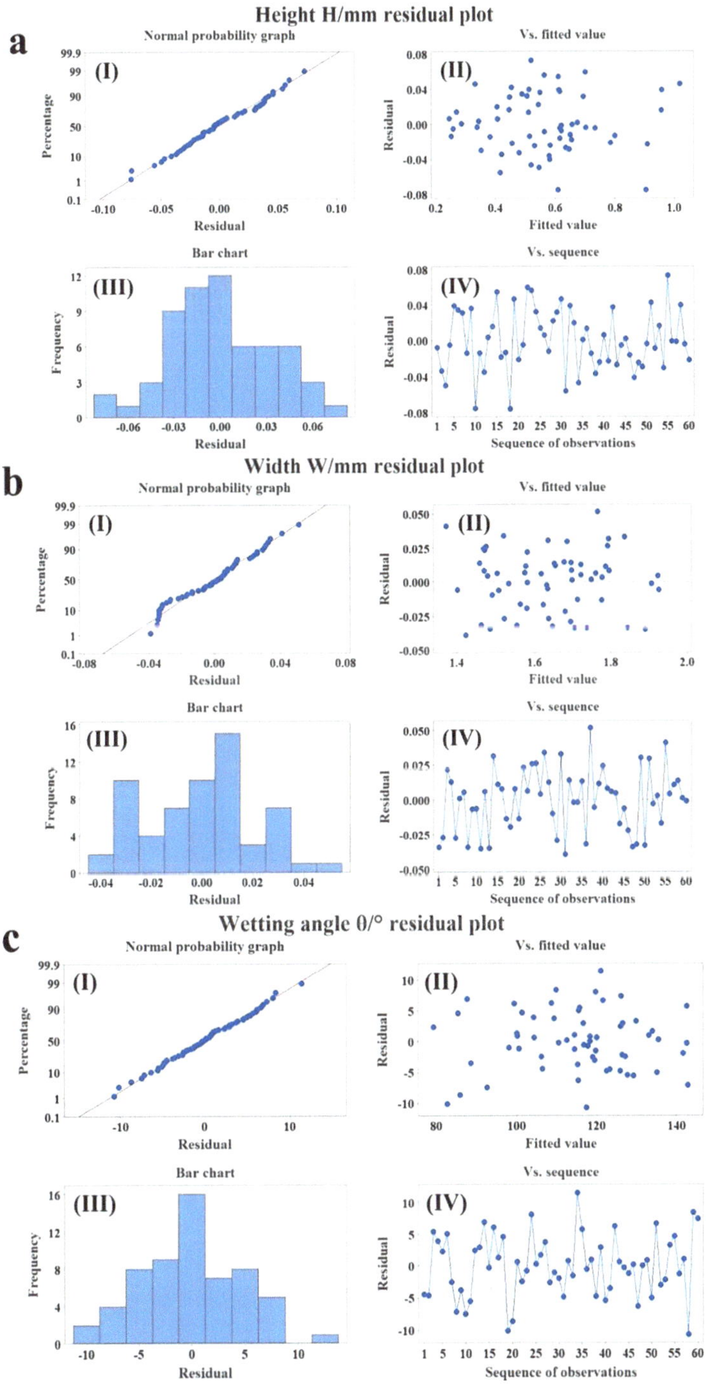

Figure 18. (**a**) Height H/mm residual plot. (**b**) Width W/mm residual plot. (**c**) Wetting angle θ/° residual plot.

A laser power of 600~750 W, scanning speeds of 9 mm/s and 11 mm/s, and a powder feeding rate of 6.67 g/min were selected to laser additively fabricate a single multi-layer, multi-channel deposition layer. In this case, the aspect ratio ranged from 0.4 to 0.5, and the wetting angle was less than 120°. As shown in Figure 19a, the cross-sectional morphology of the single-pass, multi-layer deposited layer exhibits a porous and pronounced steepening phenomenon. The wetting angle gradually decreases, and there are noticeable inclusions within the deposited layer, which will adversely affect the mechanical properties. On the contrary, Figure 19b shows that six sets of single multi-layer, multi-channel deposited layers were fabricated using direct laser deposition, with a laser power of 550 W~750 W, scanning speeds of 9 mm/s and 13 mm/s, and powder feeding rates of 2.1 g/min and 4.33 g/min. The aspect ratios ranged from 0.2 to 0.3, and the wetting angles were greater than 120°. The cross-sectional morphology of group (b) shows that the surface of the deposited layers in this value range is smooth, and the defects, such as porosity and inclusions, are significantly reduced.

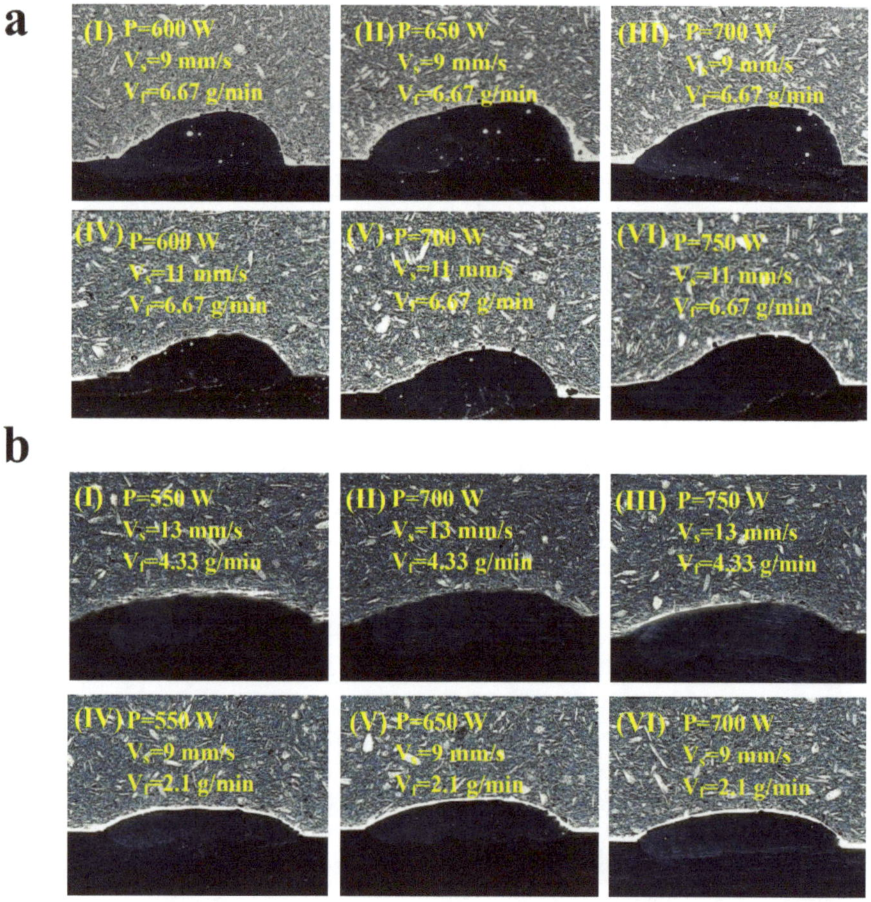

Figure 19. (**a**) Cross-sectional topography of direct laser deposition single-layer, multi-channel intercalation. (**b**) Morphology of a single-layer, multi-pass, non-inclusion cross-section for direct laser deposition.

4. Theoretical Deduction

Wettability is an important property of a material surface, and the degree of wetting is commonly measured using the wetting angle θ [22]. Young's equation describes the relationship between the contact angle and the three interfacial tensions [23]. Young's equation, also known as the wetting equation, holds a significant position in the field of material surface interfaces. It serves as the foundation for quantifying the surface wettability of a material [22] and calculating the surface tension of a solid material using the wetting angle [24]. As the research progressed, it was found that Young's equation needed to be corrected by introducing line tension when the droplets are small [25]. The cross-section of a single-pass coating exhibits morphological similarities to the lying drop and spherical crown models. Therefore, deriving a mathematical model of the relationship between the geometrical parameters of the deposited layer by combining surface tension and gravity requires the assistance of Young's equation, which is based on the laying drop and crown models.

Here, it is important to emphasize that the spherical crown model is solely a theoretical derivation of the cross-section parameter relationships for a single layer with a more regular single-channel cross-section morphology profile. Some irregular cross-sectional morphologies appear in Figures 3 and 4. These irregular morphologies indicate that this parameter combination is not within the optimal range, and this particular parameter is not applicable to the theoretical derivation of the ball-and-crown model.

The model the crown of the sphere, as shown in Figure 20, is based on the formula for the area of a circle:

$$S = \pi r^2 \tag{1}$$

The formula for calculating the volume of a spherical crown is:

$$V = \pi h^2 R - \pi h^3/3 \tag{2}$$

And the other geometric relationships:

$$(R - h)^2 + r^2 = R^2 \tag{3}$$

The associations (1)–(3):

$$V = 1/6\pi h\left(h^2 + 3r^2\right) = 1/6\pi h^3 + 1/2hS \tag{4}$$

Therefore, the crown height h, volume V, and contact area S satisfy the equation:

$$h^3 + 3Sh/\pi - 6V/\pi = 0 \tag{5}$$

Equation (5) is a cubic equation, which has a unique real root determined by solving for the roots of a cubic equation in one variable:

$$h = \sqrt[3]{\frac{3V}{\pi} + \sqrt{\frac{9V^2}{\pi^2} + \frac{S^3}{\pi^3}}} + \sqrt[3]{\frac{3V}{\pi} - \sqrt{\frac{9V^2}{\pi^2} + \frac{S^3}{\pi^3}}} \tag{6}$$

And

$$R = 1/2(h + S/\pi h) \tag{7}$$

$$\theta = \arccos(1 - h/R) \tag{8}$$

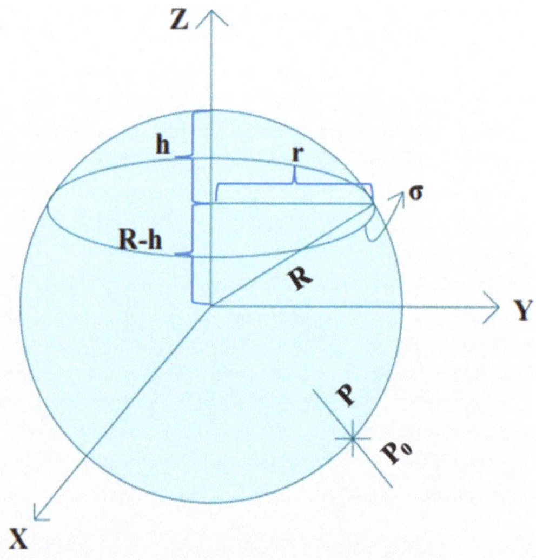

Figure 20. Spherical crown model.

The relationship between the crown height h, crown volume V, and contact area S, as well as the wetting angle θ and crown height h, is determined using the crown model.

However, in reality, the deposited layer's surface will assume a spherical shape due to surface tension, as illustrated in Figure 20. This occurs when the tension per unit length of the spherical shell is set to σ, the Z axis serves as the axis of symmetry, and there is a difference in air pressure between the inside and outside of the spherical crown, which is denoted as ΔF.

$$\Delta F = \int_0^{2\pi} d\sigma \int_0^{\alpha} (P - P_0)\cos\theta R^2 \sin\theta d\theta = \pi(P - P_0)R^2\sin^2\alpha \tag{9}$$

Once the spherical shell is in equilibrium, the difference between the internal and external pressures, ΔF, must be cancelled out by the surface tension of the shell, as shown in Figure 20.

$$\Sigma = \sigma \cdot \sin\alpha \cdot 2\pi R \sin\alpha \tag{10}$$

And

$$\Delta F = \Sigma \tag{11}$$

So

$$\pi(P - P_0)R^2\sin^2\alpha = \sigma \cdot \sin\alpha \cdot 2\pi R \sin\alpha \tag{12}$$

This can be introduced as

$$\sigma = \frac{(P - P_0)R}{2},\ R = \frac{2\sigma}{(P - P_0)} \tag{13}$$

Combining (8) with (13),

$$\theta = \arccos[1 - \frac{h(P - P_0)}{2\sigma}] \tag{14}$$

The relationship between the wetting angle θ, crown height h, internal and external pressures P and P_0, and line tension σ was determined by solving the crown model in conjunction with the surface tension.

The surface wettability of a material is influenced by three factors: the physicochemical properties of the material itself, surface tension, and gravity. A previous spherical crown model has taken into account the effect of surface tension, but it has not considered the influence of gravity on shape. When considering the effect of gravity, the analysis of the spherical crown model is performed in conjunction with Young's equation, which is also known as the wetting equation.

As shown in Figure 21, the relationship between the wetting angle θ and the interfacial tension is described by Young's equation, proposed by T. Young through mechanical analysis, i.e.,

$$\gamma_{lv}\cos\theta = \gamma_{sv} - \gamma_{sl} \tag{15}$$

where γ_{sv} represents the solid surface tension, γ_{lv} represents the liquid surface tension, and γ_{sl} represents the solid–liquid interfacial tension. Some studies [21] have explored the relationship between gravity and the wetting angle. However, further analyses were carried out based on the schematic of the spherical crown model, in conjunction with the previously described surface tension. A small section of the rectangle ABCD is intercepted from the edge of the spherical crown, as shown in Figure 22.

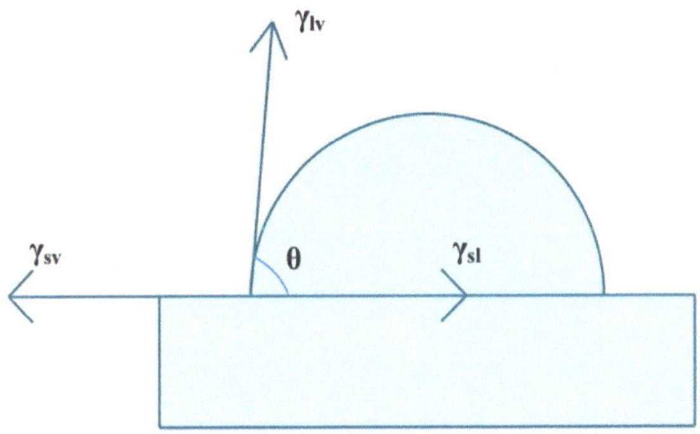

Figure 21. Schematic diagram of T. Young's equation.

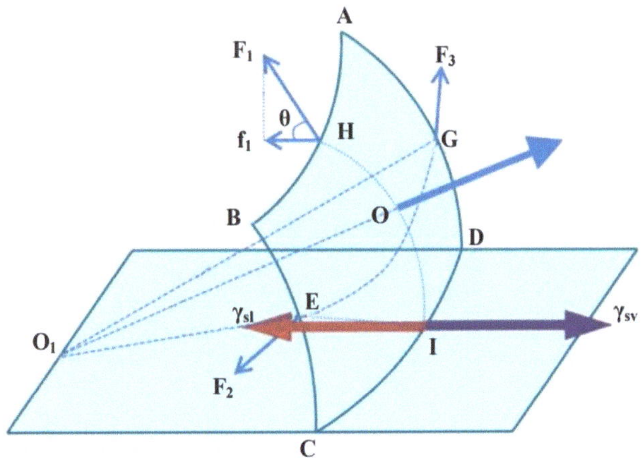

Figure 22. Tiny rectangle at the edge of the crown.

Combining Figures 21 and 22 shows that the side AB of the small rectangle ABCD is affected by surface tension:

$$F_1 = \gamma_{lv} \cdot AB \tag{16}$$

The component force in the horizontal direction:

$$f_1 = \gamma_{lv} \cdot AB \cdot \cos\theta \tag{17}$$

Similarly, sides BC and AD are subject to surface tension:

$$F_2 = \gamma_{lv} \cdot BC,\ F_3 = \gamma_{lv} \cdot AD \tag{18}$$

The component force in the horizontal direction:

$$f_2 = \gamma_{lv} \cdot BC \cdot \cos\theta, f_3 = \gamma_{lv} \cdot AD \cdot \cos\theta \tag{19}$$

The surface tension on the edge CD:

$$f_4 = \gamma_{sl} \cdot CD - \gamma_{sv} \cdot CD \tag{20}$$

Consider the case under a gravitational field and let the height of the droplet be denoted as h. The rectangle ABCD is subjected to pressure from the droplet in the direction of OO_1. The magnitude of this pressure is determined by the relationship between the force, pressure, and area:

$$F_5 = P \cdot S = \rho g h \cdot AB \cdot BC \tag{21}$$

The component force in the horizontal direction:

$$f_5 = \rho g h \cdot AB \cdot BC \cdot \sin\theta \tag{22}$$

The reason for this is that the magnitude of the combined force in the horizontal direction is zero:

$$f_6 = f_1 + f_2 + f_3 + f_4 + f_5 \tag{23}$$

$$\gamma_{lv} \cdot AB \cdot \cos\theta + \gamma_{lv} \cdot BC \cdot \cos\theta + \gamma_{lv} \cdot AD \cdot \cos\theta + \gamma_{sl} \cdot CD - \gamma_{sv} \cdot CD - \rho g h \cdot AB \cdot BC \cdot \sin\theta = 0 \tag{24}$$

The formula tidied up is as follows:

$$\gamma_{sv} \cdot CD = \gamma_{lv} \cdot AB \cdot \cos\theta + \gamma_{lv} \cdot BC \cdot \cos\theta + \gamma_{lv} \cdot AD \cdot \cos\theta + \gamma_{sl} \cdot CD - \rho g h \cdot AB \cdot BC \cdot \sin\theta \tag{25}$$

As can be seen in Figure 20,

$$AB = CD,\ BC = AD \tag{26}$$

$$\gamma_{sv} = \gamma_{lv} \cdot \cos\theta + 2 \cdot \gamma_{lv} \cdot \frac{BC \cdot \cos\theta}{CD} + \gamma_{sl} - \rho g h \cdot BC \cdot \sin\theta \tag{27}$$

Let BC = l:

$$\gamma_{sv} = \gamma_{lv} \cdot \cos\theta + 2 \cdot \gamma_{lv} \cdot \frac{l \cdot \cos\theta}{CD} + \gamma_{sl} - \rho g h \cdot l \cdot \sin\theta \tag{28}$$

From the relationship between the line tension σ and l, γ_{lv}, let $\gamma_{lv} \cdot l = \sigma$, then the equation can be written as

$$\gamma_{sv} = \gamma_{lv} \cdot \cos\theta + 2 \cdot \frac{\sigma \cdot \cos\theta}{CD} + \gamma_{sl} - \rho g h \cdot l \cdot \sin\theta \tag{29}$$

When the droplets are small and the height of the droplet is negligible, there is

$$\gamma_{sv} = \gamma_{lv} \cdot \cos\theta + 2 \cdot \frac{\sigma \cdot \cos\theta}{CD} + \gamma_{sl} \tag{30}$$

When the droplets are large and the line tension is negligible, there is

$$\gamma_{sv} = \gamma_{lv} \cdot \cos\theta + \gamma_{sl} - \rho g h \cdot l \cdot \sin\theta \tag{31}$$

Combining the relationship between the wetting angle θ, the crown height h, the internal and external pressures P and P_0, and the line tension σ, which was derived earlier based on the spherical crown model combined with the surface tension solution, and substituting it into the above equation, it can be expressed as

$$\gamma_{sv} = \gamma_{lv} \cdot [1 - \frac{h(P - P_0)}{2\sigma}] + \frac{2\sigma - h(P - P_0)}{CD} + \gamma_{sl} - \rho g h \cdot l \cdot \sin\theta \tag{32}$$

In conjunction with (15), there is

$$\tan\theta = \frac{1}{h} \cdot \frac{2\gamma_{lv}}{CD\rho g} \tag{33}$$

The novelty of this study is demonstrated by the combination of cross-sectional morphology diagrams and previously obtained data for 60 sets of process parameter combinations, in conjunction with Equation (33). The parameter values can fluctuate within different environments. When CD, ρ, g, and γ_{lv} in Equation (33) are determined, there exists a quantitative relationship between the height of the sedimentary layer h and the wetting angle θ. Based on this relationship, a set of h and θ can be determined by selecting the process parameters before conducting an experiment or engineering project. This allows us to assess whether the selected set of process parameters can produce a single-layer, single-pass morphology that meets the requirements of the latter. By selecting fewer consumables and ensuring high-level efficiency, the suitability of the process parameters can be determined to the greatest extent possible, and the wear can be minimized.

5. Conclusions

The research in this paper combines experiments and a theory. This combination aims to illustrate the relationship between the geometrical shape parameters of the single-layer, single-pass cross-section of direct laser deposition. The combination of experimentation and a theory provides a more comprehensive understanding of the relationship between the variation of cross-section parameters in layers deposited via direct laser deposition. The theory is utilized to explain the phenomenon, while the experiment serves to validate the findings. Based on this, it can be deduced that in the future, the practical application of engineering will greatly benefit from the direct laser deposition of sedimentary layer cross-section geometry. This deduction is supported by the theoretical derivation of the formula illustration. By combining these process parameters, it is expected that the efficiency and cost savings will be significantly improved. In the shortest possible time, the minimum amount of material can be verified to determine whether to proceed with the process parameters of this parameter group. This can significantly enhance industrial efficiency and save costs. This can greatly improve industrial efficiency. The conclusions of this paper are presented in the following four points:

(1) The laser power, scanning speed, and powder feeding rate have varying degrees of influence on the macroscopic forming morphology of the single-pass direct laser deposition of Inconel 738 powders. The laser power has a significant effect on the width (W) and a smaller effect on the height-to-width ratio (H/W). The scanning speed has a moderate effect on the height (H), width (W), and wetting angle (θ). The powder feeding rate has a very significant effect on the height (H) and wetting angle (θ), but it does not have a significant effect on the width (W). The laser power and scanning speed have different effects on the overall shape of the single-pass laser additive manufacturing of Inconel 738 powders.

(2) Laser power is the most important factor affecting W. As the laser power increases, W gradually increases. The scanning speed and powder feeding rate are the main factors

that influence H. As the scanning speed increases, H gradually decreases. Conversely, as the powder feeding rate increases, H gradually increases. The powder feeding rate is the most influential factor in θ. With the increase in powder feeding rate, θ gradually decreases.

(3) On the basis of Young's equation, the surface tension term and gravity term are included, indicating that the wetting angle of the material surface is influenced not only by the three interfacial tensions, but also by the surface tension and gravity. Based on surface tension and gravity, the functional relationship between H and θ is established. This relationship can be used to develop the theory of process analysis and guide the implementation of the process. It can also be used to predict and evaluate the trend and quality of deposition changes.

(4) From the experimental results, it can be seen that the single layer with multiple depositions obtained has the fewest defects and inclusions as well as a higher efficiency when the H/W value is between 0.2 and 0.3 and θ is greater than 120°. At this time, the range of laser power is 550–750 W, the range of scanning speed is 7–13 mm/s, and the range of powder feeding rate is 2.1–4.33 g/min.

Author Contributions: Conceptualization, K.Q. and C.Q.; methodology, K.Q. and P.C.; software, K.Q. and W.W.; validation, K.Q., P.C. and H.L.; formal analysis, K.Q. and C.Q.; investigation, K.Q.; resources, K.Q. and C.Q.; data curation, K.Q.; writing—original draft preparation, K.Q.; writing—review and editing, K.Q.; visualization, K.Q. and C.Q.; supervision, K.Q. and C.Q.; project administration, K.Q., W.W. and C.Q.; funding acquisition, C.Q. All authors have read and agreed to the published version of the manuscript.

Funding: This project was supported by National Natural Science Foundation of China (Grant No. 52371032).

Institutional Review Board Statement: Not applicable.

Informed Consent Statement: Not applicable.

Data Availability Statement: Data are contained within the article.

Conflicts of Interest: The authors declare no conflict of interest.

References

1. Ren, G.; Guo, S.; Zhang, B. Surface Integrity of Inconel 738LC Parts Manufactured by Selective Laser Melting Followed by High-speed Milling. *Chin. J. Mech. Eng.* **2023**, *36*, 81–95. [CrossRef]
2. Jie, H. Effect of laser power on the densification of selected zone laser melting of IN738LC alloy. *Mater. Guide* **2022**, *36*, 426–429.
3. Kastner, C.; Neugebauer, M.; Schricker, K.; Bergmann, J.P. Strategies for increasing the productivity of pulsed laser cladding of hot-crack susceptible nickel-base superalloy Inconel 738 LC. *J. Manuf. Mater. Process.* **2020**, *4*, 84. [CrossRef]
4. Taheri, M. Analysis of Solidification and Liquation Cracks in the Electron Beam Welding of IN738 Superalloy. *Metallogr. Microstruct. Anal.* **2021**, *10*, 815. [CrossRef]
5. Mostafa, A.; Masoumeh, K. Direct metal deposition of IN625 on IN738LC superalloy: Microstructure and crack analysis. *Appl. Phys. A* **2021**, *127*, 305.
6. Ponnusamy, P.; Rahman, R.A.; Masood, S.H.; Ruan, D.; Palanisamy, S. Mechanical properties of SIM-printed aluminum alloys: A review. *Materials* **2020**, *13*, 4301. [CrossRef]
7. Cheng, J.; Xing, Y.H.; Dong, B.; Zhao, L.J.; Liu, H.N.; Chang, T.Y.; Chen, M.J.; Wang, J.H.; Lu, J.W.; Wan, J. An overview of laser metal deposition for cladding: Defect formation mechanisms, defect suppression methods and performance improvements of laser-cladded layers. *Materials* **2022**, *15*, 5522. [CrossRef]
8. Zhang, Q.; Tong, W.; Chen, Z.; Yao, J.; Li, Z.; Feng, K.; Kovalenko, V.S. Influence of spot size on the geometrical characteristics of laser deep quenching hardened layer of 42CrMo steel. *Surf. Technol.* **2020**, *49*, 254–261.
9. Bax, B.; Rajput, R.; Kellet, R.; Reisacher, M. Systematic evaluation of process parameter maps for laser cladding and direct energy deposition. *Addit. Manuf.* **2018**, *21*, 487–494.
10. Ren, D.; Lin, Q.; Li, T.; Qi, H. Research on laser cladding repair process and path of cold work mold surface. *Surf. Technol.* **2018**, *47*, 54–60.
11. Zhang, R.; Zhang, R.; Chen, Y.; Liu, G. Research on the process parameters of single-pass molding process for laser cladding of iron-based alloys. *Therm. Process. Technol.* **2017**, *46*, 162–165+169.
12. Pinkerton, A.J. Advances in the modeling of laser direct metal deposition. *J. Laser Appl.* **2014**, *27*, s15001. [CrossRef]

13. Perevoshchikova, N.; Rigaud, J.; Sha, Y.; Heilmaier, M.; Finnin, B.; Labelle, E.; Wu, X. Optimisation of selective laser melting parameters for the Ni-based superalloy IN738LC using Doehlert's design. *Rapid Prototyp. J.* **2017**, *23*, 881–892. [CrossRef]
14. Mondal, S.; Bandyopadhyay, A.; Pal, P.K. Application of artificial neural network for the prediction of laser cladding process characteristics at Taguchi–based optimized condition. *Int. J. Adv. Manuf. Technol.* **2014**, *70*, 2151–2158. [CrossRef]
15. Barekat, M.; Razavi, R.S.; Ghasemi, A. Nd: YAG laser cladding of Co-Cr-Mo alloy on y-TiAl substrate. *Opt. Laser Technol.* **2016**, *80*, 145–152. [CrossRef]
16. Ansari, M.; Shoja, R.R.; Barekat, M. An empirical-statistical model for coaxial laser cladding of NiCrAIY powder on Inconel 738 superalloy. *Opt. Laser Technol.* **2016**, *86*, 136–144. [CrossRef]
17. Muthuramalingam, T. Surface quality measures analysis and optimization on machining titaniumalloy using CO_2 based laser beam drilling process. *J. Manuf. Process.* **2021**, *62*, 1–6. [CrossRef]
18. Elsheikh, A.H. Improving laser cutting quality of polymethylmethacrylate sheet: Experimental investigation and optimization. *J. Mater. Res. Technol.* **2020**, *9*, 1325–1339. [CrossRef]
19. Khoshaim, A.B. Experimental investigation on laser cutting of PMMA sheets: Effects of process factors on kerf characteristics. *J. Mater. Res. Technol.* **2021**, *11*, 235–246. [CrossRef]
20. Elsheikh, A.H. Modeling of the Transient Temperature Field during Laser Heating. *Lasers Manuf. Mater. Process.* **2021**, *8*, 97–112. [CrossRef]
21. Elsheikh, A.H. Prediction of laser cutting parameters for polymethylmethacrylate sheets using random vector functional link network integrated with equilibrium optimizer. *J. Intell. Manuf.* **2020**, *32*, 1377–1388. [CrossRef]
22. Bormashenko, E. Progress in understanding wetting transitions on rough surfaces. *Adv. Colloid Interface Sci.* **2014**, *222*, 92–103. [CrossRef] [PubMed]
23. Young, T. An essay on the cohesion of fluids. *R. Soc. Lond. Philos. Trans.* **1805**, *95*, 65–87.
24. Tavana, H.; Neumann, A.W. Recent process in the determination of solid surface tensions from contact angles. *Adv. Colloid Interface Sci.* **2007**, *132*, 1–32. [CrossRef] [PubMed]
25. Liu, Y. Research on Surface Wettability of Materials Affected by Gravity. Ph.D. Thesis, Northwestern Polytechnical University, Xi'an, China, 2017.

coatings

Article

Microstructure and Properties of Phosphorus Bronze/Brass Joints Produced by Resistance Projection Welding

Ruilin Lai [1,2], Weijun Zhang [2], Xiaofei Sheng [3], Xianjue Ye [1], Yingfeng Cai [2], Xiwei Zhang [2], Temiao Luo [2], Pinghu Chen [4], Qian Lei [1] and Yunping Li [1,*]

1 Powder Metallurgy Research Institute, Central South University, Changsha 410083, China; 133701033@csu.edu.cn (R.L.); leiqian@csu.edu.cn (Q.L.)
2 Gongniu Group Co., Ltd., Ningbo 315314, China; luotm@gongniu.cn (T.L.)
3 School of Mechanical and Electrical Engineering, Wuhan Business University, Wuhan 430058, China
4 School of Mechanical Engineering, University of South China, Hengyang 421200, China; chenpinghu1986@163.com
* Correspondence: lyping@csu.edu.cn

Abstract: In this work, we fabricated lap joints between embossed projection phosphorus bronze and flat brass through resistance projection welding (RPW). The experimental results indicated that the bronze projection moves into the softer brass without being deformed during the welding process. The tensile shear loads of the joint reached a maximum value of 273.6 N at a welding current of 5.5 kA. Under this circumstance, a reaction layer, including a columnar crystal solidification layer and a diffusion layer, is formed at the interface beside the boundary of bronze. The EDS line scan shows an elemental transition diffusion layer of about 1.5 μm between the H62 brass columnar crystal and XYK-6 phosphorus bronze. The fracture occurred on the XYK-6 side, passing through the bump instead of the welding interface, resulting in intactness of the welding interface. The results revealed that resistance projection welding is an effective method for welding copper alloys, suggesting the bright prospects of this technology in welding electrical parts.

Keywords: resistance projection welding; dissimilar joint; interface; properties; microstructure

Citation: Lai, R.; Zhang, W.; Sheng, X.; Ye, X.; Cai, Y.; Zhang, X.; Luo, T.; Chen, P.; Lei, Q.; Li, Y. Microstructure and Properties of Phosphorus Bronze/Brass Joints Produced by Resistance Projection Welding. *Coatings* **2023**, *13*, 1032. https://doi.org/10.3390/coatings13061032

Academic Editor: Alexander Tolstoguzov

Received: 17 May 2023
Revised: 31 May 2023
Accepted: 31 May 2023
Published: 2 June 2023

1. Introduction

Copper alloys, such as bronze and brass, are widely used in the field of electrical conductivity due to their excellent mechanical properties, high electrical conductivity, and high thermal conductivity [1,2]. However, copper alloys are difficult to weld, and they are prone to generate hot cracks and pores, resulting in a significant decrease in the effectiveness of their mechanical properties and electrical conductivity [3,4]. The welding methods of copper alloys mainly include fusion welding [3,4], pressure welding [5,6], and brazing [7,8]. At present, the electrical connection between the copper bars and the modules is made by brazing [9]. However, brazing makes it easy to produce false welding, leading to increased electrical resistance and decreased strength of the welding joint, endangering the safety of electricity consumption. In addition, the production process of brazing generates a lot of smoke and dust, polluting the environment and endangering health. Therefore, it is necessary to study an efficient welding method for copper alloys instead of the brazing method.

Resistance projection welding (RPW) is a type of resistance welding, which has become the most widely used method for welding sheet metals due to various advantages including easy operation, excellent adaptability for automation, high efficiency, and sound quality production [10]. The utilization of projection design can be highly effective in concentrating the welding current and heat generation. Therefore, it is perfectly applied to copper and copper alloys resistance welding, due to the generally high conductivity of these materials [11]. Recently, many studies on RPW have been reported, including its physical mechanism [10–12], parameters optimization [13–22], microstructure characterization [20–25], and numerical process simulation [16,26]. Wehle et al. [10] revealed that projection welding can be accounted for in the solid-state welding process, where the intimate weld interface is formed under plastic deformation at elevated temperatures and surface diffusion which enables re-arrangement of the interface to perfectly fit. Thomas et al. [11] pointed out that the solid-state bonds in the interface can be accomplished at extremely concise times (~milliseconds). However, weld nuggets form under longer weld times (hundreds of milliseconds); thus, the resulting joints can be formed either through solid-state or fusion processes. In the research of Chun et al. [22], the inhomogeneous microstructures, including the unmixed Al-Si coating layer and the second-phase Fe_3(Al, Si) intermetallic compound at the edge of the nugget, were responsible for the poor weld mechanical property. In Nielsen's model [25], friction was implemented between the square nut projections and the sheets during the welding process simulation, which increases the accuracy of electro-thermo-mechanical RPW modeling. Gintrowski et al. [26] reported that the RPW joint between aluminum and copper has good electrical and mechanical properties despite the brittle intermetallic AlCu layer in the interface. Saad [27] studied the effect of resistance spot welding parameters on copper and brass alloys, specifically analyzing the metallographic structures and joint properties, as well as fracture failure. However, there is limited information available on the formation mechanism of welding interfaces and their effect on joint performance. Therefore, the purpose of this research is to investigate the effect of resistance projection welding parameters on bronze/brass alloy joints and to explore the formation mechanism of the interface.

We weld phosphorus bronze and brass alloys using resistance projection welding technology in this work. Scanning electron microscopy (SEM), electron back-scattered diffraction (EBSD), and energy dispersive spectroscopy (EDS) were used to characterize the welded samples. This work aims to investigate the weld behavior and microstructure evolution of dissimilar copper alloy welded joints.

2. Materials and Methods

The base metals for the dissimilar welded joints are XYK-6 phosphorus bronze and H62 brass alloy sheets of 0.5 mm thickness. The chemical compositions of the base metals were determined by an inductively coupled plasma optical emission spectrometer (ICP-OES, Spectro Blue II, Germany), as shown in Table 1. The conductivity and mechanical properties of the base metals are presented in Table 2. It shows that the H62 brass has lower strength and conductivity compared to XYK-6 phosphorus bronze. Figure 1a shows that the melting point of H62 brass is about 900 °C, which is close to the boiling point temperature of zinc. Thus, the zinc in its structure is prone to volatilization when the brass is melted. The melting point of XYK-6 phosphorus bronze is approximately between 1060 °C and 1080 °C, which is higher than that of H62 brass (Figure 1b).

Table 1. Chemical composition of the materials.

Materials	Chemical Composition %						
	Cu	Zn	Sn	P	Fe	Ni	Pb
H62	60.5~63.5	Bal.	-	-	≤0.07	-	≤0.09
XYK-6	Bal.	1.00	1.8~2.7	0.03~0.10	0.10	0.1~0.4	0.02

Table 2. Properties of the materials.

Materials	Conductivity (%IACS)	Tensile Strength (MPa)
H62	25	350~470
XYK-6	32	480~530

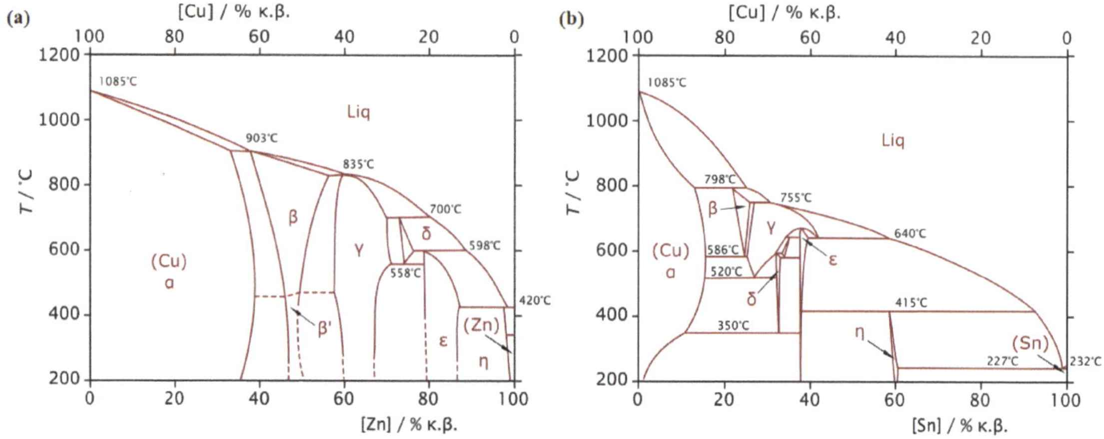

Figure 1. (**a**) Cu-Zn phase diagram and (**b**) Cu-Sn phase diagram [28].

The projection welded samples were produced using a 16 kVA mid-frequency inverter DC pedestal-type resistance-spot welding machine, as shown in Figure 2a. A high heat resistance W80Cu20 tungsten copper bar rod electrode with a flat surface diameter of 8 mm was used. Figure 2b,c shows the RPW illustration and the dimensions of the welded specimen, respectively. The projections are formed by stamping with self-made molds. Prior to welding, all coupons are cleaned by an ultrasonic cleaning machine with alcohol cleaner to avoid surface contamination. XYK-6 phosphorus bronze with an embossed projection was placed on a flat H62 brass in such a way that the projection side faced downwards as shown in Figure 2c. During the welding process, a welding current is locally applied to overlapping workpieces clamped together under pressure by the electrode.

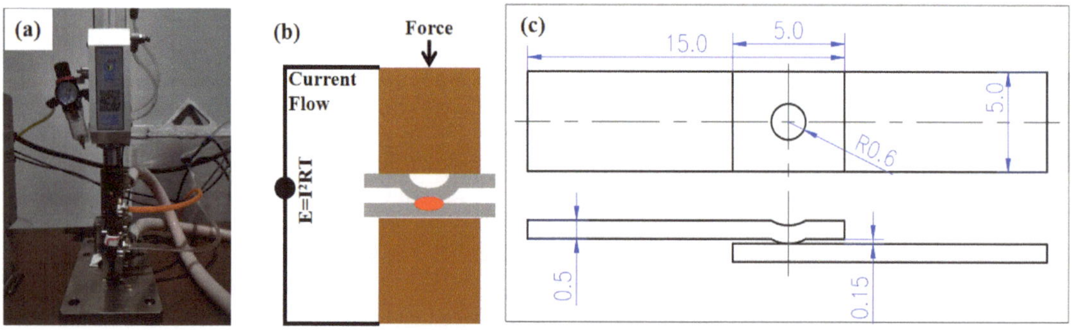

Figure 2. (**a**) Experimental equipment; schematic figures of (**b**) the projection weld illustration and (**c**) a welded specimen (all dimensions in mm).

The welding schedules are listed in Table 3. The mechanical property as the maximum tensile shear load and the electrical resistance of the RPW joints were thus determined for each specimen. The mechanical test was carried out as a tensile lap shear test (2 mm/min) on a Zwick/Roell Z010 (max. 10 kN) tensile testing machine. The electrical resistance tests were conducted using a double bridge circuit device at room temperature.

Table 3. Welding parameters used in the experiments.

Electrode Force (N)	Welding Current (kA)	Welding Time (ms)	Cooling Time (ms)
200	3.5, 4.5, 5.5, 6.5	20	10

After welding, the metallographic samples were cut into cross-sectioned joints by a wire electrical discharge machine. The samples were grounded using SiC paper with a grit size of 500, 1000, and 2000 in sequence, finally polished with 0.3 μm α-Al_2O_3 and etched in aqueous solutions of $FeCl_3$/HCl for 10 s at room temperature. The microstructures of the weld joints were characterized by a Leica EC3 optical microscope and a field emission scanning electron microscope (FESEM, FEI Quanta 650, Hillsboro, OR, USA) with a 20 kV accelerating voltage. The elemental distribution in the weld joint was determined with energy dispersive spectroscopy (EDS, Oxford Instruments, Oxford, UK). The specimens for the EBSD measurements were prepared, mechanically grounded, polished, and finally polished with Vibratory Polisher (AZoNetwork UK Ltd., Manchester, UK) for 2 h. The EBSD measurements were conducted using a field-emission scanning electron microscope (EBSD, Oxford HKL Channel 5, Oxford Instruments, Oxford, UK) at an accelerating voltage, spot size, and step size of 20 kV, 4 μm, and 0.2 μm, respectively. The micro-hardness of the sample was determined using a Vickers hardness tester under 100 g load for 10 s, and the indentations were spaced 0.1 mm apart.

3. Results and Discussion

3.1. Properties of the Weld Joints

Figure 3 shows the tensile shear load and resistance of the brass/bronze (H62/XYK-6) RPW joints under different welding currents. In the welding current range from 3.5 to 5.5 kA, the tensile shear load of the joint increased with the increasing welding current. When the welding current was 5.5 kA, it reached the maximum value of 273.60 kN. When the welding current exceeded 5.5 kA, the load capacity of the RPW joints decreased slightly as the welding current increased. The resistance of the joint decreased sharply with the increasing welding current when the welding current ranged from 3.5 to 5.5 kA. As the welding current increases, the resistance of the joint drops rapidly. However, the joint resistance decreases much more slowly when the welding current exceeds 5.5 kA. This is because more heat is generated in the weld zone as the welding current increases, resulting in a larger weld bond area and therefore lower joint resistance.

3.2. Macro- and Micrographs of the Weld Joints

Figure 4a–d shows the transverse macro cross-section of an XYK-6 projection welded onto a flat H62 sample under the welding currents of 3.5 kA, 4.5 kA, 5.5 kA, and 6.5 kA, respectively. In the case where the current was low (3.5 kA), a small discontinuous gap was observed between the interfacial boundary (see Figure 5a). Due to the low melting point of brass and the high melting point of bronze, the brass first softens during the welding process, and the bumps of the bronze squeeze the brass material. As the welding current increases, the heat generation increases, and the area where the brass softens increases. The bronze bump extruded the brass. During the welding process, the bronze bump moves into the softer brass material without being deformed. When the welding current reached 5.5 kA, a thin layer of the solidified structure appeared at the bronze interface (as shown in Figure 5b). When the welding current reaches 6.5 kA, there is not only a solidified structure at the bronze interface, but also the excessive welding heat causes more melting of the brass, and the bronze bump squeezes the molten brass so that the partially molten brass is squeezed out for welding. At the interface, splashes are formed, and it can be seen that there are hole defects in the solidified brass structure, which is due to the volatilization of zinc in the molten brass structure. During the whole welding process, the bronze did not melt. When the welding current is small, the interface does not melt and weld, but it does in solid-state diffusion welding. When the welding current increases, the interface is converted from solid-state diffusion welding to melting welding. Since zinc in brass are volatile, the welding current cannot be too large. Otherwise, the overheated welding heat will cause more melting of the brass, causing serious zinc volatilization and splashes, resulting in defective holes and weakening the joint connection performance.

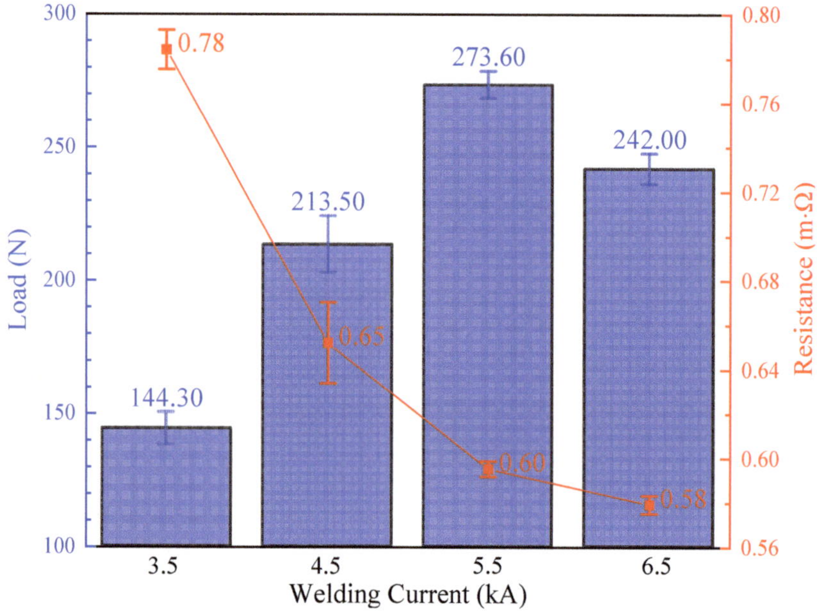

Figure 3. Effect of the welding current on the tensile shear load and resistance of the RPW joints.

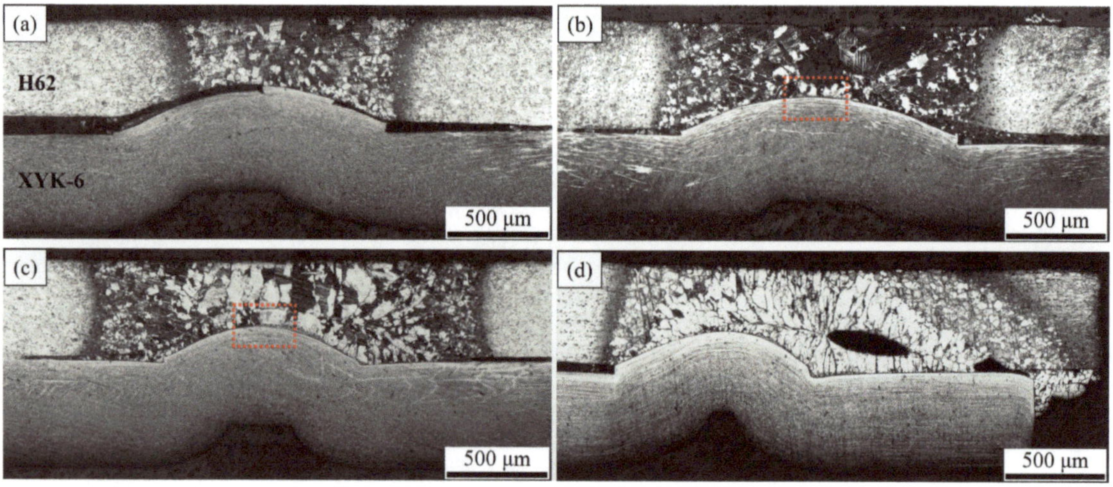

Figure 4. Traverse macro cross-section of an XYK-6/H62 projection weld: (**a**) 3.5 kA joint; (**b**) 4.5 kA joint; (**c**) 5.5 kA joint; (**d**) 6.5 kA joint.

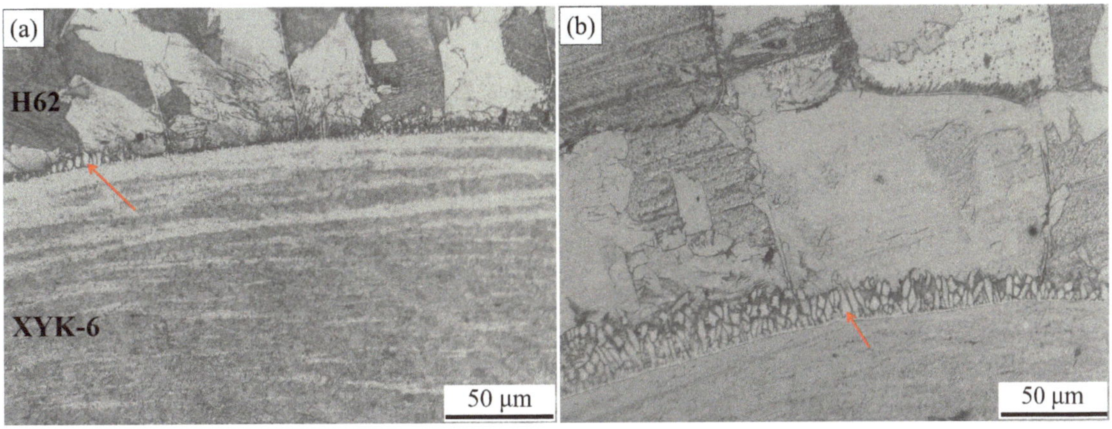

Figure 5. Optical micrograph of the H62/XYK-6 RPW interface (**a**,**b**) corresponding to the red dotted box area in Figure 4b and c, respectively.

Figure 5a shows that scattered and a few solidified columnar crystal structures are observed, indicating the onset of melting at local contact points during RPW at a welding current of 4.5 kA. Furthermore, as shown in Figure 5b, more solidified columnar crystal structures are obviously found and they are contiguous with each other, suggesting that the increasing heating accelerates the melting at the interface of H62/XYK-6 at a welding current of 5.5 kA. Therefore, a good-quality weld can be fabricated when the welding current is 5.5 kA. In the following experiments and the result analysis, these fixed parameters were chosen.

3.3. Metallurgical Characteristics

The EBSD results for H62 and XYK-6 base metals are depicted in Figures 6 and 7, respectively. The majority of grains in the H62 base metal were formed with a <111> crystal orientation, while those in the XYK-6 base metal had a prevalence of <101> crystal orientation. The average grain size in the H62 base metal and the XYK-6 base metal were 5.04 μm (Figure 6e) and 1.09 μm (Figure 7e), respectively. The fraction of high-angle grain boundaries (HAGBs), low-angle grain boundaries (LAGBs), and twin boundaries (TBs) in the H62 base metal were 77.7%, 21.6%, and 0.7% of the total grain boundaries (Figure 6f), respectively. In contrast, in the XYK-6 base metal, they were 66.9%, 32.6%, and 0.5% of the total grain boundaries (Figure 7f), respectively.

Additionally, the H62 base metal consisted of 5.2% recrystallized grains, 5.3% subgrains, and 89.5% deformed grains (Figure 6g), while the XYK-6 base metal consisted of 5.9% recrystallized grains, 12.4% subgrains, and 81.7% deformed grains (Figure 7g). In the kernel average misorientation (KAM) map of Figures 6d and 7d, the KAM values were macroscopically uniform but showed high values at the grain boundaries in both the H62 and XYK-6 base metals. Figures 6h and 7h showed that the local misorientation had a unimodal distribution with an average value of 1.80 and 1.39 in the H62 and XYK-6 base metals, respectively. The KAM value can be used to reflect the dislocation density [29], Thus, the higher KAM values observed at grain boundaries can be attributed to a higher density of geometrically necessary dislocations (GNDs) stored in the deformed structure. Based on the corresponding results shown in Figures 6c and 7c, it can be inferred that both of the H62 and XYK-6 base metals were subjected to deformation before use.

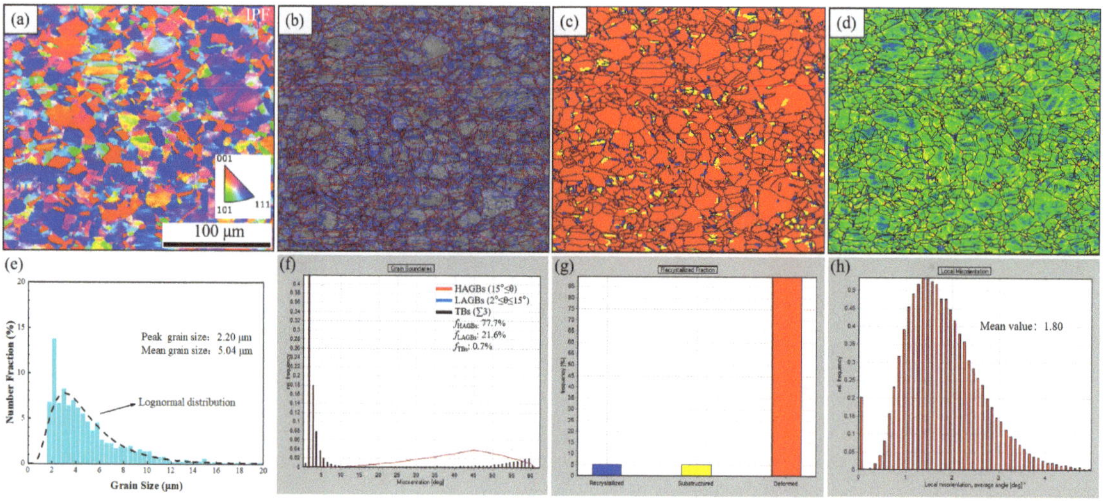

Figure 6. EBSD images of the H62 base metal: (**a**) the inverse pole figure maps; (**b**) GBs (high-angle grain boundaries (HAGBs with θ ≥ 15°)) are marked in red lines, low-angle grain boundaries (LAGBs with 2° < θ < 15°) are marked in blue lines, and twin boundaries (TBs) are marked in black lines; (**c**) recrystallized grains (in blue), subgrains (in yellow), and deformed grains (in red); (**d**) kernel average misorientation (KAM) map; (**e**) corresponding statistical distribution of grain size; (**f**) fraction of the grain boundaries; (**g**) fraction of the recrystallization; (**h**) local misorientation angle distribution.

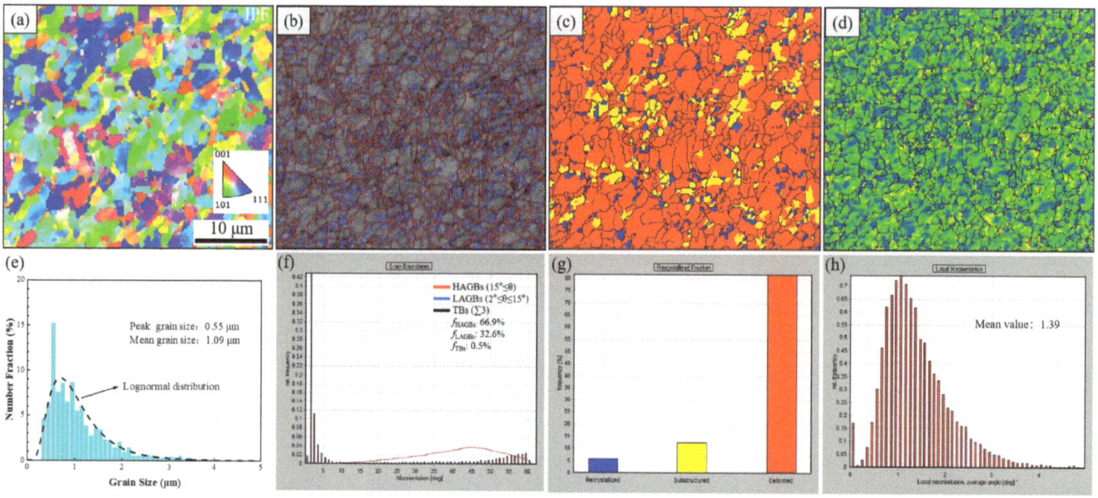

Figure 7. EBSD images of the XYK-6 base metal: (**a**) the inverse pole figure maps; (**b**) GBs (high-angle grain boundaries (HAGBs with θ ≥ 15°)) are marked in red lines, low-angle grain boundaries (LAGBs with 2° < θ < 15°) are marked in blue lines, and twin boundaries (TBs) are marked in black lines; (**c**) recrystallized grains (in blue), subgrains (in yellow), and deformed grains (in red); (**d**) kernel average misorientation (KAM) map; (**e**) corresponding statistical distribution of grain size; (**f**) fraction of the grain boundaries; (**g**) fraction of the recrystallization; (**h**) local misorientation angle distribution.

To reveal the interfacial bonding quality and mechanism, detailed microstructural characterization after welding was carried out on a typical H62/XYK-6 weld interface, as shown in Figure 8. It can be observed from Figure 8a that no significant orientation prevailed at the weld bonding interface. The grains of brass grow significantly after welding (~100 μm) compared to the grain before welding (~5.04 μm). On the other hand, the growth of the bronze grains (black dashed box) during welding is relatively slow, from 1.09 μm (before welding) to 2.82 μm (after welding). The fraction of HAGBs, LAGBs, and TBs at the weld interface were 44.2%, 54.9%, and 0.9% of the total grain boundaries (Figure 8f), respectively. In addition, the H62/XYK-6 weld interface contains 21.8% recrystallized grains, 55.9% subgrains, and 22.3% deformed grains (Figure 8g). In the KAM map of Figure 8d, the KAM was macroscopically uniform. Figure 8h shows the local misorientation is unimodal distribution. The average value is 0.71 at the H62/XYK-6 weld interface, which is lower than the H62 and XYK-6 base metals. Due to heat generation and pressure during welding, dynamic recrystallization takes place at the welding interface. As a result, the proportion of the deformed structure decreases while the recrystallized proportion increases in comparison to the two alloys before welding. This is evident from Figures 6f and 7f. Furthermore, the KAM values after welding are reduced as shown in Figure 8h.

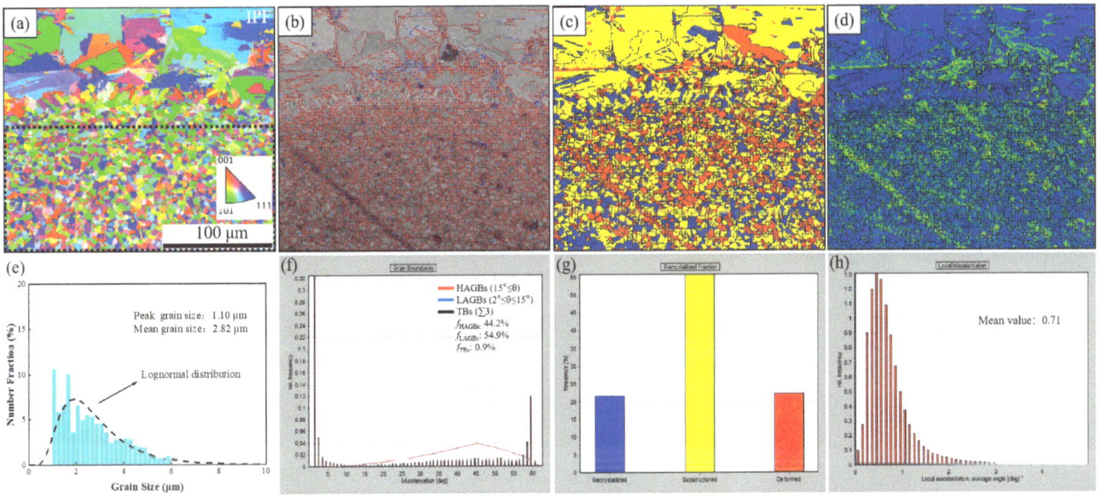

Figure 8. EBSD images of the H62/XYK-6 weld interface: (**a**) the inverse pole figure maps; (**b**) GBs (high-angle grain boundaries (HAGBs with $\theta \geq 15°$)) are marked in red lines, low-angle grain boundaries (LAGBs with $2° < \theta < 15°$) are marked in blue lines, and twin boundaries (TBs) are marked in black lines; (**c**) recrystallized grains (in blue), subgrains (in yellow), and deformed grains (in red); (**d**) kernel average misorientation (KAM) map; (**e**) corresponding statistical distribution of grain size; (**f**) fraction of the grain boundaries; (**g**) fraction of the recrystallization; (**h**) local misorientation angle distribution.

In order to better understand the formation mechanism of the interface, the elemental distributions of Cu, Zn, and Sn at the interface of two alloys after RPW processing are shown in Figure 9. It can be seen from Figure 9a that Cu is slightly rich at the bottom, and Zn is obviously rich at the top of the observed area, indicating that the composition of the columnar crystals is completely consistent with that of brass. It can be verified that the columnar crystals are all brass structures. There is no clear boundary for Sn in the elemental surface scan, possibly because its content is already low. EDS line scan analysis was performed at the welding interface to determine the elemental distribution. As shown in Figure 9b, the yellow line in the figure represents the EDS line scan position. The EDS line scan shows an elemental transition diffusion layer of about 1.5 μm between the brass columnar crystal and XYK-6. According to the metallographic structures, as shown in Figure 9, it can be inferred that the welding temperature is just over the melting point of H62 brass but never exceeded the melting point of XYK-6. As a result, the interfacial reaction in this area is mostly between liquid H62 and solid XYK-6, generating a reaction layer consisting of a columnar crystal solidification layer and a diffusion layer.

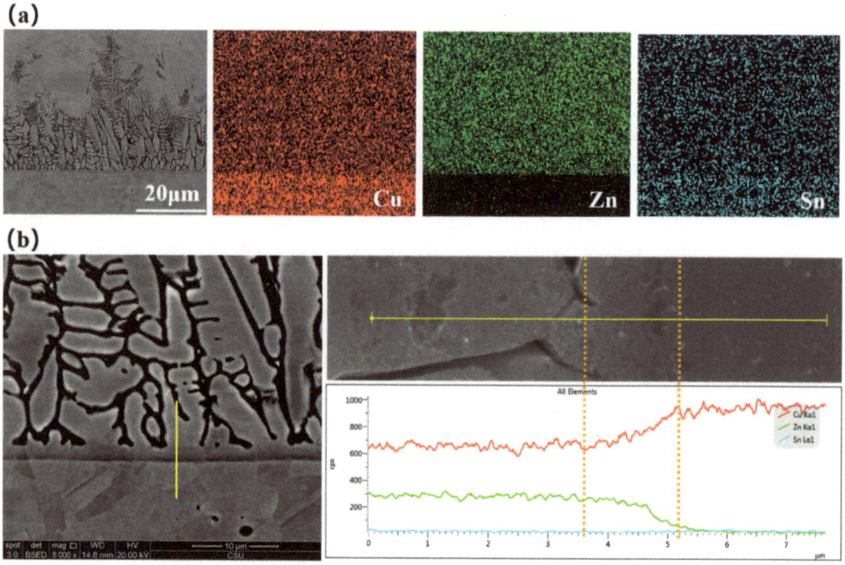

Figure 9. Elemental distribution of Cu, Zn, and Sn at the H62/XYK-6 RPW interface. (**a**) EDS mapping and (**b**) EDS line path on the SEM image and the curves of the EDS elemental analysis.

Since only a few results have been found in the literature on the formation of a diffusion layer at the surrounding area of the brass–bronze interface by RSW/RPW, we propose a hypothesis here of a three-stage reaction mechanism between near solid/solid brass and solid bronze as illustrated in Figure 10b–d. Figure 10a depicts the initial squeezing stage before welding. High compression stresses the electrode applied and arouses high strain at the H62/XYK-6 interface, producing close contact regions. When the electric current passes through the contact regions and generates Joule heat (Figure 10b), the H62 materials were softened firstly by the elevated temperature, and the XYK-6 bump extruded softened H62. Thus, the contact regions increased with diminishing gaps. Furthermore, as shown in Figure 10c, heating accelerates the softened H62 materials which fabricate the joining region by atomic interdiffusion, leading to solid-phase diffusion welding between the solid softened H62 and solid XYK-6 alloys. Additionally, as the welding current is increased, local melting occurs at the contact interface of the brass due to increased heat generation. This leads to liquid-phase diffusion between the liquid H62 brass and the solid XYK-6. The molten brass takes the interfacial diffusion layer as its crystalline surface and solidifies in the cooling direction to form dendrites (Figure 10d), thus achieving a strong welded joint.

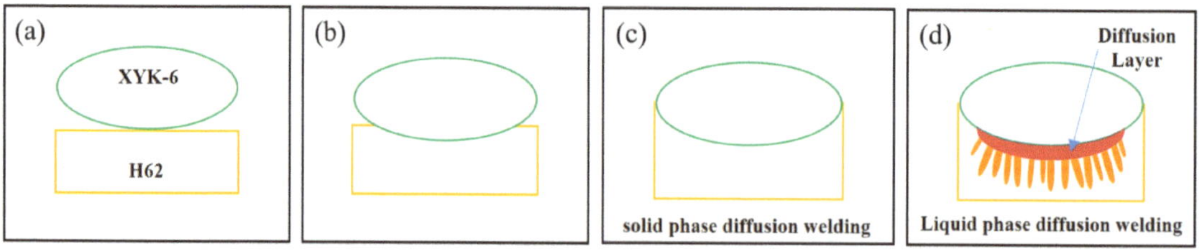

Figure 10. Schematics of the H62/XYK-6 interface at different stages during RPW: (**a**) the initial squeezing stage and (**b–d**) the welding stage.

Figure 11 shows the photograph of the fractured specimen of the RPW joint at a welding current of 5.5 kA. It can be seen from Figure 11 that the fracture occurred on the metal side of XYK-6 instead of the welding interface, resulting in the intactness of the welding interface. The schematic of the fracture mechanism is shown in Figure 11b. The interface strength is high after welding due to good interface bonding. Hence, during the tensile loading process, cracks can only develop from the interface under stress resulting in the fracture passing through the convex point between the two pieces, while the welding interface remains intact. It can be inferred that when the welding current is low, no effective bonding is formed at the interface, and its fracture will occur at the welding bump interface without passing through the bump.

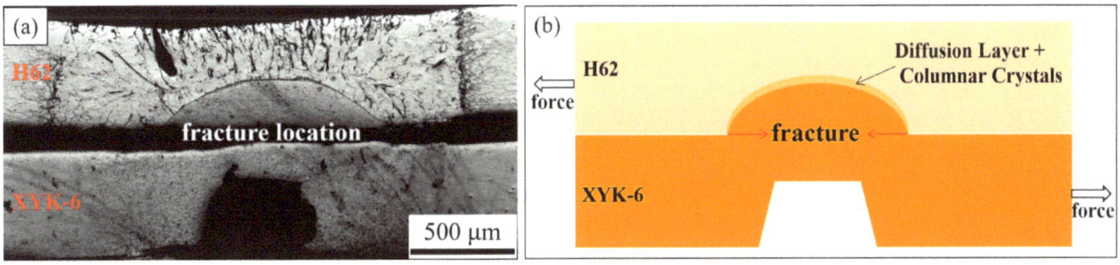

Figure 11. (**a**) Photograph of the fractured specimen and (**b**) schematic of the fracture.

In summary, the welding current has an evident effect on the interface layer. When the welding current is low, the solidified columnar crystal structures at the interface layer are fewer and discontinuous, resulting in low interface bonding strength. This is because interface bonding relies mainly on solid-phase diffusion. However, increasing the welding current leads to increased heat input, which generates more molten brass at the interface and results in the formation of continuous solidified columnar crystal structures in the interface layer. This transition from solid-phase diffusion welding to liquid-phase diffusion welding improves the interface bonding strength.

4. Conclusions

The microstructure and performance of an XYK-6 phosphorus bronze/H62 brass joint produced by resistance projection welding were investigated. The main conclusions are summarized as follows:

1. Resistance projection welding is a suitable method for producing robust welds between bronze and brass. The welded joints show a good quality tensile shear load and electrical conductivity.
2. The tensile shear load of the joint increased as the welding current increased. When the welding current was 5.5 kA, it reached the maximum value of 273.60 kN. The resistance of the joint decreased with the increase in the welding current.
3. At a welding current of 5.5 kA, a reaction layer is formed at the interface adjacent to the boundary of phosphorus bronze. This layer consists of a columnar crystal solidification layer and a diffusion layer. A tensile shear fracture occurred on the metal side of XYK-6, passing through the bump instead of the welding interface.
4. The formation mechanism of the welding interface is revealed for the first time, demonstrating a shift from solid-phase diffusion welding to liquid-phase diffusion welding, ultimately resulting in a robust welded joint. The results revealed that resistance projection welding is an effective method for welding copper alloys, suggesting the broad prospects of this technology in welding electrical parts, such as in the electrical and automotive industries.

Author Contributions: R.L. was the principal investigator of the research and wrote the original draft. X.Z. and T.L. carried out the welding tests. X.S. and P.C. analyzed the microstructure of the welded samples. Q.L. and X.Y. completed the review and editing. W.Z., Y.C. and Y.L. provided supervision. All authors have read and agreed to the published version of the manuscript.

Funding: This research received no external funding.

Institutional Review Board Statement: Not applicable.

Informed Consent Statement: Not applicable.

Data Availability Statement: Not applicable.

Conflicts of Interest: The authors declare no conflict of interest.

References

1. Freudenberger, J.; Warlimont, H. Copper and Copper Alloys. In *Springer Handbook of Materials Data*; Springer International Publishing: Cham, Switzerland, 2018; pp. 297–305.
2. Davis, J.R. *Copper and Copper Alloys*; ASM International: Materials Park, OH, USA, 2001.
3. Nayak, B.; Baliarsingh, A.K.; Biswajit, G.R. Investigation of Mechanical Properties of high Thickness Brass by TIG Welding. *Juni Khyat* **2021**, *11*, 559–564.
4. Auwal, S.; Ramesh, S.; Yusof, F.; Manladan, S.M. A review on laser beam welding of copper alloys. *Int. J. Adv. Manuf. Technol.* **2018**, *96*, 475–490. [CrossRef]
5. Nakata, K. Friction stir welding of copper and copper alloys. *Weld. Int.* **2005**, *19*, 929–933. [CrossRef]
6. Ni, Z.; Yang, J.; Li, S.; Wang, X.; Ye, F. A review on ultrasonic spot welding of copper alloys. *Mater. Sci. Technol.* **2020**, *36*, 264–283. [CrossRef]
7. Hasap, A.; Noraphaiphipaksa, N.; Kanchanomai, C. The microstructure and strength of copper alloy brazing joints. *Weld. J.* **2014**, *93*, 116–123.
8. Li, Z.; Yi, Y.; He, D.; Lai, R. Interfacial Microstructure and Shear Strength of Brazed Cu-Cr-Zr Alloy Cylinder and Cylindrical Hole by Au Based Solder. *Metals* **2017**, *7*, 247. [CrossRef]
9. Li, Y.; Chen, C.; Yi, R.; Ouyang, Y. Special brazing and soldering. *J. Manuf. Process.* **2020**, *60*, 608–635. [CrossRef]
10. Wehle, M. Basics of Process Physics and Joint Formation in Resistance Projection Welding Processes. Ph.D. Thesis, University of Stuttgart, Stuttgart, Germany, 2020.
11. Lienert, T.; Siewert, T.; Babu, S.; Acoff, S.W.P.V. Specifications, ASM handbook. In *Welding Fundamentals and Processes*; ASM International: Materials Park, OH, USA, 2011; Volume 6A.
12. Polajnar, I. *Fundamentals of Resistance Welding Processes*; Univerza v Ljubljani, Fakulteta za strojništvo: Ljubljana, Slovenia, 2007.
13. Baobuangoen, J.; Pornsing, C. *Productivity Improvement in Projection welding Process by Design and Analysis of Experiment A case of an Automobile Parts Manufacturer*; Silpakorn University: Bangkok, Thailand, 2021.
14. Paliakou, A.Y. *Effect of Electrode Pressing Force on the Amount of Energy Supplied to Interelectrode Space in Resistance Projection Welding with Constant Current Welding Machine*; IOP Conference Series: Materials Science and Engineering; IOP Publishing: Bristol, UK, 2021; p. 012034.
15. Wang, X.; Zhang, Y. Effects of Welding Procedures on Resistance Projection Welding of Nuts to Sheets. *ISIJ Int.* **2017**, *57*, 2194–2200. [CrossRef]
16. Ha, S.; Murugan, S.P.; Marimuthu, K.P.; Park, Y.; Lee, H. Estimation of lobe curve with material strength in resistance projection welding. *J. Mater. Process. Technol.* **2019**, *263*, 101–111. [CrossRef]
17. Han, G.; Ha, S.; Marimuthu, K.P.; Murugan, S.P.; Park, Y.; Lee, H. Estimation of shear fracture load in resistance projection welded sheets with dissimilar strength and thickness. *Eng. Fail. Anal.* **2021**, *120*, 105042. [CrossRef]
18. Mikno, Z.; Stepien, M.; Grzesik, B. Optimization of resistance welding by using electric servo actuator. *Weld. World* **2017**, *61*, 453–462. [CrossRef]
19. Ha, S.; Han, G.; Lee, H.; Murugan, S.P.; Park, Y. Optimal Design of Square Weld Nut by Taguchi Method and FE Analyses in Nut Projection Welding. *Trans. Korean Soc. Mech. Eng. C* **2019**, *43*, 409–418. [CrossRef]
20. Ren, K.; Qiu, R.; Ma, H.; Shi, H. Joining Aluminum Alloy to Steel by Resistance Projection-Plug Welding. *J. Mater. Eng. Perform.* **2020**, *29*, 4087–4094. [CrossRef]
21. Sejč, P.; Belanová, J.; Gábrišová, Z.; Vanko, B. The Influence of Parameters of the Resistance Projection Welding of M10 Steel Nuts to the Galvanized Steel Sheet DP 600 on Selected Joint Characteristics. *Manuf. Technol.* **2020**, *20*, 822–833. [CrossRef]
22. Chun, E.-J.; Sung-Sang, L.; Young-Tae, K.; Nam, K.-S.; Young-Min, K.; Young-Whan, P.; Murugan, S.P.; Park, Y.-D. Influence of heat-treated Al–Si coating on the weldability and microstructural inhomogeneity for hot stamped steel resistance nut projection welds. *Met. Mater. Int.* **2019**, *25*, 179. [CrossRef]
23. Van Rymenant, P. *Mechanical Characterisation and Modelling of Resistance Welding*; KU Lueven Publishing: Lueven, Belgium, 2010.
24. Luo, J.; Zhou, Z.; Cao, X.; Zou, C.; Zou, C. Microstructure and Failure Analysis of Resistance Projection Welding of Nuts to AHSS with Capacitor Discharge Welding. *ISIJ Int.* **2019**, *59*, 305–311. [CrossRef]

25. Nielsen, C.V.; Zhang, W.; Martins, P.A.F.; Bay, N. 3D numerical simulation of projection welding of square nuts to sheets. *J. Mater. Process. Technol.* **2015**, *215*, 171–180. [CrossRef]
26. Gintrowski, G.; Reisgen, U.; Schiebahn, A.; Schmachtenberg, M.; Hibert, V. Characteristics of resistance projection–welded aluminum-copper interconnects. *Weld. World* **2019**, *63*, 1593–1599. [CrossRef]
27. Mat Saad, H.N. Effect of Resistance Spot Welding Parameters on Copper and Brass. Master's Thesis, Universiti Sains Malaysia, Penang, Malaysia, 2012.
28. Baker, H.; Okamato, H. *ASM Handbook, Vol. 3, Alloy Phase Diagrams*; ASM International: Materials Park, OH, USA, 1992.
29. Jiang, J.; Britton, T.B.; Wilkinson, A.J. Evolution of dislocation density distributions in copper during tensile deformation. *Acta Mater.* **2013**, *61*, 7227–7239. [CrossRef]

coatings

MDPI

Article

Effect of Co/TiAl on Mechanical Properties of Laser Melted IN 625 on 304SS Matrix

Tong Yang [1], Wenxing Wu [1], Yuantao Lei [1], Pinghu Chen [2], Hao Liu [1], Li Zhao [1] and Changjun Qiu [1,*]

[1] School of Mechanical Engineering, University of South China, Hengyang 421200, China; yangtong202302@163.com (T.Y.)

[2] College of Mechatronics and Control Engineering, Additive Manufacturing Institute, Shenzhen University, Shenzhen 518000, China

* Correspondence: qiuchangjun@hotmail.com; Tel.: +86-139-7343-7826

Abstract: IN 625 is one of the most widely used nickel-based high-temperature alloys. However, the unstable high-temperature mechanical properties of IN625 and the difficulty of processing complex parts have limited its wider application. This work fabricated IN625 coatings with Co/TiAl (1.25, 1.55, and 1.85) on 304SS using laser melting deposition technology, with excellent high-temperature mechanical properties. The effects of Co/TiAl on the microstructure and properties of the laser-cladded resulting coatings have been carefully investigated. Compared to the IN625 coating, the addition of Co, Ti, and Al to the IN625 laser cladding coating significantly enhances its hardness and strength at room temperature, while reducing the elongation rate sharply. After heat treatment (900 °C × 10 h + 850 °C × 2 h), the hardness of the IN625 coating decreases, while the hardness of the IN625 laser cladding coating with added Co, Ti, and Al significantly increases. At a temperature of 850 °C, the yield strength and elongation rate of the IN625 laser cladding coating with Co, Ti, and Al additives exhibit an increasing-then-decreasing trend as the Co/TiAl ratio rises. The results exhibited that the coating had excellent high-temperature mechanical properties when the Co/TiAl is 1.55, and its hardness, tensile strength, yield strength, and elongation rate are 48.62 HRC, 735 Map, 665 MPa, and 11.3%, respectively.

Keywords: laser melting deposition; IN625; Co/TiAl; microstructure; high-temperature mechanical properties

Citation: Yang, T.; Wu, W.; Lei, Y.; Chen, P.; Liu, H.; Zhao, L.; Qiu, C. Effect of Co/TiAl on Mechanical Properties of Laser Melted IN 625 on 304SS Matrix. *Coatings* **2023**, *13*, 768. https://doi.org/10.3390/coatings13040768

Academic Editor: Enikö György

Received: 16 March 2023
Revised: 5 April 2023
Accepted: 13 April 2023
Published: 14 April 2023

1. Introduction

Due to the excellent high-temperature mechanical properties, good high-temperature corrosion resistance, and high-temperature oxidation resistance of IN625, it has been widely used in fields such as nuclear power, thermal power, industrial gas turbines, aerospace, and other areas [1–5]. However, the low content of the high-temperature stable γ' strengthening phase, the fact that the γ'' strengthening phase is a metastable phase, and its transformation to the δ-phase can be promoted by long-term high-temperature exposure, which will all significantly impair the high-temperature mechanical properties of IN625 [6–8]. In addition, the poor workability of the IN625 alloy further hinders its wider application, especially in applications that require high geometric complexity [9–11]. Additive manufacturing (AM) can overcome the problem of manufacturing complex alloy shapes and achieve the manufacturing of metal parts with complex geometric shapes [12–15]. The preparation of high-temperature alloy coatings with excellent mechanical properties utilizing laser melting technology has very high potential [16].

Currently, many researchers have conducted considerable work to improve the mechanical properties of AM IN625 alloy at high temperatures. Ghiaasiaan et al. [17] carried out a study on the fatigue behavior of IN625 laser additive manufactured samples at 427 °C and 649 °C, and indicated that the significant reduction in the fatigue performance of the samples can be attributed to the reduction in twinning at higher temperatures. Song et al. [18] investigated the microstructural evolution of IN625 alloy prepared by LPBF (laser powder bed fusion) and its influence on mechanical properties under high-temperature environments of 700 and 750 °C. The research revealed that the decrease in high-temperature mechanical properties was attributed to the growth of Laves phases precipitated at the grain boundaries. Zhao et al. [6] studied the microstructure and mechanical properties of the Inconel 625 alloy prepared by SLM at 870 °C. The results showed that the decrease in mechanical properties was due to the aggregation of carbide particles at the grain boundaries under high temperature. Kreitcberg et al. [19] investigated the effect of stress relief annealing, solution treatment, and hot isostatic pressing on the microstructure, room temperature, and high-temperature mechanical properties of LPBF-fabricated Inconel 625 specimens. The study found that compared to untreated specimens, the elongation rate slightly increased after stress relief annealing. This is because stress relief annealing induced an anisotropic structure characterized by a columnar δ-phase structure and particles elongated mostly along the building direction. However, the ductility decreased significantly at 760 °C compared to room temperature, owing to embrittlement caused by the precipitation of carbides at grain boundaries. Son et al. [20] studied the effect of different heat treatments (such as solution heat treatment, hot isostatic pressing, and long-term cyclic heat treatment) on the creep strength of forged IN625 and LPBF-prepared IN625 alloys at 650 and 800 °C. The study found that the LPBF-fabricated IN625 exhibited comparable or higher creep strength than forged IN625 at all temperatures, but with a sharp drop in ductility. The reason for this is that the γ'' and δ phases precipitated at 650 and 800 °C, respectively.

Up until the present time, researchers mostly focus on exploring the causes of high-temperature failure of IN625, and a small proportion of them aim to improve the high-temperature mechanical properties of laser additive IN625 alloy through heat treatment or optimizing process parameters. There have been few reports on improving the high-temperature mechanical properties of IN625 by adding intermetallic compounds. As we all know, there are three commonly used strengthening methods for nickel-based high-temperature alloys: (i) solid solution strengthening caused by high melting point elements such as Cr, Mo, Nb, W, Ta, and Co; (ii) precipitation strengthening caused by the formation of γ' phase elements such as Al, Ti, Ta, and Nb; and (iii) grain boundary strengthening induced by B and C. It has been confirmed that the γ' phase plays a crucial role in the high-temperature mechanical properties of nickel-based alloys. This is because the antiphase boundary energy of the precipitated γ' phase can hinder the movement of dislocations, thus endowing nickel-based high-temperature alloys with excellent high-temperature mechanical properties [21–24]. At the same time, Al and Ti are the main elements for the formation of the γ' phase, with Al atoms forming the $\gamma'(Ni_3Al)$ strengthening phase through precipitation strengthening, while Ti can replace Al in the $\gamma'(Ni_3Al)$ phase, forming γ'-$Ni_3(Al,Ti)$ to increase the volume fraction of the γ' phase [25,26]. In addition, it has been shown that the addition of Co elements would preferentially work in favor of the γ phase and reduce the solubility of Ti and Al in the γ phase. The remaining Ti, Al, and Co elements can enter the γ' phase and occupy the Ni lattice positions, forming ordered γ'-Ni_3 (Co,Al,Ti) and increasing its volume fraction. The element of Co can improve phase stability and reduce stacking fault energy, thereby enhancing mechanical properties [25,27].

It is also established that the element of Ti, Al, and Co can significantly promote the formation of a massive γ' phase and improve the tensile and creep properties of the traditional Ni-based alloys. Nevertheless, it has been seldom reported that the high temperature mechanical properties of IN625 can be enhanced by adding the γ' phase, forming Al and Ti. The main reason is that high-Ti and Al-content nickel-based alloys manufactured by laser additive manufacturing have a higher tendency to crack, which severely affects the mechanical properties of nickel-based high-temperature alloys during the laser additive manufacturing process [28,29]. Hence, the work described in this paper aims to ensure crack-free coatings with excellent high-temperature mechanical properties by adding AlTi intermetallic compounds and Co using laser melting. In our previous work, when preparing γ'-strengthened IN625 coatings using laser cladding technology, we found that when the content of TiAl intermetallic compound reached a certain amount (about 4.4%) and the ratio of Ti to Al was 1:1, the strengthening effect of the IN625 alloy was more significant. IN625 is a promising candidate material for the indoor and low-pressure turbine blades of gas turbines, aero engines, and rocket engines, then, with a typical working temperature of 850 °C [30]. Therefore, the work investigated in detail the effects of different Co/TiAl ratios (1.25, 1.55, and 1.85) upon the microstructure, room, and 850 °C mechanical properties of laser melted IN625 alloy coatings.

2. Materials and Methods

2.1. Materials

Prior to the laser melting process, the 304SS substrate measuring 80 mm × 50 mm × 30 mm (length, width, thickness) was sequentially ground with SiC paper of grit scale 200–2000 and cleaned using absolute ethanol. The powders with compositions presented in Table 1 were used as the melting material. Gas-atomized IN625 powders containing different Co/TiAl ratios, with a mean particle diameter of 75 μm, were deposited on the substrate as the Ni-based alloy coatings by the laser melting technique. To achieve a uniform distribution, they were processed using the QM-3SP04L ball mill at a speed of 120 r/min for two hours. Afterward, they were dried in a drying oven at 60 °C for one hour and screened through a 180 mesh before being subjected to laser melting.

Table 1. Composition of laser-melted powder.

	Mo	Ti	Al	Cr	Nb	Co	Fe	Ni
IN625	9.17	0.034	0.025	21.75	3.8	0.025	0.32	Bal.
1#:Co/TiAl = 1.25	8.27	2.20	2.20	19.61	3.43	5.5	0.288	Bal.
2#:Co/TiAl = 1.55	8.14	2.20	2.20	19.31	3.37	6.82	0.284	Bal.
3#:Co/TiAl = 1.85	8.02	2.20	2.20	19.02	3.32	8.14	0.28	Bal.

2.2. Laser Melting

Figure 1a depicts the schematic view of the Directed Energy Deposition (DED) process. To prepare the Laser additive coating with a thickness of 2 mm, an FL-1500 1.5 kW fiber laser was employed along with a synchronous powder-feeding and water-cooling system. To ensure the coatings were free from defects such as pores and microcracks, optimization of processing parameters was carried out as follows: a laser power of 650 W, spot diameter of 1.2 mm, traverse speed of 550 mm/min, energy density of 390 W/mm^2, overlap rate of 50%, and a powder delivery rate of 3.4 g/min. High-purity argon (5 Ar, 99.999%) was used as both a carrier and shielding gas at a flow rate of 10 L/min. After melting layers were deposited, the Ni-based alloy coatings with a dimension of 40 mm × 30 mm × 3 mm were fabricated successfully. Figure 1b represents the schematic view of the Sampling method. The specimen surface was smoothed and polished after preparation, and wire cutting was carried out based on the dimensions presented in Figure 1c. Preliminary testing demonstrated that the surface of all sample groups did not exhibit visible cracks.

2.3. Characterizations

Prior to the experiment, the cross-section of the laser-melted samples was ground and polished. To determine the tensile performance of the coating, a testing machine utilizing electro-hydraulic servo dynamic and static was utilized with a constant displacement rate of 0.2 mm/min. Three coating tensile samples were tested for each group. The coating-stretched sample at room temperature did not undergo heat treatment, while the 850 °C-coated stretched sample received heat treatment under the conditions of 900 °C × 10 h + 850 °C × 2 h. In addition, the hardness of the laser-melted specimens was determined via a Rockwell hardness tester, with an average value obtained by conducting five tests for each sample group, the geometry of which is shown in Figure 1c. To examine the coating microstructure, it was etched with an aqua regia solution (30 mL HCl + 10 mL

HNO_3). The phase composition of the coatings was analyzed using a Miniflex600 X-ray diffractometer (XRD) with a Cu-Kα radiation source, operating at an acceleration voltage of 35 kV and a current of 25 mA. The scanning range was set to 30°–85°, with a step size of 5°/min. The microstructure of the laser-melting coatings was examined with a MERLIN scanning electron microscope (SEM) that included an energy dispersive spectrometer (EDS). For the SEM, the working distance was approximately 10 mm and the accelerating voltage was set at 20 kV.

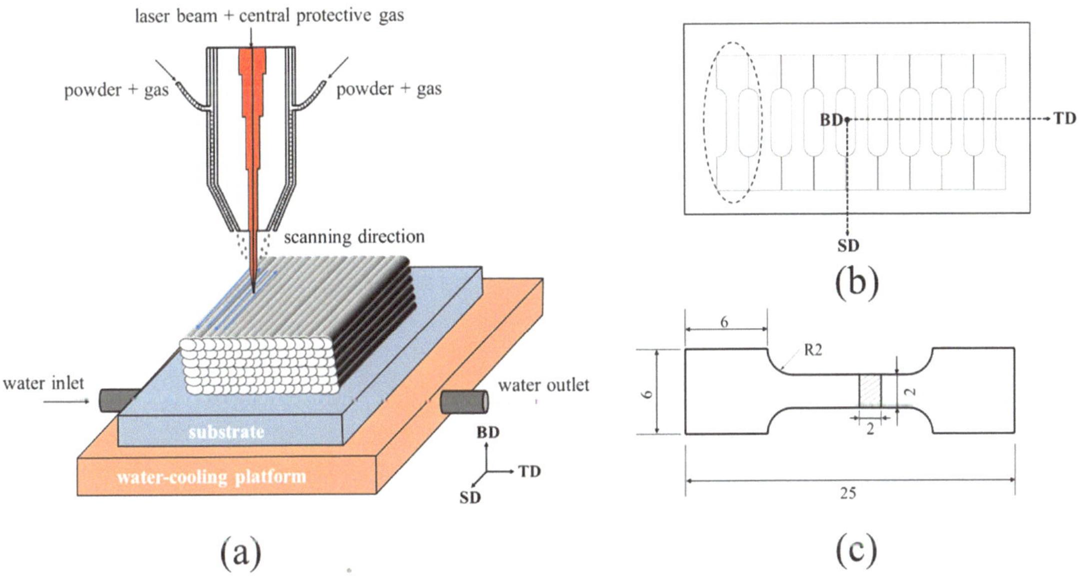

Figure 1. (**a**) Schematic diagram of sample preparation principle, (**b**) sampling methods, (**c**) tensile sample size.

3. Results and Discussion

3.1. XRD Analysis

Figure 2 shows the XRD spectra of IN625 and three distinct IN625 coatings with Co/TiAl (#1, #2, and #3) before and after heat treatment. It is evident from Figure 2a that the diffraction peaks of the Co, Al, and Ti-coated specimens in their as-deposited state are very similar to those of IN625. This is because the γ phase, γ′ phase, and γ″ phase are all mutually coherent and have similar lattice parameters, making it difficult to distinguish them. Moreover, it is difficult to observe the Laves phase and carbide diffraction peaks during testing due to the small quantity and volume fraction of precipitation phases in the unheated-treated coatings [31]. The main precipitated phase of IN625 is the matrix γ phase, which is consistent with the results of Li [32] and Ding [33] and the research of others on the fact that its microstructure is mainly the martensitic γ phase matrix at room temperature. After heat treatment at 900 °C × 10 h + 850 °C × 2 h, the coatings of IN625 and those added with Co, Al, and Ti elements in Figure 2b exhibit the diffraction peaks of Laves phase and carbide. The IN625 coating has a diffraction peak of the δ phase, which is due to the γ″ phase lattice distortion caused by high temperature (850 °C) aging, which promotes the transformation of the metastable γ″ phase into the δ phase. This is similar to the research results of Mathew et al. [34] and Di et al. [35]. The coatings added with Co, Al, and Ti elements showed the diffraction peak of the σ phase after heat treatment (Figure 2b), which is due to the formation of γ + γ′ eutectic and the enrichment of Cr, Mo, and Co elements around them during aging. This promotes the formation of the σ phase in the coatings.

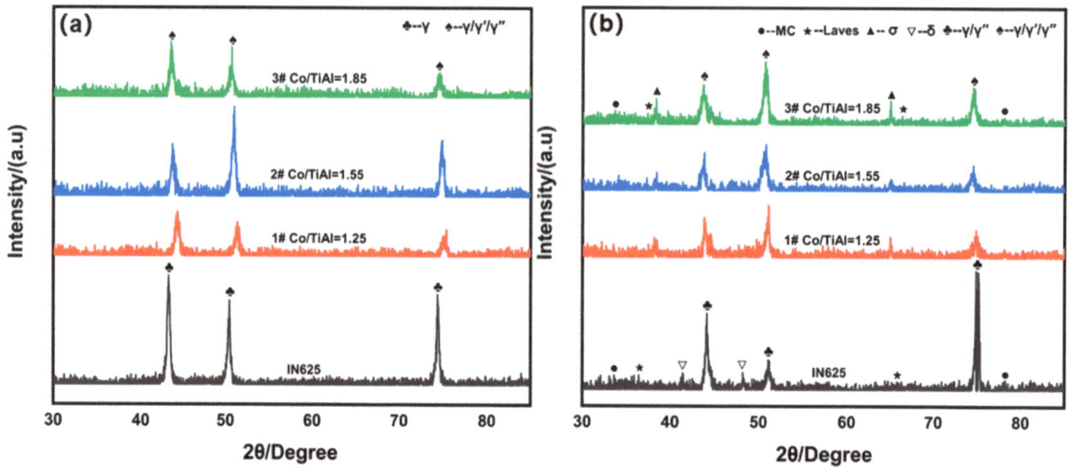

Figure 2. XRD patterns of IN625 and three different Co/TiAl IN625 coatings 1#, 2# and 3# ((**a**) Without heat treatment, (**b**) after heat treatment).

3.2. Microstructure and Structure

The microstructure and chemical composition of IN625 and three distinct Co/TiAl-coated IN625 specimens (#1, #2, and #3) were observed and analyzed through SEM and EDS. As shown in Figure 3, compared to the original IN625 coating, the addition of Ti, Al, and Co elements in the coatings results in a significant increase in the number of precipitated phases, which are mainly white spherical or plate-like precipitates. The number of precipitated phases is significantly higher for the coating with a Co/TiAl ratio of 1.55 compared to coatings with ratios of 1.25 and 1.85. EDS mapping revealed that the Nb element is significantly segregated at the grain boundaries of the coatings. The Ti element also exhibits notable segregation at the grain boundaries of #1, #2, and #3 coatings, while the Al element is only segregated in the white precipitates with uniform distribution in other areas. The segregation behavior of these elements is consistent with that observed in other nickel-based high-temperature alloys manufactured using laser additive techniques and conventionally cast alloys, as reported in previous studies [36,37]. The energy spectrum analysis of regions I, II, III, and IV in Figure 3 is presented in Figure 4. The composition analysis of region I indicates that it primarily consists of Ni, Cr, Mo, and Nb elements with a similar elemental content to the alloy powder used for preparing the coating. Thus, region I in Figure 3a can be considered as the γ-base phase. The energy spectrum analysis of regions II, III, and IV reveals substantial amounts of Ni, Ti, Al, Nb, and Co, corresponding to the formation of intermetallic compounds such as γ'-Ni_3Al, $Ni_3(Al,Ti)$, and $(Ni,Co)_3(Al,Ti)$ in the coatings. The γ'' phase primarily forms due to the $[Ni_3(Nb,Al,Ti)]$ intermetallic compound. Accordingly, the white spherical or plate-shaped precipitates observed at regions II, III, and IV in Figure 3 correspond to the γ' and γ'' phases, respectively. This observation is consistent with the findings of Zhou et al. [38].

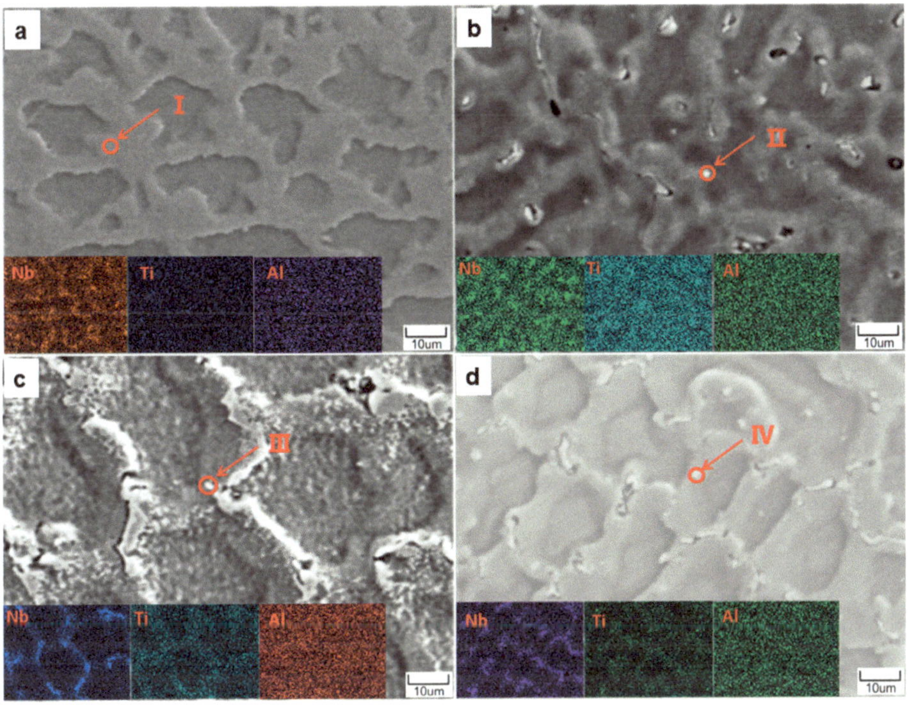

Figure 3. SEM and EDS images of the surface of the coatings without heat treatment ((**a**) IN625, (**b**) 1#, (**c**) 2#, (**d**) 3#).

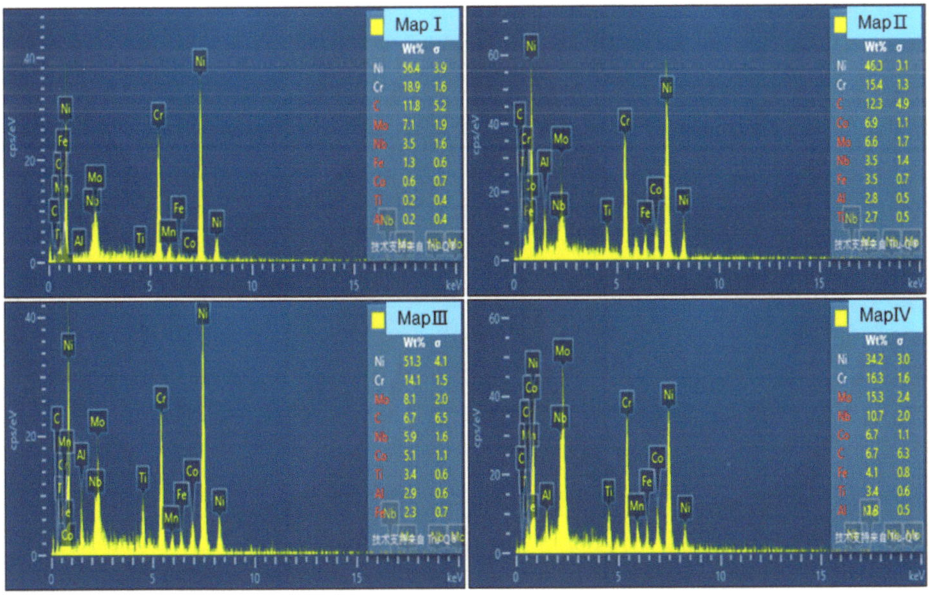

Figure 4. EDS results of the four micro-regions in Figure 3 (Map I: Region I in Figure 3; Map II: Region II in Figure 3; Map III: Region III in Figure 3; Map IV: Region IV in Figure 3).

SEM and EDS analysis of the as-heat-treated IN625 coating and the three Co/TiAl modified IN625 coatings (#1, #2, and #3) are presented in Figure 5. The SEM image in Figure 5a reveals noticeable grain boundary precipitation and coarsening in the as-heat-treated IN625 coating. Further EDS scanning analysis confirms substantial segregation of Mo and Nb elements at the grain boundaries. This is due to the relatively stable precipitation of the γ' phase within the grain during the aging process; as the aging time is increased, gradual precipitation and growth of the γ'' phase followed by transformation into the δ phase occurs. Furthermore, it is noticeable that a significant amount of Laves phases and carbides accumulate at the grain boundaries [35,36]. Figure 5b–d demonstrate that the heat-treated coatings with Co/TiAl ratios of 1.25, 1.55, and 1.85 experience lower levels of element segregation of Nb, Ti, and Al, as well as a more uniform distribution. This phenomenon can be explained by the effect of heat treatment, which reduces the element segregation of the alloy, leading to a more uniform microstructure distribution and increased refinement of the γ' and γ'' phases. Furthermore, the heat treatment dissolves some brittle phases and carbides at the grain boundaries [37,38]. It can be clearly observed from Figure 5b–d that the coatings with added Co, Al, and Ti elements feature the formation of numerous irregularly shaped $\gamma + \gamma'$ eutectic structures. When the Co/TiAl ratio is either 1.25 or 1.85, a noticeable amount of needle-shaped σ phase precipitates around the $\gamma + \gamma'$ eutectic structures (indicated by red arrows in Figure 5). It is also observed that the coating with Co/TiAl ratio of 1.85 has higher quantity and larger volume of the σ phase precipitates. However, no conspicuous precipitation of the σ phase is observed when the Co/TiAl ratio is 1.55.

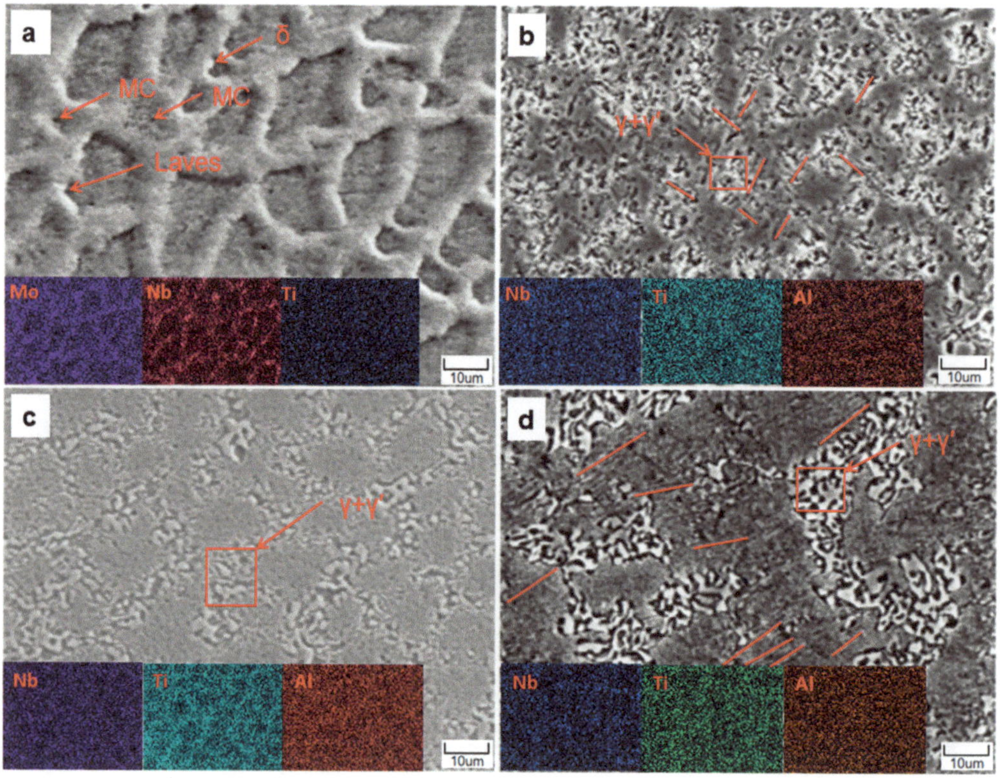

Figure 5. SEM and EDS images of the coatings surface after heat treatment ((**a**) IN625, (**b**) 1#, (**c**) 2#, (**d**) 3#).

3.3. Mechanical Property Analysis

The variations in hardness values of the IN625 coating and three Co/TiAl modified IN625 coatings, before and after heat treatment, are illustrated in Figure 6. The data presented in Figure 6 indicates that the initial hardness value of the IN625 coating is 18.32 HRC, which significantly increases with the addition of Ti, Al, and Co elements in the as-received coatings. It is worth noting that the highest hardness value (41.52 HRC) is achieved when the Co/TiAl ratio is 1.55, due to the considerable precipitation of the γ' phase that significantly contributes to the increase in hardness. This is consistent with the results reported in the study conducted by Tang et al. [39]. After heat treatment, the hardness of the IN625 coating decreased significantly as a result of the precipitation of the δ-phase at 800 °C and continuously coarsening or dissolving with increasing temperature, leading to a decrease in the hardness of the alloy [40]. In addition, carbides and Laves phases that precipitate and grow at grain boundaries are another reason for the decrease in the hardness of the IN625 coatings. Contrary to the IN625 coating, the three Co/TiAl modified IN625 coatings reveal a noticeable increase in hardness after heat treatment. The increase in hardness is mainly caused by grain strengthening arising from the increase in the number of refinements of the γ' phase in the matrix as well as the precipitation strengthening arising from the increased precipitation of the σ phase after the aging heat treatment. This result agrees with previous studies on this topic conducted by Xu et al. [36] and Kishore et al. [41].

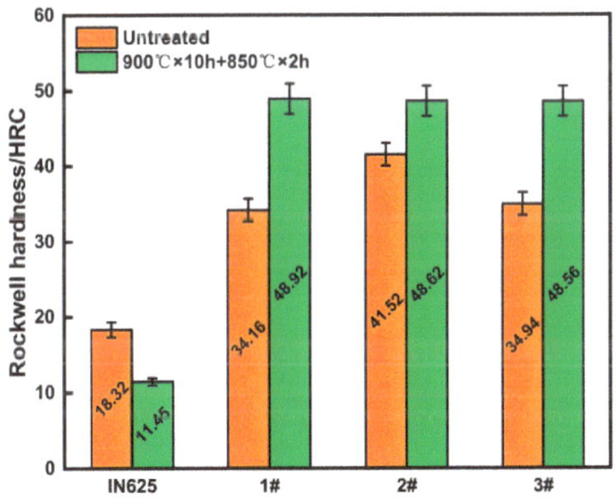

Figure 6. Hardness of four kinds coatings before and after heat treatment.

Table 2 and Figure 7 exhibit the tensile properties and stress-strain curves of both the IN625 coating and three Co/TiAl modified IN625 coatings (designated 1#, 2#, and 3#). Table 2 and Figure 7a reveal that the as-received IN625 coating has the highest elongation (27.5%) but the lowest yield strength at room temperature. As Ti, Al, and Co elements are incorporated, the strength of the samples increases notably, but at the expense of a rapid decline in elongation. This phenomenon occurs due to the preferential segregation of Co element into the γ phase, which in turn decreases the solubility of Ti and Al in the γ phase. However, the remaining Ti, Al, and Co elements can be incorporated into the γ' phase and occupy the Ni lattice sites, forming an ordered γ'-$(Ni,Co)_3(Al,Ti)$, thereby increasing its volume fraction. The L12 structure $(Ni,Co)_3(Al,Ti)$ provides a more remarkable yield strength than binary intermetallic compounds such as Ni_3Al, Co_3Ti, and so on [25,41]. A noticeable trend is observed as the Co/TiAl ratio rises, where the strength of the tensile samples initially increases before declining, while the elongation displays an opposite behavior. When the Co/TiAl ratio is 1.55, the ultimate tensile strength and yield strength reach their peaks at 1206 and 1050 MPa, correspondingly. However, the elongation diminishes drastically to merely 8.1%. The reason for this notable decline in elongation at ambient temperature is attributed to the high volume fraction of the precipitated γ' and γ'' phases.

Table 2 and Figure 7 reveal that following heat treatment, the strength and elongation of IN625 decrease dramatically at 850 °C, in contrast to the room temperature properties. This phenomenon occurs due to the growth of δ phase precipitation within the matrix during the heat treatment and 850 °C tensile process, accompanied by the accumulation of Laves phases and carbides. The growth of δ phase precipitation is prone to cause the build-up of dislocations, leading to localized stress concentration and the formation of microcracks within the matrix. Ultimately, it significantly hinders the high-temperature tensile performance [42]. The formation and growth of Laves phases and carbide particles at grain boundaries during high-temperature tensile testing can potentially undermine the stability of the grain boundaries, ultimately leading to severe embrittlement at elevated temperatures. Furthermore, the presence of carbide precipitates at grain boundaries hinders their ability to slide, thus leading to inadequate ductility at elevated temperatures [30]. The sample with a Co/TiAl ratio of 1.55 exhibits better yield strength and elongation than any of the other samples, recording values of 665 MPa and 11.3%, respectively. Nonetheless, the yield strength of the sample reduced by 385 MPa under this circumstance, which is attributed to the significant impact of the σ phase formed in the matrix post, extended aging heat treatment on the mechanical properties of the alloy [41].

Table 2. Tensile properties of the different coatings under varying conditions.

	RT			850 °C		
	UTS (MPa)	YS (MPa)	EL (%)	UTS (MPa)	YS (MPa)	EL (%)
IN625	898	608	27.5	338	320	17.9
1#	1070	880	10.8	630	554	10.4
2#	1206	1050	8.1	735	665	11.3
3#	1138	918	11.3	621	540	9.4

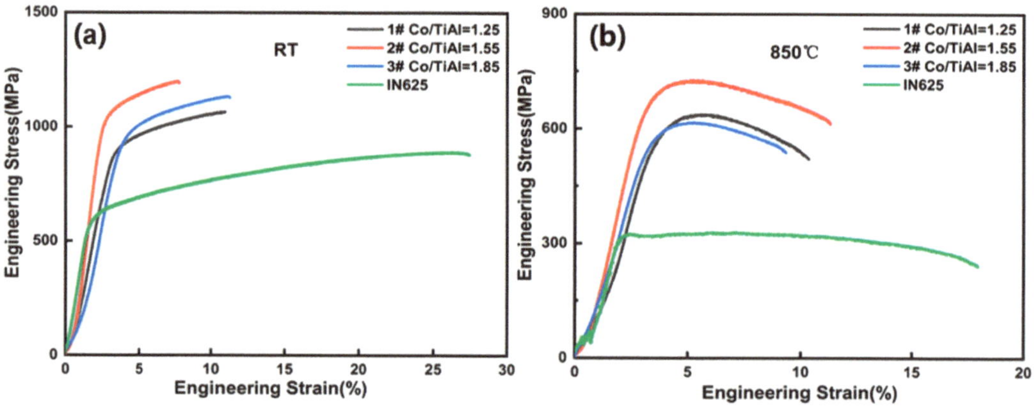

Figure 7. Tensile curves of four coatings after heat treatment and without heat treatment ((**a**) room temperature, (**b**) 850 °C).

The fracture surfaces of IN625 and various Co/TiAl IN625 (1#, 2#, and 3#) laser additive samples under ambient and 850 °C high-temperature conditions are presented in Figure 8. Figure 8a–d,(a1–d1) depict the tensile fracture surfaces of IN625, 1#, 2#, and 3# specimens under ambient temperature conditions, whereas Figure 8e–h,(e1–h1) illustrate the tensile fracture surfaces of the same specimens under high-temperature (850 °C) conditions. When subjected to ambient temperature tensile conditions, the fracture surface of IN625 specimens shows transgranular fracture and numerous dimples, which indicate ductile failure (shown in Figure 8(a1–e1)). In contrast, the fracture mode transitions to intergranular fracture under 850 °C tensile conditions, with a reduced amount of dimples compared to the room temperature case. This phenomenon is attributed to the gradual precipitation and coarsening of the δ phase along with significant carbide clustering at grain boundaries during both thermal processing and high-temperature tensile testing, which inevitably leads to decreased plasticity. It can be observed from Figure 8c. that cracks appear at the tensile fracture, and the reason for the formation may be that the γ'' phase and carbide hinder the dislocation movement during the plastic deformation process, resulting in nearby stress concentration, and microcracks will occur in the stress concentration area of the alloy when the stress concentration exceeds the tensile strength [43,44]. This is also an important reason why its elongation is lower than that of other coatings. Figure 8(b1–d1) reveals that samples containing Ti, Al, and Co also display apparent transgranular fractures, which are attributed to the precipitation of γ' and γ'' phases at the grain boundaries, leading to a stronger boundary strength than that of the interior grains, thereby promoting crack propagation from the grains. Figure 8(b1–d1) exhibit a small amount of dimples, but there are no visible dimples in Figure 8(c1), instead showing distinct cleavage surfaces, which indicates brittle fracture. Upon thermal processing, the tensile fracture surface of the specimens shifts from transgranular fracture at ambient temperature to intergranular fracture when exposed to elevated temperatures (depicted in Figure 8(f1–h1)), attributable to the formation of the σ phase at grain boundaries that are weaker than the interior grains, favoring crack propagation along the grain boundaries. As evidenced by the fracture morphology in Figure 8(g1), the grains are observed to undergo elongation, owing to the formation of the plastic $\gamma + \gamma'$ eutectic phase that is prone to elongation at high temperatures, resulting in a ductile fracture mode. Figure 8(h1) shows a considerable amount of voids, resulting from the substantial and coarse precipitation of the σ phase, resulting in cavity generation upon tensile deformation and compromising its plasticity. In addition, it is consistent with the stretch data in Table 2.

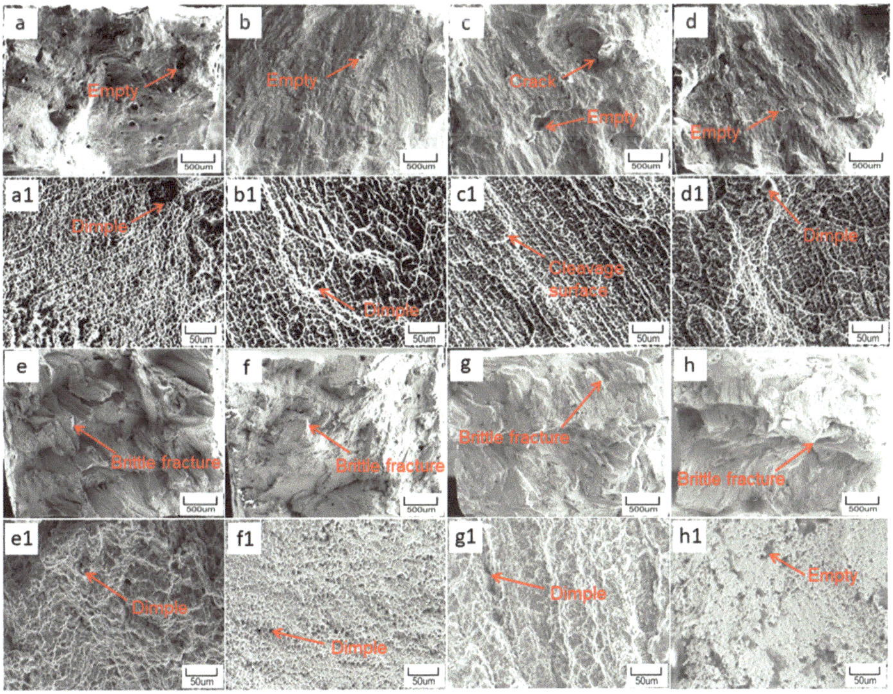

Figure 8. Tensile fracture of four coatings at normal temperature and 850 °C before and after heat treatment ((**a**,**a1**,**e**,**e1**) IN65, (**b**,**b1**,**f**,**f1**) 1#, (**c**,**c1**,**g**,**g1**) 2#, (**d**,**d1**,**h**,**h1**) 3#).

4. Conclusions

Utilizing laser-based additive manufacturing techniques, researchers fabricated IN 625 coatings with varying Co/TiAl ratios (1.25, 1.55, and 1.85) onto the surface of a 304 stainless steel substrate. Some promising results have already been obtained, as depicted below:

(1) Phase and microstructure analysis. IN 625 laser cladding layers lacking Ti, Al, and Co additives are predominantly composed of the γ phase, while the incorporation of Co, Ti, and Al during the laser cladding process for IN 625 promotes the formation of the γ' phase. Precipitation of the γ' phase reaches its maximum when the Co/TiAl ratio is 1.55. At the same time, it is worth noting that IN625 has no defects, such as cracks in the coating after adding Al, Ti, and Co. Upon thermal processing, the IN625 laser cladding layer exhibits significant coarse intergranular phase transformation precipitation, coupled with the accumulation of Laves phase and carbides. Post-heat treatment and the level of element segregation for Nb, Ti, and Al decrease while exhibiting greater homogeneity in their distribution throughout the IN 625 laser cladding layer containing Co, Ti, and Al additives. Furthermore, the formation of the $\gamma + \gamma'$ eutectic microstructure was observed, and a distinct needle-like σ phase was developed surrounding the $\gamma + \gamma'$ eutectic when the Co/TiAl ratio was set at 1.25 and 1.85.

(2) Excellent mechanical properties. In comparison to the IN625 coating, the incorporation of Co, Ti, and Al additives in the IN 625 laser cladding layer results in a substantial enhancement in coating hardness and room temperature strength, albeit accompanied with a steep reduction in ductility. At a Co/TiAl ratio of 1.55, the coating exhibits its highest yield strength and hardness values (1050 MPa, 41.52 HRC) yet records an elongation of only 8.1%. The hardness of the IN625 coating reduces upon heat treatment (900 °C × 10 h + 850 °C × 2 h). Conversely, the integration of Co, Ti, and Al into the IN 625 laser cladding layer leads to a marked surge in hardness. As the Co/TiAl ratio rises, the yield strength and elongation of the IN 625 laser cladding layer containing Co, Ti, and Al additives initially increase, but they both ultimately decline at a temperature of 850 °C. Notably, the IN625 coating with a Co/TiAl ratio of 1.55 exhibits excellent high-temperature (850 °C) mechanical properties without crack defects, exhibiting a superior combination of hardness, tensile strength, yield strength, and elongation rate, measuring 48.62 HRC, 665 MPa, and 11.3%, respectively.

The detailed mechanisms behind the variation in the overall performance of the IN625 coatings with various Al, Ti, and Co contents after heat treatment were discussed in terms of the combined effects of solid solution, microstructural refinement, and precipitation. Meanwhile, this work may have important implications for academic research as well as industrial applications of nickel-based high-temperature alloys, especially to achieve laser remanufacturing and the repair of indoor and low-pressure turbine blades for gas turbines, aero-engines, and rocket engines with high mechanical properties.

Author Contributions: Conceptualization, T.Y. and C.Q.; formal analysis, T.Y., Y.L. and L.Z.; investigation, C.Q.; writing—original draft preparation, T.Y; writing—review and editing, P.C.; visualization, W.W.; supervision, H.L. All authors have read and agreed to the published version of the manuscript.

Funding: This research was funded by National Key Research and Development Program, grant number 2017YFE0301300, National Defense Science and Technology Key Laboratory Fund Project, grant number JCKY61420052002, National Defense Science and Technology Advanced Research Fund Project, grant number 61400040204, and Hunan Province postgraduate research and innovation project, grant number 223YSC009.

Institutional Review Board Statement: Not applicable.

Informed Consent Statement: Not applicable.

Data Availability Statement: Not applicable.

Acknowledgments: The author would like to thank Tan Lirong and Guan Runjie for their help in sample preparation and tensile experiments.

Conflicts of Interest: The authors declare no conflict of interest. The funders had no role in the design of the study; in the collection, analyses, or interpretation of data; in the writing of the manuscript; or in the decision to publish the results.

References

1. Parizia, S.; Marchese, G.; Rashidi, M.; Lorusso, M.; Hryha, E.; Manfredi, D.; Biamino, S. Effect of heat treatment on microstructure and oxidation properties of Inconel 625 processed by LPBF. *J. Alloy Compd.* **2020**, *846*, 156418. [CrossRef]
2. Xu, Z.; Xie, Y.; Ebrahimnia, M.; Dang, H. Effect of B4C nanoparticles on microstructure and properties of laser cladded IN625 coating. *Surf. Coat. Technol.* **2021**, *416*, 127154. [CrossRef]
3. Marchese, G.; Parizia, S.; Rashidi, M.; Saboori, A.; Manfredi, D.; Ugues, D.; Lombardi, M.; Hryha, E.; Biamino, S. The role of texturing and microstructure evolution on the tensile behavior of heat-treated Inconel 625 produced via laser powder bed fusion. *Mater. Sci. Eng. A* **2020**, *769*, 138500. [CrossRef]
4. Ahsan, M.R.U.; Fan, X.; Seo, G.J.; Noakes, M.; Nycz, A.; Kliaw, P.; Kim, D.B. Microstructures and mechanical behavior of the bimetallic additively-manufactured structure (BAMS) of austenitic stainless steel and Inconel 625. *J. Mater. Sci. Technol.* **2021**, *74*, 176–188. [CrossRef]
5. Karmuhilan, M.; Kumanan, S. A review on additive manufacturing processes of inconel 625. *J. Mater. Eng. Perform.* **2022**, *31*, 2583–2592. [CrossRef]
6. GaoLe, Z.; Yun, J.; XiaoAn, H.; Jia, H.; Yun, W. High-temperature mechanical properties of nickel-based superalloys manufactured by additive manufacturing. *Mater. Sci. Technol.* **2020**, *36*, 1523–1533. [CrossRef]
7. Suave, L.M.; Cormier, J.; Villechaise, P.; Laigo, J. Microstructural evolutions during thermal aging of alloy 625: Impact of temperature and forming process. *Metall. Mater. Trans. A* **2014**, *45*, 2963–2982. [CrossRef]
8. Banerjee, A.; Rossin, J.; He, M.R.; Musinski, W.D.; Shade, P.A.; Cox, M.E.; Schwalbach, E.J.; Pollock, T.; Henker, K.J. Decoupling build orientation-induced geometric and texture effects on the mechanical response of additively manufactured IN625 thin-walled elements. *Mater. Sci. Eng. A* **2023**, *87*, 144826. [CrossRef]
9. Tian, Z.; Zhang, C.; Wang, D.; Liu, D.; Tian, Y. A review on laser powder bed fusion of inconel 625 nickel-based Alloy *Appl. Sci.* **2019**, *10*, 81. [CrossRef]
10. Huang, W.; Li, Y.; Yanjie, R.E.N.; Sun, J.; Xia, Z.; Zhou, L.; Li, C.; Chen, J.; Niu, Y.; Zhao, Y. Effect of scanning speed on the high-temperature oxidation resistance and mechanical properties of Inconel 625 alloys fabricated by selective laser melting. *Vacuum* **2022**, *206*, 111447. [CrossRef]
11. Parida, A.K.; Maity, K. Comparison the machinability of Inconel 718, Inconel 625 and Monel 400 in hot turning operation. *Eng. Sci. Technol. Int. J.* **2018**, *21*, 364–370. [CrossRef]
12. Balbaa, M.; Ghasemi, A.; Fereiduni, E.; Al-Rubaie, K.; Elbestawi, M. Improvement of fatigue performance of laser powder bed fusion fabricated IN625 and IN718 superalloys via shot peening. *J. Mater. Process. Technol.* **2022**, *304*, 117571. [CrossRef]
13. Sitek, R.; Molak, R.; Zdunek, J.; Bazarnik, P.; Wisniewski, P.; Kubiak, K.; Mizera, J. Influence of an aluminizing process on the microstructure and tensile strength of the nickel superalloy IN 718 produced by the Selective Laser Melting. *Vacuum* **2021**, *186*, 110041. [CrossRef]
14. Martin, E.; Natarajan, A.; Kottilingam, S.; Batmaz, R. Binder jetting of "Hard-to-Weld" high gamma prime nickel-based superalloy RENÉ 108. *Addit. Manuf.* **2021**, *39*, 101894. [CrossRef]
15. Chen, L.; Sun, Y.; Li, L.; Ren, X. Microstructural evolution and mechanical properties of selective laser melted a nickel-based superalloy after post treatment. *Mater. Sci. Eng. A* **2020**, *792*, 139649. [CrossRef]
16. Thakare, J.G.; Pandey, C.; Mahapatra, M.M.; Mulik, R.S. Thermal barrier coatings-A state of the art review. *Met. Mater. Int.* **2021**, *27*, 1947–1968. [CrossRef]
17. Ghiaasiaan, R.; Poudel, A.; Ahmad, N.; Gradl, P.R.; Shao, S.; Shhamsaei, N. High temperature tensile and fatigue behaviors of additively manufactured IN625 and IN718. *Procedia Struct. Integr.* **2022**, *38*, 581–587. [CrossRef]
18. Song, Z.; Gao, S.; Wang, Z.; Lan, L.; Hou, J.; He, B. Effects of non-equilibrium microstructures on microstructure evolution and mechanical properties of laser powder bed fusion IN625 Ni-based superalloy during long-term thermal exposure at 700 °C and 750 °C. *Mater. Sci. Eng. A* **2022**, *856*, 143883. [CrossRef]
19. Kreitcberg, A.; Brailovski, V.; Turenne, S. Elevated temperature mechanical behavior of IN625 alloy processed by laser powder-bed fusion. *Mater. Sci. Eng. A* **2017**, *700*, 540–553. [CrossRef]
20. Son, K.T.; Phan, T.Q.; Levine, L.E.; Kim, K.S.; Lee, K.A.; Ahlfors, M.; Kassner, M.E. The creep and fracture properties of additively manufactured inconel 625. *Materialia* **2021**, *15*, 101021. [CrossRef]
21. Xia, W.; Zhao, X.; Yue, Q.; Yue, L.; Wang, J.; Ding, Q.; Bei, H.; Zhang, Z. Formative and controlled mechanisms of nano-sized γ' precipitates with local phase-transition within dislocation networks of nickel-based single crystal superalloys. *Acta Mater.* **2021**, *206*, 116653. [CrossRef]
22. Zhang, Z.; Liu, Y.; Wang, Z.; Huang, J.; Chen, S.; Ye, Z.; Chen, X.; Yang, J. Influence behavior and mechanism of γ'-or γ''-precipitated phase types on damage resistance of heterogeneous interface in nickel-based superalloys. *J. Alloy Compd.* **2022**, *924*, 166559. [CrossRef]
23. Sun, B.; Zhang, T.; Song, L.; Zhang, L. Correlation between secondary γ' and high-temperature tensile behavior of a powder metallurgy nickel-based superalloy EP962NP. *Mater. Sci. Eng. A* **2023**, *866*, 144687. [CrossRef]

24. Lin, Y.C.; Wang, C.Y. Alloying-element dependence of structural, elastic and electronic properties of nickel-based superalloys: Influence of γ'volume fraction. *J. Alloy Compd.* **2020**, *838*, 155141. [CrossRef]
25. Tang, L.; Liang, J.; Cui, C.; Li, J.; Zhou, Y.; Sun, X.; Ding, Y. Precipitation and phase transformation mechanism of additive manufactured Ni-Co base superAlloy *Mater. Charact.* **2019**, *151*, 252–259. [CrossRef]
26. Tang, L.; Liang, J.; Cui, C.; Li, J.; Zhou, Y.; Sun, X.; Ding, Y. Influence of Co content on the microstructures and mechanical properties of a Ni–Co base superalloy made by specific additive manufacturing process. *Mater. Sci. Eng. A* **2020**, *786*, 139438. [CrossRef]
27. Cui, C.Y.; Gu, Y.F.; Ping, D.H.; Harada, H. Phase constituents in Ni–Al–Co–Ti quaternary alloys. *Intermetallics* **2008**, *16*, 910–916. [CrossRef]
28. Kalyanasundaram, V.; De Luca, A.; Wróbel, R.; Tang, J.; Holdsworth, S.R.; Leinenbach, C.; Hosseini, E. Tensile and creep-rupture response of additively manufactured nickel-based superalloy CM247LC. *Addit. Manuf. Lett.* **2023**, *5*, 100119. [CrossRef]
29. Griffiths, S.; Tabasi, H.G.; Ivas, T.; Maeder, X.; Leinenbach, C. Combining alloy and process modification for micro-crack mitigation in an additively manufactured Ni-base superAlloy *Addit. Manuf.* **2020**, *36*, 101443. [CrossRef]
30. Hu, X.A.; Zhao, G.L.; Liu, F.C.; Liu, W.X. Microstructure and mechanical behavior of Inconel 625 alloy processed by selective laser melting at high temperature up to 1000 C. *Rare Met.* **2020**, *39*, 1181–1189. [CrossRef]
31. Cao, Y.; Bai, P.; Liu, F.; Hou, X. Investigation on the precipitates of IN718 alloy fabricated by selective laser melting. *Metals* **2019**, *9*, 1128. [CrossRef]
32. Li, Y.M.; Zhu, R.M.; Liu, H.J.; Ding, Y.T. Evolution of precipitated phase in GH625 alloy during long term aging. *Heat Treat. Met.* **2017**, *42*, 94–99.
33. Ding, Y.T.; Wang, W.; Li, H.F.; Guo, T.B. Thermodynamic calculation of precipitated phases in alloy GH3625. *J. Lanzhou Univ. Technol.* **2016**, *42*, 1–5.
34. Mathew, M.D.; Palanichamy, P.; Rao, K. Microstructural changes in alloy 625 during high temperature creep. *Mater. Charact.* **2008**, *59*, 508–513. [CrossRef]
35. Di, X.; Xing, X.; Wang, B. Nucleation and coarsening mechanism of δ phase in Inconel 625 deposited metal. *Acta Met. Sin* **2014**, *50*, 323–328.
36. Xu, J.; Gruber, H.; Peng, R.L.; Moverare, J. A novel γ'-strengthened nickel-based superalloy for laser powder bed fusion. *Materials* **2020**, *13*, 4930. [CrossRef] [PubMed]
37. Hariharan, A.; Lu, L.; Risse, J.; Kostka, A.; Raabe, D. Misorientation-dependent solute enrichment at interfaces and its contribution to defect formation mechanisms during laser additive manufacturing of superalloys. *Phys. Rev. Mater.* **2019**, *3*, 123602. [CrossRef]
38. Zhou, S.; Hu, M.; Li, C.; Guo, Q.; Yu, L.; Ding, H.; Liu, Y. Microstructure-performance relationships in Ni-based superalloy with coprecipitation of γ' and γ'' phases. *Mater. Sci. Eng. A* **2022**, *855*, 143954. [CrossRef]
39. Tang, Y.T.; Panwisawas, C.; Ghoussoub, J.N.; Gong, Y.; Reed, R.C. Alloys-by-design: Application to new superalloys for additive manufacturing. *Acta Mater.* **2021**, *202*, 417–436. [CrossRef]
40. Gao, Y.B.; Ding, Y.T.; Meng, B.; Ma, Y.J.; Chen, J.J.; Xu, J.Y. Research progress in evolution of precipitated phases in Inconel 625 superAlloy *J. Mater. Eng.* **2020**, *48*, 13–22.
41. Kishore, K.; Kushwaha, S.; Teja, K.S.P.; Chandan, A.K.; Mukherjee, M.; Adhikary, M.; Kumar, A. Sigma phase embrittlement-induced failures of heat-resistant stainless steel traveling grate links. *Eng. Fail. Anal.* **2023**, *144*, 106979. [CrossRef]
42. Gao, Y.; Zhang, D.; Cao, M.; Chen, R.; Feng, Z.; Schleifenbaum, J.H.; Ziegler, S. Effect of δ phase on high temperature mechanical performances of Inconel 718 fabricated with SLM process. *Mater. Sci. Eng. A* **2019**, *767*, 138327. [CrossRef]
43. Marchese, G.; Lorusso, M.; Parizia, S.; Bassini, E.; Biamino, S. Influence of heat treatments on microstructure evolution and mechanical properties of Inconel 625 processed by laser powder bed fusion. *Mater. Sci. Eng. A* **2018**, *729*, 64–75. [CrossRef]
44. Hu, Y.; Lin, X.; Li, Y.; Zhang, Q.; Chen, W.; Li, W.; Huang, W. Influence of heat treatments on the microstructure and mechanical properties of Inconel 625 fabricated by directed energy deposition. *Mater. Sci. Eng. A* **2021**, *817*, 141309. [CrossRef]

coatings

MDPI

Article

Study on Cavitation Corrosion Properties of Titanium Alloy Radiation Rod with Different Roughnesses for Ultrasonic Casting

Yilong Yang [1,*], Ya Zhang [2], Xuhe Liu [1] and Haoming Zhang [1]

1 School of Mechanical Engineering, Henan University of Engineering, No. 1, Xianghe Road, Zhengzhou 451191, China; lxh_haue@163.com (X.L.); zjpzhm@sina.com (H.Z.)
2 School of Electrical and Information Engineering, Henan University of Engineering, No. 1, Xianghe Road, Zhengzhou 451191, China; henanzy1989@163.com
* Correspondence: yangyilong@haue.edu.cn

Abstract: To determine the mechanism of corrosion damage caused by cavitation, the properties of titanium alloy radiation rods with different roughnesses in 2A14 aluminum melt for ultrasonic casting were studied. The corrosion morphology, weight loss/cavitated area, reaction layer and microhardness of surface section were observed, and the collapse of a single cavitation bubble was simulated. The weight loss/cavitated area caused by the physical impact of cavitation accounted for 6.4% to 8.6% of the total weight loss/cavitated area. The corrosion product was $TiAl_3$. The reactant appeared at the Al/Ti solid–liquid interface in 4 min and the reaction layer appeared in 10 min under different roughnesses. The thickness of the work hardening layer on the surface of the material could reach 160 μm. The results show that the greater the roughness of titanium alloy in aluminum melt, the greater the rate of weight loss/cavitated area and the greater the maximum pressure in the process of cavitation bubble collapse. The evolution of the hardened layer depended on the stripping rate of the surface material caused by cavitation corrosion and the work hardening rate of the surface layer. This study provides insights to develop a new homemade Ti alloy radiation rod with better resistance to corrosion in the ultrasonic casting.

Keywords: radiation rod; $TiAl_3$; surface roughness; cavitation collapse; cavitation corrosion

Citation: Yang, Y.; Zhang, Y.; Liu, X.; Zhang, H. Study on Cavitation Corrosion Properties of Titanium Alloy Radiation Rod with Different Roughnesses for Ultrasonic Casting. *Coatings* **2023**, *13*, 1632. https://doi.org/10.3390/coatings13091632

Academic Editor: Chi Tat Kwok

Received: 29 July 2023
Revised: 13 September 2023
Accepted: 15 September 2023
Published: 18 September 2023

1. Introduction

With the development of modern industry, aluminum alloy has great application prospects in industrial applications [1–4]. However, with the continuous upgrading of industrial applications, higher requirements are put forward for the performance of aluminum alloys, and aluminum alloys produced by traditional casting methods can no longer meet this requirement [5,6]. By incorporating ultrasound into the traditional aluminum alloy casting process, the special acoustic cavitation effect of ultrasound [7–10] can refine the grain of aluminum melt and remove impurities and gases in the melt, and so aluminum alloys with stronger properties are obtained.

In the past few decades, many researchers have conducted research on the forming of aluminum alloys [11–15]. Ultrasonic melt treatment technology (UST) has significant advantages compared to other forming technologies, so this article mainly introduces ultrasonic processing technology [16–20]. Eskin [21] first proposed the application of ultrasound to lightweight alloy melts in 1997. The science and practicability of ultrasonic treatment of light alloy melt and the development of acoustic cavitation in liquid metal were discussed. The results showed that ultrasonic melt treatment could improve the degassing rate and fine filtration rate of light alloy melt, and had a great influence on the microstructure of the ingot. Li [22] prepared a new extrusion cast Al-5Cu-XLI-0.5Mn-0.3mg-0.15Ti (x = 0.3, 0.6, 0.9, 1.2 wt%) alloy (referred to as Al-5Cu-xLi alloy) using ultrasonic treatment. The addition of ultrasound could refine grain size and had a significant degassing effect. Guang [23] prepared high-strength magnesium aluminum bimetallic materials using a new type of lost foam composite casting (LFCC) method enhanced by ultrasonic vibration. Under the action of ultrasound, the Mg_2Si and $Al_{12}Mg_{17}$ phases and $Al_{12}Mg_{17}$+δ-Mg at the interface were obviously refined, and the microhardness of the entire interface became more uniform. The shear strength of the Mg/Al bimetal was significantly increased by 86.5%, reaching 69 MPa. Moussa [24] used ultrasonic melt technology to prepare high-performance nickel–aluminum bronze (NAB) alloy with finer α(Cu) and β′ phases, and the finer intermetallic κ_{II} phase particles were evenly distributed along the casting structure. The above research proved the effectiveness of ultrasound on the melt and also ignored the effect of corrosion of the ultrasonic rod. The radiation rod used for ultrasound was damaged due to the influence of ultrasonic cavitation performance during actual work [25,26]. The main reason for the damage of radiation rod was cavitation corrosion. On the one hand, cavitation corrosion reduced the service life of the radiation rod; on the other hand, the corrosion products generated were harmful to the performance of the alloy. These are important reasons limiting the popularity of the ultrasonic casting industry. Titanium alloy has excellent physical and chemical properties compared to other materials [27]. Therefore, it was necessary to study the cavitation corrosion mechanisms of the titanium alloy radiation rod. The study of cavitation corrosion mechanisms has guiding significance for future corrosion protection [28,29].

The application of ultrasonic technology was accompanied by the emergence of the cavitation phenomenon. Many scholars have conducted comprehensive and systematic research on the factors affecting the process of cavitation corrosion. These factors include the material and surface condition of the radiation rod, the characteristics of the liquid medium, and the external environment [25,30]. In the process of ultrasonic casting, the ultrasonic performance is given priority, followed by the study of reducing the cavitation corrosion of the radiation rod. Once the casting of a certain metal is determined, its physical properties cannot be changed. Therefore, the study of cavitation corrosion of radiation rod is mainly affected by the properties of the radiation rod and the working environment. These include the pressure gradient on the surface of the radiation rod, material properties [31,32], surface morphology [33,34], and others.

In this paper, the factors affecting the cavitation corrosion of a titanium alloy radiation rod in aluminum melt were further studied. However, there are few reports on the mechanism of the influence of the surface roughness of the radiation rod on the degree of cavitation corrosion. The corrosion morphology, weight loss/cavitated area, surface roughness, reaction layer of profile, and microhardness of the surface section of the radiation rod were observed before and after the experiment, and the collapse of a single cavitation bubble was simulated. Therefore, this paper deeply studies the cavitation corrosion mechanism of a radiation rod with different surface roughnesses in high-temperature aluminum melt. This provides certain technical references for the study of cavitation corrosion mechanisms of radiation rods in the process of ultrasonic casting.

2. Experimental Details

2.1. Materials

2A14 aluminum alloy is an aluminum alloy with a nominal composition of Al-0.90Si-4.20Cu-0.45Mg-0.11Zn-0.67Mn-0.10Ti-0.10Ni-0.050Fe (wt%) [35]. The titanium alloy (Ti-6Al-4V) used for ultrasonic casting is grade TC4 [36], which belongs to the (α + β) titanium alloy family. It has several advantages, such as good corrosion resistance, a small density (4.5 g/cm^3), high specific strength, high toughness, and welding performance. It has good resistance to high-temperature/pressure environments, good amplitude conductivity, and good adaptability to other external environmental factors in ultrasonic casting. Table 1 shows its composition [37].

Table 1. Contents of major solute elements in the TC4 (wt%).

Elements	Fe	C	N	H	O	Al	V	Ti
Content	≤0.30	≤0.10	≤0.05	≤0.02	≤0.20	5.5~6.8	3.5~4.5	Bal.

2.2. Sample Preparation

UST was exerted through a sonotrode [36] that was driven by a high-power ultrasonic generator (DEEPSEA, Shanghai, China, input frequency, 20 kHz; input power 2.5 kW). The output power and current were 400 W and 6 A, respectively. The sonotrode was immersed at a position of 30 mm below the melt surface. The experimental setup is shown in Figure 1, which mainly includes a resistance furnace, a graphite silicon carbide crucible, an inert gas protection device, and an ultrasonic vibration system.

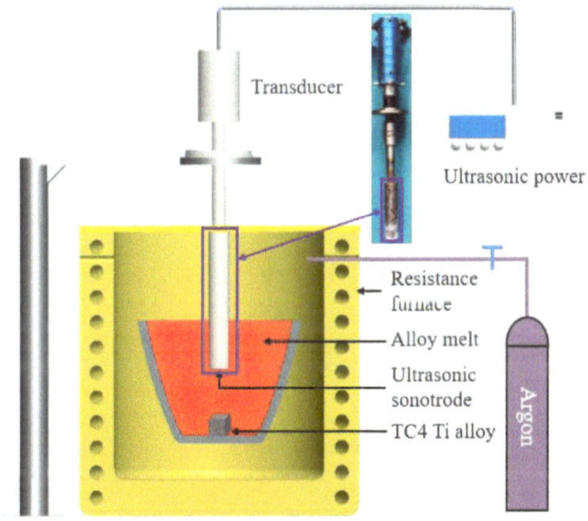

Figure 1. Schematic diagram of the experimental system.

The entire process of cavitation corrosion includes the incubation period of cavitation and acceleration period of cavitation. The duration of the cavitation incubation period increases linearly with the increase of the roughness of the material surface morphology, and the original traces generated by machining on the material surface have a significant impact on it [38]. The incubation period of cavitation corrosion is a typical period that reflects the whole process of cavitation corrosion. Therefore, the corrosion experiment of the ultrasonic radiation rod in this paper was divided into two parts. One part was a corrosion experiment on titanium alloy blocks. The corrosion morphology, weight loss/cavitated area, changes of roughness, and reaction layer of the sample were observed and counted. The other part was the corrosion experiment on the ultrasonic radiation rod. In contrast to the first part of the corrosion experiment, there was no block sample at the lower end of the radiation rod, and only cavitation corrosion experiments were conducted on titanium alloy radiation rods. In this part, the growth of the reaction layer and the microhardness change of the surface section were studied. The TC4 titanium alloy, which was the same as the radiation rod used in ultrasonic casting, was machined into a sample size of 25 mm × 25 mm × 10 mm with four different roughnesses. The samples with roughnesses from small to large were represented as A, B, C, and D. Sample A was precision-machined on a lathe, while samples B, C, and D were all rough-machined, with roughnesses of 0.4, 7.2, 9.5, and 9.8 Ra (μm), respectively. Aluminum alloy 2A14 with a weight of about 5 kG was used for each crucible experiment. The aluminum melt temperature was about 700 °C, and the experiment durations were 12, 24, 36, and 48 min, respectively. After the experiment was completed, the samples were soaked with HCl and NaOH to remove the residual aluminum on the surface, and then cleaned and dried with alcohol and acetone to calculate weight loss/cavitated area. The weight loss rate/cavitated area is given by:

$$v_t = \frac{(m - m_t)/t}{s} \quad (1)$$

where v_t, m, m_t, s, and t are defined as the weight loss rate at this moment, weight, weight at this moment, cavitated area, and experimental time, respectively.

2.3. Microstructural Characterization

A super-depth-of-field microscope (VHX5000) was used to examine the sample's surface. The original surface roughness value of the sample was tested with an optical surface profiler (Wyko, Beijing, China, Wyko NT9100). High-resolution scanning electron microscopy (TESCAN, Brno, Czech Republic, SEM: MIRA3 TESCAN) revealed the morphology of the corroded sample surface and section reaction layer. Energy-dispersive X-ray spectroscopy (Carl Zeiss, Berlin, Germany, EDS: OxfordX-Max20) and X-ray diffraction (XRD) in a Rigaku 600 X-ray diffractometer were used to perform qualitative analysis and identify the reaction layer components. The XRD was operated at a scanning speed of 0.02°/s at 40 kV using CuKα radiation (wavelength $\lambda_{K\alpha}$ = 1.54056 Å). A microhardness tester (Changzhou Sanfeng Instrument Technology Co., Ltd, Changzhou, China, HV-1000A) with a diamond indenter was used to measure the original surface microhardness of the sample. The sample size was 10 mm × 10 mm × 2 mm. Standard loads of 1.5 N were added. The operation time was 2 s. The microhardness values were obtained based on five different positions whenever possible.

3. Results and Discussion

3.1. Weight Loss Rate/Cavitated Area

Table 2 summarizes the data of initial weight, weight after ultrasonic treatment, and total weight loss of samples with different roughnesses. After 48 min of experiment, samples A: 0.4 Ra, B: 7.2 Ra, C: 9.5 Ra, and D: 9.8 Ra lost 8.1 mg, 12.7 mg, 17.3 mg, and 15.4 mg in weight, respectively. Sample A lost the least weight, while sample C lost the most weight. The rougher the surface, the greater the number of pits. A greater number of pits results in greater contact area between the pit and the aluminum melt. The aluminum melt rapidly spreads on the surface of the titanium alloy, leading to a higher rate of diffusion, chemical reaction, and recombination of Ti atoms in the aluminum melts. The change of weight loss rate/cavitated area over time of the sample under the ultrasonic condition of aluminum melt at 700 °C is shown in Table 3. Furthermore, the curve of weight loss rate/cavitated area of the block sample is drawn as shown in Figure 2.

Table 2. Weight loss/cavitated area of each sample (mg/cm^2).

Sample	Ra (μm)	Initial Weight m(g)	Weight after 48 min m_1(g)	Lost Weight/Cavitated Area (mg/cm^2)
A	0.4	27.0017	26.9936	8.1
B	7.2	27.0519	27.0392	12.7
C	9.5	27.2140	27.1968	17.3
D	9.8	26.9418	26.9264	15.4

Table 3. Weight loss rate/cavitated area of each sample (mg·min^{-1}/cm^2).

Number Rate (v_t)	12 min v_1 (mg·min^{-1}/cm^2)	24 min v_2 (mg·min^{-1}/cm^2)	36 min v_3 (mg·min^{-1}/cm^2)	48 min v_4 (mg·min^{-1}/cm^2)
A	0.168	0.152	0.185	0.169
B	0.304	0.260	0.273	0.265
C	0.460	0.349	0.421	0.359
D	0.334	0.266	0.347	0.320

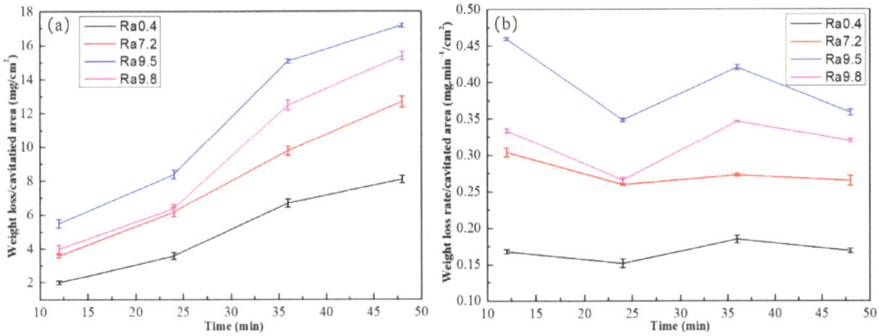

Figure 2. Curves of (**a**) weight loss/cavitated area; (**b**) weight loss rate/cavitated area.

The roughness was proportional to the weight loss rate/cavitated area, and the weight loss rate/cavitated area of the four groups of samples was relatively stable. The average weight loss rate/cavitated area of samples A: 0.4 Ra, B: 7.2 Ra, C: 9.5 Ra, and D: 9.8 Ra were 0.17 mg·min/cm^2, 0.28 mg·min/cm^2, 0.40 mg·min/cm^2, and 0.32 mg·min/cm^2. According to the weight loss rate/cavitated area, the weight loss/cavitated area of titanium alloy caused by physical impact due to the cavitation effect accounted for 6.4%~8.6% of the total weight loss/cavitated area.

3.2. *Corrosion Morphology and Corrosion Products*

3.2.1. Corrosion Morphology

Figure 3 shows the corrosion morphology of samples with different roughnesses subjected to ultrasonic vibration for 48 min. Obvious plastic deformation appeared on the surface of the samples. The etch pits formed as shown in Figure 3a were small in size, ranging from 1 to 5 μm in diameter. Each etch pit was scattered on the surface of the material in a separate form. However, the continuous expansion of some of the previously formed tiny etch pits gradually brought the original separated etch pits closer and closer, so that the walls of the etch pits were staggered together to form a continuous band of larger pits. Figure 3b shows that a continuous etch pit wall had been completely formed. The mechanical effect of the cavitation effect was concentrated at the etch pit, resulting in plastic deformation at the bottom of the slope between the bumps. The material was repeatedly pushed to the edge of the bump, resulting in a long waveform fold, and the entire surface presented a honeycomb. The average radius of the etch pit was 7–8 μm. The radius of the etch pit in Figure 3c,d continued to expand to more than 10 μm. The depth continued to deepen. Smaller etch pits merged to form larger and deeper etch pits. There were signs of reactants falling off at the grain boundary, exhibiting serious plastic deformation with astatic. The plastic deformation caused the honeycomb corrosion products to escape more easily from the radiation rod. The surface became rough and the morphology was still honeycomb. From Figure 3d, it was found that cracks began to form at the junction of the etch pits, with a trend of accelerating weight loss/cavitated area. This was caused by the impact wave formed by the cavitation effect and the continuous impact of the microjet. The continuous etch pits were affected by the cavitation effect and the continuous impact of the microjet, and fell off from the radiation rod, causing damage to the radiation rod. The evolution process of surface morphology was also different due to the influence of ultrasonic cavitation and different original surface roughness. The change process of etch pits on surfaces with small roughness lagged behind that on surfaces with large roughness.

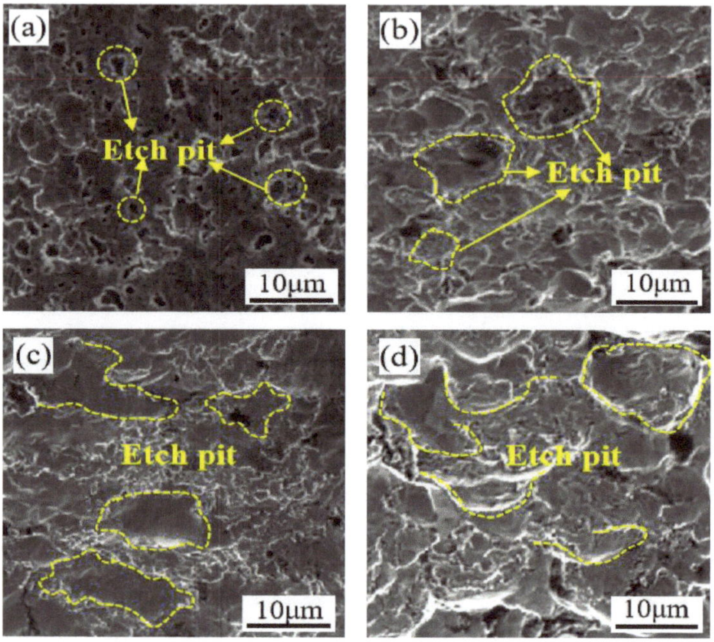

Figure 3. Surface morphologies of samples with vibration for 48 min: (**a**) A: 0.4 Ra, (**b**) B: 7.2 Ra, (**c**) C: 9.5 Ra, (**d**) D: 9.8 Ra.

3.2.2. Corrosion Product

Figure 4a shows the profile microstructure of reaction layer of sample A: 0.4 Ra after ultrasonic application for 48 min in aluminum melt at 700 °C. It can be seen from the figure that a reaction layer is formed at the Al/Ti interface. XRD and EDS were used to analyze the reaction layer of the sample profile. Figure 4b,c show the energy spectrum of point 1 and the XRD analysis of the interface, respectively. The results of EDS elemental content showed that the atomic ratios of Al and Ti at the spectral points were about 3:1, indicating that this substance was $TiAl_3$. This fund can provide an insight to develop a new homemade Ti alloy radiation rod with better resistance to corrosion in the ultrasonic melt processing.

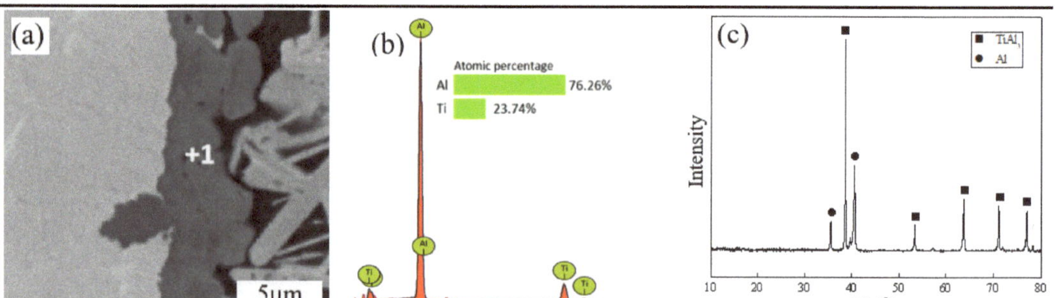

Figure 4. Sample A: 0.4 Ra: (**a**) Microstructure of the reaction layer in the profile; (**b**) energy spectrum, (**c**) XRD analysis.

3.3. Reaction Layer Morphology

Figure 5 shows the microscopic image of the Al/Ti interface of each group of samples after ultrasonic vibration for 4 min. From the beginning of 4 min, reactants began to appear at the Al/Ti interface of samples with different roughness surfaces. Very small reactants had begun to form at all surfaces, which distributed in a flocculent manner on the surface of the material. Sample A: 0.4 Ra with low roughness exhibited fewer reactants, as shown in Figure 5a. Sample D: 9.8 Ra with high roughness exhibited more reactants, as shown in Figure 5d.

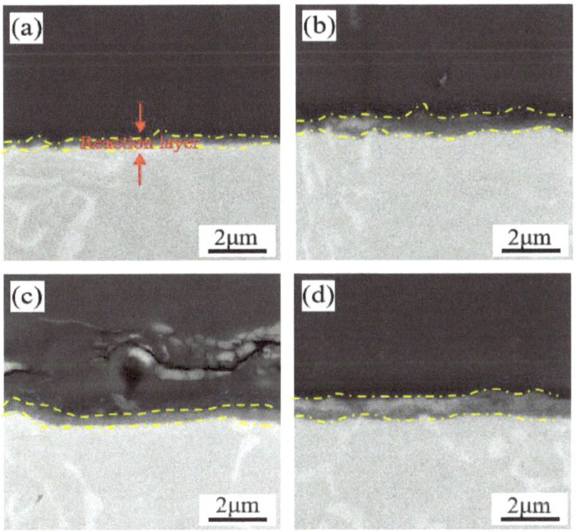

Figure 5. SEM images of the interface of samples with vibration for 4 min: (**a**) A: 0.4 Ra, (**b**) B: 7.2 Ra, (**c**) C: 9.5 Ra, (**d**) D: 9.8 Ra.

Figure 6 shows the SEM images of the interface of each sample after ultrasonic vibration for 10 min. A continuous reaction layer had been formed at the Al/Ti interface. Some granular reactants on the surface reaction layer of C: 9.5 Ra and D: 9.8 Ra titanium alloy materials with high roughness had been separated from the interface and ionized into the aluminum melt. This indicated that the cavitation effect generated by ultrasonication at this time had formed an impact on the surface and caused some weight loss/cavitated area in the interface reaction layer. EDS elemental analysis of the substance in Figure 6a,c revealed that it is composed of aluminum and titanium, with an atomic ratio of 3:1. It was determined that this substance was $TiAl_3$. The thickness of the reaction layer in Figure 6a is relatively thin, and the thickness of the reaction layer in Figure 6d is the last.

Figure 7 shows the SEM images of the interface of each sample after ultrasonic vibration for 48 min. The reaction layers of all samples thickened and exhibited an irregular arrangement. Regardless of the roughness of the sample, a large amount of free reactants appeared at the Al/Ti interface. The thickness of the $TiAl_3$ reaction layer increased to about 5 μm. There were many cracks in the reaction layers, and the cracks continued to deepen to the interior. Reactants began to form near the surface of the titanium alloy matrix. The growth of reactants might cause the adjacent matrix material to peel off as a whole, increasing the mass loss of the later material. The interface between Ti alloy matrix and reaction layer was no longer smooth, and the roughness increased obviously. This was because under the action of high-power ultrasonic, a large number of cavitation bubbles were generated in the melt. The fracture of cavitation bubble near the wall would produce vertical microjet impact on the solid surface, resulting in plastic deformation of the titanium alloy surface and increasing surface roughness. The microjets would also cause fatigue damage to the formed $TiAl_3$ reaction layer. Under the influence of mechanical impact, cracks of various sizes appeared in the reaction layer. These cracks peeled the $TiAl_3$ particles outside the reaction layers, which also verified the phenomenon that the radiation rod had a high weight loss rate/cavitated area under ultrasonic conditions.

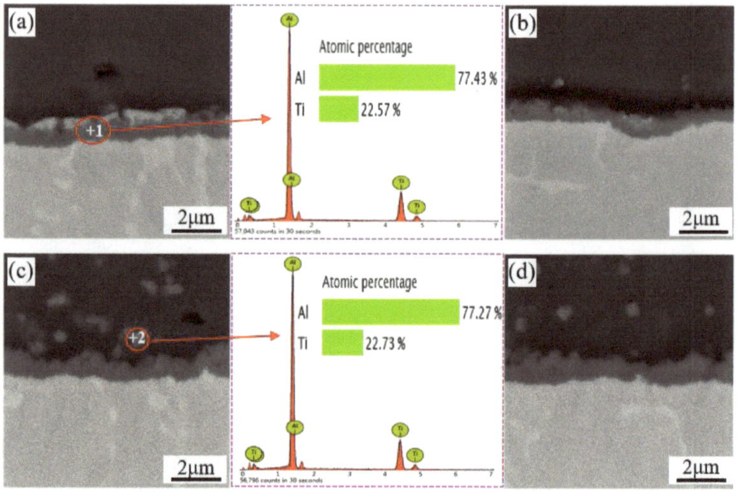

Figure 6. SEM images of the interface of samples with vibration for 10 min: (**a**) A: 0.4 Ra, (**b**) B: 7.2 Ra, (**c**) C: 9.5 Ra, (**d**) D: 9.8 Ra.

Figure 7. SEM images of the interface of samples with vibration for 48 min: (**a**) A: 0.4 Ra, (**b**) B: 7.2 Ra, (**c**) C: 9.5 Ra, (**d**) D: 9.8 Ra.

Although the surface roughness of the four groups of samples was different, the formation of Al/Ti interface reactants and the growth of the reaction layer were similar. A continuous reaction layer appeared at the Ti/Al solid–liquid interface at 10 min, and the chemical composition of the layer was determined to be $TiAl_3$. The cavitation effect occurred when the ultrasonic wave propagated in aluminum melt. The cavitation effect near the wall caused instantaneous high temperature and high pressure at the liquid–solid interface of Al/Ti. The high temperature and high pressure caused by cavitation bubble collapse increased the interdiffusion rate of Al and Ti atoms. They also increased the plastic deformation of the interface, which increased the solid–liquid contact area. This was helpful for the full diffusion of atoms. To a certain extent, the reaction rate of Al/Ti atoms was accelerated.

3.4. *Profile Microhardness*

Figure 8 shows the microhardness of four different surface roughness profiles of the radiation rod materials subjected to cavitation corrosion. In this experiment, four titanium alloy samples with different roughnesses in 700 °C aluminum melt were subjected to cavitation corrosion for 1 h, 3 h, and 5 h, respectively. Then, the microhardness test was carried out from the surface layer of the matrix every 20 μm vertically to the inside, and the test length was 200 μm.

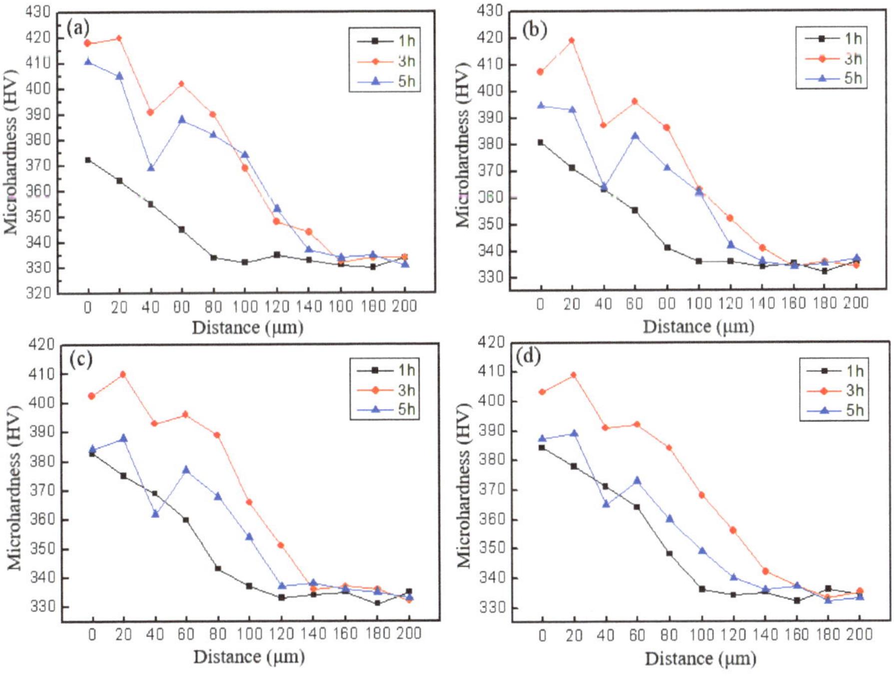

Figure 8. Profile microhardness curve of samples and application time of ultrasonic: (**a**) A: 0.4 Ra, (**b**) B: 7.2 Ra, (**c**) C: 9.5 Ra, (**d**) D: 9.8 Ra.

As an important measure of surface mechanical properties, the surface microhardness of a material can reflect the ability of the material to resist external impact. Therefore, surface microhardness is also one of the important parameters of cavitation resistance in the study of the cavitation corrosion behavior of materials. In the process of cavitation corrosion, the ability of titanium alloy material to absorb cavitation damage depends on the degree of deformation of the material surface, the depth of corrosion and the mode of destruction. The microhardness change of the material before and after cavitation corrosion is an index to measure the material's ability to absorb cavitation damage.

When the radiation rod material was damaged by cavitation corrosion for 1 h, work hardening layers of different depths were formed near the surface of each sample substrate. As cavitation corrosion progressed, the thickness of the work hardening layer increased. The microhardness variation of samples with different roughness was similar. During the incubation period of cavitation corrosion, the surface of the material began to show work hardening due to the effect of microjet shock wave. Until the end of the incubation period of cavitation corrosion, the microhardness of the near surface layer continued to rise and reached the maximum value. At this time, the cavitation energy absorbed by the plastic deformation of the material reached saturation. Then, the microhardness gradually decreased, which caused the surface material to detach from the matrix and undergo processing softening. At the same time, as the newly exposed surface was subjected to cavitation impact, the larger hardened layer of the material moved towards the interior of the matrix. The rate of surface material detachment was lower than the rate of high hardened layer hardening towards the interior. This was the reason why the work hardening layer grew over time and there was a certain thickness of work softening zone near the surface.

Table 4 shows the thickness of the work hardening layer of titanium alloy samples with different roughnesses after different cavitation times. A hardened layer of 80 μm was formed on the surface of A: 0.4 Ra at 1 h after the experiment, and the thickness increased to 160 μm at 3 h. The thickness of the hardened layer decreased to 140 μm after 5 h. However, the thicknesses of the hardened layers of the other samples with higher roughness at 1 h were all higher. This was because at the beginning of the experiment, the surface of the samples with higher roughness had higher strength of cavitation corrosion and greater plastic deformation generated by absorbing energy of cavitation corrosion. This resulted in the higher work hardening degree of the surface layer. The thickness of the hardened layer of the C: 9.5 Ra at 3 h was lower than other samples with lower roughness, indicating that the cavitation corrosion acceleration period began at this time. The surface material of the substrate gradually fell off and dispersed in the aluminum melt. Therefore, the actual thickness of the work hardening layer was lower than that of the samples with lower roughness. The thickness of the hardened layer of the four samples at 5 h was reduced by about 20 μm. This indicated that the difference between the stripping rate of the surface material caused by cavitation corrosion and the work hardening rate of the surface layer caused by cavitation corrosion was roughly equal.

Table 4. Thickness of work hardening layer (μm).

Roughness Time (t/h)	1	3	5
0.4 Ra	80	160	140
7.2 Ra	90	150	130
9.5 Ra	100	140	120
9.8 Ra	100	150	130

The formation of cavitation pits was the result of microjets and shock waves generated after the collapse of cavitation bubbles on the surface of materials. The microscopic and submicroscopic cracks on the surface of materials were the places where gas nuclei existed [39]. The gas nuclei at the crack gradually grew to collapse under the negative pressure environment formed by ultrasonic vibration, and the high-speed jet generated impacts the surface of the material. Figure 9 shows the formation, growth and collapse of cavitation bubbles in cylindrical grooves on the surface of the material. c_R and c_L represent the radial velocity of the impact point and the wave velocity of the shock wave, respectively. When $c_R > c_L$, the wave front kept in contact with the material surface, and the high-pressure region continued to exert high pressure on the solid surface. When $c_R = c_L$, the wave front gradually diverged from the surface of the material. The aluminum melt generated a radial flow along the surface of the material, with velocities pointing toward the center of the unloading wave. When the surface roughness was small, the lateral flow of aluminum melt was less hindered and the high-pressure zone between the solid and liquid interface could be completely unloaded. However, for the surface with larger roughness, the jet might directly act in the groove of a certain depth, and the lateral flow would be hindered by the two side walls. Therefore, the unloading process was difficult to complete.

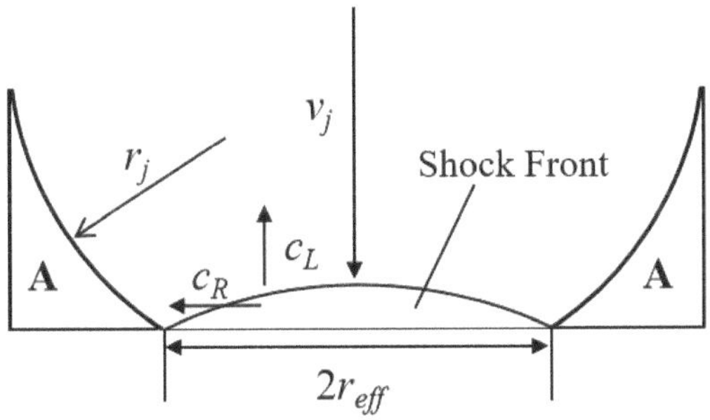

Figure 9. Schematic diagram of microjet impact grooves.

Ignoring the compression of aluminum melt in region A, Equation (4) can be obtained by using Equation (2) [40] and Equation (3) [41], which is the geometric relationship between v_j and c_L when the high-pressure region is completely unloaded.

$$r_{eff} = r_j \frac{v_j}{c_L} \tag{2}$$

$$t_r = \frac{3r_j}{2} \frac{v_j}{c_L^2} \tag{3}$$

$$v_j = 0.1673 - 0.2100c_L \tag{4}$$

where r_{eff}, t_r, r_j, v_j, and c_L are defined as the radius of the high-pressure area, the time from the start of unloading to the exact completion of the high-pressure area, radius, impact velocity, and velocity of shock wave in aluminum melt, respectively. It is assumed that the diameter and height of the cylindrical groove contour on the solid surface are equal to the diameter of the head of microjet. The bottom of the groove is an ideal plane.

If the value of c_L is 1500 m/s, then v_j is equal to 251–315 m/s. According to Equations (2) and (3), it can be seen that the high-pressure unloading completion time t_r after impact of a microjet with a jet diameter of 10 μm is 0.63–0.99 ms. The lateral flow of the microjet compresses the melt between the solid surface and the edge wall, thus prolongating the unloading time. According to the above analysis, the surface profile with reduced roughness has little obstruction and so the lateral flow of melt is not affected. The duration of high pressure is short, and the degree of damage to the substrate surface is correspondingly small. The surface with high roughness is hindered by the lateral flow of microjets, and unloading in the high-pressure area is difficult. At the same time, the surface with high roughness contains more micro-cracks, and surface tearing and convex shear will occur under the impact of high lateral flow.

3.5. Simulation of the Collapse Impact of Cavitation Bubbles near the Wall

3.5.1. Boundary

The object of study in this paper was the process of the impact of cavitation bubbles on rigid surfaces with different roughness. This model was based on the existence of spherical cavitation bubbles near solid walls [42]. The two-phase flow module in the VOF method was used to solve the problem by using Fluent software. The first phase was a 2A14 aluminum melt at 700 °C, and the second phase was a gaseous cavitation bubble.

Only half of the cavitation bubble and fluid motion process needed to be modeled due to the axisymmetric characteristics of cavitation collapse. Figure 10 shows the calculation region with an area of 1 mm × 1 mm. In this region, there was a semi-circular cavity on the Y-axis. Its radius was R. The solid wall was on the X-axis, and the minimum equilibrium position between the cavity wall and the solid wall was H. The effect of pressure pulse caused by cavitation collapse on the solid was different depending on the position of the cavity from the solid wall. In order to facilitate the following expression, the radius of cavitation bubbles R was set to 50 μm, and the ratio coefficient γ was set to:

$$\gamma = \frac{H}{R} \tag{5}$$

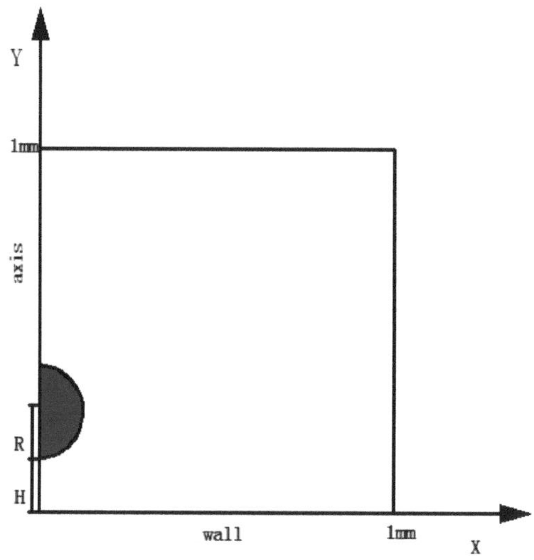

Figure 10. Simulation area.

The actual morphologies of solid surfaces with different roughnesses were modeled for simulation calculation. The maximum height difference of the 0.4 Ra, 7.2 Ra, and 9.7 Ra surface profiles was about 2 μm, 25 μm, and 40 μm, respectively. Combined with the calculation method of Ra, the absolute average value of the distance between the contour points and the equilibrium position was taken on the sampling length l:

$$\mathrm{Ra} = \frac{1}{l}\int_0^l |y(x)|dx \tag{6}$$

The approximate calculation is:

$$\mathrm{Ra} = \frac{1}{n}\sum_{y=1}^{n} |y_i| \tag{7}$$

where n and l are defined as the number of phases and sampling length, respectively. For the convenience of calculation, the solid surface was fitted to different scales of serrated morphology, and the roughness values used in actual experiments were taken as 0.4, 7, and 10, respectively. The grid of the established plane model was divided, and the minimum mesh size was 2.079×10^{-4} mm. The model required assumptions [43]: 1. ignoring the role of inertial force; 2. ignoring the surface tension of the cavity wall; and 3. assuming that the amount of gas in the cavity remained constant. The inertial force and surface tension have little influence on the whole calculation result, and the calculation amount is large, so they are ignored. In order to ensure the accuracy of simulation results, the amount of gas in the cavity must be kept constant. The left boundary of the model was set to conform to the symmetric boundary, and the following equation was established:

$$u_x = 0, \frac{\partial u_y}{\partial n} = 0 \tag{8}$$

where u_x and u_y are defined as the partial velocity in the x and y directions, respectively. The lower boundary was set as the rigid wall surface, and the upper and right boundaries were set as the infinity boundary, which satisfies

$$\frac{\partial u_y}{\partial n} = 0, \frac{\partial v_y}{\partial n} = 0 \tag{9}$$

where n is the normal direction. The gas pressure in the cavity is 2340 Pa and the fluid pressure is 5 MPa.

3.5.2. Parameter

The gas–liquid phase in the fluid satisfies the equation:

$$\frac{\partial \rho}{\partial t} + \nabla \cdot (\rho u) = 0 \tag{10}$$

The N-S equation is:

$$\frac{\partial u}{\partial t} + (u \cdot \nabla) u = -\frac{1}{\rho} \nabla p + \gamma \nabla^2 u \tag{11}$$

The VOF method was used to track the gas–liquid interface. In order to represent the proportion of liquid in the grid, the volume fraction function F_i was introduced. This function satisfied:

$$\frac{\partial}{\partial t}(F_i \rho_i) + \nabla \cdot (F_i \rho_i u_i) = 0 \tag{12}$$

$$\sum_{i=}^{2} F_i = 1 \tag{13}$$

where u, μ, ρ, and i are the velocity tensor of the fluid, the kinematic viscosity of the fluid, the density of the fluid, and the different phases in the fluid, respectively.

The density, surface tension, dynamic viscosity, and saturated vapor pressure of aluminum melt were 2450 kg/cm^3, 0.86 N/m, 1.38 $\times$ 10^{-3} m^2/s, and 2.45 kPa. The transient solution was performed by Fluent. Time and space were used in implicit format and upwind format, respectively, and pressure was calculated using the PISO algorithm. The time step was 10^{-8}, and the variable step was solved.

3.5.3. Simulation Result

Figures 11–13 show the pressure distribution diagram and the maximum pressure value of the solid interface when the ratio coefficients are equal to 0.5, 1, and 1.5, and the roughness values of the solid interface are 0.4 Ra, 7 Ra, and 10 Ra, respectively.

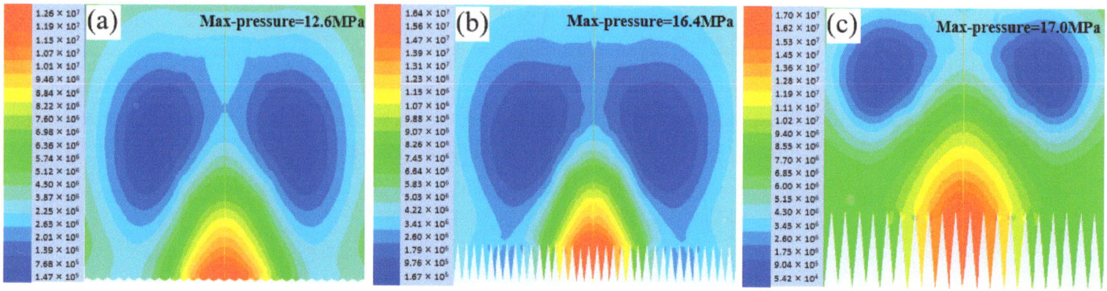

Figure 11. Cloud picture of pressure distribution at $\gamma = 0.5$: (**a**) 0.4 Ra, (**b**) 7 Ra, (**c**) 10 Ra.

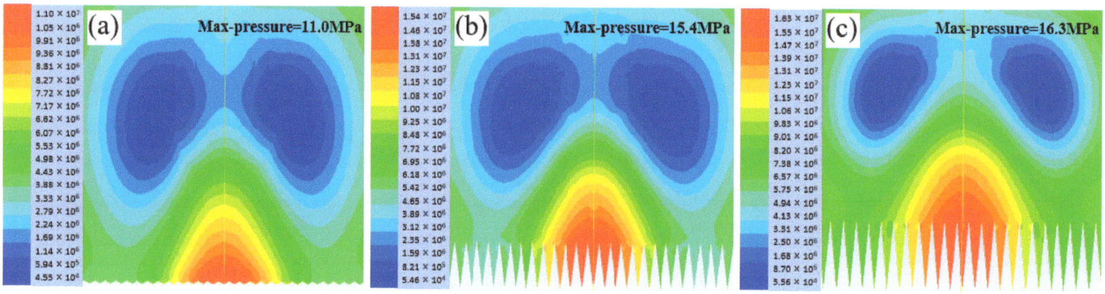

Figure 12. Cloud picture of pressure distribution at $\gamma = 1$: (**a**) 0.4 Ra, (**b**) 7 Ra, (**c**) 10 Ra.

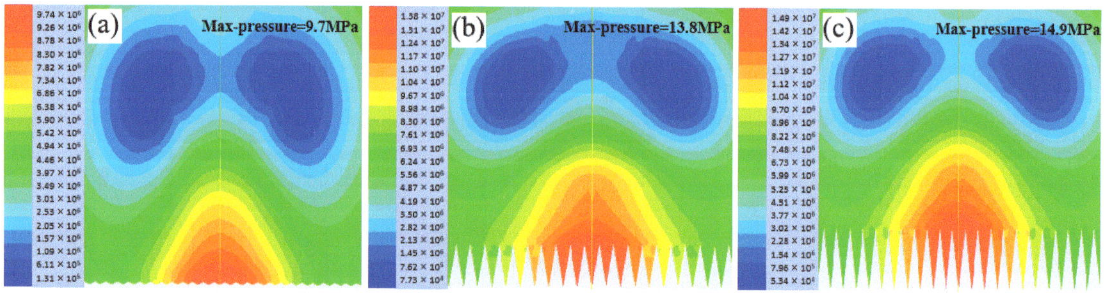

Figure 13. Cloud picture of pressure distribution at $\gamma = 1.5$: (**a**) 0.4 Ra, (**b**) 7 Ra, (**c**) 10 Ra.

When the ratio coefficient was the same, the greater the roughness of the solid interface, the greater the maximum pressure value at that point. When the roughness was constant, the smaller the ratio coefficient was, and the closer the cavity was to the solid interface, the higher the maximum pressure. The number of small profile peaks on solid surface affected by the high-pressure region was slightly different. There were five small profile peaks affecting 0.4 Ra, and about 7 Ra small profile peaks affecting 7 Ra and 10 Ra. Although these pressure values were far less than the yield strength of titanium alloys, a large and repeated impact would certainly cause damage to the solid surface. Therefore, the greater the surface roughness, the greater the maximum pressure value on the surface, and the wider the influence range of the high-pressure region. The surface with higher roughness had sharper contour peak, which was easier to form stress concentration and greater corrosion degree.

4. Conclusions

The ultrasonic corrosion behavior of titanium alloy radiation rod with different roughness was studied in aluminum melt at 700 °C. Macro- and micro morphology observation, weight loss/cavitated area detection, phase analysis, microhardness measurement, cavitation dynamics equation derivation, and single cavitation bubble collapse simulation were carried out on the experimental samples at each stage under ultrasonic vibration at different time. Our main conclusions are as follows:

(1) The corrosion rule of titanium alloy in aluminum melt was that the greater the roughness, the greater the weight loss rate/cavitated area. The weight loss/cavitated area of titanium alloy caused by cavitation effect accounted for 6.4%~8.6% of the total weight loss/cavitated area. The corrosion product was the intermetallic compound $TiAl_3$. These reactants appeared at the Al/Ti solid–liquid interface of samples with different roughness in about 4 min, and the reaction layer was formed in 10 min.
(2) With the increase of cavitation erosion time, the thickness of the work hardening layer on the surface of the material increased first and then decreased. The maximum thickness could reach 160 μm. The evolution of the hardened layer depended on the stripping rate of the surface material caused by cavitation corrosion and the work hardening rate of the surface layer.
(3) The greater the roughness of the solid interface, the greater the maximum pressure at that point. When the roughness was constant, the smaller the ratio coefficient was and the closer the cavity was to the solid interface, the maximum pressure increased. The greater the surface roughness, the wider the influence range of the high pressure zone, and the greater the degree of corrosion.

Additionally, research has thus far concentrated on the properties of titanium alloy radiation rod with different roughness in 2A14 aluminum melt for ultrasonic casting. The influence of different roughness is only one factor in the ultrasonic casting process. We can also study the effect of different power levels of ultrasound on the corrosion performance of radiation rods. This will help us understand the corrosion mechanism of titanium alloy radiation rods throughout the entire ultrasonic casting process and also provide a deeper insight to develop a new homemade Ti alloy radiation rod with better resistance to corrosion in the ultrasonic casting.

Author Contributions: Conceptualization, Y.Y.; Data curation, Y.Z.; Writing—Original draft, Y.Y.; Writing—review and editing, X.L.; Funding acquisition, H.Z. All authors have read and agreed to the published version of the manuscript.

Funding: This research was funded by the Key Scientific Research and Technological Projects in Henan Province (NO.232102230010), Key Scientific Research and Technological Projects in Henan Province (NO.222102230097), Doctoral Foundation for Henan University of Engineering (NO. D2021008), and Key Research Project Plan for Higher Education Institutions in Henan Province (NO.22B430010).

Institutional Review Board Statement: Not applicable.

Informed Consent Statement: Not applicable.

Data Availability Statement: The data presented in this study are available on request from the corresponding author.

Acknowledgments: The authors are grateful to T. Zhang and W.W. Xu of the Analysis and Testing Center of HAUE for their kind help with XRD and SEM.

Conflicts of Interest: The authors declare no conflict of interest.

References

1. Lin, R.; Liu, B.; Zhang, J.J.; Zhang, S.G. Microstructure evolution and properties of 7075 aluminum alloy recycled from scrap aircraft aluminum alloys. *J. Mater. Res. Technol.* **2022**, *19*, 354–367. [CrossRef]
2. Wan, Z.D.; Wang, Q.; Zhao, Y.; Zhao, T.Y.; Shan, J.G.; Meng, D.Y.; Song, J.L.; Wu, A.P.; Wang, G.Q. Improvement in tensile properties of 2219-T8 aluminum alloy TIG welding joint by PMZ local properties and stress distribution. *Mater. Sci. Eng. A* **2022**, *839*, 142863. [CrossRef]
3. Zhang, J.F.; Zhou, D.S.; Pang, X.Y.; Zhang, B.W.; Li, Y.; Sun, B.H.; Valiev, R.S.; Zhang, D.L. Deformation-induced concurrent formation of 9R phase and twins in a nanograined aluminum alloy. *Acta. Mater.* **2023**, *244*, 118540. [CrossRef]

4. Safyri, M.; Moshtaghi, M.; Hojo, T.; Akiyama, E. Mechanisms of hydrogen embrittlement in high-strength aluminum alloys containing coherent or incoherent dispersoids. *Corros. Sci.* **2022**, *194*, 109895. [CrossRef]
5. Miao, J.S.; Sutton, S.; Luo, A.A. Deformation microstructure and thermomechanical processing maps of homogenized AA2070 aluminum alloy. *Mater. Sci. Eng. A* **2022**, *834*, 142619. [CrossRef]
6. Guo, X.P.; Li, H.J.; Xue, P. Microstructure and mechanical properties of 600 MPa grade ultra-high strength aluminum alloy fabricated by wire-arc additive manufacturing. *J. Mater. Sci. Technol.* **2023**, *149*, 56–66. [CrossRef]
7. Balasubramani, N.; Venezuela, J.; Yang, N.; Wang, G.; Stjohn, D.; Dargusch, M. An overview and critical assessment of the mechanisms of microstructural refinement during ultrasonic solidification of metals. *Ultrason. Sonochem.* **2022**, *89*, 106151. [CrossRef]
8. Eskin, D.G.; Tzanakis, I.; Wang, F.; Lebon, G.S.B.; Subroto, T.; Pericleous, K.; Mi, J. Fundamental studies of ultrasonic melt processing. *Ultrason. Sonochem.* **2019**, *52*, 455–467. [CrossRef]
9. Priyadarshi, A.; Khavari, M.; Shahrani, B.; Subroto, T.; Yusuf, L.A.; Conte, M.; Pretice, P.; Pericleous, K.; Eskin, D.; Tzanakis, L. In-situ observations and acoustic measurements upon fragmentation of free-floating intermetallics under ultrasonic cavitation in water. *Ultrason. Sonochem.* **2021**, *80*, 105820. [CrossRef]
10. Xu, N.X.; Yu, Y.; Zhai, W.; Wang, J.Y.; Wei, B.B. A high-temperature acoustic field measurement and analysis system for determining cavitation intensity in ultrasonically solidified metallic alloys. *Ultrason. Sonochem.* **2023**, *94*, 106343. [CrossRef]
11. Lu, D.H.; Jiang, Y.H.; Guan, G.S.; Zhou, R.F.; Li, Z.H.; Zhou, R. Refinement of primary Si in hypereutectic Al-Si alloy by electromagnetic stirring. *J. Mater. Process. Tech.* **2007**, *189*, 13–18. [CrossRef]
12. Cho, W.G.; Kang, C.G. Mechanical properties and their microstructure evaluation in the thixoforming process of semi-solid aluminum alloys. *J. Mater. Process. Tech.* **2000**, *105*, 269–277. [CrossRef]
13. Wang, T.; Huang, Y.F.; Ma, Y.Z.; Wu, L.; Yan, H.Y.; Liu, C.; Liu, Y.; Liu, B.; Liu, W.S. Microstructure and mechanical properties of powder metallurgy 2024 aluminum alloy during cold rolling. *J. Mater. Res. Technol.* **2021**, *15*, 3337–3348. [CrossRef]
14. Liu, T.; Wang, Q.D.; Sui, Y.D.; Wang, Q.G.; Ding, W.J. An investigation into aluminum-aluminum bimetal fabrication by squeeze casting. *Mater. Design* **2015**, *68*, 8–17. [CrossRef]
15. Martin, J.H.; Yahata, B.; Hundly, J.M.; Mayer, J.A.; Schaedler, T.A.; Pollock, T.M. 3D printing of high-strength aluminium alloys. *Nature* **2017**, *549*, 365–369. [CrossRef]
16. Yuan, D.; Shao, S.; Guo, C.; Jiang, F.C.; Wang, J.D. Grain refining of Ti-6Al-4V alloy fabricated by laser and wire additive manufacturing assisted with ultrasonic vibration. *Ultrason. Sonochem.* **2021**, *73*, 105472. [CrossRef]
17. Jiang, R.P.; Zhao, W.H.; Zhang, L.; Li, X.Q.; Guan, S.K. Microstructure and corrosion resistance of commercial purity aluminum sheet manufactured by continuous casting direct rolling after ultrasonic melt pre-treatment. *J. Mater. Res. Technol.* **2023**, *22*, 1522–1532. [CrossRef]
18. Jang, H.S.; Lee, G.H.; Jeon, J.B.; Choi, Y.S.; Shin, S.M. Effect of ultrasonic melt treatment conditions on melt quality of Al–Mg alloy. *J. Mater. Res. Technol.* **2022**, *19*, 2645–2656. [CrossRef]
19. Xiong, Z.F.; Jiang, Y.; Yang, M.; Zhang, Y.; Lei, L. Achieving superior strength and ductility in 7075 aluminum alloy through the design of multi-gradient nanostructure by ultrasonic surface rolling and aging. *J. Alloys Compd.* **2022**, *918*, 165669. [CrossRef]
20. Sui, D.S.; Han, Q.Y. Ultrasound-assisted cast-on method: Obtaining high-quality metallurgical bonds between a bare steel insert and A354 aluminum alloy within a composite casting. *J. Mater. Process. Tech.* **2023**, *311*, 117783. [CrossRef]
21. Eskin, G.I. Principles of Ultrasonic Treatment: Application for Light Alloys Melts. *Adv. Perform. Mater.* **1997**, *4*, 223–232. [CrossRef]
22. Li, J.Y.; Pan, Y.; Yan, Z.X.; Lu, S.L.; Zhao, D.J.; Guo, W.; Wu, S. Effects of Li content on microstructure evolution and mechanical properties of squeeze-cast Al-5Cu-xLi alloy assisted with ultrasonic treatment. *J. Alloys Compd.* **2023**, *961*, 171083. [CrossRef]
23. Guang, F.; Jiang, W.M.; Wang, J.L.; Li, G.Y.; Zhang, Z.; Fan, Z.T. Development of high strength Mg/Al bimetal by a novel ultrasonic vibration aided compound casting process. *J. Mater. Process. Tech.* **2022**, *300*, 117441. [CrossRef]
24. Moussa, M.E.; Waly, M.A.; Amin, M. Effect of high intensity ultrasonic treatment on microstructural modification and hardness of a nickel-aluminum bronze alloy. *J. Alloys Compd.* **2018**, *741*, 804–813. [CrossRef]
25. Dong, F.; Li, X.Q.; Zhang, L.H. Cavitation erosion mechanism of titanium alloy radiation rods in aluminum melt. *Ultrason. Sonochem.* **2016**, *31*, 150–156. [CrossRef]
26. Jiao, L.; Zhao, Y.T.; Li, H.; Shang, H.; Zhang, Z.Z.; Xia, T.F. A High-Temperature Aluminum Liquid Corrosion Resistant Ultrasonic Amplitude Transformer and a Preparation Method Thereof. China Patent CN201210467584.4, 19 November 2012.
27. Maula, M.I.; Winarni, T.I.; Tauviqirrahman, M.; Maula, M.I.; Winarni, T.I.; Tauviqirrahman, M.; Akbar, I.; Basri, H.; Heide, E.V.D.; Jamari, J. Tresca Stress Simulation of Metal-on-Metal Total Hip Arthroplasty during Normal Walking Activity. *Materials* **2021**, *14*, 7554.
28. Thakur, A.; Kaya, S.; Kumar, A. Recent Trends in the Characterization and Application Progress of Nano-Modified Coatings in Corrosion Mitigation of Metals and Alloys. *Appl. Sci.* **2023**, *13*, 730. [CrossRef]
29. Thakur, A.; Sharma, S.; Ganjoo, R.; Assad, H.; Kumar, A. Anti-Corrosive Potential of the Sustainable Corrosion Inhibitors Based on Biomass Waste: A Review on Preceding and Perspective Research. *J. Phys. Conf. Ser.* **2022**, *2267*, 012079. [CrossRef]
30. Tian, Y.; Liu, Z.L.; Li, X.Q.; Zhang, L.H.; Li, R.Q.; Jiang, R.P.; Dong, F. The cavitation erosion of ultrasonic sonotrode during large-scale metallic casting: Experiment and simulation. *Ultrason. Sonochem.* **2018**, *43*, 29–37. [CrossRef]
31. Hong, S.; Wu, Y.P.; Zhang, J.F.; Zheng, Y.G.; Zheng, Y.; Lin, J.R. Synergistic effect of ultrasonic cavitation erosion and corrosion of WC–CoCr and FeCrSiBMn coatings prepared by HVOF spraying. *Ultrason. Sonochem.* **2016**, *31*, 563–569. [CrossRef]

32. Man, H.C.; Cui, Z.D.; Yue, T.M.; Cheng, F.T. Cavitation erosion behavior of laser gas nitrided Ti and Ti6Al4V alloy. *Mater. Sci. Eng. A* **2003**, *355*, 167–173. [CrossRef]
33. Li, D.; Kang, Y.; Wang, X.C.; Ding, X.L.; Fang, Z.L. Effects of nozzle inner surface roughness on the cavitation erosion characteristics of high speed submerged jets. *Exp. Therm. Fluid Sci.* **2016**, *74*, 444–452. [CrossRef]
34. Pitt, F.; Ramulu, M. Influence of grain size and microstructure on oxidation rates in titanium alloy Ti-6Al-4V under superplastic forming conditions. *J. Mater. Eng. Perform.* **2004**, *13*, 727–734. [CrossRef]
35. Shi, C.; Wu, Y.J.; Mao, D.H.; Fan, G.F. Effect of Ultrasonic Bending Vibration Introduced by the L-shaped Ultrasonic Rod on Solidification Structure and Segregation of Large 2A14 Ingots. *Materials* **2020**, *13*, 807. [CrossRef]
36. Li, R.Q.; Liu, Z.L.; Dong, F.; Li, X.Q.; Chen, P.H. Grain refinement of a large-scale Al alloy casting by introducing the multiple ultrasonic generators during solidification. *Metall. Mater. Trans. A* **2016**, *47*, 3790–3796. [CrossRef]
37. Li, R.Q.; Liu, Z.L.; Chen, P.H.; Zhong, Z.T.; Li, X.Q. Investigation on the Manufacture of a Large-Scale Aluminum Alloy Ingot: Microstructure and Macrosegregation. *Adv. Eng. Mater.* **2017**, *19*, 1600375. [CrossRef]
38. Zhou, Y.K.; Hammitt, F.G. Vibratory cavitation erosion in aqueous solutions. *Wear* **1983**, *87*, 163–171. [CrossRef]
39. Chahine, G.L.; Kapahi, A.; Choi, J.K. Modeling of surface cleaning by cavitation bubble dynamics and collapse. *Ultrason. Sonochem.* **2016**, *29*, 528–549. [CrossRef]
40. Kattner, U.R.; Lin, J.C.; Chang, Y.A. Thermodynamic Assessment and Calculation of the Ti-Al System. *Metall. Mater. Trans. A* **1992**, *23*, 2081–2090. [CrossRef]
41. Kennedy, C.F.; Field, J.E. Damage threshold velocities for liquid impact. *J. Mater. Sci.* **2000**, *35*, 5331–5339. [CrossRef]
42. Zhang, S.G.; Duncan, J.H.; Chahine, G.L. The final stage of the collapse of a cavitation bubble near a rigid wall. *J. Fluid. Mech.* **1993**, *257*, 147–181. [CrossRef]
43. Tomia, Y.; Robinson, P.B.; Tong, R.P.; Blake, J.R. Growth and collapse of cavitation bubbles near a curved rigid boundary. *J. Fluid. Mech.* **2002**, *466*, 259–283. [CrossRef]

coatings

MDPI

Article

Effect of Local Remelting and Recycled WC-Co Composite Reinforcement Size on Abrasive and Erosive Wear of Manual Arc Welded Hardfacings

Egidijus Katinas [1,*], Maksim Antonov [2], Vytenis Jankauskas [3] and Dmitri Goljandin [2]

1 Department of Electrical Engineering and Automation, Czech University of Life Sciences Prague, 165 00 Praha-Suchdol, Czech Republic
2 Department of Mechanical and Industrial Engineering, Tallinn University of Technology, 12616 Tallinn, Estonia
3 Department of Mechanical, Energy and Biotechnology Engineering, Vytautas Magnus University, 44248 Kaunas, Lithuania
* Correspondence: katinas@tf.czu.cz

Abstract: The hardfacings reinforced by recycled (cermet inserts of machining tools) WC-Co composite particles applied by manual arc welding can be used for protection against abrasive and erosive wear. Two categories of coatings were analyzed: with and without thermal treatment (local remelting). The hardfacings were tested under abrasive (low and medium stress) and erosive (high stress) conditions. In elevated temperature erosive conditions, coatings without heat treatment have up to two times higher wear resistance than reference material. Under low-stress conditions, coatings with the finest reinforcing particle size (180–355 μm) treated with local remelting demonstrated more than 10 times higher wear resistance than reference material. The optimal reinforcement size selection depends on the application conditions (low, average, or high stress). In most test conditions, the local remelting did not provide additional improvement. A scanning electron microscope was used for the characterization of wear mechanisms. The relation between mechanical properties and wear rates was found only for some conditions.

Keywords: WC particles; recycling; local remelting; abrasive wear; manual arc welding

Citation: Katinas, E.; Antonov, M.; Jankauskas, V.; Goljandin, D. Effect of Local Remelting and Recycled WC-Co Composite Reinforcement Size on Abrasive and Erosive Wear of Manual Arc Welded Hardfacings. *Coatings* **2023**, *13*, 734. https://doi.org/10.3390/coatings13040734

Academic Editors: Pinghu Chen, Yun Zhang and Ruiqing Li

Received: 24 February 2023
Revised: 26 March 2023
Accepted: 30 March 2023
Published: 4 April 2023

1. Introduction

The metal processing industry uses high amounts of tungsten products to process workpieces. The lifetime of metal cutting and machining tools' cermet inserts can be extremely short (sometimes just 15 min). The material loss during that time is insignificant (just a few percent), and the remaining part of the insert is discarded. Tungsten and its products (cermet inserts), according to authors [1–3], can be recycled. The inserts can be crushed into powder that can be used to reinforce thick (up to 10 mm) coatings to increase their wear resistance [4]. Such coatings are usually produced by well-controlled automated plasma and laser technologies. Their properties have been investigated [5–7], and it is possible to remelt a thermally sprayed coating to improve its inherent defects (pores, unmelted particles, poor mechanical bonding) [8]. Recycled insert powders can be used for sample preparation for the sintering process [9,10] to analyze mechanical or microstructural properties. However, information about manual arc welded hardfacings is scarce. Agriculture often uses manual arc welding to improve or repair tillage tools [11,12]. The conditions of the weld pool are less controlled, and significant or complete melting of reinforcing particles, defects (pores, slag inclusions, oxides, a lack of or low adherence between layers, etc.), as well as the generation of thermal stresses, especially in the case of thick multilayered coatings, take place [13]. According to previous investigations, the size of 135 μm WC (unused; not recycled; not composite) reinforcing particles provides a higher resistance to the thermal cracking that appears during the welding process, but the removal of coarser WC particles requires more time (and leads to higher wear resistance) [14].

The additional thermal (heat) treatment could be applied as an annealing process [15] or preheating of the substrate material before welding [16,17], remelting the applied surface by laser [8,18], or by the tungsten inert gas (TIG) method [19]. Preheating the substrate material before welding reduces thermal stress between the hardfacing and substrate material [13]. Surface remelting can also improve dilution of reinforced particles or metallurgical bonding between layer and substrate, as well as the homogeneity of the formed hardfacing [20]. However, in some cases, gas pockets can lead to microporosities near the substrate and a completely melted layer after remelting by the TIG [19]. The recycled reinforcement particles may improve the wear resistance of manual arc welded hardfacings, but welding quality must be monitored [14]. The welding defects can be removed with TIG's previously mentioned thermal treatment method. The remelting carried out by the TIG increases the dissolution and reprecipitation of reinforcing particles.

Recycling metal cutting and machining tools' cermet inserts would prevent wasting critical raw material. Manual arc welding electrodes with flux coverings incorporating crushed cermet particles can create hardfacings resistant to abrasive or erosive wear conditions. The current work aims to study the effect of remelting on the performance of multilayered manual arc welded hardfacings with various sizes of recycled reinforcement particles under low-, medium- (abrasive), and high-stress (erosive) wear testing conditions.

2. Materials and Methods

2.1. Crushing of the Inserts

The collected mixture of various machining and turning tool inserts was mainly composed of WC-Co materials, but some minor presence (less than 10%) of TiC, VC, NbC, and MoC was possible.

The inserts were crushed in two steps by disintegrators DSL-350 and DSL-175 (rough and fine crushing). Powders were sieved by Fritsch Analysette 3 into six fractions: 180–355, 355–500, 500–710, 710–1000, 1000–1400, and 1400–2000 μm (Figure 1). The powder with a size less than 180 μm was not used in this research because of the high content of adhered steel resulting from contact with the working elements of the disintegrator. It is essential to point out that the particles of such a powder are made of composite material (mainly WC grains cemented by cobalt). This might reduce the dilution of WC grains during welding and lead to higher wear rates [21].

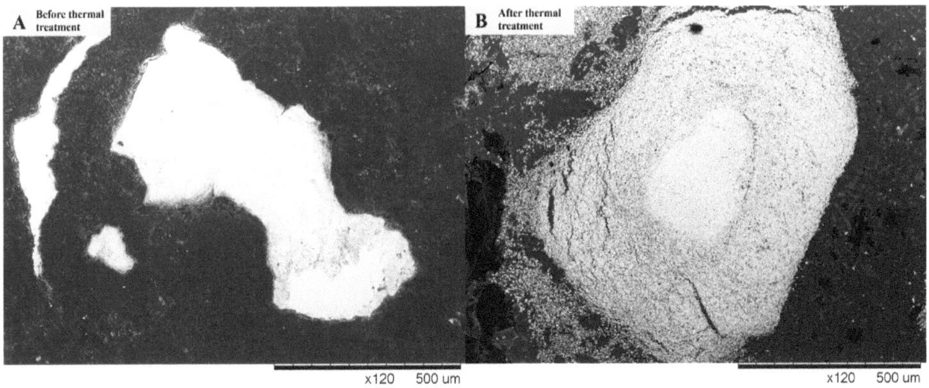

Figure 1. Crushed WC-Co particles (710–1000 μm) before (**A**) and after (**B**) thermal treatment at 1280 °C temperature.

2.2. Preparation of Manual arc Welding Electrodes

The preparation of covered electrodes and application of the manually welded arc layers was performed in cooperation with JSC "Anykščių Varis", the research center of the producer of electrodes on an industrial scale. The welding electrode for the composite coating application was composed of the central core wire covered by the flux, which had inclusions of recycled inserts (reinforcement). The diameter of the core wire was 3.2 mm, and the length was 350 mm. The grade of the low-carbon ferritic–pearlitic steel wire was SWRY–21 according to JIS G3503 or CB-08A according to GOST 2246–70.

2.3. Preparation of the Hardfacings

Two groups of hardfacings and three reference materials were investigated. The designation of sample grades consists of two digits (1X or 2X), where the first digit indicates the group (1—without; 2—with additional thermal treatment), and the second digit indicates the size of reinforcing particles (1—finest; . . . ; 6—coarsest). The list of materials is given in Table 1, along with their chemical composition.

Table 1. Chemical composition of hardfacings with reinforcing particles made from recycled inserts in wt.%.

Sample Code	C	Si	Mn	Cr	Co	W	Particle Size [1], μm	Hardness, HRC
			Hardfacings without thermal treatment					
11	3.7	2.1	4.0	5.4	3.0	29.1	180–355	58 ± 3
12	3.1	2.1	3.9	5.2	3.3	29.7	355–500	57 ± 2
13	2.7	2.2	4.0	5.5	2.8	22.7	500–710	54 ± 4
14	3.2	2.3	4.4	7.3	2.7	24.8	710–1000	56 ± 4
15	3.1	2.1	3.7	5.5	3.3	30.8	1000–1400	58 ± 4
16	3.7	2.4	3.3	5.4	4.2	39.4	1400–2000	59 ± 2
			Hardfacings with thermal treatment					
21	4.0	3.1	4.2	7.9	3.0	35.9	180–355	53 ± 4
22	2.6	1.9	3.4	5.5	3.4	32.4	355–500	59 ± 3
23	3.2	2.1	3.7	6.6	3.1	29.8	500–710	55 ± 2
24	2.5	2.6	4.5	5.5	3.3	33.5	710–1000	53 ± 3
25	2.1	2.4	4.7	5.2	3.4	32.1	1000–1400	55 ± 4
26	2.6	2.6	3.8	5.6	4.0	30.3	1400–2000	55 ± 3

Other elements composing in total 1.4–2.7% are as follows: Cu 0.2, Mo 0.5–0.8, Ni 0.1–0.6, Ti 0.5–1.0, Nb 0.1–0.3. The remainder is iron.
Chemical composition of reference materials (wt.%) is as below.
Hardox 400 (for room temperature tests): C 0.32, Si 0.7, Mn 1.6, Cr 1.4, Mo 0.6, Ni 1.5, B 0.004, P 0.02, S 0.01. Hardness, 40 ± 3 HRC.
Mn steel (Hadfield steel, for room temperature tests): C 1.2, Si 0.4, Mn 12.9, Cr 0.4, Co 0.9, S 0.05, P 0.05. Hardness, 15 ± 1 HRC.
AISI316 (for elevated temperature tests): C 0.08, Si 0.75, Mn 2.0, Cr 17.0, Mo 2.5, Ni 12.0, P 0.045, S 0.03. Hardness, 18 ± 2 HRC.
The remainder is iron.

[1] The particle size before welding and thermal treatment is indicated.

All hardfacings were welded in four layers on top of SJ 355 (EN 10025-2:2004) steel substrate plates with dimensions of 80 (length) × 40 (width) × 10 (thickness) mm and were left to cool naturally to room temperature.

Later, the thermal treatment (local melting of the coating in an argon atmosphere to reduce the extent of defects) was conducted with the second group of samples (No. 21–26).

The samples (21–26) were heated up to 300 °C and then locally heated up to 1600–1800 °C (melted) by a nonconsumable tungsten electrode in an argon gas field (*ESAB*, CADDY TIG 2200i AC/DC, North Bethesda, MD, USA). The electrode was manually moved along the longitudinal direction of the sample, and the approximately 2.5 mm wide and 80 mm long strip was treated in 10 s. After that, an electrode was shifted by approximately 2 mm, and the next strip was treated with an overlap of about 0.5 mm. Approximately 20 strips (area of 40 mm × 80 mm) were treated in close to 3 min. The bulk sample temperature was in the range of 500–1000 °C during thermal treatment. After treatment, the samples were placed in a chamber with a temperature of 350 °C, and cooling at a rate of 3 °C min^{-1} down to room temperature was performed to reduce thermal stresses (Figure 2).

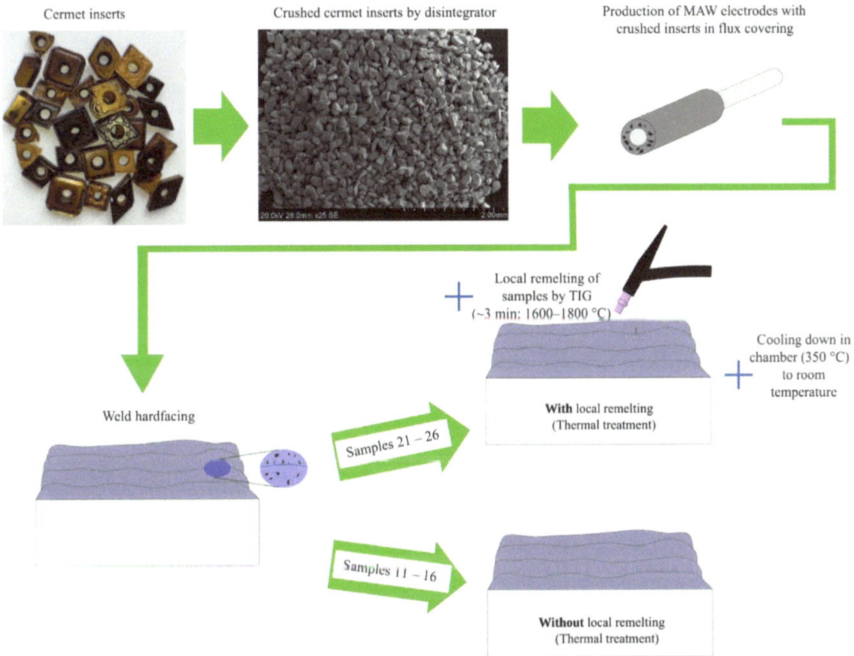

Figure 2. Schematic diagram of preparation, welding, and local remelting of the hardfacings.

Approximately 2 mm of the top surface of the hardfacing was removed during grinding. The steel substrate was partially removed by machining to provide a total thickness of samples (including coating and substrate) of 7 mm. Later, the samples were cut by *Struers* (Copenhagen, Denmark) Secotom 50 into the required shape (25 × 50 mm or 15 × 25 mm for abrasive and erosive tests, respectively).

The chemical composition was measured by BELEC compact-lab-N spectrometer. The TK-2M hardness testing machine performed the hardness measurement. At least five measurements were taken for chemical and hardness testing.

2.4. Abrasive and Erosive Wear Testing Conditions

To demonstrate the performance of hardfacings under a variety of conditions, three types of wear tests were carried out: (1) soft abrasion (low-stress) ASTM G65-04 standard (rubber wheel test); (2) medium-stress abrasion conducted similarly to ASTM G65-04 but with the assistance of a steel wheel; and (3) dynamic impacts (high-stress) by centrifugal accelerator at room and high temperature in accordance with GOST 23.201-78 standard (Figure 3). Each testing method has its application in agriculture, mining, or other industries sectors. The broader testing of the hardfacings improves the proper usage in the most familiar environment. Three testing conditions help to analyze deeper wear mechanisms specific to low-stress conditions (abrasive particles prevailing mechanism is dragging (scratching) along the sample surface), medium-stress conditions (rotation and partial fracturing of abrasive particles and higher local stresses in material's surface), and high-stress conditions (erodent particles flying after exiting a centrifugal accelerator and creating an impact with a sample surface). The wear results will show applications where the hardfacing has higher strength and a lower wear rate. A detailed description of test conditions is shown in Tables 2 and 3.

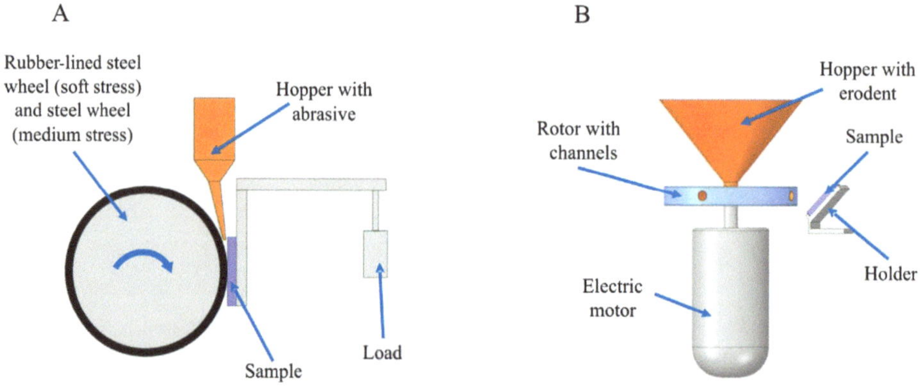

Figure 3. Schematic diagram of abrasive and erosive wear testers: (**A**) abrasive wear test with rubber-lined or steel wheel, (**B**) erosive wear test at room or high temperature.

Table 2. Abrasive wear tests conditions.

Parameter	Description	
Scheme	Block on Wheel	
Local stress applied to the abrasive particle	Soft (not broken particle)	Medium (broken particle)
Description of wheel	Rubber-lined steel wheel, diameter 228.6 mm, width 12.7 mm, Shore A hardness 60	Steel wheel (C45, EN8), diameter 228.6 mm, width 12.7 mm, hardness 165 HB
Abrasive	Quartz sand supplied by SC Anykščių kvarcas, Lithuania, size 200–425 μm, feed rate 250–300 g min^{-1}	
Circumferential velocity	2.4 m s^{-1}	
Linear abrasion (duration)	2153 m (duration 15 min)	431 m (duration 3 min)
Force again specimen	130 N	85 N
Atmosphere	Air, temperature 23 ± 2 °C, relative humidity 45 ± 5%	
Final typical wear scar size (length × width), mm	26.0 × 15.5	22.0 × 13.0

It should be mentioned that due to the deformation of the rubber wheel, the average pressure in contact changes only slightly during the test. In contrast, the steel wheel pressure is high at the beginning of the test and is reduced along with the increase in the wear scar size, and finally, it is even lower than that of the soft-stress abrasion test (Table 2). During testing with a steel wheel, there is a higher chance of abrasive particle fracture due to the metal's higher stiffness than rubber [22]. In contact with rubber, the crushing of abrasive particles is almost eliminated.

Table 3. Erosive wear test conditions (high-stress conditions due to impact).

Parameter	Description	
	Room Temperature	**High Temperature**
Erodent (weight charged into the hopper is indicated)	SiO_2 with size of 0–600 µm (average 400 µm), HV1 = 1183, 6 kg for running-in 15 kg for test at 30 m s^{-1}, 10 kg for test at 50 m s^{-1}, 8 kg for test at 80 m s^{-1}	
Impact velocity	30, 50 and 80 m s^{-1}	80 m s^{-1}
Impact angle	30° and 90°	30°
Atmosphere	Air, relative humidity 45 ± 10%	
Temperature	25	350, 450, 550, 650 °C
Heating rate	–	7 °C min^{-1} (up to 500 °C) 4 °C min^{-1} (above to 500 °C)
Cooling rate	–	7–10 °C min^{-1}
Approximate duration of erosion (influenced by flowing of erodent through the nozzle)	40 min (tests at temperature of 25, 350, and 450 °C)	

Weighing before and after the abrasive and erosive tests was performed to the nearest 1 mg using KERN EG420–3NM and Mettler Toledo ME204 balances, respectively, to determine mass loss. The erosion rate was calculated according to the procedure described in [23,24].

In order to reduce the effect of oxidation on the precision of erosive wear measurement by weight change, the samples were preliminary oxidized in an electrical chamber, *Nabertherm* (Lilienthal, Germany) L9/13 with a PID temperature controller P330, at their test temperature with heating, cooling rates, and duration of oxidation similar to those applied during the erosion test.

After wear tests, the chemical composition of samples was determined by the EDS method with an SEM Hitachi TM-1000 Tabletop Scanning Electron Microscope (SEM) (Tokyo, Japan) equipped with a Bruker XFlash® 6 | 10 detector for element analysis, which was used for imaging as well. Each type of hardfacing test was performed two times under the same testing conditions.

2.5. Hardness and Macrohardness Measurement

The personal computer-controlled Universal hardness tester *Zwick* (Ulm, Germany) BZ 25 with Vickers indenter was used to test material properties. The tests were conducted in accordance with DIN 50359-1 and EN 14577-1 [25]. A sufficiently high load was applied to determine macro-properties due to the large size of reinforcing particles, their inhomogeneous distribution due to the coating application procedure, and the difference in densities of the phases. Table 4 shows indentation testing conditions.

Table 4. Instrumented indentation testing conditions.

Parameter	Description
Load	1500 N
Rate of load application and removal	150 N s^{-1}
Duration of load application and removal	10 s
Dwell time	10 s

3. Results

Figure 4 shows typical images of the cross-sections of the hardfacings. One of the most significant cracking of the reinforcing grains is found in sample No. 14 (reinforcing particle size of 710–1000 μm; Figure 4A). These grains are inside the first applied layer, indicating that preheating is important. The substrate plate was not heated up before welding the first layer. The temperature differences between the welded layer and substrate create residual stress acting on the reinforcing particles. The possible sinking of reinforcing particles is shown in Figure 4B (sample No. 26; WC particles of 1400–2000 μm). The significantly higher density of tungsten influences sinking in comparison to steel. Due to the close position of WC-Co particles to the substrate, it is possible to assume that the dilution of the base metal by hardfacing is limited. Figure 4B also demonstrates possible cracks in the low-carbon steel binder. The uniformity of carbide distribution in manual arc welded hardfacings can be improved by applying a higher welding current (energy input) [5].

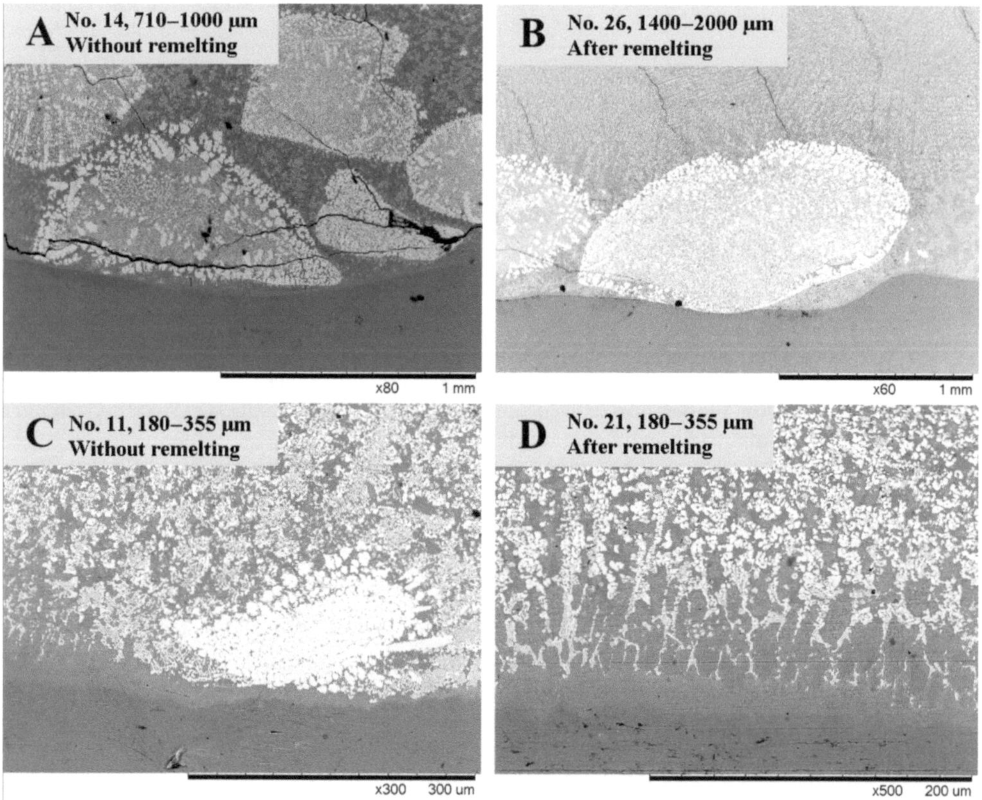

Figure 4. WC particle size variation without remelting (**A**,**C**) and after remelting (**B**,**D**): (**A**) sample No. 14 (700–1000 μm), (**B**) sample No. 26 (1400–2000 μm), (**C**,**D**) samples No. 11 and 21 (180–350 μm).

Samples with additional remelting have reduced the size of reinforcing particles (Figure 4B, No. 26, particle size 1400–2000 μm) or resulted in the complete loss of particles (Figure 4D, No. 21, particle size 180–355 μm) due to dilution. There is a broader transition zone (fusion line) between (1) hardfacing and base metal and (2) reinforcing particles and steel binder (up to 100 μm) that is usually beneficial since it has a lower gradient and lower stresses.

The following sections present the results of the samples' wear testing.

3.1. Low-Stress Abrasion Tests (with Rubber Wheel)

The results of the low-stress abrasion test indicate that additional thermal treatment (local remelting) has a generally negative effect—the average wear rate of hardfacings increases by 29 percent (Figure 5). However, not all the hardfacings had the same response to the treatment. The wear rate of the sample with the finest size of WC particles decreased by 35% (samples 11 vs. 21), while some samples with larger sizes of reinforcement particles had an extreme rise in wear (wear rate is up to two times higher; samples 23 vs. 13, 26 vs. 16). It is possible to conclude that in the case of low-stress abrasion, thermal treatment is only advantageous in the case of hardfacings with reinforcing particles with a size of 180–355 µm. The wear of Hardox 400 is significantly higher than that of hardfacings (3.7 and 2.6 times higher than coatings without or with thermal treatment, respectively). The wear of Mn steel is approximately twice as low as that of Hardox 400, while it is generally higher than that of hardfacings. Only coating 26 is less resistant to wear than Mn steel. The best coatings among their groups were 14, 21, and 24. This shows that in addition to the positive effect of thermal treatment for hardfacings with the finest reinforcement (180–355 µm, sample 21), there is an optimum particle size (coatings 14 and 24) of 710–1000 µm.

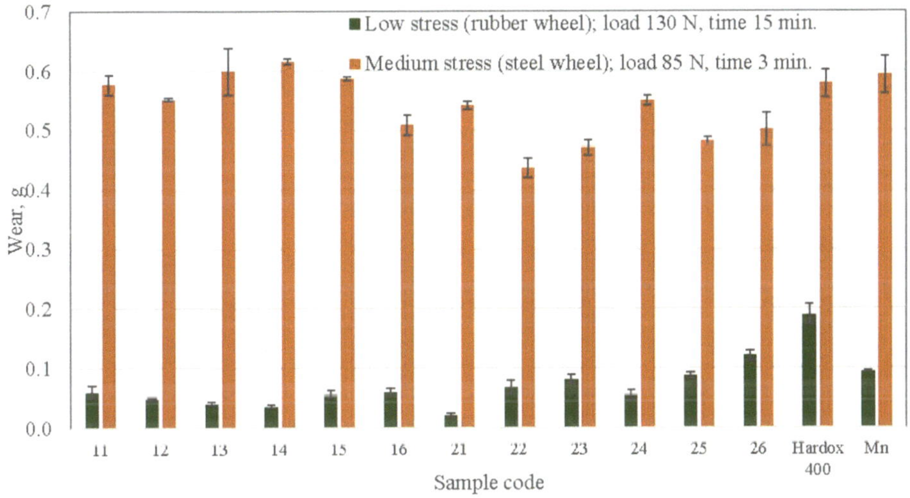

Figure 5. Wear rate of tested materials during low- and medium-stress abrasive wear test.

3.2. Medium-Stress Abrasion Tests (with Steel Wheel)

Medium-stress abrasion results (Figure 5) indicate that the increase in contact stresses (and typically observed shifting of abrasive particle behavior from scratching to rotation with indentation and fracturing) leads to a significant rise in wear rate. The wear rate during 3 min of average-stress abrasion is approximately 10 times higher than during 15 min of low-stress abrasion. The thermal treatment had a positive effect on all hardfacings. The average wear rate decreased by 13%. The lowest and most significant reduction (2 and 22%) was observed for samples 16 vs. 26 (coarsest) and 13 vs. 23, respectively. In conditions of medium-stress abrasion, the wear rates of Hardox 400 and Mn steel were similar to those of hardfacings without heat treatment, while after the heat treatment, the performance of all hardfacings was better (approximately 14%) than that of reference steels. The hardfacings with the highest hardness (16 and 22; Table 1) have the highest resistance in medium-stress abrasive conditions since they provide the highest resistance against indentation.

3.3. Solid Particle Erosion Tests at Room Temperature

As is usually expected, the wear rate of materials investigated under both angles of impact rose with the rise in impact velocity due to the increased kinetic energy of the flying erodent (Figures 6 and 7).

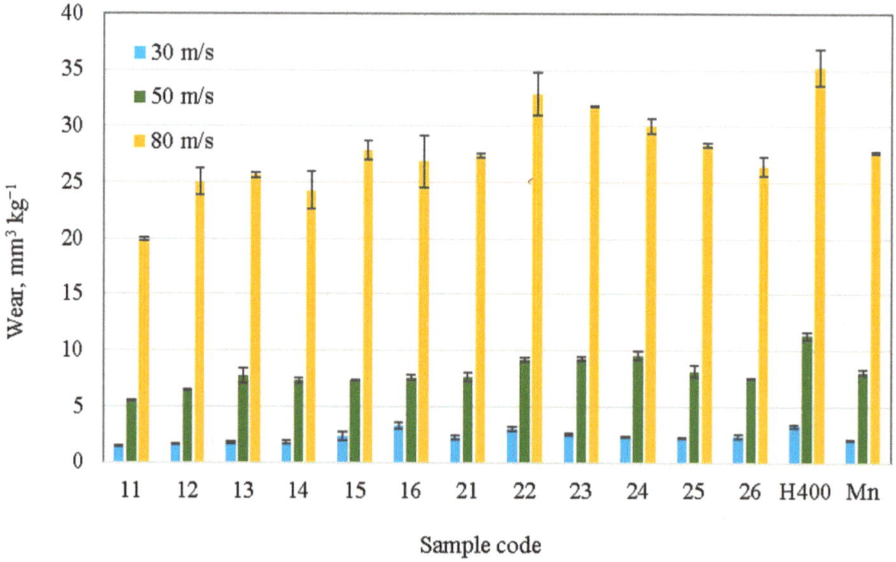

Figure 6. Effect of erodent impact velocity on wear rate of materials tested under impact angle of 30°.

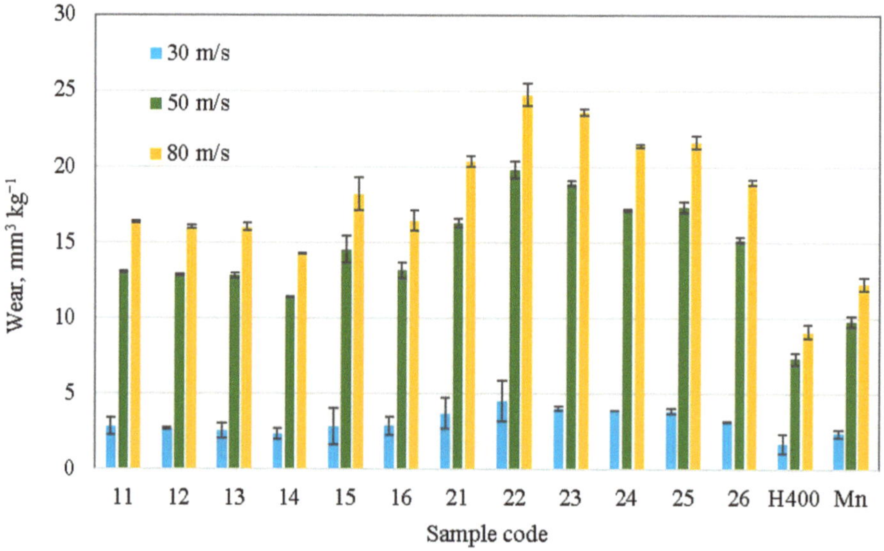

Figure 7. Effect of erodent impact velocity on wear rate of materials tested under impact angle of 90°.

The erosive wear rate results of hardfacings investigated under impact angles of 30° and 90° show (Figures 6 and 7) that heat treatment negatively affected wear resistance. Only material 16 vs. 26 experienced almost no changes, or at the slowest test speed of 30 m·s^{-1} and an angle of 30°, the wear rate decreased.

In general, hardfacings with larger reinforcement particle sizes (1000–1400 and 1400–2000 μm) had the worst wear resistance if no heat treatment was provided (Figures 6 and 7). However, after heat treatment, the coatings with the average size of hard particles (355–500 and 500–710) were the least suitable for erosion under impact angles of 30 and 90°. The coatings 11 and 21 had the lowest wear under an impact angle of 30° at all velocities, which illustrates the potential of hardfacings with small reinforcing particles.

During erosion with an impact angle of 30°, Mn steel and Hardox 400 have similar or lower (5–10% on average) wear resistance to the tested coatings, respectively; however, under normal impact angles, the steels have better wear resistance (1.7–2.7 times as compared to Hardox 400).

The sample No. 12 grain has a crack, which can appear during welding or erosion testing because of particle impact (Figure 8). Elongated grains have a lower tolerance to internal stresses than rounded ones. For example, sample No. 24 nicely incorporates a grain into the matrix. The matrix is plastically deformed and distributed over harder areas in sample No. 16 (particle size 1400–2000 μm). Reinforcing particles are washed around by the erodent particles because of different hardness between the carbide grain and the matrix. In some places, there is recognized plural splitting of the carbide grain (sample No. 25). When the reinforcing particle is examined with higher magnification, it is possible to observe its composite nature (sample No. 25), where that matrix is partially removed around the WC particles (1–5 μm) by the fine abrasive particles or fragments [19]. Microploughing on the surface of samples No. 11 and 21 is observed when erodent particles collide with the surface at the highest impact velocity of 80 m·s^{-1}.

In most cases, it is difficult to predict a cutting direction because scratching direction is various and unpredictable. The erodent particles change direction after the first impact with a surface or collision between each other. Specifically, this can be noticed after erosion with the highest impact velocity (80 m·s^{-1}), where scratches appear very chaotically.

The volumetric erosion rate can be expressed in the general form [24]

$$I = a \cdot v^m$$

where a is the coefficient that depends on the target material, impact angle, and eroding particle properties; v is impact velocity, m·s^{-1}; and m is the index. Table 5 contains the data for estimating erosion rates under both impact angles in the range of studied velocities (30–80 m·s^{-1}).

Table 5. Data for estimation of volumetric erosion rate of hardfacings and reference steels at impact angles of 30° and 90°.

Sample Code	30°			90°		
	$a \times 10^{-3}$	m	R^2	$a \times 10^{-3}$	m	R^2
11	0.2	2.68	0.99	7.4	1.81	0.86
12	0.1	2.77	0.99	6.4	1.84	0.86
13	0.2	2.75	0.99	5.1	1.89	0.86
14	0.2	2.66	0.99	5.4	1.85	0.86
15	0.4	2.53	0.99	5.0	1.92	0.85
16	2.1	2.14	0.98	7.7	1.80	0.86
21	0.4	2.55	0.99	11.5	1.75	0.86
22	0.7	2.44	0.99	14.4	1.75	0.86
23	0.4	2.58	0.99	10.0	1.82	0.86
24	0.3	2.63	0.99	12.5	1.75	0.86
25	0.3	2.62	1.00	11.0	1.78	0.86
26	0.5	2.47	0.99	7.3	1.84	0.86
H400	0.8	2.43	1.00	3.8	1.82	0.86
Mn	0.2	2.68	0.99	7.6	1.67	0.87

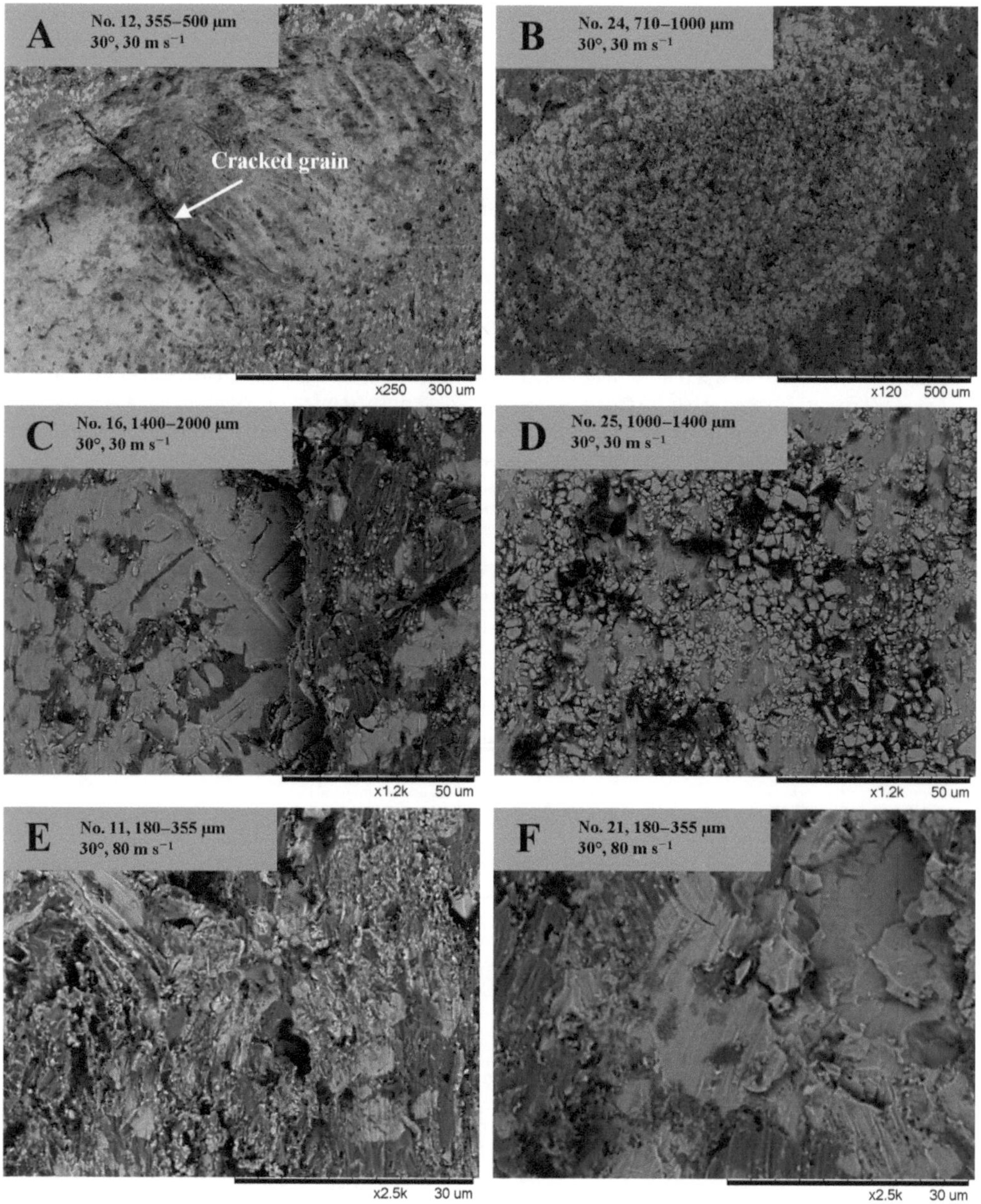

Figure 8. SEM images of the sample surfaces after erosive test at 30 (**A–D**) and 80 (**E**,**F**) m·s^{-1} particle velocity and 30° impact angle.

The energy supplied by an erodent particle to the sample during normal angle impact is higher than during oblique angle impact. As a result, it is expected to observe more severe damage. The intensity of carbide breakage (fragmentation), relocation of initial material, embedment of erodent fragments, and fatigue of the metallic matrix, i.e., the formation of a mechanically mixed layer, is increased (Figure 9). The rough surface structure can be seen on sample No. 15's surface. Compared to Figure 8 (impact angle 30°), a higher impact angle (90°) does not create as many chaotical scratches as a lower impact angle. The erodent particles at a higher impact angle have a straight path and create one impact on the sample surface, which either deforms the matrix or, after impact with a higher hardness carbide grain, bounces back from the sample surface. In comparison, a lower impact angle repeatedly creates impacts on the surface.

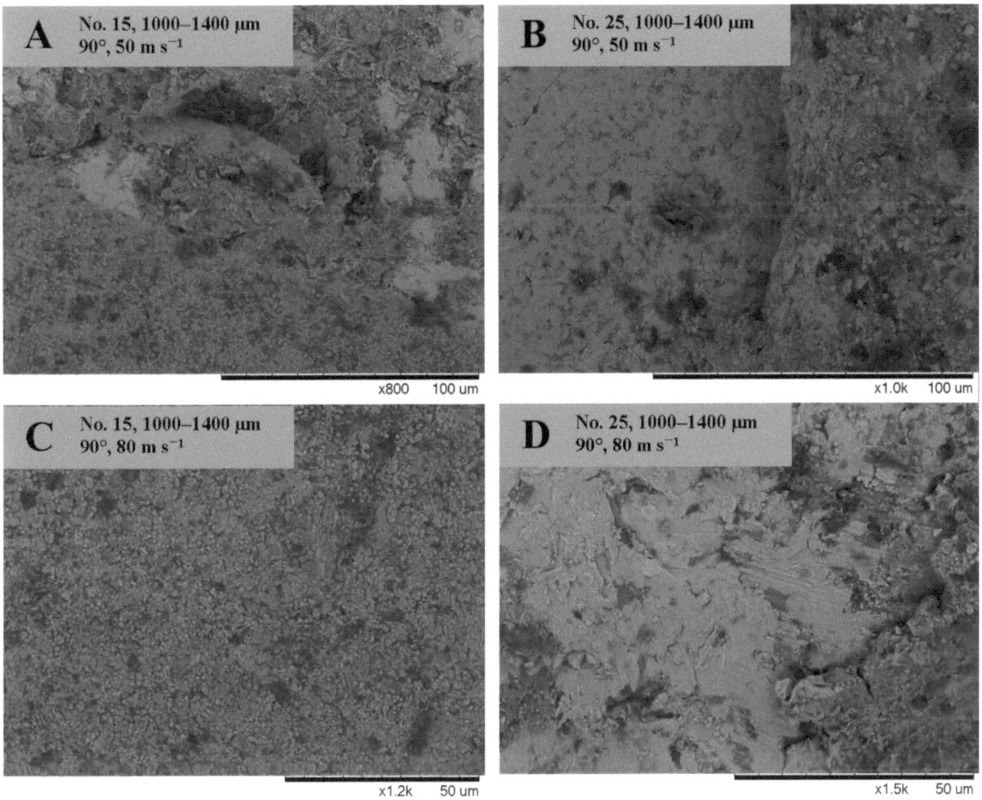

Figure 9. SEM images of the sample surfaces after erosive test at 50 (**A**,**B**) and 80 (**C**,**D**) m·s^{-1} particle velocity and 90° impact.

3.4. Solid Particle Erosion Tests at Elevated Temperatures

AISI 316 stainless steel was used as a reference material instead of Hardox 400 and Mn steel (used for room temperature tests) due to their excessive oxidation at such elevated temperatures.

The results of groups of hardfacings with and without thermal treatment are shown separately for better clarity of the images, to make it easier to track the effect of temperature on the wear rate of each material independently and its performance within its group.

The wear rate of hardfacings without thermal treatment under an impact angle of 30° and a velocity of 80 m s^{-1} at temperatures of 350, 450, and 550 °C was lower than that of reference AISI316 stainless steel. In contrast, at 650 °C, the wear of coatings 11, 13, and 15 was higher (Figure 10).

At least three parameters are influenced by the rise in temperature: (1) softening (the hardness of materials is usually reduced at high temperatures), (2) change of internal stresses, and (3) oxidation. As seen in Figure 10, the wear rates of coatings 13 and 16 are lower at 450 and 550 °C than at 350 °C and perform very well at temperatures below 650 °C. These materials were usually not the best performers during abrasive or especially erosive testing at room temperature. This enables one to conclude that their performance is improved due to reduced brittleness or internal stresses. Material 16 had the highest W

(WC) concentration and the highest hardness at room temperature (Table 1). Coating 12 performed quite well during average-stress abrasive wear testing, with an average erosion rate at 350, 450, and 550 °C and the lowest rate at 650 °C among the materials investigated.

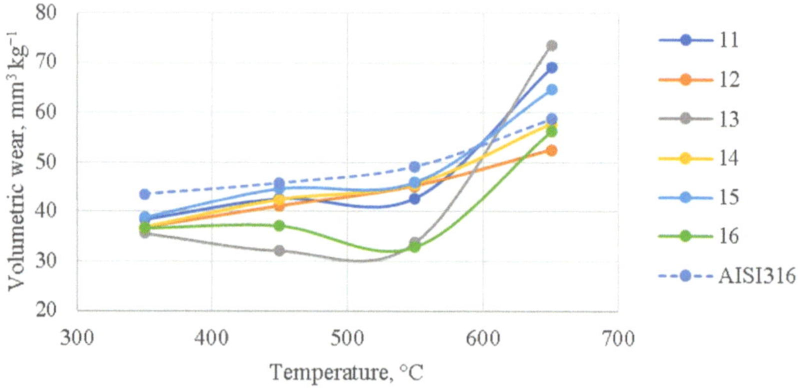

Figure 10. Effect of temperature on erosion wear of hardfacings without thermal treatment and AISI 316 steel (impact angle 30°; velocity 80 m s^{-1}).

The erosive wear rate of hardfacings after the thermal treatment (Figure 11) was quite similar to materials without thermal treatment. In addition to materials 23 and 26, material 25 also has relatively low wear rates at 350, 450, and 550 °C, which points to the conclusion that materials with average or larger reinforcement sizes may provide better erosive wear resistance at least at temperatures of 350–550 °C with an impact angle of 30°. The important fact that can also be observed from Figure 10 is that at 650 °C, all thermally treated hardfacings have lower wear resistance than reference stainless steel.

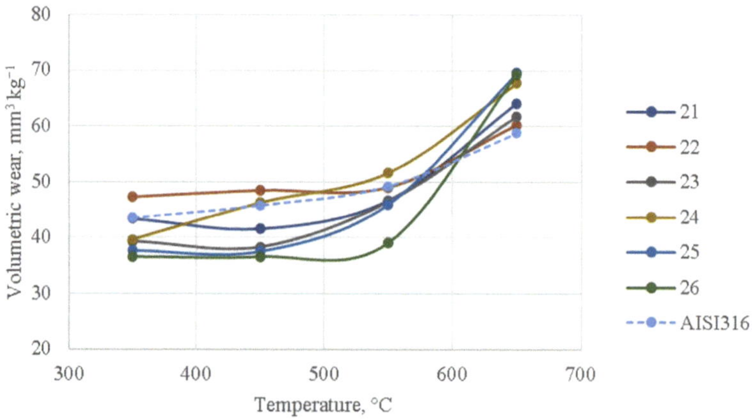

Figure 11. Effect of temperature on erosion wear of hardfacings with thermal treatment and AISI 316 steel (impact angle 30°; velocity 80 m s^{-1}).

In order to compare the average performance among hardfacings with or without local remelting and their performance vs. reference stainless steel, the differences in erosive wear rates were calculated and are presented in Table 6.

It is possible to conclude that using hardfacings without local remelting is reasonable only in the temperature range from 20 to 550 °C. At 650 °C, the reference stainless steel provides better wear resistance in erosive conditions.

Table 6. Comparison of effects (application of remelting; reference vs. hardfacing with or without remelting) providing increase (+) or decrease (−) of wear rate of studied materials (average wear rates of the groups measured at 80 m·s^{-1} with impact angle of 30° are compared).

Effects	Temperature, °C				
	20	**350**	**450**	**550**	**650**
Effect of hardfacing local remelting	+18%	+10%	+3%	+12%	+5%
Application of hardfacing without local remelting instead of reference AISI 316 stainless steel	-	−17%	−14%	−18%	+6%
Application of hardfacing with local remelting	-	−7%	−11%	−6%	+11%

3.5. Macrohardness, Young's Modulus, Plastic and Elastic Work of Indentation

Macrohardness results measured with the Vickers pyramid indenter show the same tendencies as hardness measured by the Rockwell indenter (see Table 1 and Figure 12). The general trend observed is that hardfacings without thermal treatment with the highest W (WC) content (11, 12, 15, 16) have higher hardness. This trend does not apply to heat-treated hardfacings. Material 21, which has the highest W content, had the lowest HRC hardness (Table 1) and one of the lowest hardness values determined by high-load indentation by the Vickers indenter (Figure 12). On the other hand, the coating with average W content (22) had the highest hardness according to both methods. It demonstrates that the hard reinforcement phase is not used efficiently in heat-treated coatings. The average macrohardness of hardfacings with heat treatment (Figure 12) is 14% lower than that of coatings without treatment. Possible explanations for such detrimental reinforcement behavior could be their fracturing, a change in shape from circular to dendritic, or deterioration of bonding between phases (probably due to higher internal stresses). According to a commonly used simplified approach, materials with the highest hardness should perform best, but this was not the case under most conditions tested.

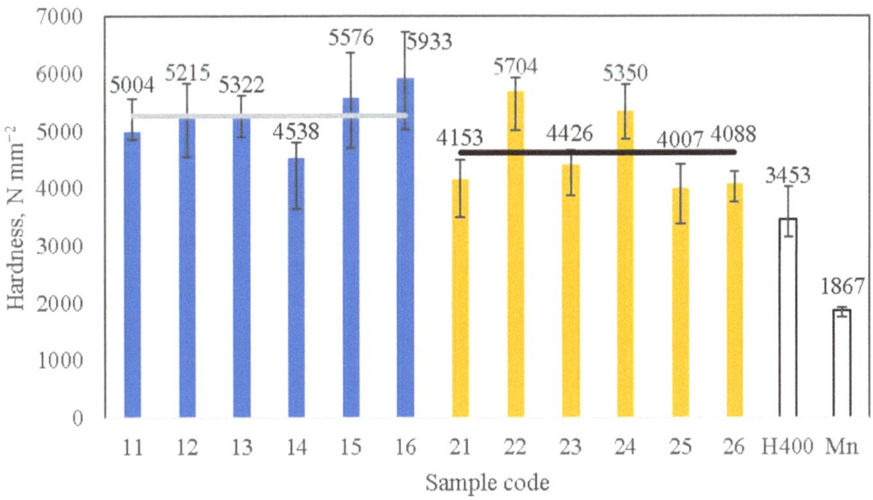

Figure 12. Universal macrohardness of tested materials determined by instrumented indentation method. Lines indicate the average values of each group (group 11–16, 5214; group 21–26, 4613 N·mm^{-2}). Color indicates the groups without (11–16), with (21–26) thermal treatment and reference materials (H400, Mn).

The results of Young's modulus testing of hardfacings shown in Figure 13 have the same tendency as the values of macrohardness (Figure 12). As such, the coatings with the highest hardness generally have the highest Young's modulus. The hardness of Hardox 400 and Mn steel is lower than hardfacings (this is expected) while Young's modulus has a similar level, showing that hard reinforcing particles with sufficiently higher Young's modulus are not performing efficiently and cannot provide their expected increase.

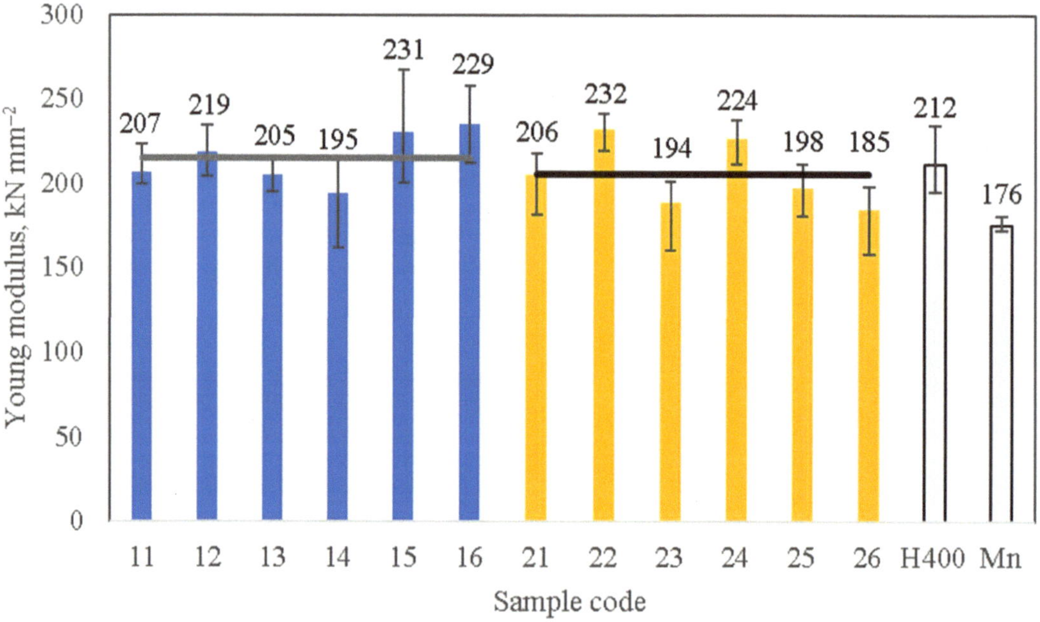

Figure 13. Young's modulus of tested materials determined by instrumented indentation method. Lines indicate the average values of each group (group 11–16, 214; group 21–26, 206 kN·mm^{-2}). Color indicates the groups without (11–16), with (21–26) thermal treatment and reference materials (H400, Mn).

The instrumented indentation testing method can provide the values of elastic and plastic work, and it is possible to compare the ratio of one to another, as shown in Figure 14. It could be concluded that hardfacings with a high portion of plastic work (14, 21, 25) can provide better wear resistance in erosive conditions at room temperature under impact angles of 30 and 90° (Figures 6 and 7). Materials 14 and 21 also have the best performance during low-stress abrasive testing. Materials 16 and 22, with the highest portion of elastic work (Figure 14), have shown the best results in the medium-stress abrasion test (Figure 5). Hardox 400 and Mn steel also have the highest portion of plastic work and superior performance at room temperature under an impact angle of 90°, while their resistance against low-stress wear is the lowest. This leads to the conclusion that comparing the portion of plastic or elastic work out of the total is more valid within the specific group than between various material groups.

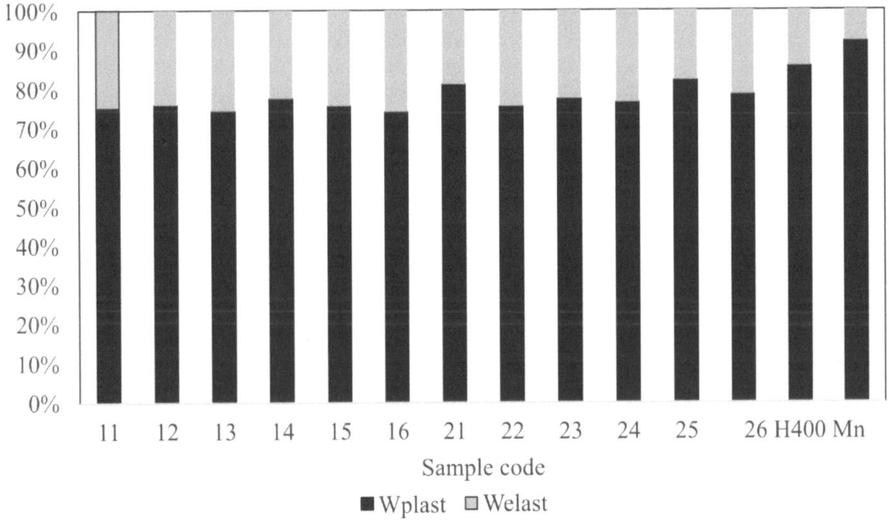

Figure 14. The ratio between plastic and elastic works performed during indentation (unloading) of materials by Vickers indenter with a load of 150 kg.

4. Discussion

The hardfacings prepared by the same manual arc welding method but with Hadfield steel as a binder and solid WC inclusions (that were not recycled and did not have composite microstructure) [26] had 2–17% higher wear at velocities of 30–80 m s^{-1} under an impact angle of 30° than hardfacings from the current work without heat treatment. The hardfacings presented in the current work with heat treatment are 2% worse than those studied in [26].

Suppose the comparison is made with hardfacings produced from solid WC particles (which were not recycled and did not have composite microstructure) and the same low-carbon or stainless steel matrix [27]. In that case, hardfacings with recycled composite reinforcement and without heat treatment (current results) have 1.1–3.4% lower wear rates depending on impact velocity (30–80 m s^{-1}) and impact angle (30° or 90°). As a result, recycled WC-Co composite reinforcing can be used without compromising wear resistance. Wear resistance can be further increased by additional chemical treatment with recycled powders (removes the impurities caused by the milling process), enhancing their bonding with a matrix of hardfacings [8].

To some extent, erosion testing with an impact angle of 30° or 90° can create similar wear conditions as abrasive wear testing with rubber or steel wheels. Both erosion testing at 30° and testing with a rubber wheel intensify scratching by abrasive particles, while erosion testing at 90° and abrasion with a steel wheel include a significant portion of indentations. On the other hand, the intensity of scratching or impacting is important. The best material without local remelting (14) from rubber wheel testing has relatively high wear rates during erosion at an impact angle of 30°. Somehow, material 21 with local remelting had meager wear rates in abrasion with the steel wheel and erosion with normal impact. This is probably since the intensity of interaction in average-stress and high-stress methods is similar, leading to the fracturing of abrasive particles and possibly embedment of fragments into the test surface and the formation of a mechanically mixed layer [22,25].

Suppose Figures 6 and 7 are compared, it is possible to conclude that, during room temperature erosion, untreated hardfacings with the smallest reinforcement size (180–355 μm) perform better, while in the case of locally remelted ones, the smallest (180–355 μm) and largest (1400–2000 μm) reinforcements experience the lower wear rates, with the highest wear rate demonstrated by hardfacings with intermediate size. It should be noted that remelting reduced the wear resistance of hardfacings in conditions of solid particle erosion, so it is more important to stress that hardfacings with intermediate size (355–1400 μm) reinforcing transformed into very unsuitable material. From one side, the enrichment of binder by harder phases (due to additional heat treatment) originating from reinforcing particles (W, Ti, Mo, Nb, and their compounds) should increase the wear resistance of binder and hardfacing in general. In addition, local remelting can improve the bonding of composite reinforcing particles. In intermediate size, however, induced additional stresses and reinforcing cracking are more significant, resulting in less wear. If bonding between phases is improved, coatings with the largest reinforcing size can perform better since such protruding particles can effectively protect the binder [9].

Typically, ductile materials (steels) have the highest wear rate close to an impact angle of 30°. In contrast, the maximum wear rate of brittle materials (for example, ceramics) is observed at 90°. We have placed the results of erosive wear tests at 30° and 90° at room temperature on the studied hardfacings in Figure 15 to compare relative wear rates.

According to Figure 15, it is clear that when the impact velocity rises from 30 to 80 m s^{-1}, the wear mechanism experienced by hardfacings during solid particle erosion changes from "brittle" to "ductile". Since WC-Co reinforcements and low-carbon steel matrix could be comparably treated as "brittle" and "ductile" phases, it could be concluded that at low velocities of impact, the performance of reinforcing particles is defining (controlling or limiting) the wear rate of the whole hardfacing. Typically, the energy of erodent particles impacting at 30 m s^{-1} is insufficient to directly (by a single impact) remove large fragments of material, including one or several reinforcing particles and binder, while the process includes selective removal of steel matrix, wear of reinforcements, and loosening of reinforcements when the matrix cannot provide support for them during the impacting by erodents. The resistance of a material to fatigue is essential. However, in the case of manual arc welded hardfacings, the brittle mechanism instead includes a combination of direct intensive fracturing of reinforcing particles (low fatigue resistance) and intensive wear of the unprotected matrix. Brittle fracturing is influenced by internal stresses (cracks), pores, and insufficient bonding between deposited weld beads (Figures 4, 8 and 9). At higher impact velocities, the ductile phase controls the wear rate. Thus, reinforcement alone has only a minor effect, while it can participate in forming a more ductile mechanically mixed layer through fracturing, relocation (shifting), and removal. The energy of the impact is enough to cause the embedment of erodent fragments. Material 16, with the highest W content, hardness, and particle size among hardfacings without heat treatment, demonstrates the inefficiency of reinforcement at a speed of 30 m s^{-1}; its behavior is shifted to the "ductile" region. There is usually no direct correlation between hardness (Table 1, Figure 12), Young's modulus (Figure 13), and wear rates (Figures 5–7, 10 and 11), probably because these measurements are relatively static. In contrast, wear is dynamic, especially during high-speed erosion. A relatively good correlation was found for hardfacings with a high portion of plastic work (14, 21, 25), resulting in better wear resistance in erosive conditions at room temperature, especially under an impact angle of 90° (Figure 7).

At elevated temperatures (Figures 10 and 11), when stainless steel is used as reference material due to the oxidation of Hardox 400 and manganese steel, the hardfacings perform sufficiently well up to 550 °C. Intensive oxidation of WC-Co reinforcement and low-carbon matrix at 650 °C limits their application temperature due to a significant rise in their wear rate. Additionally, the hardness of WC is reduced at this temperature, which influences its performance [28].

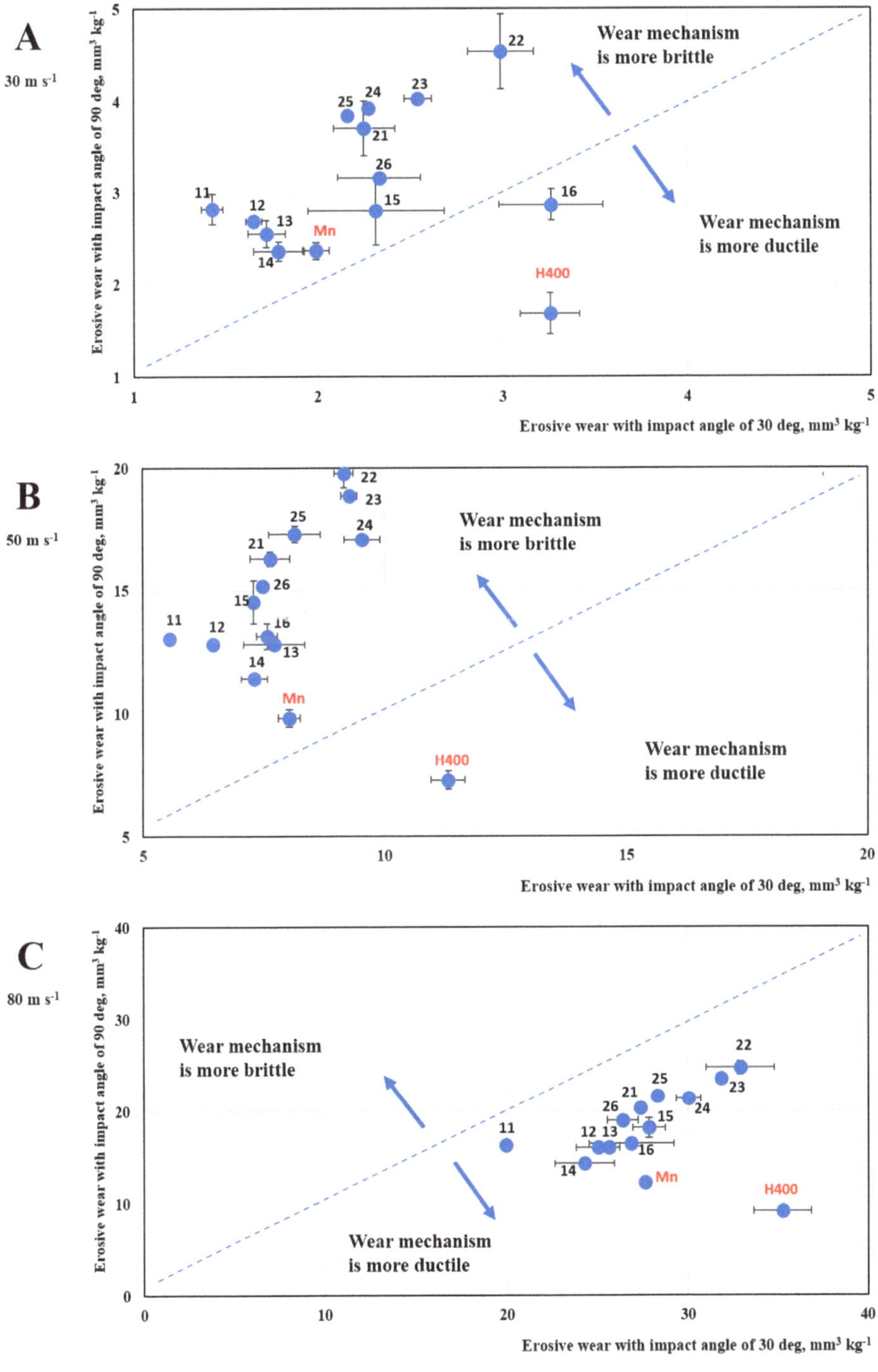

Figure 15. Evaluation of the hardfacings' erosive wear mechanism acting at various impact velocities (wear rates measured with impact angle of 30° and 90° at room temperature).

5. Conclusions

After testing the hardfacings with recycled WC-Co composite reinforcement under abrasive and erosive test conditions, we can conclude the following:

- At elevated temperature erosive conditions (temperatures 350–550 °C, impact angle 30°, velocity 80 m s^{-1}), hardfacings without heat treatment with all sizes of recycled reinforcements had up to two times the wear resistance of reference AISI316 stainless steel. At 650 °C, using developed hardfacings is not reasonable due to oxidation and softening. Data for predicting erosive wear rates at various impact velocities (temperature 20 °C, impact angle 30° and 90°) were calculated.
- The local remelting of hardfacings was beneficial only in the case of a medium-stress abrasive test (with a steel wheel). For these test conditions, it was found that materials with the highest hardness had the lowest wear rate. During low-stress (with rubber wheel) abrasive and erosive tests at all temperatures, the local remelting usually had a negative effect.
- Local remelting of hardfacings with the finest reinforcements (180–355 µm) reduced wear in low-stress abrasive conditions by more than two times. Utilization of these fine reinforcements was also efficient for protection against solid particle erosion with an impact angle of 30° at all velocities.
- The macrohardness and Young's modulus of samples decreased after local remelting. The extent of plastic work out of the total work exerted during instrumented indentation testing usually increased slightly after local remelting.
- The values of the volumetric erosion rate of hardfacings provide information for modeling erosion conditions. The extent of plastic or elastic work out of the total work, exerted by hardfacing during instrumented indentation testing, can help to predict performance in erosive or abrasive conditions.

Author Contributions: Conceptualization, V.J. and M.A.; methodology, V.J., M.A. and E.K.; validation, E.K., M.A., V.J. and D.G.; investigation, E.K. and M.A.; resources, V.J., M.A. and D.G.; data curation, E.K., M.A. and V.J.; writing—original draft preparation, E.K. and M.A.; writing—review and editing, E.K., M.A. and V.J.; visualization, E.K. and M.A.; funding acquisition, M.A. All authors have read and agreed to the published version of the manuscript.

Funding: This research was funded: by the Estonian Ministry of Education and Research, grant number M-ERA.NET DuplexCER and PRG643; and by the EIT RawMaterials grant VHE22005 by the Environmental Investment Centre (EIC) co-funding grant KIK22014.

Institutional Review Board Statement: Not applicable.

Informed Consent Statement: Not applicable.

Data Availability Statement: Data sharing is not applicable.

Conflicts of Interest: The authors declare no conflict of interest.

References

1. Zeiler, B.; Bartl, A.; Schubert, W.-D. Recycling of tungsten: Current share, economic limitations, technologies and future potential. *Int. J. Refract. Met. Hard Mater* **2021**, *98*, 105546. [CrossRef]
2. Liu, H.; Liu, H.; Nie, C.; Zhang, J.; Steenari, B.M.; Ekberg, C. Comprehensive treatments of tungsten slags in China: A critical review. *J. Environ. Manag.* **2020**, *270*, 110927. [CrossRef] [PubMed]
3. Mishra, D.; Sinha, S.; Sahu, K.K.; Agrawal, A.; Kumar, R. Recycling of Secondary Tungsten Resources. *Trans. Indian Inst. Met.* **2017**, *70*, 479–485. [CrossRef]
4. Wongsisa, S.; Srichandr, P.; Poolthong, N. Development of Manufacturing Technology for Direct Recycling Cemented Carbide (WC-Co) Tool Scraps. *Mater. Trans.* **2015**, *56*, 70–77. [CrossRef]
5. Zikin, A.; Ilo, S.; Kulu, P.; Hussainova, I.; Katsich, C.; Badisch, E. Plasma transferred arc (PTA) hardfacing of recycled hardmetal reinforced nickel-matrix surface composites. *Medziagotyra* **2012**, *18*, 12–17. [CrossRef]
6. Nurminen, J.; Näkki, J.; Vuoristo, P. Microstructure and properties of hard and wear resistant MMC coatings deposited by laser cladding. *Int. J. Refract. Met. Hard Mater.* **2009**, *27*, 472–478. [CrossRef]

7. Bartkowski, D.; Bartkowska, A. Wear resistance in the soil of Stellite-6 / WC coatings produced using laser cladding method. *Int. J. Refract. Met. Hard Mater.* **2017**, *64*, 20–26. [CrossRef]
8. Xu, H.; Huang, H. Microstructure evolution and mechanical properties of thermally sprayed coating modified by laser remelting and injection with tungsten carbide. *Ceram. Int.* **2022**, *48*, 22854–22868. [CrossRef]
9. Mégret, A.; Vitry, V.; Delaunois, F. Study of the Processing of a Recycled WC–Co Powder: Can It Compete with Conventional WC–Co Powders? *J. Sustain. Metall.* **2021**, *7*, 448–458. [CrossRef]
10. Katinas, E.; Antonov, M.; Jankauskas, V.; Tarraste, M. The Effect of Spark Plasma Sintering Thermal Cycle on Behaviour of Fe-Based Hardfacings Reinforced with WC and WC-Based Hardmetal. *Key Eng. Mater.* **2019**, *799*, 3–8. [CrossRef]
11. Królicka, A.; Szczepański, Ł.; Konat, Ł.; Stawicki, T.; Kostencki, P. The Influence of Microstructure on Abrasive Wear Micro-Mechanisms of the Claddings Produced by Welding Used in Agricultural Soil. *Materials* **2020**, *13*, 1920. [CrossRef] [PubMed]
12. Novák, P.; Müller, M.; Hrabě, P. Research of a material and structural solution in the area of conventional soil processing. *Agron. Res.* **2014**, *12*, 143–150.
13. Nagentrau, M.; Tobi, A.L.M.; Sambu, M.; Jamian, S. The influence of welding condition on the microstructure of WC hardfacing coating on carbon steel substrate. *Int. J. Refract. Met. Hard Mater.* **2019**, *82*, 43–57. [CrossRef]
14. Jankauskas, V.; Antonov, M.; Varnauskas, V.; Skirkus, R.; Goljandin, D. Effect of WC grain size and content on low stress abrasive wear of manual arc welded hardfacings with low-carbon or stainless steel matrix. *Wear* **2015**, *328–329*, 378–390. [CrossRef]
15. Xiang, Z.; Li, Z.; Chang, F.; Dai, P. Effect of Heat Treatment on the Microstructure and Properties of Ultrafine WC–Co Cemented Carbide. *Metals* **2019**, *9*, 1302. [CrossRef]
16. Majumder, P.; Sinha, A.; Biswas, A. Effect of preheating techniques on bead geometry and microhardness of weldment developed through the submerged arc welding process. *Mater. Today Proc.* **2021**, *46*, 5001–5007. [CrossRef]
17. Rajeev, G.P.; Kamaraj, M.; Bakshi, S.R. Hardfacing of AISI H13 tool steel with Stellite 21 alloy using cold metal transfer welding process. *Surf. Coat. Technol.* **2017**, *326*, 63–71. [CrossRef]
18. Karmakar, D.P.; Muvvala, G.; Nath, A.K. High-temperature abrasive wear characteristics of H13 steel modified by laser remelting and cladded with Stellite 6 and Stellite 6/30% WC. *Surf. Coat. Technol.* **2021**, *422*, 127498. [CrossRef]
19. Ghadami, F.; Sohi, M.H.; Ghadami, S. Effect of TIG surface melting on structure and wear properties of air plasma-sprayed WC-Co coatings. *Surf. Coat. Technol.* **2015**, *261*, 108–113. [CrossRef]
20. de Medeiros Castro, R.; Curi, E.I.M.; Inacio, L.F.F.; da Silva Rocha, A.; Pereira, M.; Silva, R.G.N.; de Souza Pinto Pereira, A. Laser remelting of WC-CoCr surface coated by HVOF: Effect on the tribological properties and energy efficiency. *Surf. Coat. Technol.* **2021**, *427*, 127841. [CrossRef]
21. Zikin, A. *Advanced Multiphase Tribo-Functional PTA Hardfacings*; Tallinn University of Technology: Tallinn, Estonia, 2013.
22. Antonov, M.; Hussainova, I.; Veinthal, R.; Pirso, J. Effect of temperature and load on three-body abrasion of cermets and steel. *Tribol. Int.* **2012**, *46*, 261–268. [CrossRef]
23. Kleis, I.; Kulu, P. *Solid Particle Erosion: Occurrence, Prediction and Control*; Springer: London, UK, 2008.
24. Antonov, M.; Pirso, J.; Vallikivi, A.; Goljandin, D.; Hussainova, I. The effect of fine erodent retained on the surface during erosion of metals, ceramics, plastic, rubber and hardmetal. *Wear* **2016**, *354–355*, 53–68. [CrossRef]
25. Antonov, M.; Hussainova, I.; Pirso, J.; Volobueva, O. Assessment of mechanically mixed layer developed during high temperature erosion of cermets. *Wear* **2007**, *263*, 878–886. [CrossRef]
26. Jankauskas, V.; Choteborsky, R.; Antonov, M.; Katinas, E. Modeling of Microstructures and Analysis of Abrasive Wear of Arc-Welded Hadfield Steel. *J. Frict. Wear* **2018**, *39*, 78–84. [CrossRef]
27. Katinas, E.; Antonov, M.; Jankauskas, V.; Skirkus, R. Effect of WC grain size and content on erosive wear of manual arc welded hardfacings with low-carbon ferritic-pearlitic steel or stainless steel matrix. *Key Eng. Mater.* **2016**, *674*, 213–218. [CrossRef]
28. Vornberger, A.; Pötschke, J.; Gestrich, T.; Herrmann, M.; Michaelis, A. Influence of microstructure on hardness and thermal conductivity of hardmetals. *Int. J. Refract. Met. Hard Mater.* **2020**, *88*, 105170. [CrossRef]

coatings

Article

Effect of Soil Creep on the Bearing Characteristics of Soil Slope Reinforced with CFRP and Anti-Slide Piles

Jun Wang [1,*], Lin Liu [1] and Ping Cao [2]

1 School of Building Engineering, Hunan Institute of Engineering, Xiangtan 411104, China
2 School of Resource and Safety, Central South University, Changsha 410083, China
* Correspondence: 70115@hnie.edu.cn

Abstract: In order to research the displacement characteristics and stability of a soil slope reinforced with carbon-fiber-reinforced plastic (CFRP) and anti-slide piles, the displacement composition, aging deformation and failure mode of a soil mass were analyzed. According to the Mohr–Coulomb strength criterion, a new nonlinear, accelerated creep model of soil mass was founded with the addition of a self-building M-C plastic element. Furthermore, a viscoplastic strain analytical formula of an M-C plastic element was obtained, and the tensile deformation characteristics of a CFRP sheet were also discovered under a landslide thrust creep load. According to the environmental conditions of the anti-slide pile, the CFRP was arranged along the load-bearing side of the pile to control deformation. Combining the calculation example, it is shown that the horizontal displacement of the soil slope's composite structure decreases by approximately 40% with CFRP reinforcement. Furthermore, for the first two calculation conditions, after one year, the maximum horizontal displacement decreased by 50% and increased by 10%, respectively. Simultaneously, the overall safety factor increased by 31.3% without soil creep properties. On the contrary, the overall safety factor was reduced, and the slope has a tendency toward unstable failure. Moreover, there is no through plastic zone in the slope. The stability of the reinforced slope and the bearing capacity of the pile are related to the CFRP method. Simultaneously, the structure can reduce the costs and construction difficulty of anti-slide piles in a complex environment surrounded by the soil creep effect.

Keywords: CFRP; anti-pile; nonlinear creep; displacement; stability

Citation: Wang, J.; Liu, L.; Cao, P. Effect of Soil Creep on the Bearing Characteristics of Soil Slope Reinforced with CFRP and Anti-Slide Piles. *Coatings* **2023**, *13*, 1025. https://doi.org/10.3390/coatings13061025

Academic Editor: Paolo Castaldo

Received: 5 April 2023
Revised: 16 May 2023
Accepted: 29 May 2023
Published: 1 June 2023

1. Introduction

An anti-slide pile is a type of concrete structure which can actively control the deformation of soil, enhance the stability of damaged soil and give full play to the self-bearing characteristics of soil. An anti-slide pile has the foundational characteristics of a large bearing capacity, simple structure and convenient construction. It is widely used in the construction of foundation pits, traffic slopes, mining slopes and other engineering constructions. In general, the anti-slide pile has played an important role in engineering disaster prevention and reduction [1–3]. However, adverse changes in construction conditions and geological and natural environmental factors have an important influence on the bearing stability of the anti-slide pile and can increase the sliding force and displacement of the anti-slide pile's slope structure. They can also decrease the shear strength of the soil and increase the deformation of the pile top and the phenomenon of a sliding body on top. On the other hand, unreasonable design parameters of an anti-slide pile can lead to a reduction in the bearing capacity of the anti-slide pile. Therefore, safety hazards and accidents induced by anti-slide reinforcement are everywhere [4–7]. At the same time, damaged anti-slide piles are reinforced through the following measures, which are often used in engineering: the supplementation of a second source of tension with an anchor cable and enlarging the pile's cross-sectional area, retaining wall, and pile weight before cutting slope to reduce the load. These measures often reinforce the cycle of growth, cost and increasing construction difficulty. We must find an effective and feasible method. The reinforcement of dangerous anti-slide piles has aroused widespread concern in engineering and academic regions. At present, a large number of soil creep research has focused on linear creep. A two-phase creep element model can only reflect the creep sliding failure of a soil mass. However, the use of a class model and nonlinear creep element model in soil research on anti-slide-pile reinforcement in high and steep slopes is rare, and the slope of sliding soil damage is often caused by accelerated creep. Liu et al. [3] conduced many experiments and established a nonlinear creep model. The analytical solution of the creep model is founded in a combination of the data fitting method and knowledge of elastoplastic mechanics. At the same time, a nonlinear element parameter identification method was proposed. Zhao et al. [8] established a generalized model for rock and soil materials, derived the shear and pressure transfer coefficients through the theory and obtained the coupled evolution equation of the composite fracture creep for rock and soil materials with cracks under complex load conditions and the trend of fracture failure development. Therefore, applying nonlinear creep theory to a geotechnical slope engineering stability analysis is more in line with engineering practice.

Carbon fiber reinforced plastics (CFRPs) are a new type of organic polymer materials, and viscose yarn, polyacrylonitrile fibers and asphalt silk are its basic creations. Furthermore, CFRPs are characterized by their high temperature, high elastic modulus, high strength, corrosion resistance and ease of processing. In particular, in complicated geological conditions, creep displacement and the physical and mechanical properties of the material's stability are presented. It can also prolong the service lives of rock and soil structures [3] as a new and functional material that is widely used in engineering reinforcement. For example, Zhang et al. [9] studied the fracture-bearing characteristics of CFRP-reinforced concrete structures and broadened the application of pile structures in geotechnical engineering. Xiong et al. [10] proposed a bond model of a CFRP and concrete aggregate and revealed the coupling distribution characteristics of the load and stress of the CFRP, concrete and the surrounding rock soil according to the anchoring principle of the soil–anchor structure. However, although there are numerous geotechnical reinforcements methods for anti-slide pile engineering, few consider the instantaneous intensity of soil and the unilateral increase in concrete strength, and the intensity of carbon-fiber-reinforced plastics and their thickness and volume. In addition, the inherent accelerated creep of the soil mechanical properties of an anti-slide pile's load-bearing characteristics is ignored. It is clear that the reinforcement parameters are not reasonable for maintaining the effect on the stability of the slope as a whole. There are even accidents of engineering failure of the anchorage design. The research conclusion and reinforcement effect are limited. The research results show that the nonlinear creep model, the identification of the parameters of the soil pile and CFRP composite structure under disbond and discontinuous deformation conditions, the effect of CFRP yield deformation on crack growth failure, the safety factor and the strength at each point in the composite structure have not been solved, which are bound to be studied further in the future. Therefore, in the present article, three stages of soil pile creep properties are investigated, taking a high and steep anti-slide pile reinforcement and using the soil slope as the subject. The use of CFRP to improve the pile's bearing capacity is investigated, and the deformation of the soil is determined. It can be deduced that this method can control plastic zone development and strengthen the guard slope's stability characteristics for the application of CFRP in infrastructure construction.

2. Creep Behavior of Soil

2.1. Aging Deformation of Cemented Slope Soil

The reinforcement of a soil slope with an anti-slide pile is mainly used for high and steep slopes formed by artificial activities. In the process of the anti-slide pile's construction and soil excavation, soil stress is released. In fact, the sliding displacement of the adjacent slope often occurs under the action of the dead weight load and pile anchor stress. In the process of sliding deformation, due to the internal friction characteristics of the soil and the favorable factors retaining the pile structure, the speed of the sliding body gradually slows down, and the slope tends to be stable. However, a retained soil slope that belongs to a permanent structure is often influenced by rainfall, blasting vibrations and artificial disturbances, combined with the properties of the soil creep mechanics that are inherent in sliding displacement. In this case, to reduce the strength of the soil mass, the supported slope structure appears to have a large deformation and displacement mutation. Then, the development of displacement over time will continue. It can be observed that the slope finally shows instability failure. The aging deformation characteristics of the potential sliding body of the slope are shown in Figure 1.

$$u = u_0 + u_c \tag{1}$$

where u is the total displacement of slope soil, m, u_0 is the instantaneous displacement, m, and u_c is the creep displacement, m.

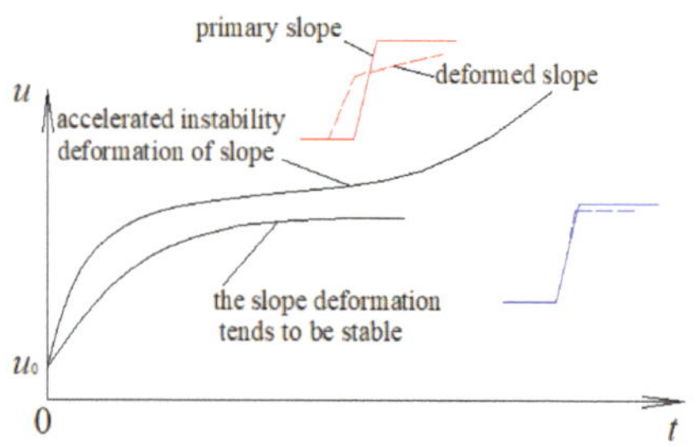

Figure 1. Time deformation of slope soil.

At the same time, in the pile reinforcement project, the slope deformation and the state of its stability are analyzed with a lack of real-time tracking. The evolution of the instability of a high and steep soil slope is hard to predict. In order to strengthen the monitoring of the displacement of a slope's soil mass, its aging characteristics were discovered, and a precautionary system of landslide prediction was also established. Improved engineering slope reinforcement is an important topic and a form of disaster prevention.

2.2. Nonlinear Creep Model

It is obvious that soil is a type of porous and granular material with low strength. Under the action of an external load, the pore structure and volume of soil are easy to adjust and change. It is easily shown that slippage and the creep phenomenon are caused. When the creep load increases and the shear strength of the soil exceeds the shear strength, the soil presents accelerated creep displacement and damage. Furthermore, experiments show that creep is an inherent mechanical property of soil mass, and a certain amount of creep displacement often occurs in soil under a creep load [10]. When the creep load is less than the soil's shear strength, there are two stages of creep, and the creep is stable. In contrast, the three stages of accelerated creep are not stable. However, a significant amount of soil creep research focuses on linear creep. A two-phase creep element model can only reflect the creep sliding failure of a soil mass. In addition, the class model and nonlinear creep element model are used for high and steep slopes. Research on the anti-slide pile reinforcement of soil slopes is rare, and sliding soil damage to slopes is often caused by accelerated creep. Therefore, a nonlinear creep theory regarding geotechnical slope engineering stability is more in line with engineering practice.

It is known that the difficulty in studying the nonlinear creep of pile-anchored soil is to search for accurate nonlinear elements to reflect plastic deformation. Unlike traditional metal materials, the strength of soil is not only related to its own material properties but also to external stress conditions. The analysis of plastic deformation is more complicated than that of metal materials. For a soil structure reinforced by a concrete anti-slide pile, the load is mainly carried by the embedded anchorage section. When the soil pressure is high, the accelerated creep displacement effect of the soil has great influence on the internal force distribution of the anchorage section and the plastic zone of the retaining soil structure [11,12]. Therefore, the classical creep element model cannot accurately evaluate the nonlinear bearing capacity of the anti-slide pile. In this paper, according to the displacement characteristics of the soil pile components, the Mohr–Coulomb strength criterion of soil mass was applied. Based on the empirical nonlinear element model theory, the self-built M-C plastic yield element was added to analyze the accelerated creep displacement characteristics of the soil mass. This aspect, named the M-C plastic element, is mainly used to solve the accelerated plastic strain. The entire creep process of the retaining soil was simulated by a parallel connection between the M-C plastic yield element and the three elements of a generalized Kelvin body. Meanwhile, the elastic element was used to simulate the loaded section of the pile, and the viscoplastic element and elastic element were connected in parallel to simulate the anchorage section of the pile. The self-built nonlinear creep model is shown in Figure 2.

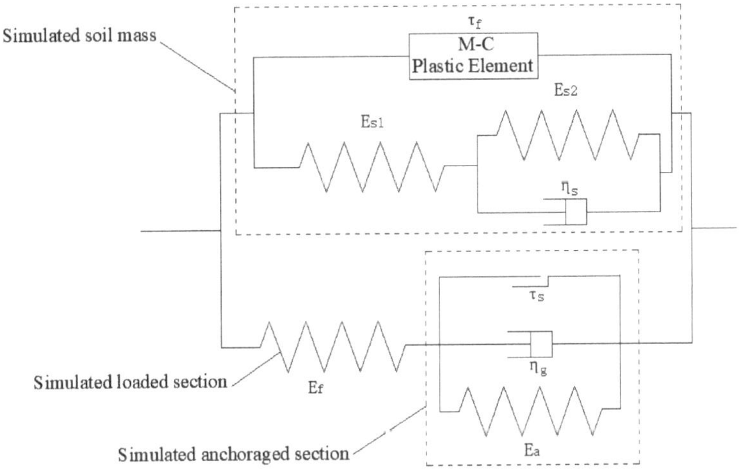

Figure 2. Nonlinear creep model of slope reinforcement.

In order to facilitate the nonlinear creep analysis of the loaded section, anchorage section and soil mass, the following assumptions were made:

(1) The soil of the retaining slope is a continuous, homogeneous and isotropic material without groundwater seepage.
(2) The creep load comes from the active and passive earth pressures of the retaining slope soil without earthquake, construction and ground overloading.

Using Figure 2, according to the theory of creep mechanics, the creep synergy equation of a soil pile structure is obtained in Equations (2) and (3).

$$\varepsilon_s(t) = \begin{cases} \sigma_s[\frac{t}{\eta_g} + \frac{1}{E_{s1}} + \frac{1}{E_{s2}}(1 - e^{-\frac{E_{s2}}{\eta_s}t})] & (\tau < \tau_s) \\ \sigma_s[\frac{t}{\eta_g} + \frac{1}{E_{s1}} + \frac{1}{E_{s2}}(1 - e^{-\frac{E_{s2}}{\eta_s}t})] + \varepsilon_{MC} & (\tau \geq \tau_s) \end{cases} \tag{2}$$

$$\sigma_s = \int_0^{L_a} \tau(x)\pi D \mathrm{d}x \tag{3}$$

where $E_{s1}, E_{s2}, \eta_s, \eta_g, \tau_s$ are the elastic modulus, MPa, viscoelastic modulus, MPa, viscoelastic coefficient, MPa·d, viscosity coefficient, MPa·d and the shear yield strength of simulated soil creep, kPa, respectively, E_f, σ' simulates the elastic modulus and creep stress of the pile under a load, MPa, $\tau(x)$ is the shear stress distribution of the soil- pile, kPa ε_{MC} is the plastic strain reflecting the accelerated creep, and L_a is the length of the loaded section, m.

The plastic strain ε_{MC} of the nonlinear model can be obtained via the Mohr–Coulomb yield criterion, which is shown in Equation (4).

$$F = -\frac{1}{2}\sigma_s + \frac{1}{2}\sigma_s \sin\varphi + c\cos\varphi \tag{4}$$

According to the conclusion of the creep experiment in Refs. [13,14], the viscoplastic strain rate of the M-C nonlinear creep element can be derived in Equation (5).

$$\dot{\varepsilon_{MC}} = \eta_s \left[\frac{-\frac{1}{2}\sigma_s + \frac{1}{2}\sigma_s \sin\varphi + c\cos\varphi}{F_0} \right] \left(-\frac{1}{2} + \frac{1}{2}\sin\varphi \right) \tag{5}$$

where c, φ are the shear strength index of the soil, kPa and °, and F_0 is the initial yield strength of the slope soil.

Therefore, the viscoplastic strain of the nonlinear creep element can be obtained via a differential solution. The solution is written as:

$$\varepsilon_{MC} = \int_{t_p}^{t_F} \eta_s [\frac{\frac{1}{2}\sigma_s - \frac{1}{2}\sigma_s \sin\phi - c\cos\phi}{F_0}](\frac{1}{2} - \frac{1}{2}\sin\phi)dt \tag{6}$$

Combining the plastic yield principle of soil structure, yield failure occurs when accelerated creep is present. Then, the failure time of the accelerated creep can be calculated as follows:

$$t_F = t_0 + \frac{1}{[\int_0^{l/2} \frac{k\varepsilon_0}{d} \exp(\pi dx) - \tau_0}} \tag{7}$$

where t_0 is the initial time of the accelerated creep, h, τ_0 is the initial bonding strength of the reinforcement and anchorage agent, kPa, ε_0 is the initial strain of the accelerated creep, t_P is the plastic yield time, h, and t_F is the accelerated failure time, h.

3. Load-Bearing Characteristics of Soil Pile Carbon Fiber Reinforcement

3.1. Tensile Creep Test of Carbon-Fiber-Reinforced Plastic Sheet

Generally, a CFRP-reinforced anti-slide pile mainly bears a tensile effect. Under the action of landslide thrust, the deformation of the CFRP and pile are coordinated. At the same time, with the increase in the creep deformation of the soil behind the pile, the tensile deformation of the CFRP also increases with time. A carbon fiber epoxy resin composite rectangular sheet was used in the CFRP aging and tensile deformation experiment in this paper. In the following testing parameters, the carbon fiber was a Cymax-L-C type and an XH 130 A/B type epoxy resin adhesive with A:B = 3:1. Regarding the landslide thrust, the tensile load was 400 kN and the creep tensile test time was 45 h. The experimental results of the aging tensile deformation test are shown in Figure 3.

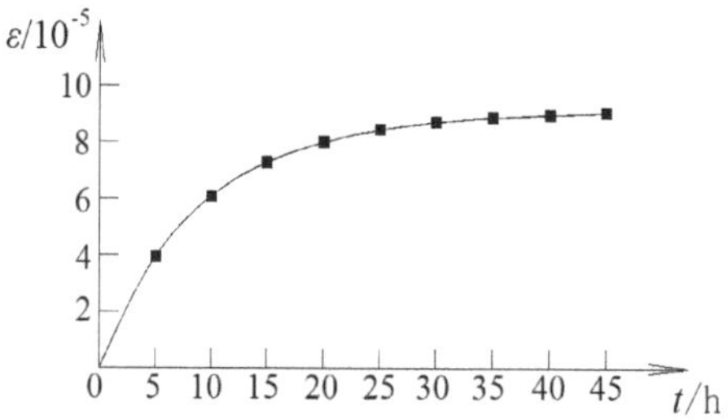

Figure 3. Time deformation of sheet material.

It can be seen from Figure 3 that under the creep load, the tensile deformation of the CFRP sheet exhibits stable creep deformation characteristics. In addition, the maximum tensile strain is 8×10^{-5}, which is smaller than that of the slope soil. On the other hand, the sheet deformation rate attenuated at the initial stage of tensile deformation and remained stable after 20 h.

3.2. Load-Bearing Characteristics of Soil Pile–CFRP Composite

Concrete engineering is mainly reinforced with CFRPs for building beam–column structures. Generally, in the form of a reinforcement and sheet, the composite reinforcement structure is formed through external sticking, embedding and concrete. Therefore, according to the pile geometry size, buried conditions, stress characteristics and engineering construction environment, the CFRP placement in the pile is generally under the external load side to enhance the stability of the soil, especially for the reinforcement of the damaged pile.

Under the action of earth pressure, bending deformation and the horizontal displacement of the loaded section of the concrete anti-slide pile occur, which are mainly manifested in the compression of the concrete outside the pile body and the tensile damage of the concrete inside the pile body. Meanwhile, when the load on the top of the loaded section is transferred to the intersection of the loaded section and the anchoring section, the pile body is broken. Therefore, improving the concrete strength of the tensile side of the in-service anti-slide pile or the damaged anti-slide pile has an obvious effect on controlling the further deformation of the soil and pile. CFRP can be closely bonded with concrete via phenolic resin to enhance the tensile strength of the concrete, and it cannot not be corroded by soil and the water in the soil. Therefore, it has a good application prospect for strengthening unstable rock and soil structures. Inside, an anti-slide pile can appear as a composite structure throughout the CFRP reinforcement with concrete, reinforced soil and steel. At the same time, the load-bearing characteristics are above. The strength of the carbon-fiber-reinforced plastic is generally greater than pile strength of the concrete, so the elastic modulus and yield strength of the composite structure are increased. Moreover, the bearing capacity increases and can reduce the creep displacement of the soil mass. Therefore, the stability of the retaining slope is enhanced. A structural diagram of an anti-slide pile reinforced by CFRP is shown in Figure 4.

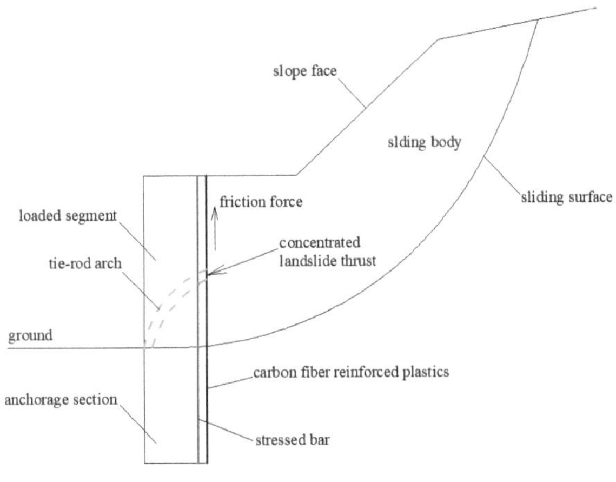

Figure 4. Reinforced slope by CFRP anti-pile.

$F_{si}(i = 1, 2, 3)$ is the frictional force, written as:

$$F_{si} = \int_0^A q_{si} dA \tag{8}$$

where A and q_{si} are the frictional area and the frictional intensity, m^2 and kPa, respectively.

In order to study the influence of the load section on the sliding force and control deformation better, the shear/span ratio variable is cited in this paper [13,14]. According to the influence of landslide thrust on pile deformation, a tie-rod arch is formed, and then the deformation of the pile under the loaded section is obtained according to the tie-rod arch deformation characteristics. We incorporated the anti-slide pile into the tie-rod arch model (Figure 5):

$$\lambda = \frac{\sum M}{Qh_0} = \frac{(P \cdot a \cdot \cos\alpha + F_{s1} \cdot l + F_{s2} \cdot b - P \cdot b \cdot \sin\alpha)}{(P\cos\alpha + F_{s1}) \cdot h_0} \tag{9}$$

where λ, Q, P, M, h_0, a, l, b and α are the shear/span ratio, the shear stress, kPa, the concentrated load, kN, the bending moment, kN·m, the effective height, m, the distance between the sliding force and the base, m, the distance between the frictional curve and slide curve, m, the width of the anti-slide pile, m, and the angle determined by the equivalent concentrated thrust and horizontal direction.

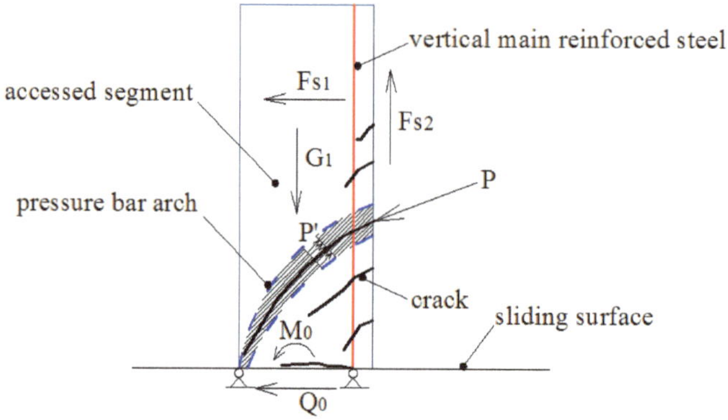

Figure 5. Transmission mode and destruction pattern from Ref. [14], copyright 2021, Springer. Q0, M0, Fs1, Fs2, P and Gs1 are the shear force on the sliding surface, the bending moment on the sliding surface, the lateral friction force, the lateral frictional force at the bearing segment, the sliding force, and the gravity load of the bearing segment, respectively.

Equation (9) shows that the shear/span ratio is equal to $\frac{a}{h_0}$ without considering the lateral friction force, F_{s1}. Thus, the reasonability of Equation (9) is verified. The thickness distribution of the sliding body and the mechanical monitor results indicate that the distance between the location of the sliding force, P, and the slide surface is about (1/4~1/3)h1. The shear/span ratio ranges from 1 to 5. Thus, shear pressure and inclined tensile failures occur.

When cracks occur in the loaded section, greater deformation and stress redistribution occur. The compression failure occurs in the loaded section. Furthermore, the crack expands further. Finally, the fracture failure of the pile body occurs. There is an obvious correlation between the shear/span ratio and the diagonal dip angle. The crack propagation angle decreases nonlinearly with the increase in the shear/span ratio. This relation is shown in Figures 3 and 6.

The fitting power function equation is:

$$\lambda = m47.017\theta^{-0.8453n} \tag{10}$$

where m and n are experimental constants, and λ and θ are the shear/span ratio and diagonal inclination angle, °, respectively.

Clearly, the shear/span ratio decreases the diagonal dip angle (α in Figure 1). In addition, the decrease rate slightly decreases with the increase in the shear/span ratio. The larger variation in the arch curvature affects the shear strength more significantly.

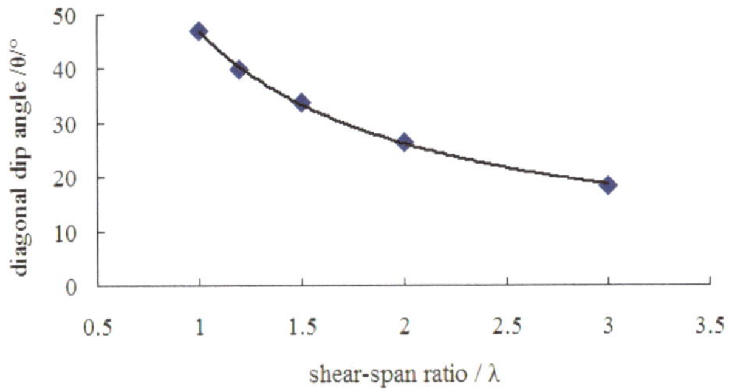

Figure 6. Relation between shear/span ratio and diagonal dip angle of arch from Ref. [14], copyright 2021, Springer.

4. The Example Analysis

4.1. Calculation Range and Calculation Parameters

In order to verify the composite structure system of the CFRP and the anti-pile slope, it should have the advantages of a high efficiency and reasonable ability to control soil deformation, improving the pile strength, reducing the plastic zone and improving safety factor of the retaining side slope. In the present article, a concrete anti-slide pile was selected to reinforce a high, steep homogeneous soil slope. The calculation range of the blocked side slope is as follows: the calculated length is 67 m, the calculated height is 44 m, the slope height is 22 m, and the slope top width is 28 m. The calculated dimensions of the slope height and width meet the requirements of a finite difference numerical simulation calculation. The cohesion is 18 kPa, the internal friction angle is 10°, the bulk density of soil mass is 18 kN/m^3, the foundation coefficient of the soil is 3.2 $\times$ 10^4 kN/m^3, the elastic modulus of the soil is 8 MPa, Poisson's ratio is 0.35, the viscoelastic modulus of the soil is 1.25 MPa and the viscoelastic coefficient is 8 MPa·d. The viscosity coefficient of the soil is 18 MPa·d, and the internal force calculation of the anti-slide pile adopts M method, $m = 6 \times 10^3$ kN/m^4. There is no groundwater, the environment is class I and the safety grade of slope is class I. The size of the anti-slide pile is 2.5 m $\times$ 2 m, the length of the pile body is 18 m, the load-bearing section is 8 m, the anchoring section is 10 m, C30 concrete was used for pouring, the vertical stress reinforcement is HRB400 and the side protective layer inside the pile body is 80 mm. Due to the relatively large earth pressure, there is a large bending deformation in the loaded section of the pile. The use of a CFRP sheet to reinforce the inside side of the pile is proposed. The tensile strength of the epoxy resin binder is 30 MPa, the elastic modulus is 3 GPa, the tensile strength of the CFRP is 2 GPa and the elongation is 1.5%. The tangential strength of the bond layer with epoxy resin is 45 MPa. The calculation range of the blocked edge slope is shown in Figure 7, and the grid mesh is shown in Figure 8.

Through the powerful 3D finite difference numerical software, which is based on the Powerstaion platform, the addition of the self-built nonlinear creep constitutive model was applied, which embedded the plastic deformation constitutive model via the horizontal displacement reflection. Using the CFRP composite material model, the soil pile's penal horizontal displacement, the plastic zone and the safety coefficient of the slope were analyzed. In the retaining structure, a brick element was used for soil, a pile element was used for the anti-slide pile, and a shell element was used for carbon-fiber-reinforced plastic sheet. An interface was used for interface bonding. The calculated width was the unit length, and the calculation model had 4896 nodes and 22,334 units. The other boundaries were free. In order to explain the influence of the CFRP and the creep effect on the displacement, the plastic zone and the safety factor distribution characteristics of the composite structure of the retaining soil, a numerical calculation was carried out under the following three conditions: (1) the soil slope was not reinforced with anti-slide piles; (2) the soil slope was reinforced by an anti-slide pile without CFRP; (3) the soil slope was reinforced by a composite reinforcement of an anti-slide pile and carbon-fiber-reinforced plastic. Non-creep and horizontal creep displacement after one year were considered. The calculation results are shown in Section 4.2.

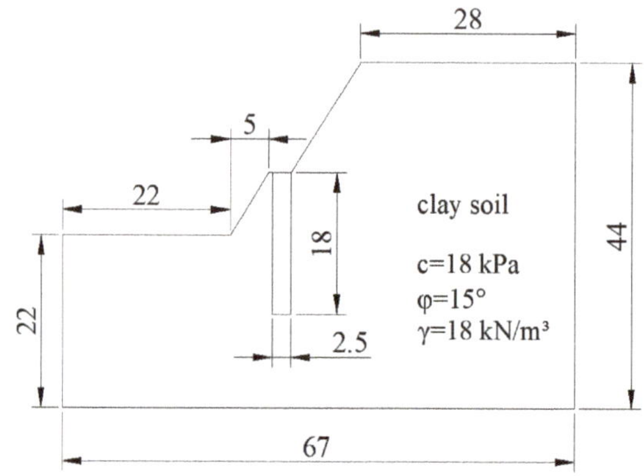

Figure 7. Calculation scope of reinforced slope (unit: m).

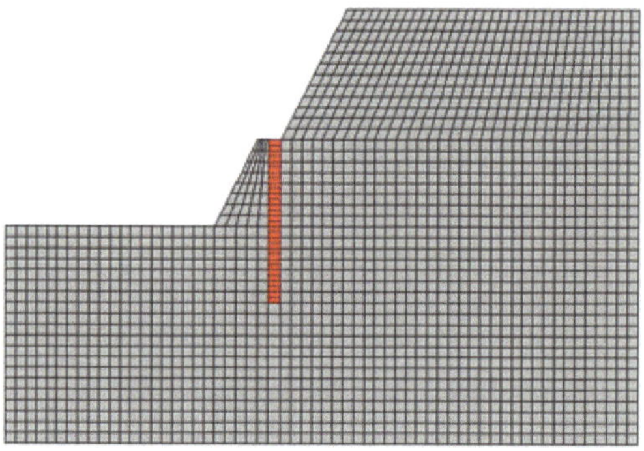

Figure 8. The reinforced slope FD calculation mesh.

4.2. The Calculation Results

As shown in Figures 9 and 10, under the condition of an unreinforced slope for a numerical calculation of a short duration, the excavation platform was not considered. It is denoted that the horizontal displacement points to the plane negative displacement. The maximal displacement is about 10 cm, and it is mainly distributed in the slope's shoulder position. In the meantime, the minimum displacement is 1.0 cm, and ot is mainly distributed in the local area. Thus, the slope toe produced obvious horizontal displacement and exceeded the standard value. At the same time, a shear plastic zone from the top to the foot of the slope appeared in the shallow part of the slope and extended into the slope. The overall safety factor of the slope is 0.99. Therefore, the slope is in an unsteady state.

As shown in Figure 11, by applying a pile to reinforce the soil slope, the horizontal displacement was obviously improved. Meanwhile, the maximum horizontal displacement is 5.0 cm, which is mainly distributed in the pile top local. The displacement is reduced by nearly 50% when compared to the first case. Therefore, the anti-slide pile reinforcement effect is obvious. However, under the influence of soil displacement creep, the soil pile head is in hidden danger, and is still sliding at the same time as the least-deep horizontal displacement occurs in the slope body figure. Therefore, the deformation tends to be stable.

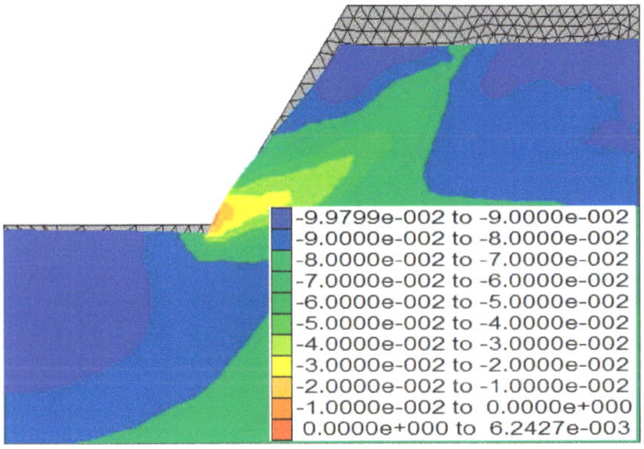

Figure 9. Contour of horizontal displacement without reinforcement with anti-slide piles (unit: m).

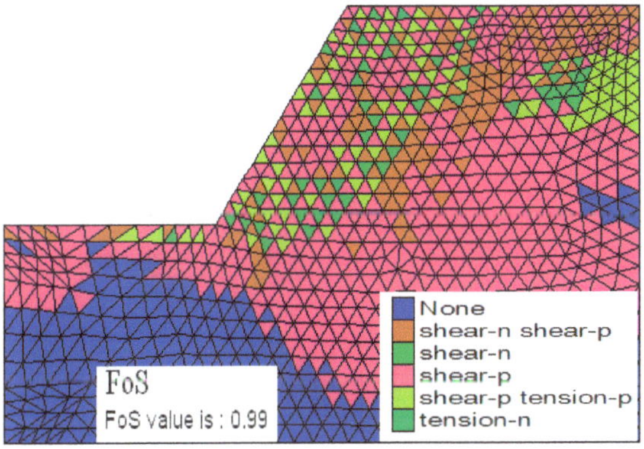

Figure 10. Plastic state and safety factor without reinforcement with by anti-slide piles.

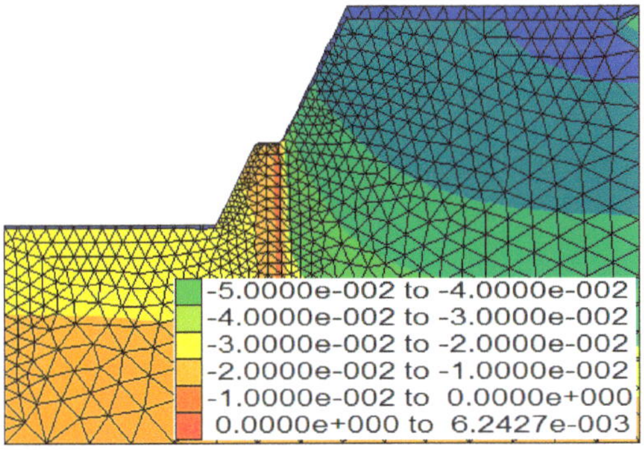

Figure 11. Contour of horizontal displacement without CFRP (Unit: m).

In order to further control the creep displacement of the pile top, the pile-bearing capacity was improved to prevent hidden danger. For the CFRP in the side of the pile body, the horizontal displacement was further improved by the block slope (see Figure 12), and the maximum horizontal displacement is 4.0 cm. In addition, the maximum horizontal displacement area shrinks, and the maximum horizontal displacement of the pile-bearing section is 1.8 cm. Meanwhile, its deformation is within the allowed value.

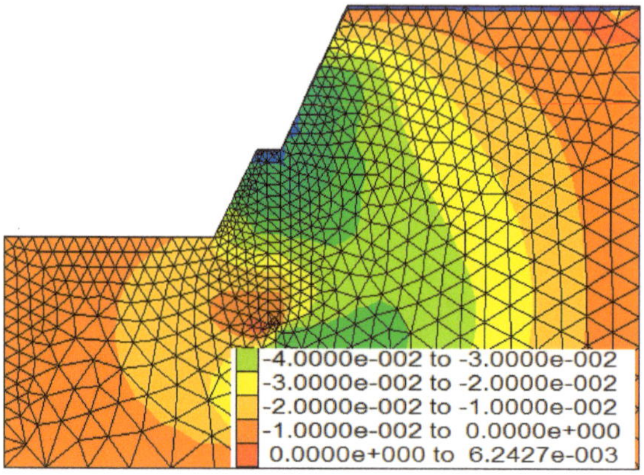

Figure 12. Contour of horizontal displacement with anti-slide piles and CFRP (Unit: m).

As shown in Figure 13, concentrating on the shallow surface of the pile's top slope, the plastic zone decreases with CFRP composite reinforcement of the anti-slide pile. There is no transfixable plastic zone appearing in other positions. The overall safety factor of the slope is 1.3. Thus, the slope is in a stable state.

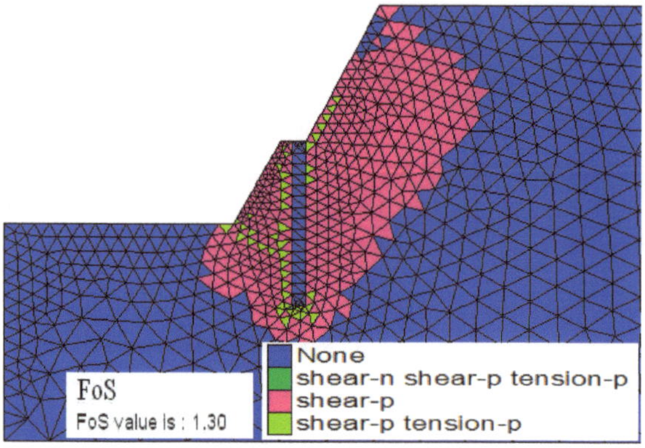

Figure 13. Plastic state and safety factor with anti-slide piles and CFRP.

Specifically, the effect of creep on the horizontal displacement of the retaining structure is not considered in the above calculation results. In order to verify the influence of soil creep properties on the bearing capacity of the composite structure system, the horizontal displacement of the soil after one year is obtained by combining the nonlinear creep model in this paper, as shown in Figure 14.

Figure 14 shows that through the pile of the CFRP-composite-reinforced soil slope, the maximum horizontal displacement is about 5.5 cm one year later. Meanwhile, the maximum horizontal displacement of the reinforcement moment increased by 10%, and the maximal displacement and pile top position distribution widened. Thus, the tendency of the increasing displacement leads to a landslide hazard. Therefore, the soil creep effect should not be neglected.

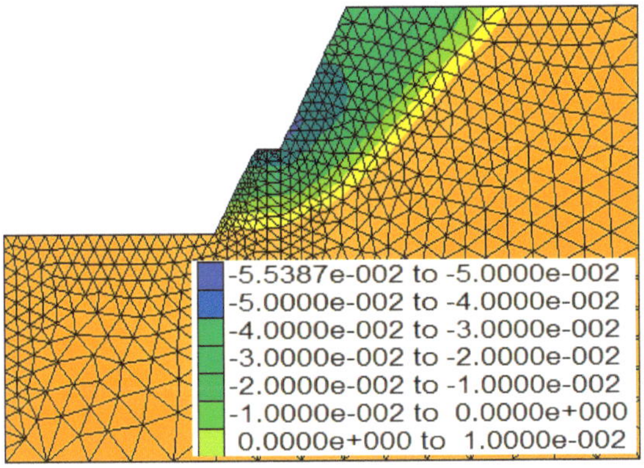

Figure 14. Contour of horizontal displacement with anti-slide piles and CFRP under creep (Unit: m).

5. Discussion

(1) At present, only the instantaneous strengths of geotechnical materials are considered in the design parameters of geotechnical engineering reinforcements. Based on the external conditions and these parameters, in combination with the design code, the type selection, geometric size and structural design of the reinforcement structure are carried out to ensure the stability of the project. Although the method has summarized many practical experiences and can be applied well to engineering practice, it ignores the inherent creep characteristics of a rock and soil mass and the change in load conditions after stress redistribution. Furthermore, time-dependent deformation, even accelerated deformation, and the weakening of the strength parameters of the composite material, interface debonding and the fracture of the retaining structure and the plastic zone of the perforated rock–soil structure will still occur. Once the deformation exceeds the allowable value, the rock and soil structure will have the possibility of failure due to instability.

(2) Considering the creep effect of soil mass, the slope displacement increases, and the overall safety factor decreases. At the same time, the displacement at the top of the pile increases most significantly. The main reason for this is that the position is located on the slope berm and faces an empty and unsupported structure. Furthermore, the upper slope's mass landslide thrust is larger with the concentration of stress. Thus, the displacement is at the maximum value. However, the displacement at the bottom of the slope is minimal because of the anti-slide piles, which disturb the deformation of the slope. Simultaneously, the limitations of the model presented herein are as follows: ① discontinuous deformation occurs after the debonding of pile–soil interface; ② the pile-soil structure produces accelerated deformation and creep of residual deformation after failure; ③ the soil mass structure changes easily on the heterogeneous macadam soil. The plastic deformation of the anti-slide pile occurs easily when the strength of the loading section is low. In view of these factors, these problems need to be researched further.

6. Conclusions

In this work, CFRP is a new kind of reinforcing material for use in a soil slope reinforced by an anti-slide pile. Its application broadens the reinforcement form of concrete structures and promotes the development of modern geotechnical reinforcement technology. In geotechnical reinforcement engineering, it can reduce the dead weight of an anti-sliding structure, improve the bearing capacity and enhance the stability of soil, especially for the reinforcement of an anti-sliding pile structure.

This paper finds advantages in a soil slope reinforced by an anti-slide pile, and a nonlinear creep model which can reflect three stages creep, attenuation creep, stable creep and accelerated creep, was established. Meanwhile, the creep model and viscoplastic strain analytical formula were found, and the tensile deformation characteristics of the CFRP sheet under landslide thrust creep load were obtained. Combining with numerical examples, it is verified that the reinforcement scheme of laying CFRP on the load-bearing side of the anti-slide pile can significantly control the further deformation of the soil and pile, reduce the penetration of the plastic zone and improve the construction's overall safety factor. It should be noted that the deformation trend of the CFRP composite reinforcement of an anti-slide pile was also discovered while considering the creep effect. In the third case, compared with the first and second cases, after one year, the maximum horizontal displacement decreased by 50% and increased by 10%, respectively. Simultaneously, the overall safety factor increased by 31.3% without soil creep properties. On the contrary, the overall safety factor is reduced, and the slope has a tendency toward unstable failure. Thus, the soil creep effect must be paid attention to in engineering practice.

Author Contributions: Data curation, J.W. and P.C.; formal analysis, P.C. and L.L.; project administration, J.W. and P.C.; supervision, J.W. and L.L.; writing original draft, J.W. All authors have read and agreed to the published version of the manuscript.

Funding: The research was funded by the Natural Science Foundation of Hunan Province (No.2019JJ40056), scientific Research Foundation of Hunan Province Education Department (No.18A345, 18B391, 22C0765),and the construct program of applied specialty disciplines in Hunan province.

Institutional Review Board Statement: Not applicable.

Informed Consent Statement: Not applicable.

Data Availability Statement: All data used in this paper are listed in the tables and figures of this manuscript.

Acknowledgments: The authors gratefully acknowledge the support from the Natural Science Foundation of Hunan Province (No. 2019JJ40056), the National Natural Science Foundation of China (No.51804110), the Scientific Research Foundation of Hunan Province Education Department (No.18A345, 18B391, 19B124), and the construction program of applied specialty disciplines in Hunan province. These works are gratefully acknowledged.

Conflicts of Interest: The authors declare no conflict of interest.

References

1. Nechita, P.; Roman, M. Review on polysaccharides used in coatings for food packaging papers. *Coatings* **2020**, *10*, 566. [CrossRef]
2. Sorin, M.A.; Claudia, B.; Claudia, V.P.; Alina, O.; Lonut, M.; Georgeta, T. Edible and functionalized films/coatings-performances and perspectives. *Coatings* **2020**, *10*, 687.
3. Liu, C.G.; Bi, H.J.; Wang, D.; Li, X.N. Stability reinforcement of slope using vegetation considering the existence of soft rock. *Appl. Sci.* **2021**, *11*, 9228. [CrossRef]
4. Chen, Y.; Feng, R.; Xu, J. Flexural behaviour of CFRP strengthened concrete-filled aluminium alloy CHS tubes. *Constr. Build. Mater.* **2017**, *142*, 295–319. [CrossRef]
5. Lee, H.; Jun, W.T.; Chung, W. Bond behavior of near surface mounted CFRP rods under temperature cycling. *Eng. Struct.* **2017**, *137*, 67–75. [CrossRef]
6. Wang, Y.X.; Guo, P.P.; Ren, W.X. Laboratory investigation on strength characteristics of expansive soil treated with jute fiber reinforcement. *Int. J. Geomech.* **2017**, *17*, 04017101. [CrossRef]
7. Zhang, C.Y.; Pu, C.Z.; Cao, R.H. The stability and roof-support optimization of roadways passing through unfavorable geological bodies using advanced detection and monitoring methods among others in the Sanmenxia Bauxite Mine in China's Henan Province. *Bull. Eng. Geol. Environ.* **2019**, *78*, 5087–5099. [CrossRef]
8. Zhao, Y.L.; Zhang, C.S.; Wang, Y.X.; Lin, H. Shear-related roughness classification and strength model of natural rock joint based on fuzzy comprehensive evaluation. *Int. J. Rock Mech. Min. Sci.* **2021**, *137*, 104550. [CrossRef]
9. Zhang, H.X.; Zhu, H.S.; Sun, L.; Li, J. Experimental study on bond-slie behavior between FRP bars and concrete. *J. Shenyang Jianzhu Univ.* **2008**, *24*, 989–992.
10. Xiong, C.N.; Shao, Y.B.; Chen, Z.M. Analysis on yield strength of CFRP-strengthened concrete filled circular steel tubularunder initial loading state. *J. Build. Struct.* **2021**, *42*, 73–83.

11. Yuan, H.P.; Zhao, P.; Wang, Y.X.; Zhou, H.L.; Luo, Y.H.; Guo, P.P. Mechanism of deformation compatibility and pile foundation optimum for long-span tower foundation in flood-plain deposit zone. *Int. J. Civ. Eng.* **2017**, *15*, 887–894. [CrossRef]
12. Zhao, Y.L.; Wang, Y.X.; Wang, W.J. Modeling of non-linear rheological behavior of hard rock using triaxial rheological experiment. *Int. J. Rock Mech. Min. Sci.* **2017**, *93*, 66–75. [CrossRef]
13. Li, J.; Jiang, X.G.; Wang, H.Z. Analytical element for Timoshenko beam on elastic foundation. *Eng. Mechancis* **2018**, *35*, 221–229.
14. Wang, J.; Liu, J.; Liang, Q. Internal force of the anti-slide pile in soil based on the deep beam model. *KSCE J. Civ. Eng.* **2021**, *25*, 782–792. [CrossRef]

coatings

MDPI

Article

Experimental Study of Electrolytic Processing of Discharge-Assisted Jet Masks

Chaoda Chen [1], Shaofang Wu [1,*], Tong Zhang [2], Yage Wang [2], Xiaoqiang Shao [1] and Fengrong Mo [1]

[1] School of Naval Architecture and Ocean Engineering, Guangzhou Maritime University, Guangzhou 510725, China; cdchen_edu@163.com (C.C.); jjyuedu@126.com (X.S.); llzhongedu@126.com (F.M.)
[2] College of Mechanical and Electrical Engineering, Guangdong University of Science and Technology, Dongguan 523000, China; gugu0769@126.com (T.Z.); wangyagegdust@163.com (Y.W.)
* Correspondence: sfwu@gzmtu.edu.cn or sfwu_edu@126.com

Abstract: There has been some research on jet electrolytic processing at home and abroad, and the phenomenon of discharge during the process has been reported, but there has been little research on the mode of jet electrolysis with the aid of discharge. A jet mask electrolytic processing experiment was set up to prepare a blue oil mask on the surface of the workpiece using photolithography; two processing modes were achieved using different tool electrodes, the workpiece was processed by two types of motion, the processing micro-pits were observed morphologically using an optical microscope, and the test data were analyzed by plotting graphs. Experiments show that a blue oil mask with a thickness of 50 μm covers the workpiece to strengthen the fixity, and that jet electrolytic discharge machining can effectively improve the depth-to-width ratio by increasing the contribution to depth by 30%–38% and the contribution to width by 2%–18%, compared to jet electrolytic machining. The former has less island effect than the latter, with a flatter bottom and better-machined shape.

Keywords: jet electrolytic machining; in-line dressing of electrical discharges; workpiece morphology analysis

Citation: Chen, C.; Wu, S.; Zhang, T.; Wang, Y.; Shao, X.; Mo, F. Experimental Study of Electrolytic Processing of Discharge-Assisted Jet Masks. *Coatings* **2023**, *13*, 1280. https://doi.org/10.3390/coatings13071280

Academic Editor: Tadeusz Hryniewicz

Received: 7 June 2023
Revised: 17 July 2023
Accepted: 20 July 2023
Published: 21 July 2023

1. Introduction

Jet electrolytic machining (Jet-ECM) is a special form of electrochemical machining (ECM) that uses an electrolytic jet as a tool. It offers advantages over other electrochemical machining processes such as good domain fixation, high machining speeds, and machining position flexibility [1,2]. The use of a displacement stage to drive the nozzle cathode to follow a pre-defined path enables rapid and precise fabrication of complex structures on the workpiece surface [3–5]. Hinduja and Pattavanitch [6] electrochemically machined a simple square pocket and a pocket with a humanoid protrusion, using tools with round and square electrodes. Hackert-Oschätzchen et al. [7] used a continuous electrolytic free jet to obtain a high degree of positioning in the removal area and electrochemically processed cavities with a depth of about 180 μm and a width of about 190 μm. Kawanaka and Kunieda [8] investigated the effect of current conditions and nozzle movement speed on the surface roughness of stainless steel, and achieved a mirror-like surface with Rz surface values of less than 0.2 μm. It was shown that the surface roughness of the cavities decreased as the nozzle translation speed increased, and that for constant speeds the surface roughness of the cavities decreased and increased again after reaching a minimum value [8–10].

Electrochemical discharge machining (ECDM) is considered to be a hybrid machining method. Material removal is based on two phenomena: electrochemical dissolution of the material and corrosion due to hot inter-electrode discharge [11,12]. Kim et al. [13] proposed an ECDM drilling method for glass workpieces, where a working fluid containing potassium hydroxide facilitated the process. Kobayashi [14] described a solution for obtaining shaped cavities by ECDM and electro-corrosion, using an electrolyte based on sodium nitrate. Liu, J.W. et al. [15] proposed a discharge mechanism analysis of electrochemical discharge machining of particle-reinforced metal matrix composites in electrolytic machining by dividing the time experienced by the discharge phenomenon of the output of the ECM process into three phases: electrolytic machining under the condition of power supply, electrolytic composite machining by arc discharge, and electrolytic machining under the condition of capacitive power supply. Yu, W. [16] mentioned in an experimental study of electrohydraulic beam machining of microstructures that spark discharge was generated between the nozzle and the workpiece at an initial machining gap of 0.5 mm, a feed rate of 0.1 mm/min, an electrolyte pressure of 0.3 MPa, an electrolyte solution of 18% $NaNO_3$ and a machining voltage of 280 V. For Cui, J. [17], the study of the motion characteristics, electromagnetic characteristics, and oscillation of charged particles in the discharge channel is an important part of the research on the mechanism of EDM. Liu, H. and Xieeryazidan, A. [18] investigated the effect of energy distribution on electrochemical discharge machining, and experiments showed that the optimal combination of process parameters at 50 V voltage, 20 kHz frequency, and 80% duty cycle achieved a more uniform distribution of discharge energy, which was able to efficiently machine pairs of hard and fragile materials. Chuang et al. [19] utilized the interaction of electrochemical and EDM compensating techniques for the machining of CrNiFe alloy material machining; the machining accuracy was characterized by electrical pulse signal matching, processing time was reduced by 80%, and the efficiency of the equipment was improved. Sambathkumar and Arunagirinathan [20] investigated the working gap (Wg) between the tool and the workpiece in ECDM affecting the process parameters, and the results showed that even a difference of 2 μm in the Wg significantly affects the ECDM process in terms of quality and quantitative metrics. Zhao et al. [21] optimized the process without the loss of cathode tools by constructing a coupled multiphysics field model for electrochemical mask processing and combining it with numerical simulation analysis to realize the distribution state of multiphysics fields. Zhan and Zhao [22] proposed a plasma-assisted electrochemical machining method, where a higher voltage is applied to promote cathodic hydrogen precipitation, inducing a plasma discharge that leads to rapid heating of the electrolyte around the working electrode, which facilitates subsequent anodic etching by enhancing the kinetics of the electrochemical reaction. The material removal rate was increased by about 20 times, compared to ECM. Jui et al. [23] used electrochemical discharge machining to process microvias on glass with a high aspect ratio, of 11. Wang et al. [24]. proposed a novel electrochemical spark grinding composite machining method, which has lower discharge energy compared to conventional electrochemical discharge machining and solves the edge collapse and breakage problems.

Jet electrolytic discharge machining (Jet-ECDM) is a Jet-ECM with superimposed discharge effects and a special form of ECDM using a jet as a tool and using a mask for positional constraints; the model is unique and novel, most of the research is still in the early exploratory stage, and there are few production applications in practice.

The experimental setup of jet electrolytic machining with a mask covering the workpiece is established, and two models of Jet-MECM and Jet-MECDM are experimentally compared, using jet mask electrolytic machining and jet mask electrolytic discharge machining. The two modes of motion are used to machine the workpiece and analyze the morphological excavation characteristics of the machined workpiece, which is of great significance to the in-depth study of Jet-ECDM. This paper is dedicated to the above, providing a new method and theoretical basis for Jet-ECM.

2. Materials and Experimental Methodology

The experimental setup for jet mask electrolytic processing is shown in Figure 1. The machine mainly consists of a linear module for the machine, a working feeder for the servo motor, and a work table. The machine is mounted on a rigid marble base which has a large vibration resistance, poor thermal conductivity, and a coefficient of thermal expansion. To prevent small vibrations in the surroundings from being transmitted to the machine, an adjustable damping pad is installed under the base. The electromechanical control section consists mainly of Mitsubishi servo motors, a Renishaw micrograting system, and a motion control card from Shenzhen Raytheon. The human–machine interface is programmed using LabVIEW 2022, the NI virtual instrument development platform.

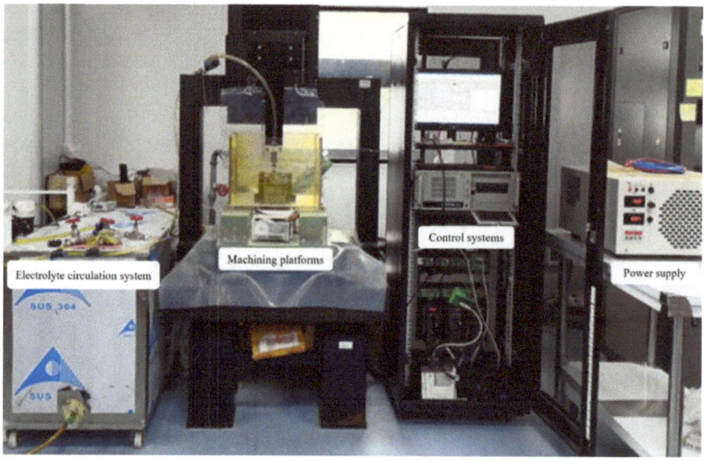

Figure 1. Diagram of the experimental setup for electrolytic processing of jet masks.

The electrolyte circulation system consists of circulation tubes, pressure control valves, pressure gauges, and reservoirs, and is an important part of the laboratory processing equipment. The high-voltage pulse power supply has excellent characteristics, such as reliable precision and outstanding stability. Under the control of human–machine interaction, the motion platform can move along the x, y, and z axes to complete the precise machining of the workpiece. At the same time, fluctuations in current and voltage during online processing can be monitored in real time with a digital oscilloscope.

The main parameters of the machine tool equipment for jet mask electrolytic processing are shown in Table 1.

Table 1. Main parameters of the machine tool equipment for jet mask electrolytic processing.

Name	Description	Unit
External dimensions	1000 × 750 × 2230	mm
X/Y direction travel	200 × 200	mm
Z-directional travel	300	mm
X/Y/Z axis positioning accuracy	5	μm
Control system software	Labview	-

The self-developed jet mask electrolytic processing machine tool equipment, after parameter testing, reached the expected target, and the equipment control was a precise and stable operation, which laid a good foundation for experimental research.

Jet mask electrolytic machining technology is for making skeleton patterns on the mask and covering the mask on the surface of the anode workpiece; it uses $NaNO_3$ electrolyte to spray processing on the workpiece, to make the workpiece surface obtain the same geometry as the mask skeleton pattern process method. This method solves the problem of complex, microfine tool electrodes and nozzles that are difficult to manufacture and assemble, and thus is widely used in the manufacture of surface micronano-structures, whose processing schematic diagram is shown in Figure 2.

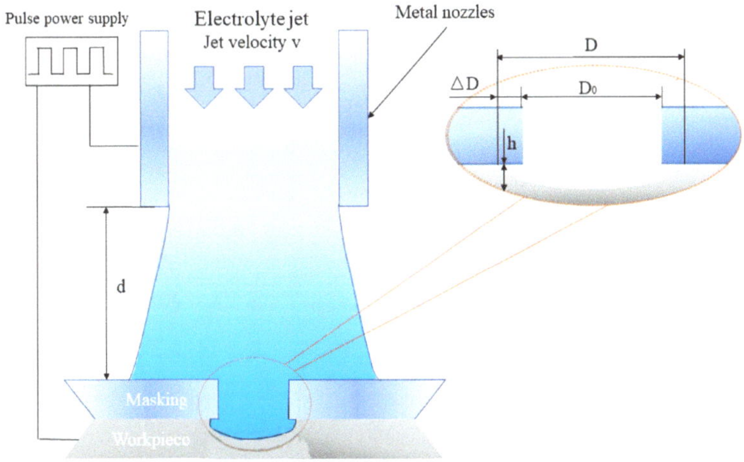

Figure 2. Principle diagram of mask jet electrolytic processing.

As shown in Figure 2, the workpiece is connected to the positive pole of the power supply, and the metal nozzle is connected to the negative pole of the power supply; the electrolyte is driven by the pump, and rushes at high speed to the surface of the mask with the skeleton pattern. The high-speed charged electrolyte forms an array of liquid electrodes under the constraint of the high-resolution mask, and carries out mass etching of the anode workpiece material (the part covered by the mask is not processed, while the part not covered by the mask is processed), thus processing a structure similar to the mask skeleton pattern. At the same time, the use of the CNC motion stage enables the cathode nozzle to achieve scanning motion, increasing the flexibility of the process. The use of mask plates solves the problem of complex microfine tool electrodes that are difficult to manufacture and assemble, while the jet electrolytic processing and scanning motion not only avoids the need to design cavity fixtures to meet specific flow field requirements, but also increases processing flexibility by enabling the processing of many different types of the workpiece without changing the process method or fixture. The mature lithography process provides micro- and nanoscale insulating masks in a variety of patterns, enabling the process method to meet the processing requirements of a variety of applications. A comparison experiment was carried out between jet mask electrolytic discharge machining (Jet-MECDM) and jet-mask electrolytic machining (Jet-MECM), using a metal nozzle for the Jet-MECDM tool and a plastic nozzle for the Jet-MECM tool, with other processing conditions remaining unchanged. The actual process diagram for the comparison experiment is shown in Figure 3.

Figure 3a shows that discharge occurs inside the jet electrolytic flow injection and producing a superposition effect. Figure 3b shows that no discharge occurs in the jet and it is conventional electrolytic processing. Whether or not the discharge occurs, inevitably has a different effect on the machining effect. The experimental sample is the initial sample with low surface roughness (i.e., stainless steel workpiece), with the size specification of $70 \times 50 \times 1$ mm^3. To facilitate the measurement, the spacing between holes is set to be 4 mm, i.e., the distance between a and b is 4 mm, thus obtaining the machining dimensions of the mask as shown in Figure 4.

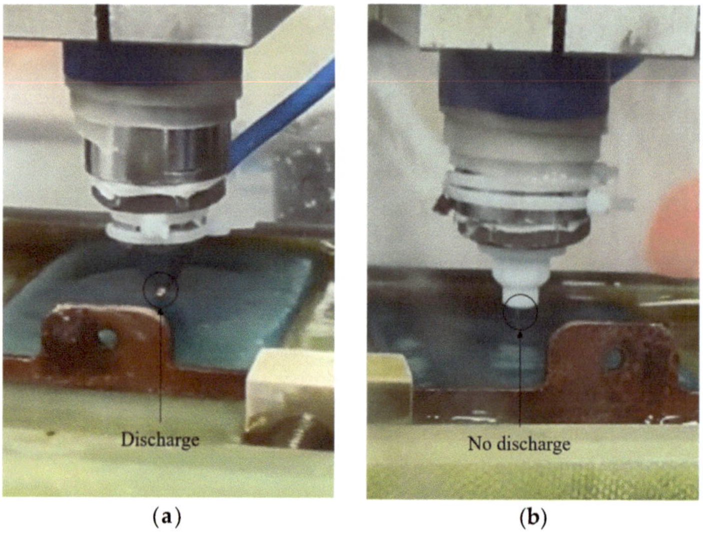

Figure 3. Comparison diagram of the actual machining process. (**a**) Jet-MECDM (Discharge); (**b**) Jet-MECM (No discharge).

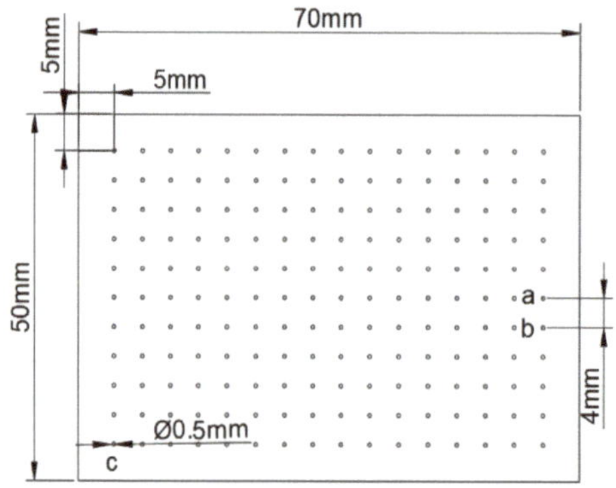

Figure 4. Processing dimensions of the mask.

A blue oil mask with a thickness of 50 μm was prepared on the surface of the stainless steel workpiece using the photolithography process, as shown in Figure 5.

Figure 5. Masked mirror stainless steel workpiece.

The corrosion factor (EF) is used to describe the fixed-domain nature of masked jet electrolytic processing, and is expressed as

$$EF = \frac{H}{\Delta D} = \frac{H}{D - D_0} \quad (1)$$

In Equation (1) H is the crater depth, ΔD is the difference between the crater diameter and the mask hole diameter, D is the crater diameter and D_0 is the mask hole diameter. The smaller ΔD is, the closer the crater diameter is to the mask hole diameter, and the larger H is, the deeper the crater depth is.

From Equation (1), it can be seen that the larger the EF value, the better the fixation and the better the definiteness, and the shape of the processed crater is closer to the skeletonized pattern; the smaller the E value, the worse the fixation. In the machining effect, the crater depth (H), diameter (D), material removal rate (MRR), and corrosion factor (EF) indicators are used to evaluate the machining effect.

3. Experimental Results and Discussion

The experiment is set up, the mask workpiece is prepared, and the differences between the two processes are analyzed through comparative experiments. Two types of processing are investigated: one is the stationary method, where the jet is sprayed right at the workpiece and the workpiece is processed through the mask hole. The second is a scanning and moving method, where the jet is scanned in a line at a certain speed, and the effect of nozzle movement on the shape of the workpiece is investigated.

3.1. Stationary Approach

As shown in Figure 2, the electrolyte is sprayed onto the workpiece in a jet state under a certain pressure, and the workpiece is machined by the jet through the mask hole in the mode of no feed on the X, Y, and Z axes, and a crater is formed at the location of the mask hole after a period of machining. The experimental conditions are shown in Table 2. $NaNO_3$ electrolyte pressure is 0.03 MPa, and the concentration is 15%.

Table 2. Specific parameters for single-well experimental conditions (Stationary Approach).

Process Parameters	Symbols	Unit	Parameter Values
Peak voltage	U	V	400
Voltage frequency	f	Hz	1000
Voltage duty cycle	r	%	50
Nozzle-to-workpiece distance	d	mm	2.5
Processing time	-	s	10
Mask plate diameter	-	μm	500
Masking board types	-	-	Blue oil mask

The experimental data on the depth H (depth) and diameter D (diameter) of the single holes (pits) machined by Jet-MECDM and Jet-MECM under the same machining conditions are shown in Table 3, with the units of depth H and width D in μm.

Table 3. Depth and width of comparative experiments (Stationary Approach).

Test No. 1	Jet-MECDM Depth	Jet-MECM Depth	Jet-MECDM Diameter	Jet-MECM Diameter
1	760	590	570	530
2	880	590	620	520
3	850	640	640	490
4	860	600	600	480
5	825	610	590	525

The experimental data on the depth H and diameter D of the hole (crater) are an important basis for the analysis of the efficiency of machining and whether the shape of the machining can be maintained.

As shown in Figure 6, the primary longitudinal axis represents the depth of a single hole (pit), and H (depth), and is represented by a bar graph. The secondary longitudinal axis represents the width of the single hole (crater), and D (diameter), and is represented by a line graph.

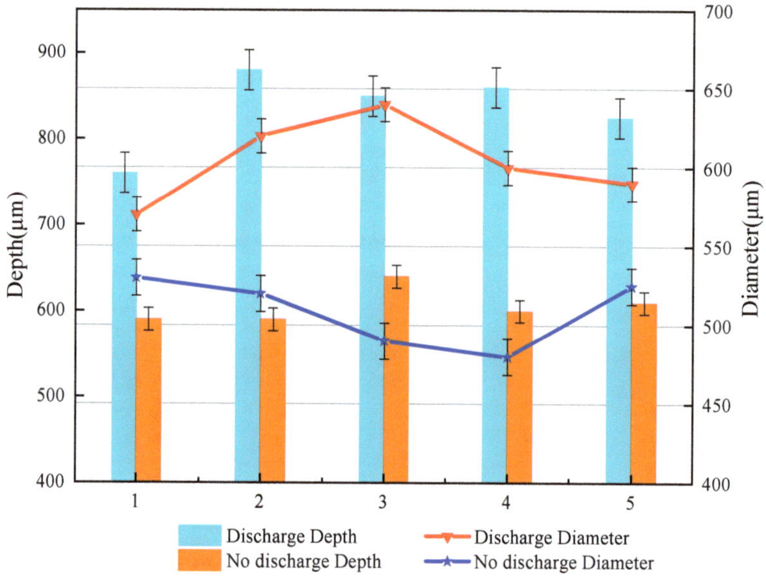

Figure 6. Experimental diagram comparing the crater depth and width of the two models (Stationary Approach).

From Table 3 and Figure 6, we can see that (1) the Jet-MECDM method processes single-hole depths of between 760 and 880 μm, with a mean value of 835 μm. The Jet-MECM method processes single-hole depths of between 590 and 640 μm, with a mean value of 606 μm. The difference between the two means is 229 μm, and the Jet-MECDM contribution to depth increases by 38%. (2) The Jet-MECDM method processes hole widths between 570 and 640 μm, with a mean value of 604 μm. The Jet-MECM method processes hole widths between 480 and 530 μm, with a mean value of 509 μm. The difference between the two mean values is 95 μm, and the contribution of Jet-MECDM to width increases by 18%, compared to Jet-MECM. (3) Jet-MECDM contributes more to depth (38%) than width (18%), when compared to Jet-MECM. The aim of using the mask is to constrain the width, i.e., to obtain a greater increase in material removal, as far as the processed morphology allows.

To further investigate its machined morphology, a single hole was observed in 3D using a laser scanning confocal microscope, as shown in Figure 7.

As can be seen from Figure 7, the shape of the single hole machined by the Jet-MECDM method is as regular as that of the single hole machined by the Jet-MECM method, and the inner wall of the single hole machined by both is equally smooth. To investigate the role of the mask as a constraint, the single hole with the mask was observed in 2D using a laser scanning confocal microscope, as shown in Figure 8.

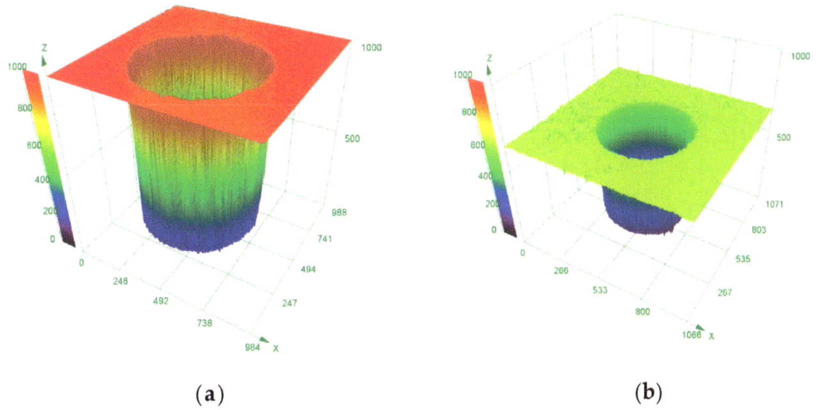

Figure 7. Contrasting experimental machining 3D morphology. (**a**) Jet-MECDM machining 3D profile; (**b**) Jet-MECM machining 3D profile.

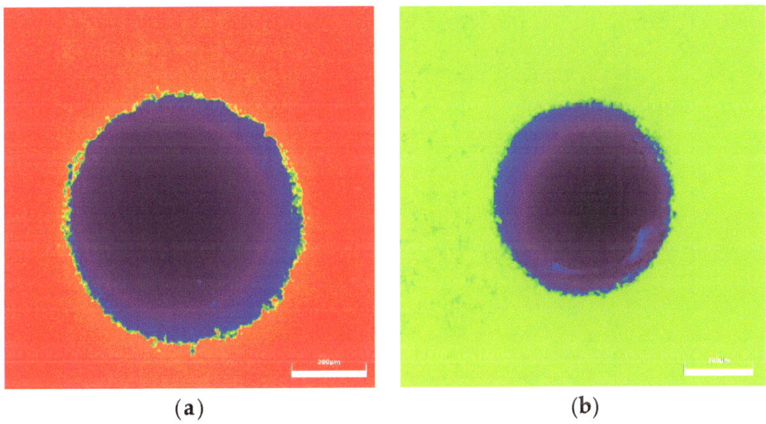

Figure 8. Comparison of experimental machined 2D profiles. (**a**) Jet-MECDM machining 2D profile; (**b**) Jet-MECM machining 2D profile.

As can be seen in Figure 8, the Jet-MECDM does not have any significant damage to the mask due to the discharge effect, and the mask is well constrained. The comparison in Figure 8 also shows that the color of the bottom of the Jet-MECM hole is not consistent, indicating a large deviation in the bottom morphological profile. This is further confirmed by the side measurements of the hole cross-sections for both machining methods, as shown in Figure 9.

As can be seen from Figure 9, the shape of the Jet-MECDM machined single hole (pit) is close to cylindrical, while the shape of the Jet-MECM machined single hole (pit) is closer to an inverted trapezoid. The steepness of the inner wall of the Jet-MECDM machined single hole (pit) is steeper and almost vertical, with less side etching. In contrast, the Jet-MECM process has a different amount of lateral corrosion at the top and bottom of the single hole (crater), creating an inverted trapezoid shape with a large top and a small bottom. Both Jet-MECDM and Jet-MECM process the bottom of the single hole (pit) without the island effect of normal mask electrolytic processing. However, the bottom of the Jet-MECDM process is flatter.

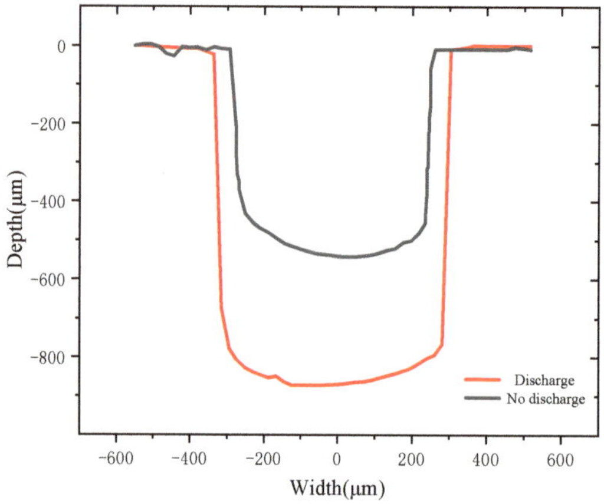

Figure 9. Single-hole (crater) depth and width measurement diagram (Stationary Approach).

3.2. *Scanning Movement Method*

As shown in Figure 2, the Y and Z axes are not fed, the X axis moves at a certain speed of motion (scanning), and the parameters of the experimental conditions are shown in Table 4. $NaNO_3$ electrolyte pressure is 0.03 MPa, and the concentration is 15%.

Table 4. Specific parameters for single-well experimental conditions (Scanning Movement).

Process Parameters	Symbols	Unit	Parameter Values
Peak voltage	U	V	400
Voltage frequency	f	Hz	1000
Voltage duty cycle	r	%	50
Nozzle-to-workpiece distance	d	mm	2.5
Scanning speed	-	mm/s	0.15
Mask board diameter	-	μm	500
Masking board types	-	-	Blue oil mask

The experimental data on the depth H (depth) and width D (diameter) of the single holes (pits) machined by Jet-MECDM and Jet-MECM under the same conditions are shown in Table 5, with the units of depth H and width D being μm.

Table 5. Depth and width of comparative experiments (Scanning Movement).

Test No. 1	Jet-MECDM Depth	Jet-MECM Depth	Jet-MECDM Diameter	Jet-MECM Diameter
1	116	85	530	510
2	121	87	559	550
3	119	95	560	561
4	112	95	580	574
5	117	85	614	583

The experimental data for the depth H and width D of the holes (pits) are an important basis for the analysis of machining efficiency and quality. For ease of observation, the experimental data in Table 5 is plotted as a relationship, as shown in Figure 10.

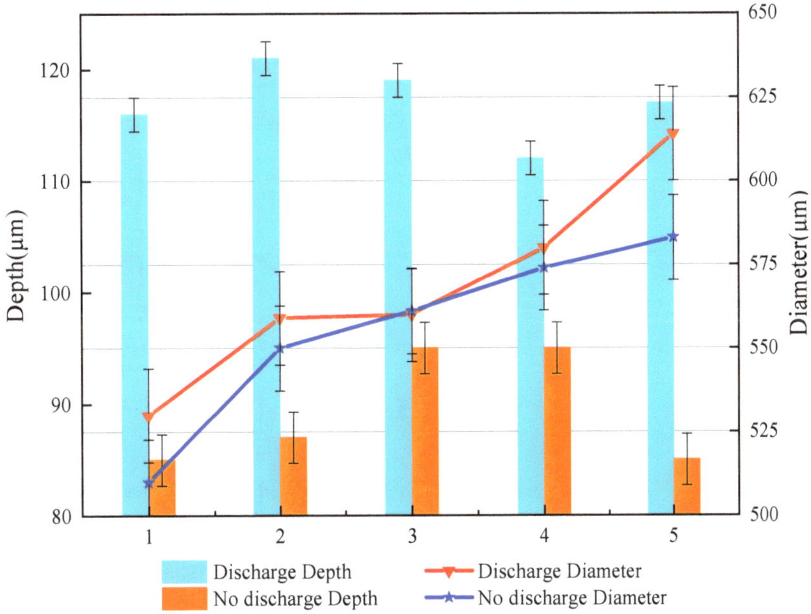

Figure 10. Experimental diagram comparing the crater depth and width of the two models (Scanning Movement).

As shown in Figure 10, the primary longitudinal axis represents the hole (pit) depth H (depth) and is represented by a bar graph. The secondary longitudinal axis represents the width D (diameter) of a single hole (pit), and is represented by a line graph.

From Table 5 and Figure 10, it follows that

(1) The Jet-MECDM method processes single-hole depths between 112 and 121 μm, with a mean value of 117 μm. The Jet-MECM method processes single-hole depths between 85 and 95 μm, with a mean value of 90 μm. The difference between the two means is 27 μm, with Jet-MECDM contributing 30% more to depth than Jet-MECM.

(2) The Jet-MECDM method processes single-hole widths between 530 and 614 μm, with a mean value of 569 μm. The Jet-MECM method processes single-hole widths between 510 and 583 μm, with a mean value of 556 μm. The difference between the two means is 13 μm, and the contribution of Jet-MECDM to width increases by 2% over Jet-MECM.

(3) A comparison of Jet-MECDM and Jet-MECM in the scanned condition reveals that Jet-MECDM machining depths increase and material removal rates improve for almost equal amounts of side etching.

The single-hole (crater) depth and width measurement diagram is shown in Figure 11.

From Figure 11, it can be seen that the inner wall of the Jet-MECDM machined single hole (pit) is steeper and more vertical, and the island of the machined single hole (pit) is 15 μm higher. It can be seen that the side etching of the upper and lower part of the Jet-MECM machined single hole (pit) is different, and the island of the machined single hole (pit) is 30 μm higher, the island of the normal mask electrolytic machining. The island effect is smaller in Jet-MECDM than in Jet-MECM, and the bottom is flatter, resulting in a better shape.

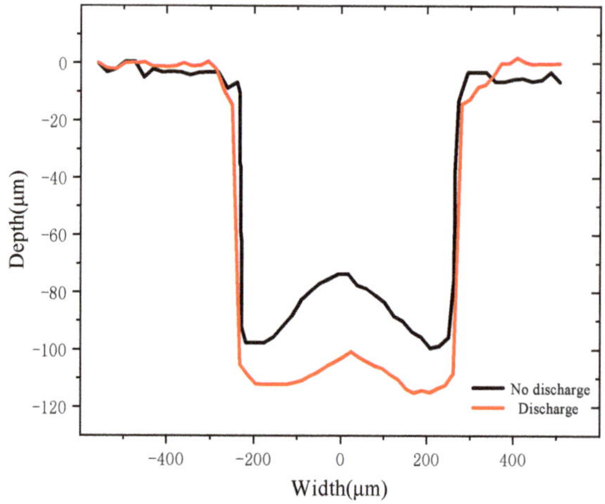

Figure 11. Single-hole (crater) depth and width measurement diagram (Scanning Movement).

4. Conclusions

This paper investigates the experimental electrolytic processing of discharge-assisted jet masks, and draws the following conclusions.

A comparison experiment between Jet-MECDM and Jet-MECM was carried out, and the results showed that

(1) The experimental setup for jet electrolytic processing was set up to complete the preparation of mask workpieces. Under other processing conditions, the use of metal nozzles produced a discharge compound effect, while the use of plastic nozzles did not.

(2) In the stationary method, Jet-MECDM increases the contribution to depth by 38% and the contribution to width by 18% over Jet-MECM, i.e., the machining efficiency is improved to a greater extent than the machined shape allows. The bottom of the single hole (crater) machined by both methods does not show the island effect of normal electrolytic machining.

(3) In the scanning movement method, Jet-MECDM contributes 30% more to the depth and 2% more to the width than Jet-MECM, i.e., the width is essentially the same but the depth increases significantly and the machining efficiency improves dramatically. The island effect of normal electrolytic machining occurs at the bottom of single holes (pits) machined by both methods, with the former having less island effect than the latter, a smoother bottom, and a better-machined shape.

Using the Jet-MECDM process, jet electrolysis is used throughout the machining process, which can improve the machining accuracy, while the discharge effect can increase the material removal capacity and effectively improve the depth-to-width ratio (effectively improving the machining efficiency). In the next stage, the process will be studied in depth in terms of different masks, different workpieces, and different processing voltages.

Author Contributions: Conceptualization, C.C.; methodology, C.C.; software, C.C.; validation, C.C., and S.W.; formal analysis, C.C. and T.Z.; investigation, Y.W.; resources, C.C.; data curation, X.S. and F.M.; writing—original draft preparation, C.C.; writing—review and editing, S.W.; visualization, C.C.; supervision, S.W.; project administration, S.W.; funding acquisition, C.C. All authors have read and agreed to the published version of the manuscript.

Funding: This research was funded by the National Natural Science Foundation of China (grant NO. 52075104), and Guangzhou Higher Education Quality and Reform Project (2023CJRHJD002).

Institutional Review Board Statement: Not applicable.

Informed Consent Statement: Not applicable.

Data Availability Statement: Data are contained within the article.

Conflicts of Interest: The authors declare no conflict of interest.

References

1. Kawanaka, T.; Kunieda, M. Mirror-Like Finishing by Electrolyte Jet Machining. *CIRP Ann.* **2015**, *64*, 237–240. [CrossRef]
2. Zhang, X.; Song, X.; Ming, P.; Li, X.; Zeng, Y.; Cai, J. The Effect of Electrolytic Jet Orientation on Machining Characteristics in Jet Electrochemical Machining. *Micromachines* **2019**, *10*, 404. [CrossRef] [PubMed]
3. Fan, G.; Chen, X.; Saxena, K.K.; Liu, J.; Guo, Z. Jet Electrochemical Micromachining of Micro-Grooves with Conductive- Masked Porous Cathode. *Micromachines* **2020**, *11*, 557. [CrossRef] [PubMed]
4. Liu, Y.; Qu, N. Investigation on the performance of macro electrochemical machining of the end face of cylindrical parts. *Int. J. Mech. Sci.* **2020**, *169*, 105333. [CrossRef]
5. Fu, X.Q.; Kang, M.; Zheng, Q.Y. Research on Flow Field Simulation and Experiment of Numerical Control Electrochemical Machining. key Eng. *Mater.* **2011**, *458*, 63–68.
6. Hinduja, S.; Pattavanitch, J. Experimental and numerical investigations in electro-chemical milling. *CIRP J. Manuf. Sci. Technol.* **2016**, *12*, 79–89. [CrossRef]
7. Hackert-Oschätzchen, M.; Meichsner, G.; Zinecker, M.; Martin, A.; Schubert, A. Micromachining with continuous electrolytic free jet. *Precis. Eng.* **2012**, *36*, 612–619. [CrossRef]
8. Mitchell-Smith, J.; Speidel, A.; Clare, A.T. Advancing electrochemical jet methods through manipulation of the angle of address. *J. Mater. Process. Technol.* **2018**, *255*, 364–372. [CrossRef]
9. Mitchell-Smith, J.; Speidel, A.; Gaskell, J.; Clare, A.T. Energy distribution modulation by mechanical design for electrochemical jet processing techniques. *Int. J. Mach. Tools Manuf.* **2017**, *122*, 32–46. [CrossRef]
10. Chu, W.S.; Kim, C.S.; Lee, H.T.; Choi, J.O.; Park, J.I.; Song, J.H.; Jang, K.H.; Ahn, S.H. Hybrid manufacturing in micro/nanoscale: A review. *Int. J. Precis. Eng. Manuf. Green Technol.* **2014**, *1*, 75–92. [CrossRef]
11. Singh, T.; Dvivedi, A. Developments in electrochemical discharge machining: A review on electrochemical discharge machining, process variants and their hybrid methods. *Int. J. Mach. Tools Manuf.* **2016**, *105*, 1–13. [CrossRef]
12. Singh, T.; Appalanaidu, B.; Dvivedi, A. Improvement in energy channelization behaviour during micro hole formation in Y-SZ ceramic with magnetic. *Measurement* **2022**, *194*, 111079. [CrossRef]
13. Kim, D.J.; Ahn, Y.; Lee, S.H.; Kim, Y.K. Voltage pulse frequency and duty ratio effects in an electrochemical discharge microdrilling process of Pyrex glass. *Int. J. Mach. Tools Manuf.* **2006**, *46*, 1064–1647. [CrossRef]
14. Kobayashi, F. Electrochemical discharge drilling methode. *Pantent JP.* **1997**, *32*, 64–69.
15. Liu, J.W.; Yue, T.M.; Guo, Z.N. An analysis of the discharge mechanism in electrochemical discharge machining of particulate reinforced metal matrix. *Int. J. Mach. Tools Manufacture.* **2010**, *50*, 86–96. [CrossRef]
16. Yu, W. Experimental study on micro-structure machining by electrohydraulic beam. Master's Thesis, Dalian University of Technology, Dalian, China, 2013.
17. Cui, J. Basic Law and Simulation Research of Micro EDM. Ph.D. Thesis, Harbin Institute of Technology, Harbin, China, 2007.
18. Liu, H.; Xieeryazidan, A. Study of Gas Film Characteristics in Electrochemical Discharge Machining and Their Effectson Discharge Energy Distribution. *Micromachines* **2023**, *14*, 1079. [CrossRef]
19. Chuang, M.-C.; Jan, C.-M.; Wang, Y.-J.; Hsu, Y.-L. Intelligent Monitoring and Compensation between EDM and ECM. *Appl. Sci.* **2023**, *13*, 927. [CrossRef]
20. Sambathkumar, S.; Arunagirinathan, R.S. A Simple Technique for the Precise Establishment of the Working Gap in an Electrochemical Discharge Machining Process and Some Experimental Results Thereof. *Micromachines* **2022**, *13*, 1367. [CrossRef]
21. Zhao, R.; Huang, L.; Zhao, H.; Cao, Y.; Tian, W.; Wang, N. Study of Mask Electrochemical Machining for Ring Narrow Groove under the Action of Multiple Physical Fields. *Coatings* **2022**, *12*, 605. [CrossRef]
22. Zhan, S.; Zhao, Y. Plasma-assisted electrochemical machining of microtools and microstructures. *Int. J. Mach. Tools Manuf.* **2020**, *156*, 103596. [CrossRef]
23. Jui, S.K.; Kamaraj, A.B.; Sundaram, M.M. High aspect ratio micromachining of glass by electrochemical discharge machining (ECDM). *J. Manuf. Process.* **2013**, *15*, 460–466. [CrossRef]
24. Wang, T.; Liu, Y.; Wang, K.; Lv, Z. Investigation on a sustainable composite method of glass microstructures fabrication—Electrochemical discharge milling and grinding (ECDM-G). *J. Clean. Prod.* **2023**, *387*, 135788. [CrossRef]

coatings

Article

Structural Transformations on the Surface of Al-Ti Cathodes Subjected to Vacuum Arc Heating

Gennady Pribytkov *, Irina Firsina and Victoria Korzhova

Institute of Strength Physics and Materials Science SB RAS, 634055 Tomsk, Russia; iris1983@ispms.ru (I.F.); vicvic@ispms.ru (V.K.)
* Correspondence: gapribyt@mail.ru

Abstract: AlTiN nitride coatings on the surfaces of metal-working tools can greatly extend their service life. The coatings are deposited from plasma flows generated by vacuum arc burning on the cathode surface. The elemental and charge composition of the plasma flows, as well as the content of metal drops, depend on the cathode's structure. In this paper, the microstructure, elemental, and phase compositions of the surface layer of Al-Ti cathodes subjected to vacuum arc heating were studied. These cathodes had similar elemental compositions (Ti + 50 at.% Al) but differed from one another in their phase composition and microstructure (grain size, porosity). The cathodes were studied by X-ray diffraction analysis, scanning electron microscopy, and electron probe analysis. It was found that during vacuum arc heating, surface fusion or thermal cracking of the cathode's surface layer occurs. The thickness, structure, and phase composition of the modified layer were controlled by the thermal conductivity of the cathode material, which, in turn, depended on the phase composition and porosity of the cathodes. The maximum thickness of the modified layer (up to 400 μm) was observed on the surface of the sintered cathode due to the lower thermal conductivity of the porous structure of the cathode. The obtained results can be used for the development of coating deposition technology based on vacuum arc sputtering of multicomponent cathodes.

Keywords: multicomponent cathodes; vacuum arc; titanium; aluminum; titanium aluminides; modified layer; thermal conductivity; thermal cracking

Citation: Pribytkov, G.; Firsina, I.; Korzhova, V. Structural Transformations on the Surface of Al-Ti Cathodes Subjected to Vacuum Arc Heating. *Coatings* **2023**, *13*, 1906. https://doi.org/10.3390/coatings13111906

Academic Editors: Pinghu Chen, Yun Zhang and Ruiqing Li

Received: 7 September 2023
Revised: 12 October 2023
Accepted: 18 October 2023
Published: 7 November 2023

1. Introduction

Vacuum arc evaporation of metal and metallic alloy cathodes in reactive gas media is widely used for the deposition of wear-resistant coatings on the surfaces of metal-working tools [1–3]. The surface of the cathodes, used for the coating's deposition by vacuum arc evaporation, is heated to high temperatures exceeding the melting point of the cathode material. Surface melting leads to an increased content of drops in the products of the arc erosion, and the lower the melting temperature of the cathode, the greater the number of droplets [4,5]. Heating of the surface layer of the composite cathodes produced by forging a mechanical mixture of elemental powders is accompanied by structural and phase transformations [6,7]. The duration of the reactions between the powder components in the mixture depends on the dispersity of the powders [7,8]. As a result, the mechanical mixtures of elemental powders turn into refractory compounds [9,10]. The heating rate of the cathode, along with the magnitude of the current, depends on the thermal conductivity of the cathode material, which, in turn, is determined by the phase composition and structure. The thickness of the modified layer of the cathode depends on two factors: thermal conductivity and melting temperature. As a rule, the thermal conductivity of the intermetallic compounds is less than that of metals. Thus, the surface temperature of an intermetallic cathode of similar elemental composition is expected to be higher than that of a composite cathode represented by a powder mixture. On the other hand, due to the higher melting point of metal aluminides compared to the low melting point of aluminum in composite cathodes, the thickness of the melted surface layer on composite cathodes can be thicker than that on aluminide cathodes. Due to the high hardness, poor plasticity, and low thermal conductivity of the intermetallic compounds, thermal stress cracks emerge on the surface layer of the intermetallic cathode [10]. The thermal cracking leads to the spalling of the modified layer and an enhancement in the rate of the cathode erosion [10].

The physical properties of the cathode's material influence the emission rate, charge, and elemental composition of the ejected plasma flows. In [11–13], a correlation was found between the cohesive energy of the cathode material, the combustion voltage, and plasma parameters. The influence of the structure and composition of the cathode on the drop ejection, charge, and energy of ions were studied and discussed in [14–19]. The structure state of the cathode's surface layer that was subjected to vacuum arc heating depends on both the elemental composition and pressure of the gaseous media in the technological chamber. A redeposition of the ejected drops on the cathode surface occurs under elevated gas pressure [20]. Another effect related to the gaseous media is the poisoning of the cathode surface as a result of interaction with the reactive gases [21,22].

TiAlN coatings deposited from vacuum arc plasma in a nitrogen environment remain the most popular among wear-resistant coatings for metal-working tools due to their high durability [23–25]. To enhance the properties of TiAlN coatings, cathodes of different designs are used: mosaic [26], composite [6,26], or cast [6]. The study of the phase composition of the surface layer of composite and Ti-50 at.% Al-cast cathodes were subjected to vacuum arc treatment in [6] using the X-ray diffraction method. According to X-ray diffraction patterns, the ratio of Ti_3Al, TiAl, Ti_2Al_5, and $TiAl_3$ aluminides in the surface layer of both cathodes depends on the distance of the analyzed spot from the center of the round cathode. Ti_3Al and TiAl aluminides have melting points of 1475 °C and 1447 °C, respectively, which are higher than the melting point of aluminum (660 °C), which is the basis of the powder composite cathode. In the present work, comparative studies of the surface layer of Al-Ti cathodes modified by a vacuum arc were fulfilled. The cathodes have the same elemental composition but differ in phase composition. Contrary to [6], we studied the cross-section of the modified layer. The research results can be useful in the development of advanced industrial technologies for the production of metal-working tools with high durability.

2. Materials and Experimental Technique

Four cathodes of similar compositions (Ti + 50 at.% Al) obtained by different technologies (Table 1) were analyzed. Since the production technology of commercial cathodes is not disclosed by producing companies, the cathode production technology was determined based on the results of microstructural studies (Table 1). The sintered cathode was obtained by mixing titanium powders (fraction less than 125 μm) and $TiAl_3$ intermetallic compound (fraction less than 50 μm), pressing under a pressure of 0.3 GPa, vacuum sintering at 1250 °C and isothermal holding for 4 h.

Table 1. Supplier and production technology of the Ti-Al cathodes.

Cathode	Supplier	Production Technology
No. 1, sintered	ISPMS SB RAS, Tomsk, Russia	Sintering
No. 2, "Polema"	JSC "Polema", Tula, Russia	Hot pressing of Al + Ti mixture
No. 3, "Umicore"	"Umicore Materials AG", Hanau-Wolfgang, Germany	Hot pressing of Al + Ti mixture
No. 4, cast	-	Casting

The starting phase composition of the cathodes was studied on its back surface layer, which was not exposed to the vacuum arc. The working surfaces of cathodes with a diameter of 70 mm were treated for 60 min with an arc current of 90 A in N_2 gas under a pressure of 0.065 Pa. Structural studies of the cathodes were carried out using X-ray diffraction (XRD)

(Co K_α radiation, DRON-7, St. Petersburg, Russia) and scanning electron microscopy (LEO EVO 50, Oberkochen, Germany). ASTM data files and PDWin 4.00 software were used for phase identification and characterization. The elemental composition of the modified layer was determined by X-ray spectroscopy (EDS). The microstructure and phase composition of the cathodes were studied on cross-sections polished and etched with Keller's reagent (2.5 mL HNO_3 + 1.5 mL HCl + 1 mL HF + 95 mL H_2O). The microhardness of the samples was measured on etched sections using a PMT-3 microhardness tester under a load of 0.98 N.

3. Results

3.1. Starting Structural State of Cathodes

The X-ray patterns of the back surface layer of the cathodes in the starting structural state and from the working surface exposed to the vacuum arc are shown in Figure 1. Table 2 presents the phase composition of the cathodes, estimated from the sum of the intensities of X-ray reflections by the processing of X-ray diffraction patterns (Figure 1). Due to the superposition of lines of different phases as well as due to the high background level, this method does not allow us to determine the exact content of the constituents. However, according to the data in Table 2, the sintered cathode in the starting state consists mainly of the Ti_3Al phase and a small amount of the γ-TiAl intermetallic compound. Hot-pressed cathode No. 2 consists almost entirely of Ti and Al. In addition to Ti and Al, there are intermetallic compounds TiAl and Ti_3Al in the structure of hot-pressed cathode No. 3. The cast cathode consists almost entirely of the Ti_3Al intermetallic compound.

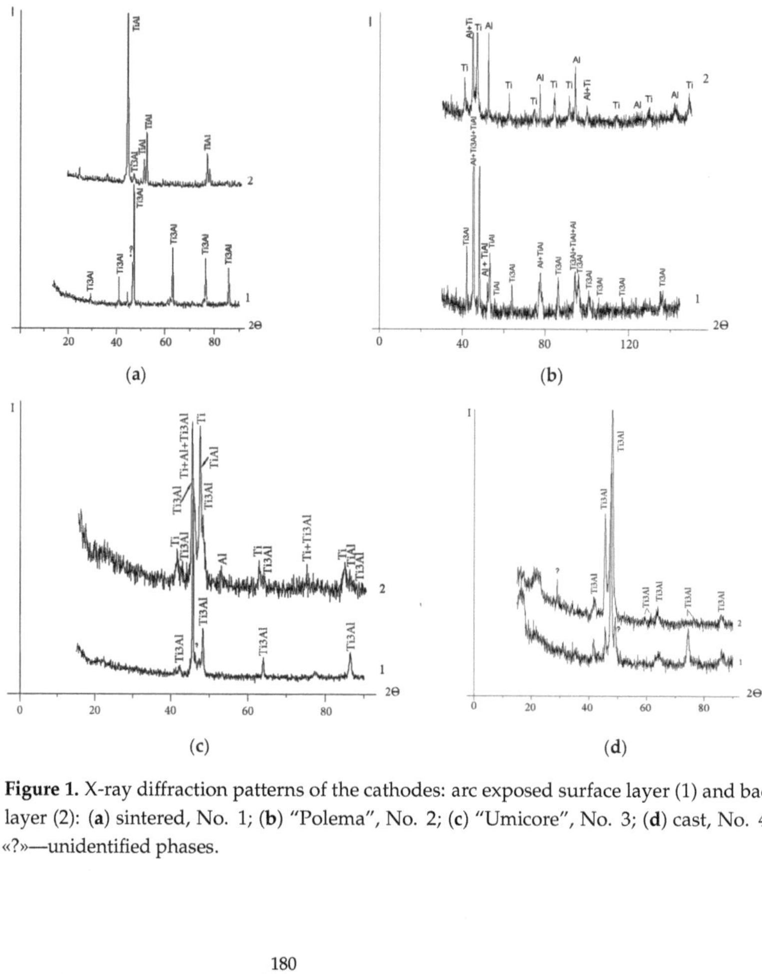

Figure 1. X-ray diffraction patterns of the cathodes: arc exposed surface layer (1) and back surface layer (2): (**a**) sintered, No. 1; (**b**) "Polema", No. 2; (**c**) "Umicore", No. 3; (**d**) cast, No. 4. Symbol «?»—unidentified phases.

Table 2. X-ray diffraction analysis of phase composition of cathodes.

Cathode	Analyzed Site	Content of Phases, Vol.%	
		Major Constituents	Other Phases
No. 1. Sintered	Arc exposed surface layer	95% Ti_3Al	5% unidentified
	Back surface layer	72%–82% γ-TiAl	21%–12% Ti_3Al, traces of Al_2O_3, 7% unidentified
No. 2. "Polema"	Arc exposed surface layer	59%–78% Ti_3Al	21%–41% TiAl
	Back surface layer	41% Ti, 57% Al	2% Ti_3Al, 2% unidentified
No. 3. "Umicore"	Arc exposed surface layer	89% Ti_3Al	11% TiAl + traces of $TiH_{0.71}$
	Back surface layer	47%–74% Ti	23%–12% Ti_3Al, 24%–12% TiAl, 7%–8% Al
No. 4. Cast	Arc exposed surface layer	81% Ti_3Al	11% $TiH_{0.71}$, 8% unidentified
	Back surface layer	99% Ti_3Al	1% unidentified

The results of the X-ray phase analysis are in accordance with the data of optical metallography of the cathode material in the starting state (Figure 2). In the etched section of the sintered cathode (Figure 2a), two constituents are presented. According to structural studies [27], a ring structure is formed as a result of the sintering of powder mixtures of Ti and $TiAl_3$ intermetallic compounds, which have an integral elemental composition close to TiAl. The ring structure consists of Ti_3Al aluminide in the peripheral part and a compound of $TiAl_3$ inclusions. With the increasing duration of isothermal sintering, the width of the rings increases. Accordingly, the volume fraction of Ti_3Al also increases. This two-phase ring structure is especially seen in backscattered electron images (Figure 3a). In the cathode structure sintered at 1250 °C for 4 h, the Ti_3Al phase occurs in the form of several inclusions inside TiAl grains (Figure 2a). The $TiAl_3$ phase after vacuum sintering and isothermal holding is completely consumed in the reaction $TiAl_3 + Ti \rightarrow TiAl\ (Ti_3Al)$. The relatively low content of the Ti_3Al phase in the sintered cathode is confirmed via calculations based on the sum of the intensities of X-ray reflections (Figure 1a, Table 2).

The microstructure of the "Umicore" and "Polema" hot-pressed cathodes (Figure 2b,c) indicates the presence of structurally isolated pure elements—Al (light phase) and Ti (dark gray phase). However, hot-pressed cathodes from these two suppliers differ significantly in microstructure and phase composition. The "Umicore" cathode (Figure 2c) has a coarser structure. The size of the Ti particles (105 ± 46 μm) is nearly an order of magnitude larger than in the "Polema" cathode (12.2 ± 6.0 μm). In addition, intermetallic interlayers are clearly seen at the interphase boundaries of titanium and aluminum in the "Umicore" cathode. According to the X-ray diffraction data (Table 2), this cathode contains 30%–40% intermetallic phases. In the "Polema" cathode, the content of Ti_3Al is only 2%, and there are no intermetallic interlayers at the boundaries. The cast cathode has a lamellar structure typical of a titanium alloy, with rare defects in the form of pores and non-metallic inclusions (Figure 2d).

More detailed information about the starting structure of cathode materials was obtained by studying backscattered electron images (Figure 3). In contrast to the optical micrographs in Figure 2b,c, titanium grains appear as lighter areas against the background of a dark aluminum matrix in the backscattered electron images. Accordingly, the grains of intermetallic compounds of the Ti-Al system in the images get darker as the amount of aluminum they contain increases.

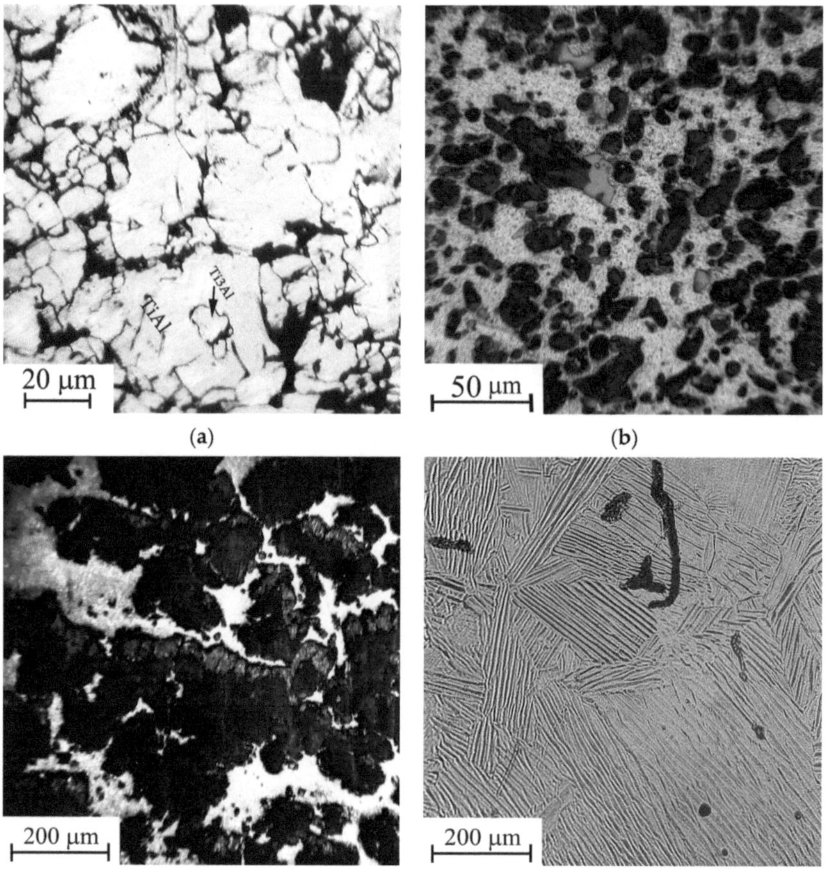

(**a**) (**b**)

(**c**) (**d**)

Figure 2. Starting microstructure of Ti-Al cathodes (back surface layer): (**a**) sintered, No. 1; (**b**) "Polema", No. 2; (**c**) "Umicore", No. 3; (**d**) cast, No. 4. (Optical metallography, etching with Keller's reagent).

The elemental composition of the phases presented in the sintered cathode corresponds to the results of X-ray diffraction analysis. Local analysis of the aluminum matrix of the "Polema" cathode disclosed the titanium content up to 3.6 at.% compared to approximately an order of magnitude lower equilibrium solubility of Ti in solid aluminum [28]. This discrepancy most likely has methodological reasons. One of the possible reasons is associated with the insufficiently small size of the electron probe for the dispersed structure of the "Polema" cathode, as well as with the possible presence of titanium grains in the analyzed subsurface layer. The validity of this explanation is confirmed by local analysis data. The content of Ti in the aluminum matrix of the "Umicore" cathode with a coarse structure is almost three times less than that in the "Polema" cathode (Table 3). The differences in the microhardness of aluminum in the "Polema" and "Umicore" cathodes are also associated with different dispersities of the mechanical mixture of Ti and Al (Table 3).

Electron-backscattered images of the cross-section of the "Umicore" cathode (Figure 3) present large titanium particles within the black color aluminum matrix. Single Ti_3Al inclusions occur in Ti particles near the boundary with the aluminum matrix. In some areas of the boundaries of intermetallic inclusions, the local elemental analysis reveals interlayers of aluminum solid solution in titanium. In the areas selected for local analysis, the Ti_3Al compound present in the starting cathode according to X-ray diffraction analysis was not detected. This is an additional confirmation of the strong homogeneity of the cathode structure, which is presented in Figure 2c.

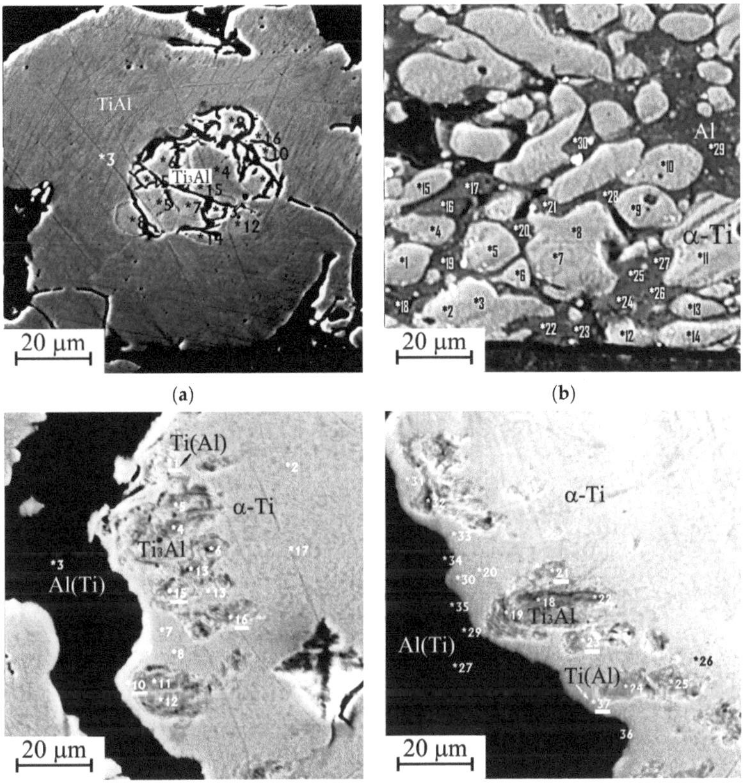

(a) (b) (c) (d)

Figure 3. Starting microstructure of the cathodes in backscattered electrons: (**a**) sintered, No. 1; (**b**) "Polema", No. 2; (**c**,**d**) "Umicore", No. 3. Spots of local elemental analysis are indicated with the numbers (Table 3).

Table 3. Starting elemental composition of the cathodes.

Cathode	Analyzed Site	Composition, at.%		No. of Points in Figure 3	Phase	H_μ, MPa
		Ti	Al			
No. 1, sintered	Core	70.9 ± 5.8	29.1 ± 5.8	1, 5–12, 15, 16 (Figure 3a)	Ti_3Al	5188 ± 794
	Periphery	52.1 ± 1.3	48.4 ± 0.8	2–4 (Figure 3a)	TiAl	3333 ± 172
No. 2, "Polema"	Light grains	100.0	0.0	1–15 (Figure 3b)	α-Ti	2385 ± 191
	Gray matrix	1.8 ± 0.7	97.5 ± 1.1	16–30 (Figure 3b)	Al(Ti)	555 ± 24
No. 3, "Umicore"	Inclusions in the gray phase	77.65 ± 6.65	22.35 ± 6.65	4–6, 11–13 (Figure 3c), 18, 19,22, 24, 25, 32 (Figure 3d)	Ti_3Al	1287 ± 126
	Gray phase	100.0	0.0	2, 7–9, 14, 17 (Figure 3c), 20, 26, 31, 33 (Figure 3d)	α-Ti	2163 ± 319
	Black phase	0.74 ± 0.67	99.25 ± 0.65	3 (Figure 3c), 27, 29, 34, 35, 36 (Figure 3d)	Al(Ti)	440 ± 4
	At the border of gray phases and inclusions	91.95 ± 5.45	8.5 ± 5.45	1, 10, 15, 16 (Figure 3c), 21, 23, 37 (Figure 3d)	Ti(Al)	-
No. 4, cast	Zone 2 (Figure 4)	-	-	-	-	4640 ± 1

3.2. Structural-Phase State of Arc Modified Surface Layer of the Cathodes

The results of X-ray diffraction analysis of a thin (up to 10 μm) layer adjacent to the working surface of the cathodes (Figure 1) are shown in Table 2. The data on the elemental composition of the modified layer and phase composition obtained based on EDS microanalysis are presented in Tables 3–5 along with the hardness of the structure constituents.

Table 4. Results of elemental analysis of the modified surface layer sintered cathode.

Zones	Description of Structural Components (Figure 5)	Composition, at.%		No. of Points in Figure 5	Phases
		Ti	Al		
1	Continuous grey layer	64.82 ± 1	35.18 ± 1	1–8	Ti_3Al
2	Base	66.28 ± 1.5	33.44 ± 1.5	9, 11, 14–19	Ti_3Al
	Light grey spots	57.26 ± 1.4	42.5 ± 1.4	10, 12, 13, 18, 20	Ti_3Al + TiAl

Table 5. Results of elemental analysis of the modified surface layer "Polema" cathode.

Zones	Description of Structural Components (Figure 6)	Composition, at.%		No. of Points in Figure 6	Phases
		Ti	Al		
1	Continuous grey layer	57.56 ± 1.77	42.44 ± 1.77	1–10	Ti_3Al + TiAl
	Light gray inclusions	27.0 ± 7.75	73.0 ± 7.75	7, 11, 13	$TiAl_3$
2	Light grains	100.0	0.0	3, 8, 12, 15, 18, 24, 25, 27	α-Ti
	Taupe matrix	8.31 ± 6.02	91.69 ± 6.02	14, 16, 17, 19–23, 26, 28–30	Al(Ti)

3.2.1. Cast Cathode No. 4

The cracks of a length up to 500 μm are found in zone 2, adjacent to the cathode's surface layer (zone 1), subjected to arc treatment (Figure 4). The crack density and crack tip opening displacement decrease with distance from the cathode's surface layer. The most probable reason for the surface cracking is a temperature gradient resulting in the thermal stress. A thin layer is seen adjacent to the working surface. The layer has a thickness of 20–50 μm and, according to X-ray diffraction data, mainly consists of Ti_3Al. Along with Ti_3Al, there is a small amount of $TiH_{0.71}$ hydride. The appearance of a hydride may be associated with the presence of hydrocarbon and, possibly, water molecules in the vacuum chamber. No other phases were found in the modified layer. Since the thin surface layer is separated from the multitracked zone by a sharp boundary, it can be argued that it was formed as a result of the crystallization of a thin melt film on the working surface of the cathode.

3.2.2. Sintered Cathode No. 1

The modified layer on the sintered cathode (Figure 5, zone 1) has of the greatest thickness (250–400 μm). The porosity of the layer is minimal, in contrast to the starting material, which has high porosity in the form of dark areas in the carcass structure, corresponding to the two-phase Ti_3Al + TiAl region on the equilibrium diagram (Figure 5, zone 2). The occasional cracks are directed mainly in parallel with the cathode surface. The average elemental composition of the layer corresponds to the Ti_3Al phase. The structure of the modified layer is apparently formed during the crystallization of the melted layer on the cathode surface. The main reason for the great thickness of the melted layer is the low thermal conductivity with the high cathode porosity. Due to the thermal effects, the phase composition and structure of the cathode material adjacent to the modified layer differ from the original one (Figure 3a). The thickness of the melted layer on the surface of the porous sintered cathode is several times greater than that of the melted layer on the surface of the non-porous cast cathode.

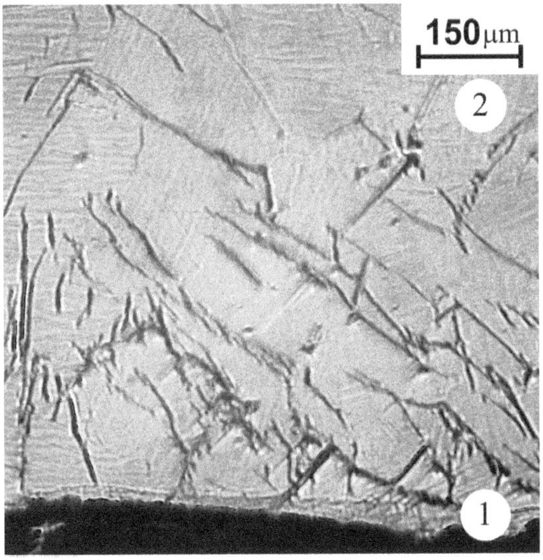

Figure 4. The microstructure of the cast cathode near the surface layer subjected to the arc treatment (optical metallography, etching Keller's reagent). 1—arc exposed layer; 2—back surface layer.

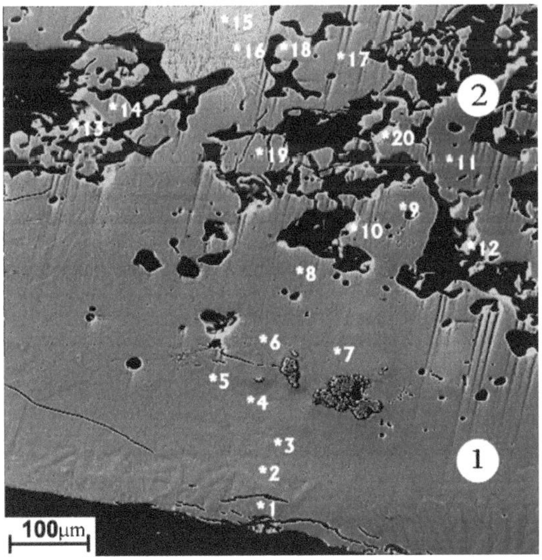

Figure 5. The image of the sintered cathode modified layer in backscattered electrons. 1—arc exposed layer; 2—adjacent zone. Spots of local elemental analysis are indicated with the numbers (Table 4).

3.2.3. Cathode No. 2, "Polema"

The thickness of the modified surface layer on the "Polema" cathode is 50–100 μm (zone 1 in Figure 6). Approximately half of the thickness is occupied by a layer adjacent to the arc-exposed surface layer with defects in the form of cracks and crumbled areas. The elemental composition of the continuous layer (Table 5) corresponds to the two-phase region Ti_3Al + TiAl. This is consistent with the results of the X-ray phase analysis (Table 2). The origin of the layer is undoubtedly associated with the interaction of Ti and Al in the pressed powder mixture during arc heating. In some areas of this layer, there are small aluminum-enriched inclusions (points 7, 11, 13 in Figure 7). Between the continuous surface layer and the underlying starting material, there is a transition region with cavernous defects. The formation of the cavities in the transition region can be explained by the partial melting of low-melting aluminum and the migration of liquid Al between Ti particles under the action of a temperature gradient. It leads to a nonuniform distribution throughout the volume and the formation of voids during crystallization.

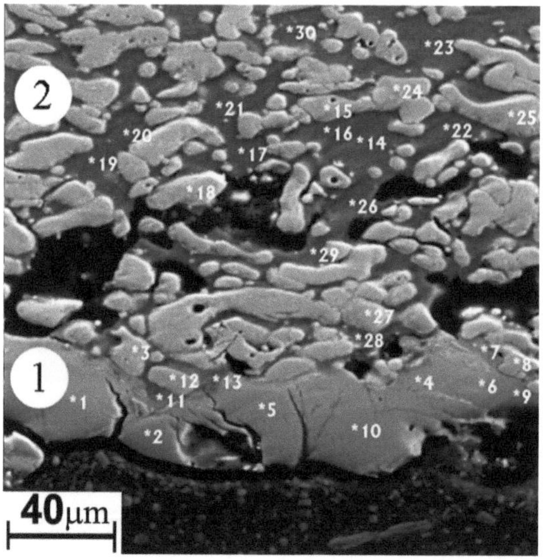

Figure 6. Image of the modified layer "Polema" cathode in backscattered electrons. 1—arc exposed surface layer; 2—back surface layer. Spots of local elemental analysis are indicated with the numbers (Table 5).

Table 6. Results of elemental analysis of the modified surface layer of the "Umicore" cathode.

Zones	Description of Structural Components	Composition, at.%		No. of Points in Figure 7a,b	Phases
		Ti	Al		
1	Surface layer	71.09 ± 5.11	28.91 ± 5.11	5 (Figure 7a), 12, 15 (Figure 7b)	Ti_3Al
		49.9 ± 0.5	50.1 ± 0.5	1, 3 (Figure 7a)	TiAl
		57.28 ± 3.25	42.72 ± 3.25	2, 4–6, 8, 9 (Figure 7a), 11, 13, 14 (Figure 7b)	Ti_3Al + TiAl
2	Light gray grains	100.0	0.0	10 (Figure 7b), 1–10 (Figure 7c)	α-Ti
	Black matrix	0.22 ± 0.25	99.78 ± 0.25	11–20 (Figure 7)	Al(Ti)

Zone 2 in Figure 6 has the structure of a mechanical mixture of Ti grains in an aluminum matrix and is visually no different from the structure of the material on the back side of the cathode (Figure 2b).

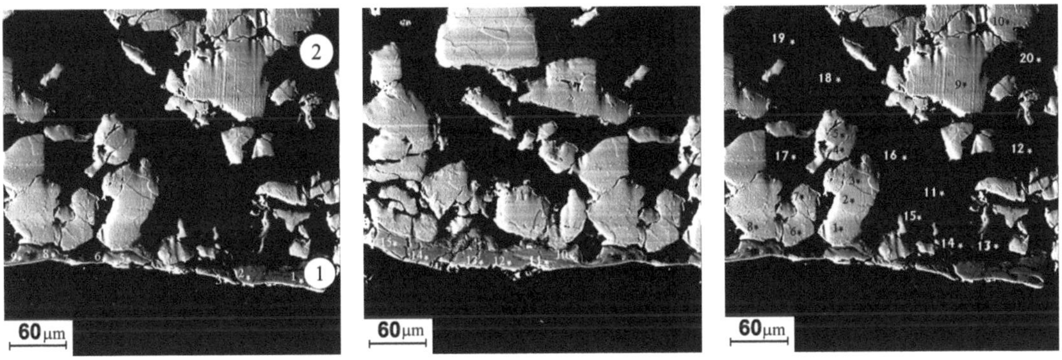

(a) (b) (c)

Figure 7. Image of the modified layer "Umicore" cathode in backscattered electrons (**a**–**c**). 1—arc exposed surface layer; 2—back surface layer. Spots of local elemental analysis are indicated with the numbers (see Table 6).

3.2.4. Cathode No. 3, "Umicore"

The thickness of the modified layer on the "Umicore" cathode (zone 1) is significantly less than that on the "Polema" cathode and varies widely. The formation of the intermetallic compounds' layer in the modified layer of the "Umicore" cathode originates from the same mechanism as for the "Polema" cathode. It is associated with the interaction of Ti and Al at the interphase boundaries of particles. However, due to the large size of Ti particles at the "Umicore" cathode and the small specific surface area of the Ti-Al interphase boundaries, the intermetallic layer has a small and irregular thickness, up to zero in some areas of the modified layer (Figure 7). According to the results of local elemental analysis, the visually unchanged structure of the cathode material adjacent to the surface intermetallic layer consists of large Ti particles in an aluminum matrix (Figure 7, Table 6, zone 2).

4. The Discussion of the Results

The thermodynamic driving force of structural transformations in the surface layer of cathodes is explained by the decrease in the free energy of the system during its transition from the initial non-equilibrium state to the equilibrium one. The greater the degree of non-equilibrium of the initial structural state, the greater the magnitude of the driving force. The "Polema" cathode compacted from the mechanical mixture of fine Ti and Al powders, has the most non-equilibrium structure of all the studied cathodes. The starting structure of the "Umicore" cathode is located nearer to the equilibrium than that of the "Polema" cathode since it is compacted from coarser powders and contains titanium aluminides. The non-equilibrium structure of the sintered cathode is determined by the presence of significant porosity and islands of the residual $TiAl_3$ phase. The single-phase structure of the cast cathode with the minimum porosity is closest to equilibrium.

The transition rate of the starting structure to the equilibrium state also depends on the temperature. The temperature on the arc-treated surface of the cathode in a stationary mode of arc treating and the thickness of the modified surface layer depend on the rate of heat removal from the working surface of the cathode through its body towards the back water-cooled surface. The heat transfer rate is defined by the thermal conductivity of the cathode material, which, in turn, depends on its structure and phase composition.

The structural changes on the surface of the cast cathode consist of a slight surface melting and cracking of the surface layer under vacuum arc heating. The main reason for melting and cracking is probably the low thermal conductivity of titanium aluminides (26.4 W/(m·K)) for TiAl, which differs little from the thermal conductivity of pure Ti (22 W/(m·K)) [29]. The rate of heat removal from the working surface of the sintered cathode is significantly lower due to its increased porosity, which led to a thickness of the melted layer on the cathode surface of 400 μm. The presence of a relatively thick film of liquid metal on the cathode surface at high arc currents leads to the formation of drops and a deterioration in the quality of the deposited coating.

The thermal conductivity of the hot-pressed "Polema" and "Umicore" cathodes, which consist of a mechanical mixture of Ti and Al powders, is several times higher than the thermal conductivity of cast and sintered cathodes. The reason is that hot-pressed cathodes contain aluminum with a thermal conductivity of 237 W/(m·K) [29], which is an order of magnitude higher than that of the titanium aluminides. Despite the high thermal conductivity of the hot-pressed cathodes, the temperature in the modified layer exceeds the Al melting point.

The higher surface temperature of the "Polema" cathode resulted in the thicker melted layer can be associated with additional heat influx from exothermic reaction in the dispersed mechanical mixture of Ti and Al particles with the formation of titanium aluminides. In the modified layer of the "Umicore" cathode, compacted from coarse powders, titanium aluminides are formed only in a thin surface layer of irregular thickness (Figure 4).

Let us compare the studied cathodes from the point of view of the best applicability in the coating technology. Sintered and cast cathodes appear to be unsuitable for the following reasons. The thick layer of the melt generated by the vacuum arc on the surface of the sintered cathode is the source of metal drops. The drops deposit on the coating surface resulting in bad roughness. Thermal cracks in the modified layer of the cast cathode can result in destruction and rapid erosion of the cathode. The most suitable for practical use seems to be cathode No. 2 ("Polema") with a homogeneous structure, which consists of fine Ti grains in an Al matrix.

5. Conclusions

The structural transformations occur in the modified surface layer of Ti-Al cathodes when they are heated by a vacuum arc. The intensity and consequences of such structural transformations depend on the initial structure and phase composition of the cathode material. The most significant changes were revealed in the modified layer of the "Polema" cathode, in which a dispersed mechanical mixture of Ti and Al particles transforms into a continuous intermetallic layer up to 100 μm thick. The "Polema" cathode seems to be the most suitable for use in vacuum arc coating technology due to the homogeneous structure and moderate thickness of the modified surface layer.

Author Contributions: G.P.: conceptualization, methodology, supervision, experimental design, manuscript writing, project administration; I.F.: formal analysis, investigation, data handling, visualization, translation; V.K.: formal analysis, investigation, conducting XRD-analysis, data handling. All authors have read and agreed to the published version of the manuscript.

Funding: The work was performed according to the Government research assignment for ISPMS SB RAS, project FWRW-2021-0005.

Institutional Review Board Statement: Not applicable.

Informed Consent Statement: Not applicable.

Data Availability Statement: The data presented in this study are available on request from the corresponding author.

Conflicts of Interest: The authors declare no conflict of interest.

References

1. Coll, B.F.; Sanders, D.M. Design of vacuum arc-based sources. *Surf. Coat. Technol.* **1996**, *81*, 42–51. [CrossRef]
2. Sanchette, F.; Ducros, C.; Schmitt, T.; Steyer, P.; Billard, A. Nanostructured hard coatings deposited by cathodic arc deposition: From concepts to applications. *Surf. Coat. Technol.* **2011**, *205*, 5444–5453. [CrossRef]
3. Veprek, S.; Veprek-Heijman, M.G.J.; Karvankova, P.; Prochazka, J. Different approaches to superhard coatings and nanocomposites. *Thin Solid Films* **2005**, *476*, 1–29. [CrossRef]

4. Nurnberg, A.W.; Fang, D.Y.; Bauder, U.H. Temperature dependence of the erosion of Al and TiC by vacuum arcs in a magnetic field. *J. Nucl. Mater.* **1981**, *103–104*, 305–308. [CrossRef]
5. Pribytkov, G.; Korzhova, V.; Korosteleva, E.; Krinitcyn, M. Effect of Silicon on the Surface Modification of Al-Cr Powder Cathodes Subjected to Vacuum Arc Treatment. *Coatings* **2022**, *12*, 958. [CrossRef]
6. Rafaja, D.; Polzer, C.; Schreiber, G.; Polcic, P.; Kathrein, M. Surface modification of Ti-Al targets during cathodic arc evaporation. *Surf. Coat. Technol.* **2011**, *205*, 5116–5123. [CrossRef]
7. Syed, B.; Zhu, J.; Polcik, P.; Kolozsvari, S.; Hakansson, G.; Johnson, L.; Ahlgren, M.; Joesaar, M.; Oden, M. Morphology and microstructure evolution of Ti-50 at.% Al cathodes during cathodic arc deposition of Ti-Al-N coatings. *J. Appl. Phys.* **2017**, *121*, 245309. [CrossRef]
8. Zhu, J.Q.; Johansson-Jöesaar, M.P.; Polcik, P.; Jensen, J.; Greczynski, G.; Hultman, L.; Odén, M. Influence of Ti–Si cathode grain size on the cathodic arc process and resulting Ti–Si–N coatings. *Surf. Coat. Technol.* **2013**, *235*, 637–647. [CrossRef]
9. Koller, C.M.; Hahn, R.; Ramm, J.; Kolozsvári, S.; Mayrhofer, P.H. Microstructural modifications in powder-metallurgically produced $Al_{0.675}Cr_{0.275}Fe_{0.05}$ targets during cathodic arc evaporation. *J. Vac. Sci. Technol. A* **2016**, *34*, 021603. [CrossRef]
10. Zöhrer, S.; Golizadeh, M.; Koutná, N.; Holec, D.; Anders, A.; Franz, R. Erosion and cathodic arc plasma of Nb-Al cathodes: Composite versus intermetallic. *Plasma Sources Sci. Technol.* **2020**, *29*, 025022. [CrossRef]
11. Anders, A.; Yotsombat, B.; Binder, R. Correlation between cathode properties, burning voltage and plasma parameters of vacuum arcs. *J. Appl. Phys.* **2001**, *89*, 7764. [CrossRef]
12. Eriksson, A.O.; Zhirkov, I.; Dahlqvist, M.; Jensen, J.; Hultman, L.; Rosen, J. Characterization of plasma chemistry and ion energy in cathodic arc plasma from Ti-Si cathodes of different compositions. *J. Appl. Phys.* **2013**, *113*, 163304. [CrossRef]
13. Syed, B.; Joesaar, M.J.; Polcek, P.; Kolozsvari, S.; Hakansson, G.; Johnson, L.; Ahlgren, M.; Oden, M. Effect of work function and cohesive energy of the constituent phases of Ti-50 at.% Al cathode during arc deposition of Ti-Al-N coatings. *Surf. Coat. Technol.* **2019**, *357*, 393–401. [CrossRef]
14. Rosen, J.; Anders, A.; Hultman, L.; Schneider, J.M. Temporal development of the composition of Zr and Cr cathodic arc plasma streams in a N_2 environment. *J. Appl. Phys.* **2003**, *94*, 1414–1419. [CrossRef]
15. Franz, R.; Polcik, P.; Anders, A. Element- and charge-state-resolved ion energies in the cathodic arc plasma from composite AlCr cathodes in argon, nitrogen and oxygen atmospheres. *Surf. Coat. Technol.* **2015**, *272*, 309–321. [CrossRef]
16. Zhirkov, I.; Landalv, L.; Gothelid, E.; Ahlgren, M.; Eklund, P.; Rosen, J. Effect of Si on DC arc plasma generation from Al-Cr and Al-Cr-Si cathodes used in oxygen. *J. Appl. Phys.* **2017**, *121*, 083303. [CrossRef]
17. Zhirkov, I.; Petruhins, A.; Polcik, P.; Kolozsvári, S.; Rosen, J. Effect of Ti-Al cathode grain size on plasma generation and thin film synthesis from a direct current vacuum arc plasma source. *AIP Adv.* **2019**, *9*, 045008. [CrossRef]
18. Zhirkov, I.; Oks, E.; Rosen, J. Effect of N_2 and Ar gas on DC arc plasma generation and film composition from Ti-Al compound cathodes. *J. Appl. Phys.* **2015**, *117*, 213301. [CrossRef]
19. Zhirkov, I.; Eriksson, A.O.; Petruhins, A.; Dahlqvist, M.; Ingason, A.S.; Rosen, J. Effect of Ti-Al cathode composition on plasma generation and plasma transport in direct current vacuum arc. *J. Appl. Phys.* **2014**, *115*, 123301. [CrossRef]
20. Zhu, J.; Eriksson, A.; Ghafoor, N.; Johansson, M.P.; Sjolen, J.; Hultman, L.; Rosen, J.; Oden, M. Characterization of worn Ti-Si cathodes used for reactive cathodic arc evaporation. *J. Vac. Sci. Technol. A* **2010**, *28*, 347–353. [CrossRef]
21. Franz, R.; Mendez-Martin, F.; Hawranek, G.; Polcik, P. Erosion behavior of composite Al-Cr cathodes in cathodic arc plasmas in inert and reactive atmospheres. *J. Vac. Sci. Technol. A* **2016**, *34*, 021304. [CrossRef]
22. Pohler, M.; Franz, R.; Ramm, J.; Polcik, P.; Mitterer, C. Cathodic arc deposition of $(Al,Cr)_2O_3$: Macroparticles and cathode surface modifications. *Surf. Coat. Technol.* **2011**, *206*, 1454–1460. [CrossRef]
23. Kameneva, A.; Antonova, N.; Pesin, M.; Makarov, V.; Nikitin, S.; Bublik, N. Structural and phase transformations control in Ti and Al cathode materials, WC-Co substrate, and $Ti_{1-x}Al_xN$ coating to improve their physico-mechanical and wear properties. *Int. J. Refract. Met. Hard Mater.* **2022**, *102*, 105726. [CrossRef]
24. Holzapfel, D.M.; Music, D.; Hans, M.; Wolff-Goodrich, S.; Holec, D.; Bogdanovski, D.; Arndt, M.; Eriksson, A.O.; Yalamanchili, K.; Primetzhofer, D.; et al. Enhanced thermal stability of (Ti,Al)N coatings by oxygen incorporation. *Acta Mater.* **2021**, *218*, 117204. [CrossRef]
25. Koller, C.M.; Glatz, S.A.; Kolozsvári, S.; Bolvardi, H.; Mayrhofer, P.H. Thermal stability and oxidation resistance of architecturally designed Ti-Al-N- and Ti-Al-Ta-N-based multilayers. *Surf. Coat. Technol.* **2020**, *385*, 125444. [CrossRef]
26. Schalk, N.; Weirather, T.; Polzer, C.; Polcik, P.; Mitterer, C. A comparative study on $Ti_{1-x}Al_xN$ coatings reactively sputtered from compound and from mosaic targets. *Surf. Coat. Technol.* **2011**, *25*, 4705–4710. [CrossRef]
27. Pribytkov, G.A.; Andreeva, I.A.; Korjova, V.V. Bulk changes and structure formation in solid phase sintering of Ti–$TiAl_3$ powder mixtures. *Powder Metall. Met. Ceram.* **2008**, *47*, 687–692. [CrossRef]
28. *Phase Equilibrium Diagrams, Datebase*, Version 3.1; American Ceramic Society: Columbus, OH, USA, 2006.
29. Binnewies, M.; Milke, E. *Thermochemical Data of Elements and Compounds*; Wiley-VCH Verlag GmbH: Weinheim, Germany, 2002; 919p.

coatings

MDPI

Article

Fabrication and Characterization of Titanium Borides by Electron Beam Surface Alloying

Fatme Padikova [1,2,*], Daniela Nedeva [2], Vladimir Dunchev [3], Borislav Stoyanov [4], Maria Ormanova [1], Nikolay Nedyalkov [1] and Stefan Valkov [1,2,*]

1 Institute of Electronics, Bulgarian Academy of Sciences, 1784 Sofia, Bulgaria; m.ormanova@ie.bas.bg (M.O.); nnn_1900@yahoo.com (N.N.)
2 Department of Mathematics, Informatics and Natural Sciences, Technical University of Gabrovo, 4 Hadji Dimitar Street, 5300 Gabrovo, Bulgaria; dnedeva1419@abv.bg
3 Department of Material Science and Mechanics of Materials, Technical University of Gabrovo, 4 Hadji Dimitar Street, 5300 Gabrovo, Bulgaria; v.dunchev@tugab.bg
4 Department of Industrial Design and Textile Engineering, Technical University of Gabrovo, 4 Hadji Dimitar Street, 5300 Gabrovo, Bulgaria; stoyanov_b@mail.bg
* Correspondence: f.padikova@yahoo.com (F.P.); stsvalkov@gmail.com (S.V.)

Abstract: This study shows the possibility of the fabrication of titanium borides by an alloying of a titanium substrate with boron powder via a scanning electron beam. During the electron beam alloying experiments, the speed movement of the specimens was varied, where it was 4 and 6 mm/s. The thickness of the alloyed zone formed by the lower velocity of the movement of the workpiece is greater than that of the coating obtained by the higher speed movement. The phase composition of the coatings is in the form of the TiB_2 phase, as well as some amount of undissolved boron in both considered cases. In the case of the lower speed of the movement of the sample, the undissolved boron is within the whole volume of the alloyed zone, while at the higher speed movement, it is on the top of the specimen. The hardness of the obtained coatings by the higher speed of the specimen movement reached values of about 4500 HV. Considering the values of the surface alloy fabricated via the lower velocity movement of 4 mm/s, it is about 2600 HV, which is lower than that of the one obtained by the 6 mm/s speed of the sample movement. The result obtained for the friction coefficient (COF) for the specimen alloyed by the speed of the specimen motion of 4 mm/s is 0.40; the value for the coating obtained at a speed movement of 6 mm/s is 0.34. In both cases, these values are lower than that of the titanium substrate.

Keywords: electron-beam alloying; titanium borides; hardness; phase composition; structure

Citation: Padikova, F.; Nedeva, D.; Dunchev, V.; Stoyanov, B.; Ormanova, M.; Nedyalkov, N.; Valkov, S. Fabrication and Characterization of Titanium Borides by Electron Beam Surface Alloying. *Coatings* **2023**, *13*, 1941. https://doi.org/10.3390/coatings13111941

Academic Editors: Pinghu Chen, Yun Zhang and Ruiqing Li

Received: 18 October 2023
Revised: 8 November 2023
Accepted: 12 November 2023
Published: 14 November 2023

1. Introduction

Due to their excellent functional properties, Ti-based alloys are used in various branches of industry, such as aerospace, automotive, railway, marine, cutting and forming tools, and others. However, some disadvantages could be noted, such as low hardness and wear resistance [1,2]. There are various methods to overcome them, the most suitable being the techniques for modifying the structure and properties of the materials on the surface, since they are most actively exposed to exploitation.

There exist a number of technologies for surface treatment and modification, including thin film deposition [3], ion implantation [4], surface treatment by high energy fluxes (laser or electron beams) [5–7], etc. Nowadays, the electron beam surface alloying (EBSA) methods are among the most promising. They have many advantages over conventional ones because of the very short time of the process, the uniform distribution of the energy of the electron beam, and the possibility of very accurate repeatability. The process of EBSA involves moving the specimen with a constant velocity where the workpiece is irradiated by accelerated electrons. The kinetic energy of the accelerated electrons is transferred into heat, leading to the formation of a temperature field from the surface to the depth of the material [8]. The heating and cooling rates can reach quite high values (10^4–10^5 K/s), resulting in a transformation in the microstructure, phase and chemical composition, formation of a preferred crystallographic orientation, changes in the surface topography, etc. [9,10]. The technological conditions of the process can be selected so that the treated surface is melted—the formation of a melt pool. When alloying elements are introduced into the liquid phase, followed by subsequent solidification, composite layers and coatings are formed with significantly enhanced functional characteristics compared to the initial material [11,12]. A lot of examples exist in the scientific literature aiming to improve the surface properties of Ti substrates employing the electron beam surface alloying technology by forming surface coatings and structures such as Cr [13], Al [14], carbon [15], etc. In all cases, composite Ti-based structures with improved functional properties were obtained, namely titanium aluminides, titanium carbides, etc.

On the other hand, titanium borides are characterized by very high hardness values and are extremely promising for the improvement of the mentioned mechanical characteristics [16]. Typical values of the hardness of the bulk titanium borides reach values of 25 GPa [17] and can exceed 40 GPa when the compound is in the form of coatings [18]. This makes them very promising to be applied as reinforcement materials or in the form of protective films where high hardness is required. In [19], a Ti-TiH_2 material was reinforced by titanium diboride powder and the results obtained showed a significant improvement in the mechanical properties. Similarly, in [20] the possibilities of modifying composite titanium matrices by introducing TiB_2 particles by laser cladding technology were researched. The results presented in [20] showed a significant rise in the hardness of the cladded zone after the application of the cladding procedure. The authors of [21] described the mechanical properties of titanium borides fabricated on Si substrates by direct current magnetron sputtering and the results showed maximal values of the hardness of 23 GPa. However, the results published in [22] show that the hardness of titanium borides fabricated on high-speed steel by DC magnetron sputtering can reach more than 40 GPa. Similarly, Kelesoglu et al. [23] investigated the mechanical characteristics of titanium borides deposited on a molybdenum sheet by DC magnetron sputtering where the influence of the external magnetic field applied during the deposition procedure was studied. The results reported in [23] pointed to hardness values ranging from 2200 to 6900 HV, where it was found that the application of the external magnetic field has a major impact on the enhancement of the mechanical properties. Nevertheless, in all considered cases, all measured values for the hardness of the fabricated titanium boride coatings and thin films significantly exceed that of the substrate materials used for experimental purposes.

Following the evaluation of the scientific literature, the application of titanium boride coatings leads to a great improvement in the mechanical characteristics of different substrates. A number of investigations consider the application of TiB compound in the form of reinforcement particles aiming to improve the functional properties of titanium-based materials [19,20]. Also, the formation of titanium borides in the form of thin films and coatings for protective purposes by magnetron sputtering technologies was also extensively investigated [21–23]. However, the fabrication of Ti–B hard composite surface structures and coatings on Ti substrate by an electron beam surface alloying technology is currently not yet studied. For this reason, the present paper aims to study the possibilities of the fabrication of composite coatings in the system of Ti–B with a significantly increased hardness and improved coefficient of friction compared to the titanium substrate achieved via the EBSA technology.

2. Materials and Methods

In this work, coatings in the titanium–boron system were fabricated by an electron-beam surface alloying technology. For the needs of the experimental work, 4 mm thick substrates of commercially pure titanium (with a purity of 99.8%) were used, having a circular shape with an area of 314 mm^2. A total of 50 mg pure boron powder (with an average size of the particles of 10 ± 2 μm and purity of 99.5%) and 0.05 mL acetone were mixed and uniformly applied to the surface of the Ti substrate and, after drying and evaporation of the acetone, a homogeneous boron film was obtained. The alloying experiments were realized by means of an EVOBEAM Cube system. During the alloying procedure, the specimens are irradiated by accelerated electrons while moving in a single direction with a constant velocity. The applied technological conditions were as follows: accelerating voltage—60 kV; electron beam current—18 mA; frequency of scanning of the electron beam—200 Hz. The electron beam was focused on the top of the surface of the modified specimens. During the process, the speed of movement of the specimens varied between 4 and 6 mm/s. The experiments were realized in the conditions of circular scanning geometry where the diameter of the circle figure was 20 mm. The technological conditions presented above were optimized in order to melt the modified surface and achieve a molten pool. In this approach of electron beam modification, the trajectory of the electron beam overlaps. This leads to a longer lifetime of the melt pool and successful introduction of the alloying elements. The application of a circular manner of scanning tends to a more homogeneous distribution of the boron particles. In Figure 1, a scheme of the experimental work is presented.

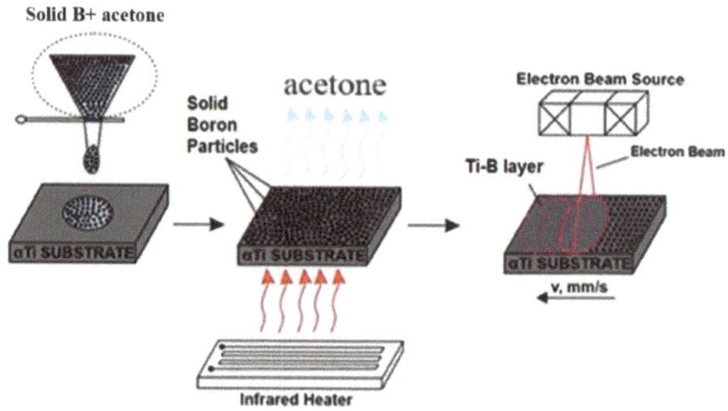

Figure 1. A scheme of electron beam surface alloying procedure.

X-ray diffraction (XRD) measurements were used for the evaluation of the phase composition of the Ti–B surface alloys. The experiments were realized in symmetrical (Bragg–Brentano) mode with Cu Kα characteristic X-ray radiation (1.54 Å). The measurements were carried out from 25 to 85 degrees at a 2θ scale at a step of 0.01° and a counting time of 1 s per step.

The microstructure and chemical composition were examined by scanning electron microscopy (SEM), equipped with a detector for EDX (energy dispersive X-ray spectroscopy) experiments. Backscattered and secondary electrons were used during the SEM experiments. An accelerating voltage of 20 kV was used.

The Vickers microhardness of the specimens obtained was studied. A load of 200 g was used. The applied load was realized for a times of 10 s with a gradient of 20 g/s. The experiments were realized according to the ISO 6507-1 standard [24].

A mechanical wear test (ball-on-flat) was conducted to study the coefficient of friction (COF) of the obtained coatings using a micro-tribotester (UMT-2, Bruker, CA, USA) where the sliding ball was coated with Cr. The experiments were performed for 300 s with a load of 2 N at room temperature.

3. Results and Discussion

The XRD patterns of the fabricated surface alloys in the Ti–B system are shown in Figure 2. Both diffractograms, corresponding to specimen movement speeds of 4 and 6 mm/s, respectively, are typical for polycrystalline materials. Diffraction maxima of pure α–Ti, pure B, as well as TiB_2 are present in both cases. The ICDD (International Center for Diffraction Data) database was used for the evaluation of the phase composition of the fabricated coatings, ICDD PDF# 44 1294 for α–Ti, ICDD PDF# 31 0207 for B, and ICDD PDF# 35 0741 for TiB_2. Also, the reference peak positions, according to the ICDD database of the aforementioned phases, are shown in Figure 2. The results show good agreement between the positions of the experimentally obtained diffraction maxima and those available in the crystallographic database. Concerning the pure B phase, some peaks do not appear in the experimentally obtained diffraction patterns which, however, are characterized by very low intensities in the crystallographic database. This, together with the small amount of the boron content could be considered as the reason that some diffraction maxima of the B phase disappear from the experimentally obtained diffractograms. Nevertheless, according to the ICDD database, the main boron peaks located between 33 and 39° at a 2θ scale are well visible, meaning that some amount of undissolved boron exists in both considered cases. The intensities of the TiB_2 peaks are the highest. This means that the titanium diboride phase is the main one detected on the surfaces of the samples. It should be noted that no amorphous-like halos can be observed at the lower Bragg angles, meaning that the obtained surface alloys and coatings are fully crystalline. Additionally, the phase composition remains unchanged, which is proof that the movement speed of the specimens does not affect the phase formation during the process of EBSA within the presented technological conditions. It should be noted that the ratios between the intensities of the peaks of the TiB_2 phase are in good agreement with the ICDD database. Moreover, no changes in the intensities of the diffraction peaks were observed as a function of the movement speed of the specimens. This means that changing this technological parameter does not lead to a change in the texturing and formation of a preferred crystallographic orientation of the samples. Such texture changes and the formation of preferred crystallographic orientation significantly influence the resultant functional properties of the formed surface alloys and coatings [25]. At the same time, no significant changes in the peak positions of the aforementioned phase are observable. The observed shift between the peaks of the coatings formed by the speed of the movement of the workpiece of 4 and 6 mm/s, respectively, is less than 0.01° at a 2θ scale, meaning that the lattice parameters and unit cell volumes remain unchanged or their deviation from each other is insignificant. This again could be considered very important for the resultant functional properties of the fabricated Ti–B surface alloys and coatings [26,27].

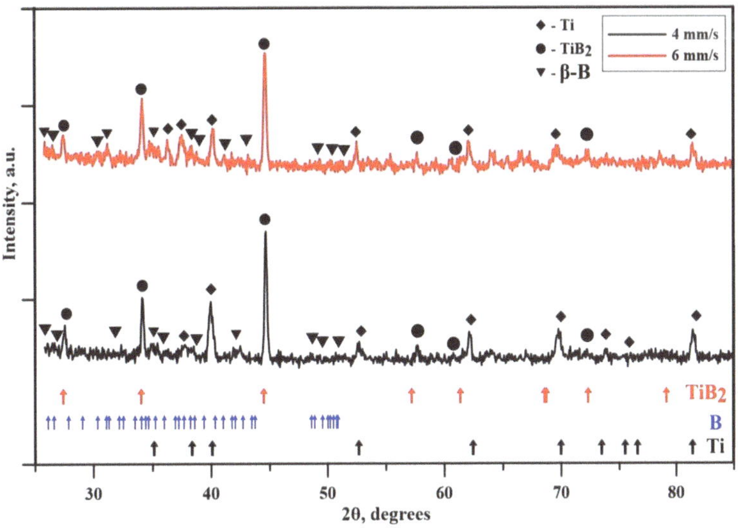

Figure 2. X-ray diffraction patterns of the fabricated titanium borides by the speeds of the specimen movement of 4 mm/s and 6 mm/s.

Figure 3 presents a cross-sectional SEM micrograph of the specimen alloyed by the lower movement speed (i.e., 4 mm/s). The alloyed zone is marked as A, while the Ti substrate is indicated as B. The chemical composition of the specimen is studied by energy-dispersive X-ray spectroscopy (EDX) experiments. The points where the chemical composition was investigated are marked as 1 and 2. The experimentally obtained EDX spectra are presented in Figure 4 and the results are summarized in Table 1. Figure 4a corresponds to Point 1 marked in Figure 3 while Figure 4b is associated with Point 2. The measured average thickness of the alloyed zone (i.e., Zone A from Figure 3) is about 50 μm. The results obtained show that the applied technological conditions, particularly the selected movement speed of the sample of 4 mm/s, led to the formation of a clearly separated alloyed zone and base titanium substrate. Considering the structure of the obtained surface alloy, it is obvious that two regions can be distinguished—darker and brighter. The chemical composition of both zones was investigated where the darker region corresponds to Point 1 while the brighter one is to Point 2. The results revealed that Point 1 consists of undissolved boron meaning that it has not been completely dissolved into the Ti matrix. However, considering the brighter zone, the amount of Ti is 36.4 at% and the concentration of B is 63.6 at%. The binary Ti–B phase diagram shows that the analyzed brighter zone fully corresponds to the TiB_2 compound [28]. Also, it should be noted that some defects, like pores, cracks, etc., exist (marked with black arrows in Figure 3). They could be attributed to the existence of two different compounds (TiB_2 and pure boron) with different thermo-physical properties. The presence of two different elements within the alloyed zone could be considered as a prerequisite for the formation of such defects which can be formed during the process of miscibility between them. The results obtained by the SEM and EDX experiments are in agreement with those of XRD measurements confirming that the phase composition of the formed surface alloy in the system of Ti–B is in the form of a TiB_2 structure with some amount of undissolved boron. It is clearly visible that the undissolved boron element is not only on the surface but also within the depth of the formed surface alloy, meaning that the undissolved boron element penetrated during the alloying procedure but did not dissolve into the Ti matrix.

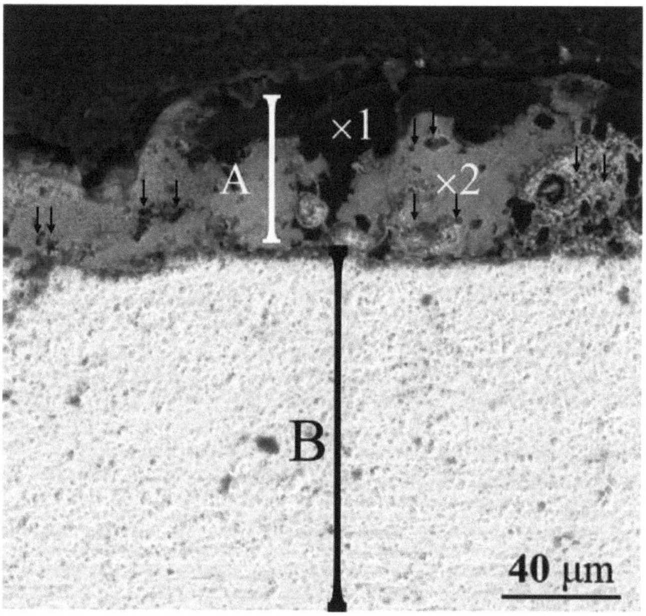

Figure 3. A cross-sectional SEM image of the specimen alloyed with a speed motion of 4 mm/s.

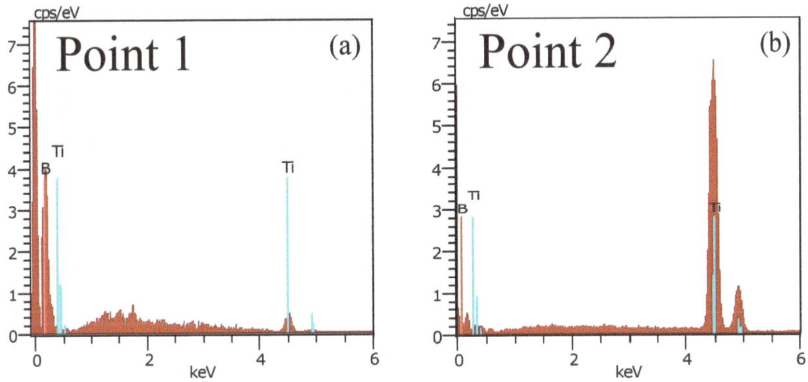

Figure 4. EDX spectra taken from (**a**) Point 1 and (**b**) Point 2 marked in Figure 3.

Table 1. Chemical composition of each measured point marked on the cross-sectional images presented in Figures 3 and 5.

Point	Ti, at%	B, at%
1	0.3 ± 0.1	99.7 ± 1.8
2	36.4 ± 1.9	63.6 ± 4.8
3	0.7 ± 0.1	99.3 ± 1.6
4	32.6 ± 1.9	67.4 ± 5.4

Figure 5 presents a cross-sectional SEM micrograph of the sample processed by the higher movement speed (i.e., 6 mm/s). The alloyed region is marked as Zone A, while the Ti plate is indicated as B. The average thickness of the alloyed zone formed with a movement speed of 6 mm/s is about 40 μm, which is slightly less compared to the alloyed zone produced with a movement speed of 4 mm/s. According to [29], the speed of the samples during the electron-beam processing of the materials has an impact on the temperature of the surface, heat input, and depth of the alloyed zone. It was shown that the temperature on the surface increases with the decrease in the movement speed of the sample in all cases. This correlates with the results achieved in this study where the lower movement speed of 4 mm/s led to the formation of a thicker alloyed zone as compared to that obtained with the higher movement speed of 6 mm/s. The EDX results of the chemical composition of the fabricated coatings are shown in Figure 6 and summarized in Table 1. Figure 6a corresponds to Point 3 while Figure 6b corresponds to Point 4 marked on the cross-sectional image presented in Figure 5. The results show that the darker regions (Point 3 from Figure 5) of the alloyed zone fully correspond to undissolved boron. The chemical composition of the brighter zone (i.e., Point 4 indicated in Figure 5) consists of Ti and B, indicating that a composite phase in the system of Ti–B has been successfully formed. It was found that the amount of boron is 67 at% while the titanium amount is 33 at%, corresponding to an ideal 2:1 ratio. The binary Ti–B phase diagram shows that the considered zone fully corresponds to the TiB_2 compound [28]. When comparing the two alloyed zones, no noticeable difference between their phase compositions can be established, apart from a slightly higher concentration of pure undissolved boron in the structure of the specimen treated with the higher movement speed. This is undoubtedly caused by the reduced input temperature per unit of time. This temperature was apparently not enough to completely melt the solid boron particles atop the surface of the substrate. However, under the undissolved B particles, the top of the titanium substrate was melted, and some amount of the boron particles were introduced into the molten material, leading to the formation of the TiB_2 compound under the undissolved boron. Also, no or a very small number of structural imperfections, such as pores and cracks, can be observed, which could be attributed to the fact that the undissolved boron remains on the top of the surface alloy, not within the whole alloyed zone.

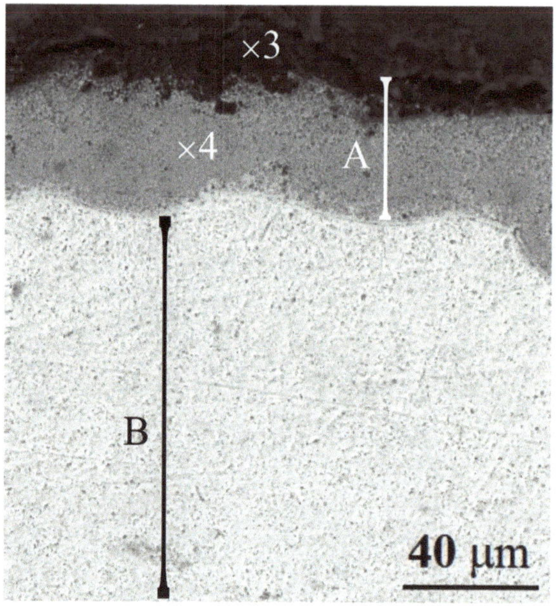

Figure 5. A cross-sectional SEM micrograph of the sample processed by the velocity movement of 6 mm/s.

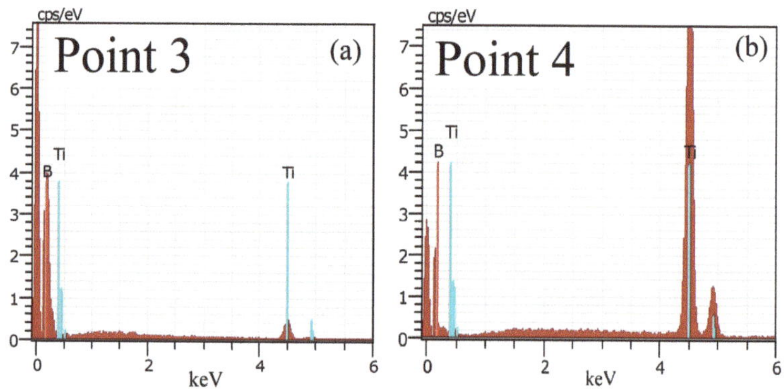

Figure 6. EDX spectra taken from (**a**) Point 3 and (**b**) Point 4 marked in Figure 5.

The results of the micro-hardness measurement of the fabricated coatings, as well as that of the pure titanium substrate, are shown in Figure 7 and are presented in Table 2. Micrographs of the experimentally obtained indentation imprints are also shown above the points. They indicate the micro-hardness values as proof of the very large difference of the studied mechanical characteristic between both types of coatings (i.e., obtained by the different speeds) and the substrate. The hardness of the substrate is about 210 $HV_{0.2}$. This value is typical for pure titanium. The average hardness of the surface alloy in the system of Ti–B obtained at the specimen's movement speed of 4 mm/s is 2610 $HV_{0.2}$, or it is more than 12 times greater in comparison to that of the Ti substrate. The hardness of the specimen fabricated by the speed movement of 6 mm/s is even higher. The measured values reached more than 4540 $HV_{0.2}$ which is more than 20 times greater than the Ti plate. The hardness of the surface alloy formed by the velocity of the movement of the specimen of 4 mm/s is lower than the fabricated one by the velocity of 6 mm/s. This could be attributed to the significantly larger number of defects, such as pores, cracks, etc., resulting in the reduced density of the fabricated coating. As already mentioned, the fabricated coating by the velocity of 4 mm/s is characterized by much more structural imperfections leading to a decrease in the hardness, as well as to a much higher scattering in the measured results. Nevertheless, the values obtained for the specimen alloyed by the lower speed of the sample movement are 12 times greater than that of the titanium substrate, which is considered a significant improvement of the surface properties as well. As mentioned in the introduction, the authors of [21] described the possibility of fabrication of thin TiB_2 films by DC magnetron sputtering where the bias voltage applied to the substrate was varied. It was shown that the hardness of the films reached 13–23 GPa (1300–2300 HV) which are values lower than those obtained in the present study. It was mentioned that the highest values for the hardness were measured at the film with the highest density which was achieved when the bias was changed from positive to negative. Moreover, the lower hardness values were attributed to the influence of the substrate on nanoindentation measurements, as well as to the selection of the radius of the indentation tip [21]. However, ref. [22] shows results for the hardness of TiB_2, fabricated by DC magnetron sputtering where the substrate was rotated or was in a stationary form. The measured values are in the range from 23 to 42 GPa (2300 to 4200 HV) where the highest hardness was measured on the samples fabricated at the stationary conditions. It was discussed that the enhancement in the mentioned mechanical characteristic is due to the formation of the preferred crystallographic orientation, as well as the much denser structure of the deposited coating when the substrate has not been rotated. The results published in [22] are completely in accordance with those presented in this study. Similarly, other authors [30,31] reported results for the hardness of TiB_2 ranging from 4000 to 5000 HV, which is completely in accordance with those presented in this study. This means that the fabrication of titanium boride coatings by both DC magnetron sputtering [22] and electron-beam surface alloying can lead to the formation of surface structures with very similar values for hardness, although the two technologies are very different. In [22], the deposition of TiB_2 films was realized on high-speed steel substrate by a sputtering of a TiB_2 target, while in the present investigation, the Ti–B coatings were formed by the alloying of pure Ti substrate with B powder by a scanning electron beam. As already mentioned, in the case of the DC magnetron sputtering, the control of the structure and properties was carried out by the rotation of the sample or being in a stationary form [22]. In this study, the control of the structure and properties of the TiB-based coatings fabricated by the EBSA was realized by the control of the speed movement of the specimen under the electron beam. As a result, the thicknesses of the coatings obtained by the DC magnetron sputtering range from 1.4 to 2.5 μm, while in this study the coatings are much thicker—from 40 to 50 μm due to the difference between both technologies. The fabrication of titanium borides by the laser beam surface alloying was shown in [32]. The results showed that the hardness and the friction coefficient are a bit lower than the values obtained in the present study. It was found that the hardness is in the range of 800–3500 HV, and the coefficient of friction was 0.39–0.41, depending on the power of the laser beam [32]. The modification of Ti substrates by a scanning electron beam where alloying elements have not been added leads to an increase in the hardness of only about 15%, which is much lower than the values obtained in this study [33]. This means that the modification of Ti and Ti-based materials by the fabrication of surface alloys in the binary Ti–B system leads to a great enhancement of the surface characteristics as compared to the modification without adding alloying elements.

Figure 8 presents the distribution of the microhardness from the surface to the depth of the samples (i.e., fabricated by the speed movement of 4 and 6 mm/s). Figure 8a presents the results of the coating obtained by 4 mm/s while Figure 8b exhibits the results of the coating obtained by 6 mm/s. The hardness of the specimen fabricated by the lower speed movement of 4 mm/s is in the range of 2400–2700 HV at depths of up to 50 μm, followed by a decrease to values ranging from 300 to 370 HV at depths from 50 to 100 μm. The measured values at depths higher than 100 μm are about 200 HV. This is a typical value for the Ti substrate. The obtained values for the sample fabricated by the higher speed motion are in the range of 4100–4900 HV at depths of up to 40 μm. With an increase in the depth of the specimen, the hardness decreases significantly as it is in the range from 300 to 360 HV within the distances from 40 to 70 μm from the surface to the bulk. At higher depths, the hardness drops further to about 210 HV. These results are fully consistent with those for the microstructure and thickness of the obtained coatings in the system of Ti–B where the measured hardness remains relatively unchanged within the whole coating in both cases. After that, it decreases significantly but is still a bit higher than the titanium material because in both cases a heat-affected zone with improved mechanical properties exists. The existence of such a region is typical for electron-beam-processed materials [34]. Below this heat-affected zone, the measured microhardness decreases to values that are typical for the titanium substrate.

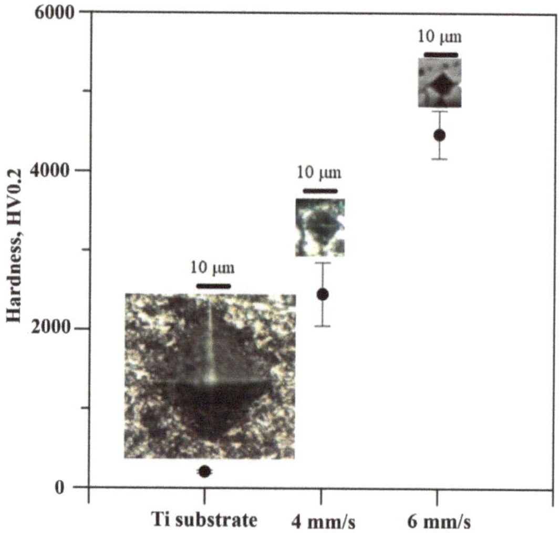

Figure 7. Hardness of the fabricated Ti–B surface alloys and Ti substrate.

Table 2. Hardness and coefficient of friction (COF) of the titanium substrate and Ti–B coatings formed by the speed movement of 4 and 6 mm/s.

Specimen	Hardness, HV0.2	COF
Substrate	210 ± 12	0.48 ± 0.12
4 mm/s	2610 ± 590	0.40 ± 0.16
6 mm/s	4540 ± 380	0.34 ± 0.13

The coefficient of friction (COF) of the Ti–B coatings formed by the speed motion of 4 and 6 mm/s as well as of the pure Ti substrate is shown in Figure 9 and is presented in Table 2. The measured values on the surface of the titanium substrate exhibit a value of 0.48. The coefficient of friction of the Ti–B coating formed by the velocity of the sample motion of 4 mm/s is 0.40 or it is lower in comparison with the Ti substrate. The values for the COF of the coating obtained by the speed movement of 6 mm/s are further reduced where the average measured value is 0.34. In all considered cases, the manufactured coatings in the system of Ti–B exhibit lower values of the coefficient of friction, whereas at the higher speed of the sample movement, it is the lowest one. This means that in all cases, the fabrication of coating in the system of Ti–B leads to a reduction in the COF, meaning that the tribo-properties on the surface of the coatings are better than that of the base Ti substrate. The measured values are lower than those reported in [23] where the TiB_2 coatings were fabricated by DC magnetron sputtering, meaning that the fabricated surface alloys in the present study (i.e., by the electron-beam surface alloying technology) have better tribological properties. In the case of the deposition of the coatings by DC magnetron sputtering, the adhesion is significantly worsened in comparison with that obtained by the alloying by high energy fluxes, such as EBSA. This could be considered a prerequisite for the formation of surface alloys and coatings with a lower COF and, therefore, better tribo-properties. However, as already pointed out, the friction coefficient measured at the specimen alloyed by the higher velocity movement exhibits lower values than that obtained by the higher speed of the sample movement. This difference could be attributed to the significant number of structural imperfections, such as pores as well as undissolved boron material within the volume of the coating fabricated by 4 mm/s speed movement. As already mentioned, the undissolved B element within the coating formed by the speed movement of 4 mm/s is distributed within the whole volume of the surface alloy while in the case of fabrication of coatings in the system of Ti–B, it is only on the top of the alloyed layer. According to the authors of [35], the larger number of structural imperfections as well as undissolved particles plays a major role in the deterioration of the friction properties of the materials. A comparison of the hardness and coefficient of friction between titanium borides fabri-cated via different technologies is shown in Table 3.

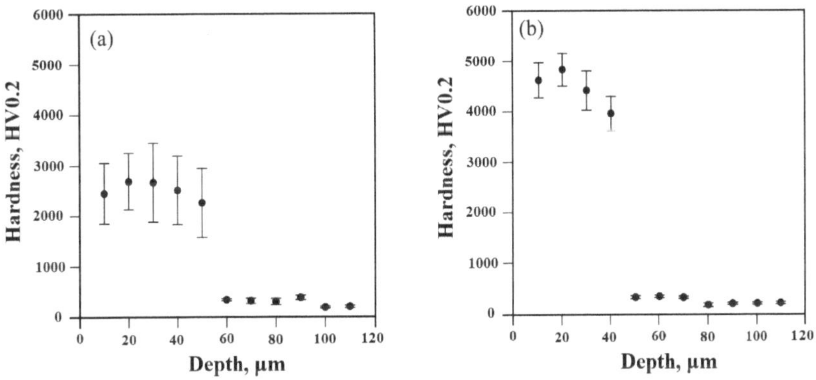

Figure 8. Hardness distribution from surface to the depth of the specimen fabricated by the speed movement of (**a**) 4mm/s and (**b**) 6 mm/s.

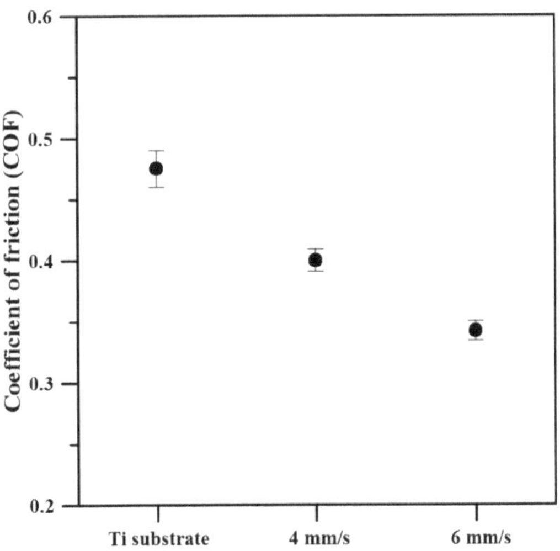

Figure 9. Coefficient of friction of Ti substrate and Ti–B coatings formed by electron-beam surface alloying using 4 and 6 mm/s.

Table 3. Hardness and coefficient of friction (COF) of titanium borides fabricated via different technologies.

No.	Coating	Technology	Hardness, HV	COF	Ref.
1	Titanium boride (TiB_2)	Electron-beam alloying	2610–4540	0.34–0.40	This study
2	Titanium boride (TiB_2)	DC magnetron sputtering	1300–2300	–	[21]
3	Titanium boride (TiB_2)	Magnetron sputtering (DC and RF)	2300–4200	–	[22]
4	Titanium boride (TiB_2)	DC magnetron sputtering	2200–6900	0.50–0.65	[23]
5	Titanium boride (TiB_2)	DC magnetron sputtering	4000–5000	–	[30]
6	Titanium boride (TiB_2)	DC magnetron sputtering	4500–5300	–	[31]
7	Titanium borides (TiB, TiB_2)	Laser alloying	800–3500	0.39–0.41	[32]

This study presents the fabrication of surface alloys and coatings in the binary Ti–B system on Ti substrate by the electron-beam surface alloying technique. It was shown that the alloying procedure by the aforementioned technological conditions led to the formation of a surface alloy where the main phase is TiB_2 and some amount of undissolved boron is also observable. In the application of the lower velocity of the specimen motion of 4 mm/s, the undissolved B element is distributed within the whole alloyed zone, while the application of the higher velocity of 6 mm/s remains on the surface. The incorporation of alloying elements within an alloyed zone formed by high energy fluxes (such as electron or laser beams) is carried out by the so-called Marangoni flows formed due to the very high thermal gradient in the molten material [36,37]. It is suggested that the influence of the Marangoni flows on the miscibility between the materials and the homogeneous distribution is significant and can be characterized by the so-called surface tension number and can be expressed by (1).

$$S = \frac{(d\sigma/dT)qd}{\mu u_0 k} \quad (1)$$

In (1), S is the surface tension number, $(d\sigma/dT)$ is the temperature coefficient of surface tension, q is the net heat flow, d is the beam diameter, μ is the viscosity, u_0 is the velocity of the movement of the specimen, and k corresponds to the thermal conductivity. In cases where the surface tension number is low, the convection flows are negligible and the incorporation of alloying elements and melt homogenization become insufficient. Moreover, the miscibility between the alloying elements and the base material is very low. In cases of high values of the surface tension number, strong convection flows exist which are responsible for the integration of the elements within the melt pool and the homogenization and miscibility between the alloying elements and the substrate are significant [37]. This is in accordance with our results. It was found that the decrease in the speed of the specimen motion causes the fabrication of the surface alloy where the undissolved boron is within the whole alloyed zone. In contrast, the use of the higher velocity of the motion of the specimen leads to the formation of a surface alloy where some amount of undissolved boron appears on the surface. Following Equation (1), it is obvious that the speed of the movement of the samples is inversely proportional to the surface tension number and, therefore, the lower speed corresponds to a higher surface tension number and stronger convection flows which are responsible for the incorporation and homogeneity distribution of the B element within the molten Ti. However, in the present particular case, the application of the lower velocity of the sample motion leads to the formation of a much more inhomogeneous structure than that of the specimen fabricated by the higher velocity movement. In the case of the higher speed, the B powder was not completely melted, and a composite Ti–B structure was formed under the thin unmelted boron layer. In this case, the Ti substrate is melted and the molten titanium material heats further the lower part of the boron film, leading to the introduction of some B amount into the substrate and forming the composite Ti–B structure under the undissolved B film. With the application of 4 mm/s speed (i.e., the lower velocity of the motion of the specimen), both Ti and B materials were completely melted, and the boron element was successfully introduced into the Ti matrix. The existence of a significant amount of undissolved B within the fabricated surface alloy by the lower velocity of the sample's movement has a great impact on the measured surface hardness and coefficient of friction and could be considered a prerequisite for the worse values of the discussed mechanical characteristic. Nevertheless, in both considered cases, some amount of undissolved boron exists. This means that further optimization of the technological conditions of the alloying procedure by a scanning electron beam has to be performed. Considering the velocity of the motion of the samples, at higher than 6 mm/s values, the amount of undissolved B element will be even higher. At the same time, the use of a speed lower than 4 mm/s will result in a significant modification of the morphology on the surface and a significant amount of evaporation will be predominant. Therefore, further optimizations are needed in light of the power of the electron beam and the frequency of scanning.

4. Conclusions

The possibilities of fabrication of titanium borides by electron-beam alloying were presented. During the experiments, the speed of the specimen's movement was chosen to be 4 and 6 mm/s. The following conclusions can be drawn:

- The thickness of the alloyed zone formed by the lower speed of movement is greater than that of the alloyed zone obtained by the higher speed of movement.
- The phase composition of the surface alloys in the system of Ti–B is in the form of a TiB_2 phase, as well as some amount of undissolved boron in both considered cases. In the case of the lower speed of movement of the sample, the undissolved boron is within the whole alloyed zone, while at the application of the higher speed of movement, it is on the top of the specimen.
- The hardness of the surface alloy fabricated via the higher speed of movement of 6 mm/s significantly exceeds that of the base titanium substrate. The measured values of the obtained coatings using the speed of movement of the specimen reached about 4500 HV. The values of microhardness of the surface alloy fabricated via the lower speed of motion of 4 mm/s are about 2600 HV.
- The coefficient of friction (COF) for the specimen alloyed by a speed of motion of 4 mm/s is 0.40; the value for the coating obtained by a speed of motion of 6 mm/s is 0.34. In both cases, these values are lower than that of the titanium substrate.

Author Contributions: Conceptualization, F.P. and S.V.; methodology, F.P., D.N., V.D., B.S., M.O., N.N. and S.V.; validation, F.P., V.D. and B.S.; formal analysis, F.P., V.D., M.O. and S.V.; investigation, F.P., D.N., V.D., B.S., M.O. and S.V.; data curation, F.P., D.N., V.D., B.S., M.O. and S.V.; writing—original draft preparation, F.P. and S.V.; writing—review and editing, F.P., S.V. and N.N.; visualization, F.P. and D.N.; supervision, S.V.; project administration, S.V. and M.O.; funding acquisition, S.V. All authors have read and agreed to the published version of the manuscript.

Funding: This work was supported by the Project of OP "Science and Education for Smart Growth", "Creation and development of centres of competence"—BG05M2OP001−1.002−0023−C01 "INTELLIGENT MECHATRONICS, ECO- AND ENERGY-SAVING SYSTEMS AND TECHNOLOGIES" (IMEEST).

Institutional Review Board Statement: Not applicable.

Informed Consent Statement: Not applicable.

Data Availability Statement: Data are contained within the article.

Conflicts of Interest: The authors declare no conflict of interest.

References

1. Kartal, G.; Timur, S.; Urgen, M.; Erdemir, A. Electrochemical boriding of titanium for improved mechanical properties. *Surf. Coat. Technol.* **2010**, *204*, 3935–3939. [CrossRef]
2. Makuch, N.; Kulka, M.; Dziarski, P.; Przestacki, D. Laser surface alloying of commercially pure titanium with boron and carbon. *Opt. Laser. Eng.* **2014**, *57*, 64–81. [CrossRef]
3. Lou, B.-S.; Lin, Y.-C.; Lee, J.-W. Mechanical Properties and Corrosion Resistance of AlCrNbSiTiN High Entropy Alloy Nitride Coatings. *Coatings* **2023**, *13*, 1724. [CrossRef]

4. Dou, Z.; Guo, Y.; Zhang, F.; Zhang, D. Effect of Low-Energy Nitrogen Ion Implantation on Friction and Wear Properties of Ion-Plated TiC Coating. *Coatings* **2021**, *11*, 775. [CrossRef]
5. Osintsev, K.; Gromov, V.; Ivanov, Y.; Konovalov, S.; Panchenko, I.; Vorobyev, S. Evolution of Structure in AlCoCrFeNi High-Entropy Alloy Irradiated by a Pulsed Electron Beam. *Metals* **2021**, *11*, 1228. [CrossRef]
6. Nevskii, S.; Sarychev, V.; Konovalov, S.; Granovskii, A.; Gromov, V. Formation Mechanism of Micro- and Nanocrystalline Surface Layers in Titanium and Aluminum Alloys in Electron Beam Irradiation. *Metals* **2020**, *10*, 1399. [CrossRef]
7. Findik, F. Laser cladding and applications. *Sustain. Eng. Innov.* **2023**, *5*, 1–14. [CrossRef]
8. Valkov, S.; Ormanova, M.; Petrov, P. Electron-Beam Surface Treatment of Metals and Alloys: Techniques and Trends. *Metals* **2020**, *10*, 1219. [CrossRef]
9. Petrov, P.; Dechev, D.; Ivanov, N.; Hikov, T.; Valkov, S.; Nikolova, M.; Yankov, E.; Parshorov, S.; Bezdushnyi, R.; Andreeva, A. Study of the influence of electron beam treatment of Ti_5Al_4V substrate on the mechanical properties and surface topography of multilayer TiN/TiO_2 coatings. *Vacuum* **2018**, *154*, 264–271. [CrossRef]
10. Gromov, V.; Konovalov, S.; Ivanov, Y.; Shliarova, Y.; Vorobyov, S.; Semin, A. Structure and properties of the CrMnFeCoNi high entropy alloy irradiated with a pulsed electron beam. *J. Mater. Res. Technol.* **2022**, *19*, 4258–4269. [CrossRef]
11. Meisner, S.; Yakovlev, E.; Semin, V.; Meisner, L.; Rotshtein, V.; Neiman, A.; D'yachenko, F. Mechanical behavior of Ti-Ta-based surface alloy fabricated on TiNi SMA by pulsed electron-beam melting of film/substrate system. *Appl. Surf. Sci.* **2018**, *437*, 217–226. [CrossRef]
12. Weglowski, M.; Blacha, S.; Phillips, A. Electron beam welding–Techniques and trends—Review. *Vacuum* **2016**, *130*, 72–92. [CrossRef]
13. Zhang, L.; Peng, C.; Yao, X.; Guan, Q.; Lu, R. Surface alloying of Cr on Ti6Al4V alloy induced by high-current pulse electron beam. *Surf. Coat. Technol.* **2019**, *370*, 288–297. [CrossRef]
14. Mei, X.; Fu, J.; Li, X.; Rotshtein, V.; Koval, N.; Ma, T. Surface alloying of Al films/Ti substrate based on high-current pulsed electron beams irradiation. *Rare Met.* **2014**, *33*, 155–160. [CrossRef]
15. Valkov, S.; Nedeva, D.; Dunchev, V.; Padikova, F.; Ormanova, M.; Stoyanov, B.; Nedyalkov, N. Fabrication and Characterization of Ti/TiC Composite Layers by an Electron-Beam Surface Modification. *Coatings* **2023**, *13*, 951. [CrossRef]
16. Zhang, X.; Lü, W.; Zhang, D.; Wu, R.; Bian, Y.; Fang, P. In Situ Technique for Synthesizing (TiB+TiC)/Ti Composites. *Scr. Mater.* **1999**, *41*, 39–46. [CrossRef]
17. Munro, R. Material Properties of Titanium Diboride. *J. Res. Natl. Inst. Stand. Technol.* **2000**, *105*, 709–720. [CrossRef] [PubMed]
18. Fuger, C.; Hahn, R.; Hirle, A.; Kutrowatz, P.; Weiss, M.; Limbeck, A.; Hunold, O.; Polcik, P.; Riedl, H. Revisiting the origins of super-hardness in TiB_2+z thin films—Impact of growth conditions and anisotropy. *Surf. Coat. Technol.* **2022**, *446*, 128806. [CrossRef]
19. Bravo Barcenas, D.I.; Chávez Aguilar, J.M.; Jiménez Alemán, O.; Olmos Navarrete, L.; Flores Jiménez, M.F.; González Albarrán, M.A.; Farias Velázquez, I.G. Microstructure and Mechanical Properties of Ti-TiH_2 Based Matrix Composites Reinforced with xTiB_2 Particles Processed by Powder Metallurgy. *Coatings* **2023**, *13*, 587. [CrossRef]
20. Lin, Y.; Wang, H.; Zhang, M.; Lin, H.; Yan, D.; Lin, Q.; Kang, X.; Wang, X. Gradient Coating of Laser Cladding TiB2/Ti-based Alloy on Titanium Alloy Surface. *Coatings* **2023**, *13*, 743. [CrossRef]
21. Sanchez, C.; Rebollo Plata, B.; Maia Da Costa, M.; Freire, F., Jr. Titanium diboride thin films produced by dc-magnetron sputtering: Structural and mechanical properties. *Surf. Coat. Technol.* **2011**, *205*, 3698–3702. [CrossRef]
22. Panich, N.; Sun, Y. Effect of substrate rotation on structure, hardness and adhesion of magnetron sputtered TiB_2 coating on high speed steel. *Thin Solid Films* **2006**, *500*, 190–196. [CrossRef]
23. Kelesoglu, E.; Mitterer, C. Structure and properties of TiB_2 based coatings prepared by unbalanced DC magnetron sputtering. *Surf. Coat. Technol.* **1998**, *98*, 1483–1489. [CrossRef]
24. *ISO 6507-1*; Metallic Materials Vickers Hardness Test Part 1: Test Method. ISO: Geneve, Switzerland, 2018.
25. Ackerland, D.R. *The Science and Engineering of Materials*; PWS Engineering: Boston, MA, USA, 1985.
26. Zhou, Y.; Niinomi, M.; Akahori, T. Effects of Ta content on Young's modulus and tensile properties of binary Ti–Ta alloys for biomedical applications. *Mater. Sci. Eng. A* **2004**, *371*, 283–290. [CrossRef]
27. Lee, Y.; Welsch, G. Young's modulus and damping of Ti6Al4V alloy as a function of heat treatment and oxygen concentration. *Mater. Sci. Eng. A* **1990**, *128*, 77–89. [CrossRef]
28. Yazlak, M.; Christ, H.; Yang, W.; Hasemann, G.; Kruger, M.; Gorr, B. Thermodynamic modelling of the V–Ti–B system. *Calphad* **2022**, *79*, 102477. [CrossRef]
29. Angelov, V.; Ormanova, M.; Kaisheva, D.; Lazarova, R.; Dimitrova, R.; Petrov, P. Selective electron beam surface alloying of aluminum with TiCN nanoparticles. *Nucl. Instrum. Methods Phys. Res. B* **2019**, *440*, 88–94. [CrossRef]
30. Ye, J.; Ulrich, S.; Sell, K.; Leiste, H.; Stüber, M.; Holleck, H. Correlation between plasma particle fluxes, microstructure and properties of titanium diboride thin films. *Surf. Coat. Technol.* **2003**, *174–175*, 959–963. [CrossRef]
31. Berger, M.; Larsson, M.; Hogmark, S. Evaluation of magnetron-sputtered TiB_2 intended for tribological applications. *Surf. Coat. Technol.* **2000**, *124*, 253–261. [CrossRef]
32. Monisha, K.; Shariff, S.; Sekar, A.; Raji, R.; Manonmani, J.; Senthilselvan, J. Titanium boride coating by high power diode laser alloying of amorphous boron with titanium and its surface property investigations. *Opt. Laser Technol.* **2024**, *170*, 110159. [CrossRef]

33. Ormanova, M.; Nikolova, M.; Angelov, V.; Petrov, P. Investigation of the microstructure and the thermal processes in a Ti6Al4V alloy surface-modified by scanning electron beam. *J. Phys. Conf. Ser.* **2020**, *1492*, 12065. [CrossRef]
34. Ren, X.; Wang, R.; Wei, D.; Huang, Y. Effect of scanning electron beam surface tungsten alloying on microstructure and hardness. *Mater. Lett.* **2022**, *308*, 131190. [CrossRef]
35. Gerald, O.J.; Wenge, L.; Tao, Z.Y.; Long, L.C.; Qiang, L. Influence of plasma spraying current on the microstructural characteristics and tribological behaviour of plasma sprayed Cr_2O_3 coating. *Boletín Soc. Española Cerámica Vidr.* **2021**, *60*, 338–346. [CrossRef]
36. Chan, C.; Mazumder, J.; Chen, M. A two-dimensional transient model for convection in laser melted pool. *Metall. Trans. A* **1984**, *15*, 2175–2184. [CrossRef]
37. Almeida, A.; Petrov, P.; Nogueira, I.; Vilar, R. Structure and properties of Al–Nb alloys produced by laser surface alloying. *Mater. Sci. Eng. A* **2001**, *303*, 273–280. [CrossRef]

coatings

MDPI

Article

Influence of Initial Structural Dimensions of Plates on Welding Distortion

Nan Guo [1], Hao Zhang [1], Xiaojie Tang [2,*], Xiqiang Ma [1,3] and Xiao Wang [1]

1 School of Mechatronics Engineering, Henan University of Science and Technology, Luoyang 471003, China; maxiqiang@haust.edu.cn (X.M.)
2 School of Computer Science, Heze University, Heze 274021, China
3 Longmen Laboratory, Luoyang 471003, China
* Correspondence: tang1xiaojie@163.com

Abstract: Aiming at the complex full-field deformation problem that easily occurs when welding plates, this paper adopts the elastic–plastic finite element method with heat-force coupling to study the deformation law of plates in different initial states. First, a rectangular plate finite element model with an initial radius and Gaussian heat source model was established to obtain the welding temperature field and deformation field of the plate; then, the method based on digital image correlation technology was used to detect the full-field dynamic deformation of the plate to verify the accuracy of the finite element model; finally, the influence of the initial structural dimensions of the plate on the weld deformation was investigated. The study shows the following: the thermoelastic–plastic finite element model proposed in this paper has high accuracy in both static and dynamic deformation; plates with the same curvature, and different lengths and widths of the initial structure of the plate welding deformation are saddle-shaped, and the edge effect of the welding of the plate is evident, independent of the length of the plate; and the maximum out-of-face deformation of the welding of the plate is linearly related to the length and the closer the aspect ratio of the plate is to 1, the smaller the out-of-face deformation is.

Keywords: finite element; initial structure of sheet; edge effect; metamorphosis; plastic strain

Citation: Guo, N.; Zhang, H.; Tang, X.; Ma, X.; Wang, X. Influence of Initial Structural Dimensions of Plates on Welding Distortion. *Coatings* **2023**, *13*, 2039. https://doi.org/10.3390/coatings13122039

Academic Editor: Bohayra Mortazavi

Received: 26 October 2023
Revised: 22 November 2023
Accepted: 27 November 2023
Published: 4 December 2023

1. Introduction

Sheet metal is widely used in automobile, ship, aerospace, and other fields; however, during the welding process, the welding deformation of the plate has complex geometric non-linear problems, which is an important factor affecting the machining accuracy, external shape, and structural performance of the welded products. It is an urgent issue that needs to be addressed in industrial production.

Suman S et al. [1] conducted a study on the deformation and stress of weld seams using double-sided submerged arc welding. They applied the initial processing deformation of the plate as a structural load to the finite element model of the same welding joint. The study revealed that there is a stable residual stress distribution in the welding zone and the heat-affected zone, and a compressive residual stress distribution in the weld seam area. Nikhil, R et al. [2] studied the tensile deformation behavior in the heat-affected zone (HAZ) of the Mod. 9Cr-1Mo, and the effects of local deformation properties of HAZ and material parameters of the intrinsic model on the deformation behavior of the HAZ interface were discussed and analyzed through experiments. Zhu Z-K et al. [3] used the finite element method to numerically simulate the temperature field, residual stress field, and welding deformation of multi-pass welding of Q690D thick plates under different welding heat input and groove angle, and studied the peak temperature, residual stress, and welding deformation of welded parts during welding. Liu Zuguo [4,5] et al. proposed a laser welding method with cold air-assisted heat dissipation, established a thermoelastic–plastic multi-physical field coupled numerical model of laser welding, and studied the effects of the parameters, such as the applied position of the heat sink, the strength, and the distance from the laser beam on the weld deformation. Shen Wei [6] et al. considered the effects of initial deformation and geometrical nonlinearity of the plate, established the stress amplification coefficient model at the weld joints of T- and cross-shaped joints, and quickly evaluated the notch stress field and fatigue strength of plate welded joints. Qiu, Y [7] et al. considered the critical stress amplification factor of the initial deformation of the plate connection, established the limit state equation of a typical welded joint, and adopted the JC method to evaluate the fatigue reliability of the welded joint of the plate. Their study shows that the reliability index of the welded joint of the plate is closely related to the plate thickness.

There is also more research on the structural and technological factors affecting welding deformation. Meng D [8] studied the effects of constraints and structural parameters on the welding deformation of unequal-thickness spliced plates, and obtained the welding deformation distribution law of unequal-thickness butt joints, which showed that the welding gap has the greatest influence on the welding deformation, followed by the angle of the beveled weld, the thickness of the thick plate, and the thickness of the thin plate. Wei S [9] et al. considered the effect of welding angle deformation and the initial welding deformation on the butt joints of the plate, and proposed a modified stress amplification factor calculation formula and verified the accuracy of their formulae. Hashiguchi T [10] et al. studied the characteristics of out-of-plane deformation and residual stress caused by patch welding through numerical analysis and patch welding tests. The research shows that the size of welding deformation depends on the size of the patch size. The tensile residual stress of the weld toe increases with the size of the patch. He Z-T [11] et al. predicted the residual stress distribution and deformation of dissimilar metal welded members, and found that with the increase in groove angle, the lateral and longitudinal residual stress near the heat-affected zone increased. The transverse shrinkage and angular deformation of the joint also increase significantly, but the residual stress of the weld and fusion line decreases. Ryu, Hyunsu [12] et al. established a quantitative relationship between welding deformation and temporary pieces setting of temporary weld parts through tests, and proposed a numerical method for welding deformation considering the setting of temporary parts, which can be used for rational use of temporary parts in the plate assembly stage and greatly improve productivity. Ma X-Q [13] et al. proposed a method based on the three-dimensional thermal digital image correlation (DIC) method to measure the three-dimensional full-field dynamic deformation of the curved surface of sheet metal, studied the evolution law of in-plane and out-of-plane deformation in the process of surfacing welding and cooling of the thin curved plate, and established the relationship between curvature of the curved plate and weld contraction. Zhou Z [14] et al. studied the welding deformation characteristics of coplanar double lap-joints by using the thermoelastic–plastic finite element method, and analyzed the influence of welding direction on welding deformation and the mechanism of welding deformation of coplanar double lap-joint. Forcellese A [15] et al. studied the high-speed deformation behavior of AA6082-T6 aluminum alloy friction stir welded sheet under biaxial balanced tension, and discussed the influence of loading rate on the deformation and fracture mechanism of the weld. Zhou, Hong [16,17] et al. combined the thermoelastic–plastic finite element method with the inherent deformation finite element method, accurately predicted the out-of-plane buckling behavior of typical marine plate girder welded structure with ortho-cross reinforcement, and studied the influence of welding sequence on the outward welding deformation. Chino T [18] et al. used the inherent strain method to quickly predict the deformation of resistance spot welding, and effectively predicted the deformation of automobile parts with 23 resistance spot welding within about 90 min. Compared with the test measurement, the prediction model has good accuracy. Xia Xiaowei [19] et al. proposed a simulation method for deformation control of large electron beam welding. By introducing inherent strain into the welding simulation, the optimal welding sequence and clamping conditions were obtained, and the reliability of the simulation was verified through contour measurement tests.

In summary, this paper proposes a thermodynamically coupled elastoplastic finite element method to predict the welding deformation of the Q235 plate, adopts a full-field dynamic deformation detection method based on digital image correlation technology to verify the accuracy of the model, and reveals the influence of the initial structural dimensions of Q235 plate on the deformation law of the outside/inside of the welding surface.

2. Welded Plate Coupling Modeling and Validation

2.1. Establishment of Thermodynamic Coupling Finite Element Model

In this paper, the thermal coupling finite element method was used to predict welding deformation. Firstly, thermal analysis was carried out, geometric models were established, unit types, material properties, etc., were defined, the requirements of structural analysis were taken into account when dividing the grid, and the finer grid at the place of stress concentration, that is, the weld was divided. Then, the heat source model was applied and the boundary conditions and load steps were determined to analyze the temperature field. Finally, the thermal unit was transformed into the corresponding structural unit, the temperature of the node was read, and the thermal load was loaded into the thin curved plate structure for stress field analysis and calculation.

To ensure the accuracy of the finite element calculation and improve the calculation efficiency, the following assumptions were adopted: (1) The material follows the Mises yield criterion. (2) The mechanical behavior in the plastic zone is subject to the rheological law. (3) The thermal physical properties and mechanical properties of the material change with temperature. (4) The influence of viscosity and creep is not considered. (5) Material isotropy.

The size of the thin plate used was 300 mm × 200 mm × 3 mm, the initial radius was 500 mm, and the material was Q235. Figure 1 shows the mesh division of the finite element model of the plate. The size of the element at the weld was 3 mm × 1 mm × 1 mm, and the whole model had 26,244 nodes and 15,000 units. When the welding heat source moved in the direction of x, the stress and deformation were calculated, and considering that there was no constraint on the plate placed around the welding table in the deformation detection test, three direction displacement constraints (x, y, z), two direction displacement constraints (x, y), single direction displacement constraints (z), and no constraints were applied to the four points on the plate grid model in turn.

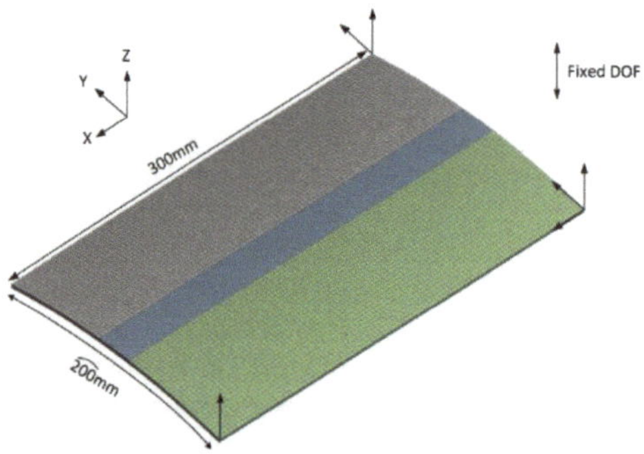

Figure 1. Mesh division of plate finite element model.

When calculating the temperature field, radiation heat dissipation was ignored in the calculation in this paper, and only convective heat transfer of materials and air was considered. Convective edge conditions were applied to all surface nodes exposed to air in the finite element model. The convective coefficient used varies with temperature, as shown in Table 1.

Table 1. Convective heat transfer coefficient.

Temperature/°C	20	100	300	500	750	1000	1590	5000
Convective heat transfer coefficient/W·K·m^{-2}	2.5	5.4	7	7.7	8.2	8.5	9.1	10

The temperature field determined by the welding heat source is the main driving force of welding deformation. In this paper, the Gaussian heat source model was used to describe the heat distribution of manual arc welding and tungsten argon arc welding, and the calculation error was small.

The heat flow distribution formula of Gaussian heat source model is as follows:

$$q_r = q_m \exp(-3\frac{r^2}{R^2}) \tag{1}$$

where, r is the distance from any point in the heat source area to the center of the heat source, R is the radius of the arc area, q_r is the surface heat flow at radius r, and q_m is the maximum heat flow at the center of the heat source, whose value can be calculated by the following formula:

$$q_m = \frac{3}{\pi R^2}\eta UI \tag{2}$$

In the formula, η is the welding thermal efficiency, the thermal efficiency of argon arc welding is generally selected at the value 0.7–0.8, U is the welding voltage, and I is the welding current. The actual welding temperature field can be obtained by adjusting the thermal efficiency, effective heating radius, and loading mode, and the accuracy of calculation can be improved.

The Gaussian heat source was applied to the weld surface unit as a load, and the arc moved by constantly changing the heating area. Each time the center of the heat source moved towards a certain distance, the unit in the range of the heat source was selected and the heat flow value was applied. Then, the corresponding nonlinear transient calculation was performed until the heat source moved to the position of the end of the welding. After the heat source was loaded, the heat source load was removed, and the convection was loaded until the thin curved plate was cooled to room temperature.

The temperature calculation model was converted into a structural calculation model, whereby the calculation results of each step of the transient thermal analysis process were loaded onto the structural calculation model, and the action time step was consistent with the time step of the thermal calculation. Then, the calculation of welding deformation was completed.

2.2. *Welding Coupling Model Accuracy Verification*

2.2.1. Welding Deformation Detection Methods

In this paper, an inspection method based on digital image correlation technology was used to detect the out-of-face deformation of welded plates. The digital image correlation method is a measurement and analysis method to calculate the correlation between the reference scatter image and the deformed scatter image to obtain the coordinates of the point to be matched after displacement.

The measurement principles for 3D full-field strain dynamic detection for plate welding deformation are shown in Figure 2, respectively, and the measurement process involves 2n images of n (n $\geq$ 2) states during the deformation process. First, scattering features were prepared on the surface of the welded plate to be measured, and two industrial CCD cameras (Basler, Ehrensburg, Germany) were used to acquire scattering images in real time during the plate forming process. Then, the computational regions were marked on the images, and the improved digital image correlation algorithm was utilized to perform correlation matching computation on the marked image regions. During the whole matching calculation process, all the left images of the deformed state were correlated with the left image of the undeformed state as the reference image, and all the right images of the deformed state were correlated with the left image of the same state as the reference image. After matching, for the left and right images of any state, the 3D coordinates of all points in the calculated area could be reconstructed by using the internal and external parameters of the camera obtained via calibration through the principle of triangulation. On this basis, the 3D displacement field and strain field of the measured surface could be obtained by further calculation.

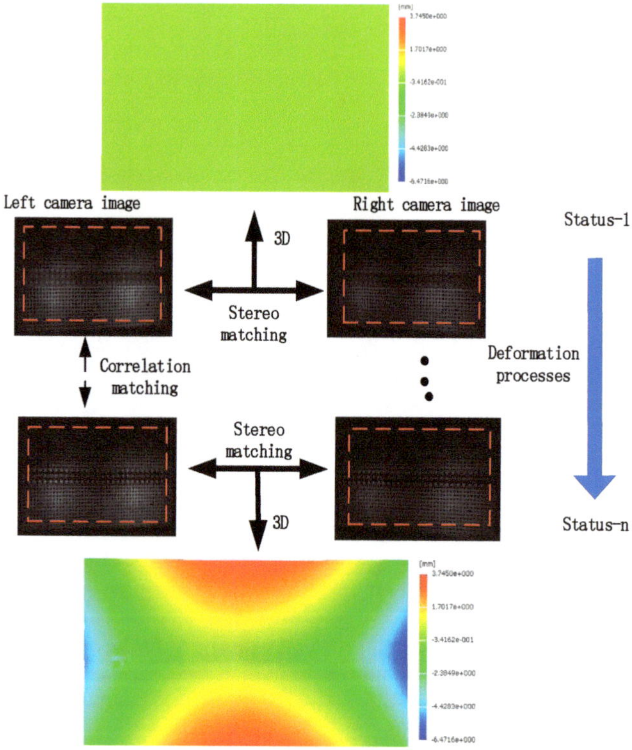

Figure 2. Three-dimensional detection method of plate welding deformation.

2.2.2. Welding Deformation Detection Test

In order to avoid the influence of arc light of TIG welding on the deformation detection device, in the welding process, the plate was placed horizontally on the welding platform (supporting platform), the welding torch is surfacing from the longitudinal center of the sheet from left to right, and the welding torch deformation detection device operated below the test plate, as shown in the schematic diagram of the welding test setup in Figure 3. The deformation detection device mainly includes 2 CCD cameras, two LEDs, etc., a control system, a post-processing system, and auxiliary devices. Test plate structure size and welding process parameters were the same as the numerical simulation.

Before the test, black/white high-temperature paint was sprayed on the inspection surface as a scattering spot that could be recognized by the inspection device, and the inspection device was set to collect one image per second. During the test, the welding robot and the deformation detection device worked simultaneously, the welding speed was 5 mm/s, the plate was cooled in the air after the welding was completed, and the inspection device ended the operation after 60 s. After the test, the deformation images acquired by the left and right cameras were processed in the deformation detection software (ANSYS18.0, ANSYS, Canonsburg, PA, USA).

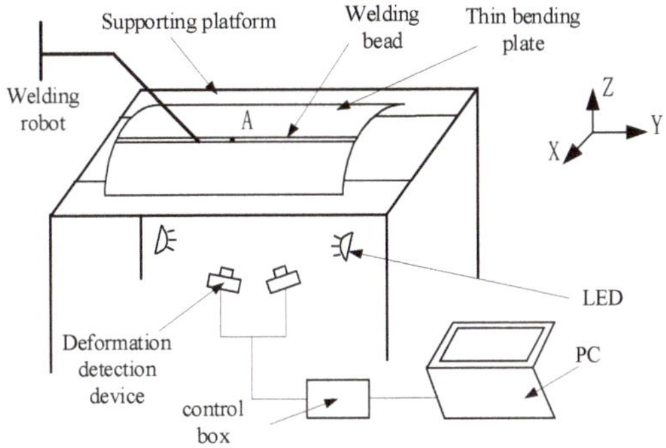

Figure 3. Schematic diagram of welding and deformation detection of thin bending plate.

Figure 4 shows the dynamic deformation curve of key point A. During the whole welding and cooling process, the Z-direction deformation trend of the test and simulation of key point A was consistent, but in the cooling process, the rebound rate generated by the test was lower than the simulation, because the simulation model was more ideal, and the test plate had an unavoidable initial deformation and initial stress. Figure 5 shows the contour plot of the out-of-face deformation of the plate after cooling, from which it can be seen that the saddle-shaped deformation occurred in both the test and simulation, with the maximum deformation of 4.329 mm in the test and 4.689 mm in the simulation.

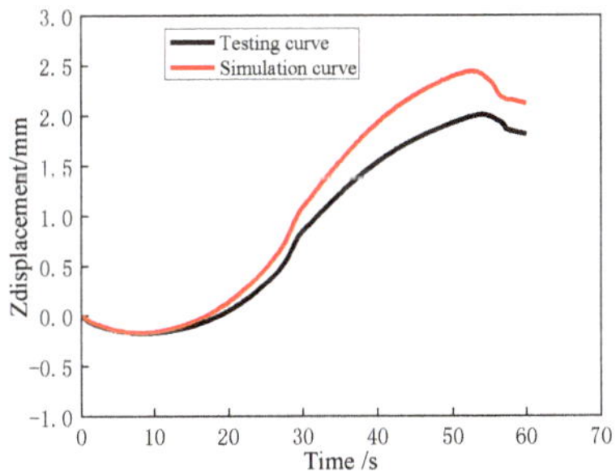

Figure 4. Dynamic deformation curve of key point A.

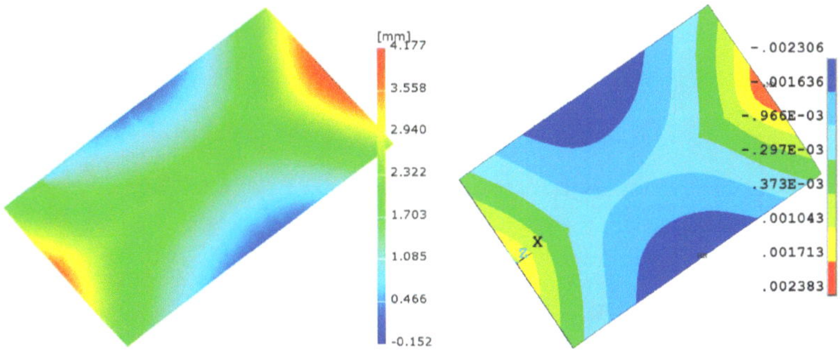

(**a**) Experimental (**b**) Simulation

Figure 5. Contour map of external deformation after cooling of the plate.

By comparing the dynamic deformation curves at critical points and the residual out-of-plane deformations from tests and simulations, it can be seen that the thermoelastic–plastic finite element method based on thermal coupling predicted the welding deformations with high accuracy and could be used to study the influence law of the initial structural dimensions on the welding deformations.

3. Effect of Initial Structural Dimensions on Weld Distortion

3.1. Effect of Plate Length on Welding Distortion

In order to study the effect of plate length on welding deformation, the size parameters of the welding plate used were as follows: radius of 500 mm, width of 200 mm, thickness of 3 mm, and length of 200/250/300/350/400. The welding position was convex, the welding method was TIG surface cladding, the welding current was 170 A, the welding voltage was 15 V, the welding speed was 5 mm/s, and the thermal efficiency was 0.8.

3.1.1. Full-Field Deformation Analysis

Figure 6 shows the contour plot of welding residual deformation of plates with different lengths, from which it can be seen that the out-of-face deformation of the plate after cooling is saddle-shaped with longitudinal downward concavity and transverse upward convexity, with the maximum out-of-face deformation of 2.990, 3.845, 4.698, 5.744, and 6.830 mm, respectively, for the five plates; this suggests that, in the case of the aspect ratio of 1~2, the maximum out-of-face deformation increases with the increase in aspect ratio. In addition, the trend of maximum out-of-face deformation in the longitudinal direction (weld direction) is faster, while the trend of maximum out-of-face deformation in the transverse direction (perpendicular to the weld direction) is relatively slower.

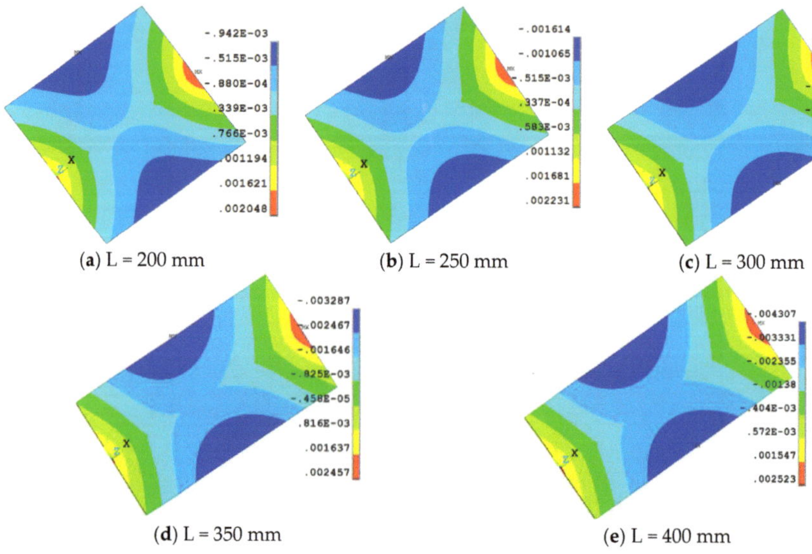

Figure 6. Contour maps of out-of-plane deformation of bending plates of different lengths.

Figure 7 shows the relationship between the maximum out-of-plane deformation of a plate with a radius of 500 mm and the maximum out-of-plane deformation of a flat plate and the plate length. It can be seen from Figure 7 that with the increase in the aspect ratio of the plate, the maximum out-of-plane deformation of curved and flat plates shows a linear trend of increase; this is because the length of the plate is increased, and the greater the residual deformation produced by the longitudinal stress, the deformation is also gradually increased. Thin flat plate out-of-face deformation increased significantly faster because compared with the flat plate, the curvature of the plate in the initial structure has a greater rigidity; accordingly, in the same length and width, the curvature of the plate with greater rigidity has a greater resistance to deformation, and therefore the out-of-face deformation is smaller. However, when the length and width ratio is close, the out-of-face deformation of the curvature plate is slightly lower than that of the thin flat plate. Furthermore, it can be seen that when the length and width dimensions of the plate reach a certain value, the deformation of the curvature plate with greater stiffness is larger instead. The maximum out-of-face deformation of the curvature plate was linearly fitted, and the relationship between the maximum deformation of the curvature plate and its maximum deformation could be obtained using the following equation:

$$\delta_{max} = 0.017L - 0.41 = 0.017(L - 24) \quad (3)$$

where, δ_{max} is the maximum out-of-plane deformation of the curved plate, and L is the length of the curved plate.

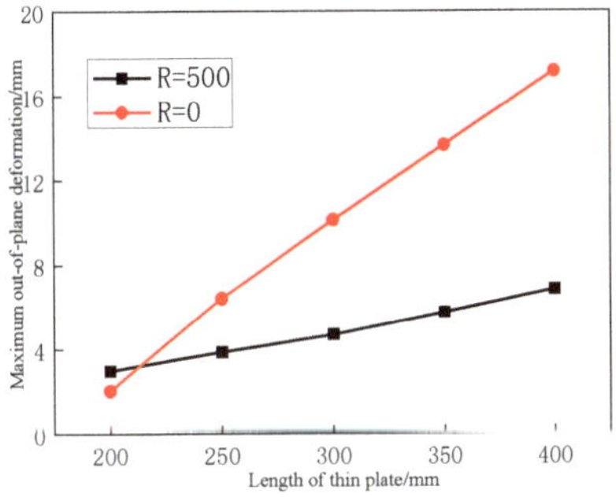

Figure 7. Relationship between maximum out-of-plane deformation and plate length.

The above formula shows that, under the premise of ensuring that other parameters remain unchanged, the maximum out-of-plane deformation of a curved plate is linearly related to its length. Therefore, in practical engineering applications, choosing the appropriate aspect ratio can effectively control the deformation of curved plate weldments with large lengths.

3.1.2. In-Plane Deformation Analysis

Figure 8 shows the longitudinal/transverse plastic strain distribution along the weld for different lengths of bending plate; it can be seen that the plastic strain distribution along the weld for different lengths of bending plate has an edge effect, and the scope of the role is the same, that is, the plastic strain size changes sharply in the range of about 50 mm from the ends of the weld, the plastic strain distribution in the middle region is relatively stable, and the length of the longitudinal strain and the transverse strain is very small. This is because the change in length does not affect the surrounding metal constraints on the weld, so the longitudinal contraction and transverse contraction force per unit length are the same. Because the length of the change does not affect the surrounding metal constraints on the weld, in the same heat input, the weld area unit length of the longitudinal contraction force and transverse contraction force is the same, and thus the stability of the longitudinal contraction and transverse contraction remains unchanged. Therefore, the length of the bending plate weld plastic strain size and edge effect have little effect on the scope of action.

The relationship between the in-plane shrinkage of the bent and flat plates and the length of the test plate is shown in Figure 9. As shown in Figure 9a, with the increase in aspect ratio, the longitudinal shrinkage of both bent and flat welded joints increases linearly. When the aspect ratio increases from 1 to 2, the increment of the longitudinal shrinkage of the bent plate is about 0.1 mm, which is much smaller than the increment of the longitudinal shrinkage of the flat plate. As shown in Figure 9b, the transverse shrinkage of the curved plate slightly decreases with the increase in length, and the change trend is opposite to that of the flat plate. In addition, when the aspect ratio increases from 1 to 2, the transverse shrinkage decreases within 0.1 mm, indicating that the length has little effect on the transverse shrinkage of the bending plate. In general, the effect of length on the inner shrinkage of curved plates is less evident than that of flat plates.

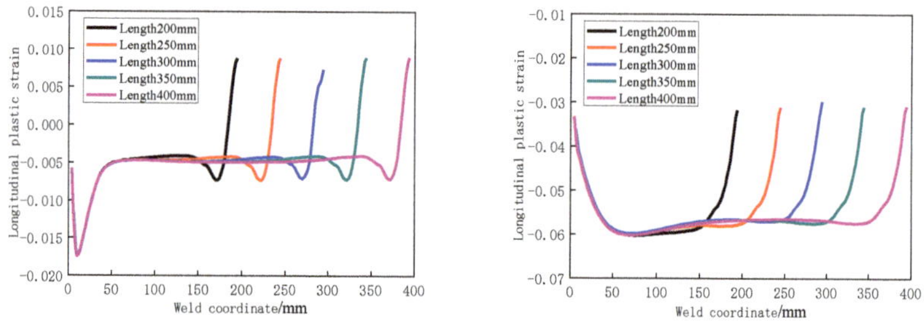

(**a**) Longitudinal strain (**b**) Transverse strain

Figure 8. Strain distribution of welds of different lengths of test plate.

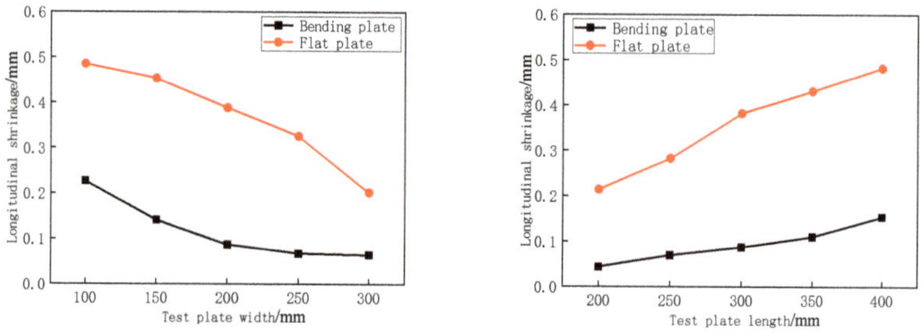

(**a**) Longitudinal contraction (**b**) Transverse contraction

Figure 9. In-plane shrinkage of curved plates of different lengths.

3.2. *Effect of Plate Width on Welding Deformation*

In order to study the effect of plate width on welding deformation, the size parameters of the welding plate used were as follows: radius of 500 mm, length of 300 mm, thickness of 3 mm, length of 100/150/200/250/300. The welding position was convex, the welding method was TIG surface cladding, the welding current was 170 A, the welding voltage was 15 V, the welding speed was 5 mm/s, and the thermal efficiency was 0.8.

3.2.1. Full-Field Deformation Analysis

Figure 10 shows the full-field out-of-face deformation contour plot of different width bending plates and the out-of-face deformation of different width bending plates after complete cooling is saddle-shaped; with the increase in width, the maximum out-of-face deformation decreases, in which the transverse deformation decreased from 9.09 mm to about 0.02 mm, the longitudinal deformation increased from 1.64 mm to about 3.82 mm, and the transverse out-of-face deformation is much larger than that in the longitudinal direction; therefore, width mainly affects the transverse out-of-plane deformation of the curved plate.

Figure 11 shows the relationship between the maximum out-of-plane deformation and the width of the bending plate and flat plate. The influence of the width on the maximum out-of-plane deformation of the bent plate is opposite to that of the flat plate. When the length is fixed, the maximum out-of-plane deformation of the bent plate decreases with the increase in the width; when the width increases to a certain extent, the maximum out-of-plane deformation tends to be stable, and the aspect ratio of the bending plate is about 1. Therefore, the selection of a bending plate with a relatively small length–width ratio can effectively reduce welding deformation.

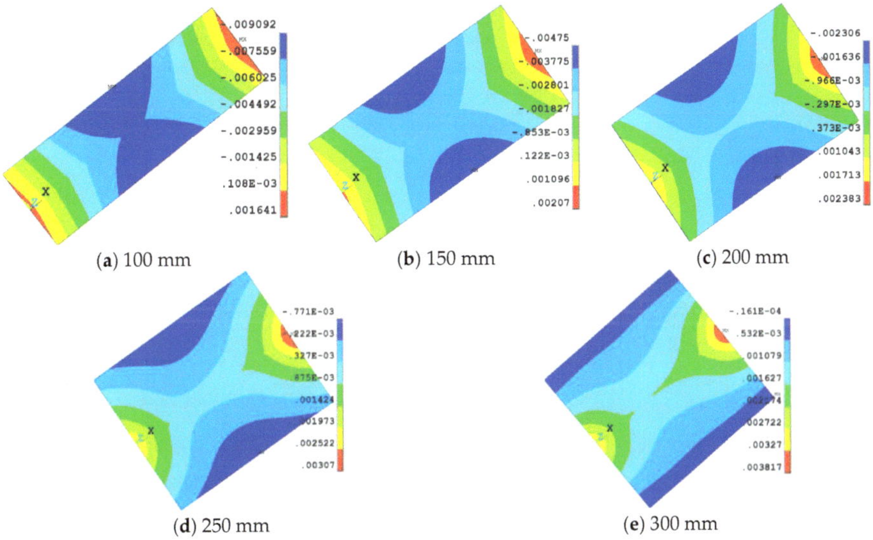

Figure 10. Contour map of out-of-plane deformation of curved plates with different widths.

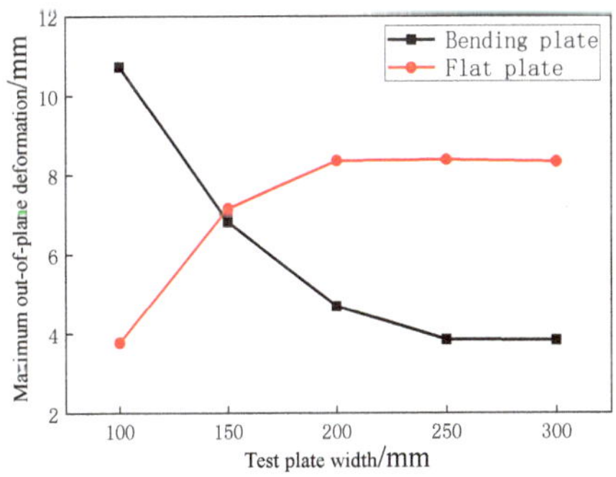

Figure 11. Relationship between the maximum out-of-plane deformation and the width of the test plate.

3.2.2. In-Plane Deformation Analysis

The longitudinal/lateral plastic strain distribution along the weld of bending plates with different widths is shown in Figure 12. Under the same welding process and bending plate curvature, the change in width has a certain influence on the stable zone of plastic strain distribution along the weld. With the increase in width, the longitudinal plastic strain along the bending plate weld gradually decreases, while the lateral plastic strain slightly increases. However, the plastic strain distribution within a range of 50 mm from both ends of the weld is almost consistent. The range of the edge fluctuation zone and the middle stable zone also remains essentially consistent, indicating that the width of the bending plate does not affect the distribution range of the edge effect of the weld plastic strain, but has a certain effect on the shrinkage deformation in the stable region.

Figure 13 shows the in-plane shrinkage of bent and flat plates in relation to the width of the test plate. The longitudinal shrinkage of the bending plate gradually decreases with the increase in width, and when the length–width ratio is close to 1, the longitudinal shrinkage tends to stabilize, and the transverse shrinkage increases with the increase in width, but the increase rate decreases gradually. It shows that when the length–width ratio is large, the effect of width on the in-plane shrinkage is evident. Since the heat input is fixed, the temperature gradient of the welded seam and its adjacent area in different widths are almost the same, and so is the contraction stress, therefore, the changing trend of the above in-plane shrinkage is closely related to the longitudinal stiffness of the bending plate and the transverse constraint.

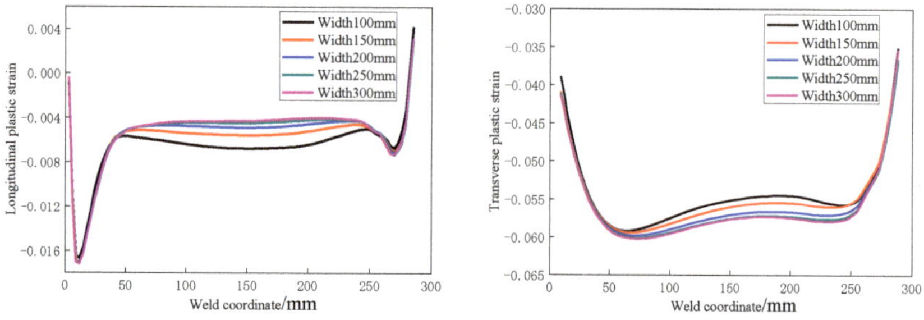

(**a**) Longitudinal strain (**b**) Transverse strain

Figure 12. Strain distribution of welds of different widths of bending plates.

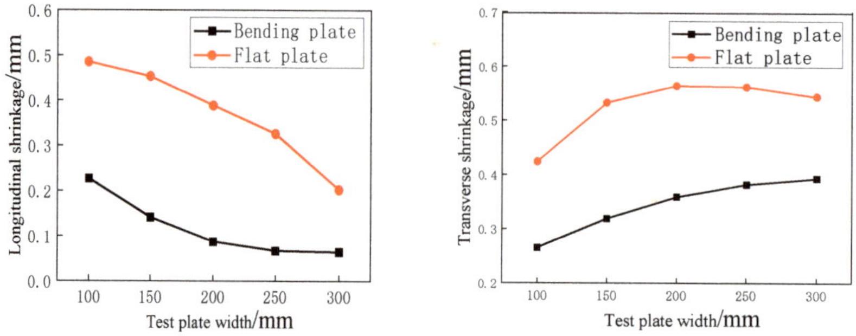

(**a**) Longitudinal contraction (**b**) Transverse contraction

Figure 13. In-plane shrinkage of curved plates of different widths.

It can be seen from the above analysis that the influence of size on bending plate welding deformation is much smaller than that of a flat plate, and the smaller length–width ratio can effectively reduce the welding deformation of bent plate weldments.

4. Conclusions

In this paper, the thermal-force coupled thermoelastic–plastic finite element method was used to investigate the influence law of initial structural dimensions of Q235 plate on welding residual deformation. The following main research results were obtained:

(1) A finite element numerical simulation model of plate welding was established to obtain the welding temperature field and deformation field, and the accuracy of the finite element model calculations was verified by comparing the calculation results with the experimental measurements, which can be used to study the prediction of welding deformation.

(2) The welding deformation of the plate is saddle-shaped, and the contraction deformation of the bent plate along the weld direction has an edge effect. The influence area of the edge effect has no evident relationship with the plate structure, and its distribution range is fixed, about 50 mm from both ends of the weld.

(3) From the perspective of welding deformation analysis, curved panels are less affected by initial dimensions compared to flat panels. When designing the dimensions of welded curved panels, using a smaller aspect ratio can effectively reduce welding deformation. Simultaneously, the greater the length–width ratio of the panel, the greater the longitudinal shrinkage and the smaller the transverse shrinkage within the plane. The out-of-plane deformation of the curved panel has a linear relationship with the length of the panel.

Author Contributions: Conceptualization, N.G.; methodology, N.G. and H.Z.; software, X.T.; validation, N.G. and X.T.; investigation, X.M.; resources, X.W.; data curation, N.G.; writing—original draft preparation, X.T.; writing—review and editing, H.Z.; supervision, X.M.; project administration, X.W.; funding acquisition, N.G. All authors have read and agreed to the published version of the manuscript.

Funding: This research was funded by the Natural Science Foundation of Henan Province of China (Grant No. 232300421336), the National Key R&D Program of China (Grant No. 2021YFB2011000), and the Wind Tunnel Industry Project of Longmen Laboratory (Grant No. LMFKCY2023001).

Institutional Review Board Statement: Not applicable.

Informed Consent Statement: Not applicable.

Data Availability Statement: The data presented in this study are available in this article.

Conflicts of Interest: The authors declare no conflict of interest.

References

1. Suman, S.; Biswas, P.; Patel, S.K.; Singh, V.P.; Kumar, A.; Kuriachen, B. Measurement of residual stresses in submerged arc welded P91 steel using surface deformation. *Mater. Today Proc.* **2020**, *21*, 1707–1712. [CrossRef]
2. Nikhil, R.; Krishnan, S.A.; Moitra, A.; Vasudevan, M. Tensile Deformation Study on Heat Affected Zone of Mod. 9Cr-1Mo Steel Weld. *J. Mater. Eng. Perform.* **2023**, *32*, 2288–2297. [CrossRef]
3. Zhu, Z.; Yang, H.; Zhang, Z.; Zhang, Y.; Zhou, L. Numerical simulation of residual stress and deformation for submerged arc welding of Q690D high strength low alloy steel thick plate. *China Weld.* **2021**, *30*, 49–58. [CrossRef]
4. Liu, Z.; Jin, X.; Zhang, J.; Hao, Z.; Li, J.; Chen, H. Numerical and experimental investigation on the mechanism of synchronous trailing cold air heat sink in eliminating the deformation during laser welding SUS301L thin sheet. *Opt. Laser Technol.* **2022**, *153*, 108258. [CrossRef]
5. Liu, Z.; Jin, X.; Li, J.; Hao, Z.; Zhang, J. Numerical simulation and experimental analysis on the deformation and residual stress in trailing ultrasonic vibration assisted laser welding. *Adv. Eng. Softw.* **2022**, *172*, 103200. [CrossRef]
6. Shen, W.; Qiu, Y.; Li, X.; Han, X.; Berto, F.; Hu, D. Stress magnification effect of initial deformation on the notch stress field and fatigue strength of thin plate welded joints. *Mar. Struct.* **2021**, *78*, 102999. [CrossRef]
7. Qiu, Y.; Shen, W.; Yan, R.; Xu, L.; Liu, E. Fatigue reliability evaluation of thin plate welded joints considering initial welding deformation. *Ocean Eng.* **2021**, *236*, 109440. [CrossRef]
8. Meng, D.; Zhang, L.; Li, Y.; Zhang, G. Welding deformation of unequal thickness tailor welded blanks under multiple factors. *IOP Conf. Ser. Earth* **2020**, *508*, 012223. [CrossRef]
9. Shen, W.; Qiu, Y.; Xu, L.; Song, L. Stress concentration effect of thin plate joints considering welding defects. *Ocean Eng.* **2019**, *184*, 273–288. [CrossRef]
10. Hashiguchi, T.; Hirohata, M.; Jármai, K. An Investigation on the Features of Deformation and Residual Stress Generated by Patch Welding with Different Plate Sizes. *Processes* **2022**, *10*, 1312. [CrossRef]
11. He, Z.; Huang, B.; Chen, L.; Zheng, J.; Yang, J.; Wang, S. Effect of Groove Angle on Stress Field and Deformation of Q345/316 Dissimilar Metal Welded Joints. *J. Mater. Eng. Perform.* **2022**, *31*, 6891–6903. [CrossRef]
12. Ryu, H.; Kang, S.; Lee, K. Numerical Analysis and Experiments of Butt Welding Deformations for Panel Block Assembly. *Appl. Sci.* **2020**, *10*, 1669. [CrossRef]
13. Ma, X.; Guo, N.; Yang, F.; Liu, C.; Guan, Z. Deformation Evolution Law of Surfacing Welding on Thin Bending Plates Based on the Three-Dimensional Thermal Digital Image Correlation Method. *Coatings* **2022**, *12*, 1084. [CrossRef]
14. Zhou, Z.; Zhan, X.; Gao, Z.; Yan, T.; Liu, J. Effect of welding direction on deformation of Ti6Al4V alloy coplanar double lap-joint produced by dual laser beam bilateral synchronous welding. *Opt. Laser Technol.* **2020**, *131*, 106447. [CrossRef]
15. Forcellese, A.; Simoncini, M. High-speed deformation of pinless fswed thin sheets in aa6082 alloy. *Metals* **2019**, *10*, 15. [CrossRef]
16. Zhou, H.; Wang, J. Accurate FE computation for Out-of-plane welding distortion prediction of fillet welding with considering self-constraint. *J. Ship Prod. Des.* **2019**, *35*, 317–327. [CrossRef]
17. Zhou, H.; Wang, J.; Zhang, H.; Liu, J.; Mo, Z. Prediction and mitigation of out-of-plane welding distortion of a typical block in fabrication of a semi-submersible lifting and disassembly platform. *Mar. Struct.* **2021**, *77*, 102964. [CrossRef]
18. Chino, T.; Kunugi, A.; Kawashima, T.; Watanabe, G.; Can, C.; Ma, N. Fast prediction for resistance spot welding deformation using inherent strain method and nugget model. *Materials* **2021**, *14*, 7180. [CrossRef] [PubMed]
19. Xia, X.; Wu, J.; Liu, Z.; Ma, J.; Ji, H.; Lin, X.; Gao, X. Prediction and Validation for Welding Deformation of the Upper Port Stub. *IEEE Trans. Plasma Sci.* **2022**, *50*, 4453–4458. [CrossRef]

coatings

MDPI

Article

Axial Compression Bearing Capacity of Bolted Drum-Shaped Spherical Shell Joints: Experimental and Numerical Analysis

Yajie Yan [1,2], Maoqing Liu [3], Zichun Zhou [4,*] and Xingpeng Ma [5]

1 College of Intelligent Construction, Shanxi Electronic Science and Technology Institute, Linfen 041000, China; tylaoxi@163.com
2 College of Civil Engineering, Taiyuan University of Technology, Taiyuan 030024, China
3 Shanxi 6th Construction Group Co., Ltd., Taiyuan 030024, China; 13613517102@163.com
4 College of Architectural Engineering, Shanxi Vocational University of Engineering Science and Technology, Taiyuan 030000, China
5 Faculty of Architecture, Civil and Transportation Engineering, Beijing University of Technology, Beijing 100000, China; mxp3262@163.com
* Correspondence: zhouzichun@sxgkd.edu.cn

Abstract: Bolted drum-shaped spherical shell joints (BDSSJs) represent a type of joint applicable to space grid structures. These joints merge the benefits of both bolted spherical joints and welded hollow spherical joints, embodying features such as a compact size, favorable centerline alignment with members, a high degree of adjustability, and high installation efficiency. Through unidirectional axial compression tests on specimens of BDSSJs, this study examines the stress distribution, force transmission pathways, ultimate bearing capacity, and failure modes of the joint, thereby determining its bearing capacity and presenting a bearing capacity calculation formula for such joints. By establishing a finite element model with parameters identical to the experimental specimens, this study analyzes the force and deformation of BDSSJs under unidirectional compression, identifying the high-stress areas during the compression process of BDSSJs. The findings of this study provide a basis for the practical engineering application of such joints, as well as theoretical support for subsequent dynamic performance into BDSSJs.

Keywords: space grid structure; bolted drum-shaped spherical shell joint; bearing capacity; theoretical analysis

Citation: Yan, Y.; Liu, M.; Zhou, Z.; Ma, X. Axial Compression Bearing Capacity of Bolted Drum-Shaped Spherical Shell Joints: Experimental and Numerical Analysis. *Coatings* **2024**, *14*, 229. https://doi.org/10.3390/coatings14020229

Academic Editor: Vincent Ji

Received: 4 January 2024
Revised: 3 February 2024
Accepted: 12 February 2024
Published: 16 February 2024

1. Introduction

In space grid structures, joints serve to connect members and transfer loads. A failure in a joint could lead to a redistribution of stresses, possibly resulting in the collapse of the entire structure. Hence, studying the mechanical properties of joints is crucial. Bolted spherical joints (BSJs) and welded hollow spherical joints (WHSJs) are widely employed joint types in space grid structures. BSJs are notable for their high degree of prefabrication and ease of installation, while WHSJs are characterized by simplistic stress states and a high bearing capacity. The details of the BSJ and the WHSJ are shown in Figure 1a,b. Both are extensively used in various public and industrial building sectors like stadiums, airport terminals, train stations, and industrial factories. In recent years, numerous researchers have extensively researched both types of joints.

The BSJs were developed by the German MERO company in 1942 [1], and in the domain of BSJs, a variety of focused studies have been conducted to comprehend their mechanical behavior and performance under different conditions. Initial inquiries examined the fundamental properties of BSJs. For instance, Ebadi Jamkhaneh et al. [2] studied the effects of the BSJ bolt tightening levels on the tensile and compressive axial behavior. Ghasemi et al. [3] conducted experiments on individual BSJs, discovering the load–displacement relationship under tensile loads and various tightening levels. Yaser Doaei [4] investigated the impact of bolt sizes on the stress levels within BSJs. Xu Jing [5] proposed an active sensing method using piezoelectric sensors to monitor the tightness of BSJ connections. Fan et al. [6] analyzed the initial bending stiffness and bending moment capacity under axial compression, laying the groundwork for understanding the basic mechanical behavior of BSJs. Building upon this, Si et al. [7] extended the investigation to the tensile performance of corroded BSJs and those with members, elucidating the degradation law of the tensile properties over time. Moreover, Ding et al. [8] explored the tensile performance at varying screwing depths of high-strength bolts, which was furthered by Yang et al. [9], who introduced a new form of BSJs to tackle the issue of an insufficient high-strength bolt screwing depth, conducting tensile performance experiments on this new variant. Transitioning into specialized conditions, Huang et al. [10] examined the tensile performance of bolted spheres under high-temperature conditions, deriving the tensile bearing capacity of BSJs across a spectrum of high temperatures. On the fatigue front, Tian et al. [11] analyzed the rod element's ultra-low-cycle fatigue performance in a grid structure with BSJs, paving the way for understanding fatigue behavior under cyclic loading. Extending the fatigue analysis to high-cycle regimes, Xu et al. [12], Qiu et al. [13], and Zhou et al. [14] each explored the high-cycle fatigue performance of M20, M30, and M60 high-strength BSJs, respectively, charting the corresponding S-N curves to illustrate the fatigue life of these joints. These assorted investigations not only broaden the understanding of BSJs' mechanical behavior under various conditions but also contribute to optimizing their design and application in engineering projects. In relation to the threads in BSJs, Landeta et al. [15] described the process parameters and testing methods used in experiments, as well as the metallographic analysis of the workpiece after thread formation. Bustillo et al. [16] discussed the potential application of friction drilling technology in joining different materials, particularly in seeking the possibility of nut-free connections in low-load structures.

(**a**) (**b**)

Figure 1. Typical joint types in space grid structures: (**a**) bolted spherical joint and (**b**) welded hollow spherical joint.

In the domain of WHSJs, several researchers have undertaken distinct yet interrelated studies. Initially, the focus has been directed towards understanding the basic performance under various loading conditions. For instance, Xing et al. [17] investigated the failure mechanisms and design strength under a uniaxial load, while Liu et al. [18] examined the axial performance of WHSJs with H-beams under tension and compression. Following this, a deeper understanding of fatigue behavior was sought. Zhang et al. [19] embarked on analyzing the fatigue behavior at the weld toe in the steel tube of WHSJs through 20 constant-amplitude fatigue tests, utilizing an infrared thermal imager to predict the location of fatigue failure based on surface temperature variations. Similarly, Duan et al. [20] conducted fatigue tests on 16 full-scale WHSJ specimens and explored the reinforcing effect of CFRP on their fatigue performance, discovering a shift in the fatigue fracture location from the tube–sphere connection to the tube–endplate connection post-reinforcement. With a foundational understanding of fatigue behavior, researchers then explored more specialized aspects and external factors affecting WHSJs. Yan et al. [21] undertook bending tests to examine the influence of welding residual stress on the ultimate bearing capacity and bending stiffness of WHSJs. Zhao et al. [22] took it a step further by investigating the impact of the tension force on the probability distribution of the bending moment capacity of WHSJs with multiple random corrosion pitting, introducing a method to analyze the bending capacity statistically. On a different note, Huang et al. [23] studied the residual performance after high-temperature exposure to understand the residual mechanical behavior and failure modes. Lastly, Shu et al. [24] derived the failure mode of WHSJs with external triangular ribs through axial compression tests, shedding light on the stress distribution on the spherical surface, thereby adding a new dimension to the understanding of WHSJs' performance under varied conditions.

However, the aforementioned joints have specific drawbacks. The weight of traditional BSJs is directly proportional to the cube of their diameter, causing a dramatic weight increase with scaling, complicating on-site installation, and leading to the false tightening of bolts. WHSJs rely heavily on weld quality for performance and safety, necessitating standard-compliant welding, which is hard to control on construction sites. This paper introduces a new type of joint—the bolted drum-shaped spherical shell joint (BDSSJ), which connects nodes and members using bolts and sleeves, eliminating the need for on-site welding. Meanwhile, with its hollow drum shape, the joint simplifies force transfer and reduces self-weight.

This paper examines BDSSJs' failure mechanisms and design strength values through unidirectional axial compression tests and numerical simulation. A comparative analysis with WHSJs is conducted to propose a failure criterion suitable for these joints and elucidate the stress distribution in the "drum body" under axial loads. A reliable finite element model (FEM) of the joint is established to analyze the stress distribution and allows for the comparison with the experimental results to determine the joint's ultimate bearing capacity. This analysis reveals the joint's failure mechanism, providing a reference for studying the joint's stress state under cyclic loads in subsequent research.

2. Structural Design of BDSSJs

2.1. Design and Functional Attributes of the Joint

The BDSSJ primarily consisted of high-strength bolts for the grid structure, sleeves, curved European nuts, curved washers, hollow drum-shaped spherical shells, and related members for grids; a detailed assembly of the joint is depicted in Figure 2. The load path aligned with that of the traditional BSJs. When members in the spatial grid were under tension, the load path was as follows: steel tube → cone/seal plate → bolt → hollow drum-shaped spherical shell → tension. Conversely, under compression, the load path was as follows: steel tube → cone/seal plate → sleeve → hollow drum-shaped spherical shell → compression.

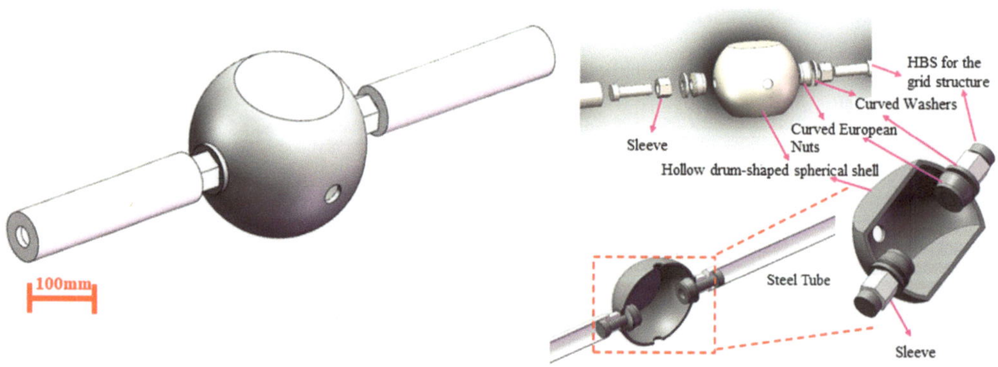

Figure 2. Bolted drum spherical shell joint.

The main innovations of this new type of BSJ include the following: (1) it combined the advantage of the efficient assembly installation of BSJs with the evident force-bearing characteristics of WHSJs; (2) the joint adopted a hollow drum shape, reducing its self-weight; (3) it can be used at the joints of buildings, site of the installation equipment, and the points supporting a roof amongst others, to achieve the advantage of saving building space.

2.2. Fabrication of the Joint

The BDSSJ can be processed based on the fabrication of welded hollow spheres, with bolt holes and installation holes set on the drum surface. The fabrication of the BDSSJ began with the cutting and forging of the materials, shaped into a hemispherical form. Attention needed to be given to the cooling rate during forging to maintain a uniform temperature. Welding then followed, involving groove welding techniques to ensure structural integrity, after which shot peening was performed to enhance the surface properties. Weld flaw detection was then carried out to ascertain the quality of the welds. Subsequently, drum surface welding was executed, followed by additional processing and shaping to refine the joint to its final form. The specimens then underwent annealing heat treatment to relieve internal stresses, followed by surface grinding to remove rust and other residues. They were, finally, coated with anti-rust oil for protection. The fabrication process is illustrated in Figure 3.

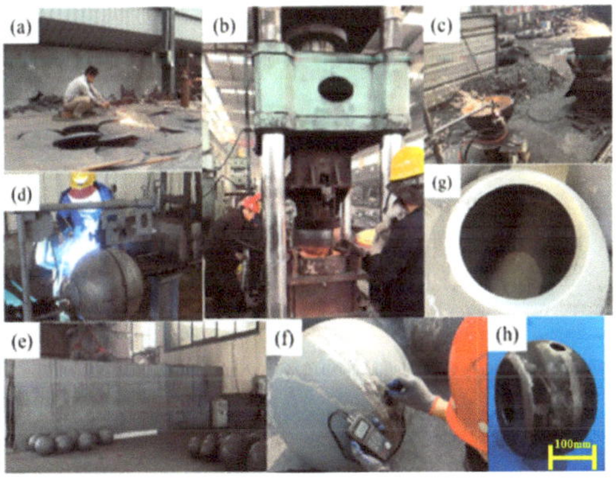

Figure 3. The fabrication process of the BDSSJ: (**a**) cutting, (**b**) forging and shaping, (**c**) hemispherical surface processing, (**d**) welding and shaping, (**e**) shot peening, (**f**) weld flaw detection, (**g**) drum surface welding, (**h**) processing, shaping, and treatment.

3. Experimental Investigation

3.1. Specification and Design of Specimen

3.1.1. High-Strength Bolts and Sleeves

Based on the standard "Bolted Spherical Joints of Space Grid Structures" [25] (JGT 10-2009) and experimental conditions, this experiment used M20 high-strength bolts, with a nominal length of 73 mm and a performance grade of 10.9S. They were made of 35CrMo steel, and their threads were of regular rolling with a pitch of 2.5 mm. The surface of the bolt was treated with oxidative blackening treatment, and its effective section area was 245 mm^2. The sleeves used in this experiment were selected according to "High-strength Bolts for Space Grid Bolted Spherical Joints" (GB/T 16939-2016) and were required to fit well with the M20 bolt.

3.1.2. Curved European Nuts and Curved Washers

This paper utilized curved European nuts, one end of which was cut to have a convex surface with a diameter identical to the inner diameter of the spherical shell, ensuring a tight fit; the other end was optimized to reduce the nut's volume and weight to save steel. A single-sided spherical washer was placed between the sleeve and the shell's connection surface, with one side milled to match the shell surface's concave curvature. Therefore, the diameter of the spherical washer's curvature equaled the spherical shell's outer diameter. The detailed dimensions and actual images of the curved European nuts and curved washers are shown in Figure 4.

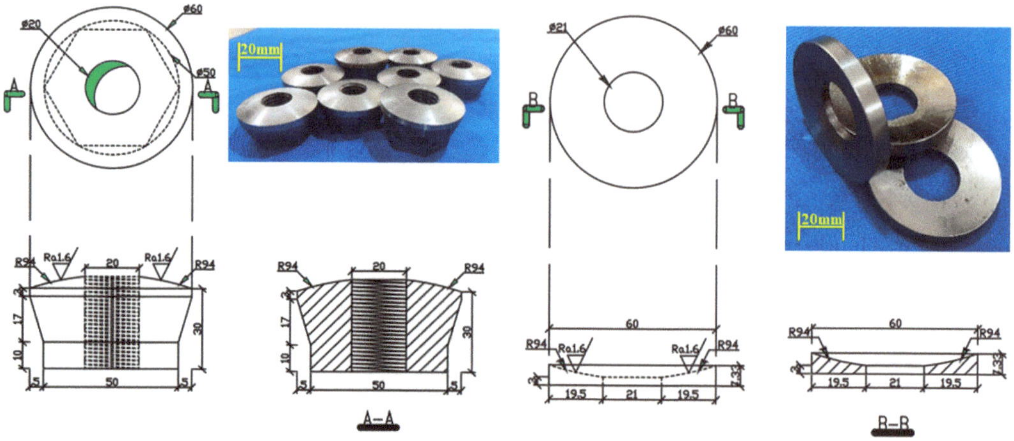

Figure 4. Curved European nuts and curved washers (unit: mm).

3.1.3. Drum-Shaped Spherical Shell

This study used the Q345B drum-shaped spherical shell, with the fabrication process outlined in Section 2.2, and detailed dimensions depicted in Figure 5.

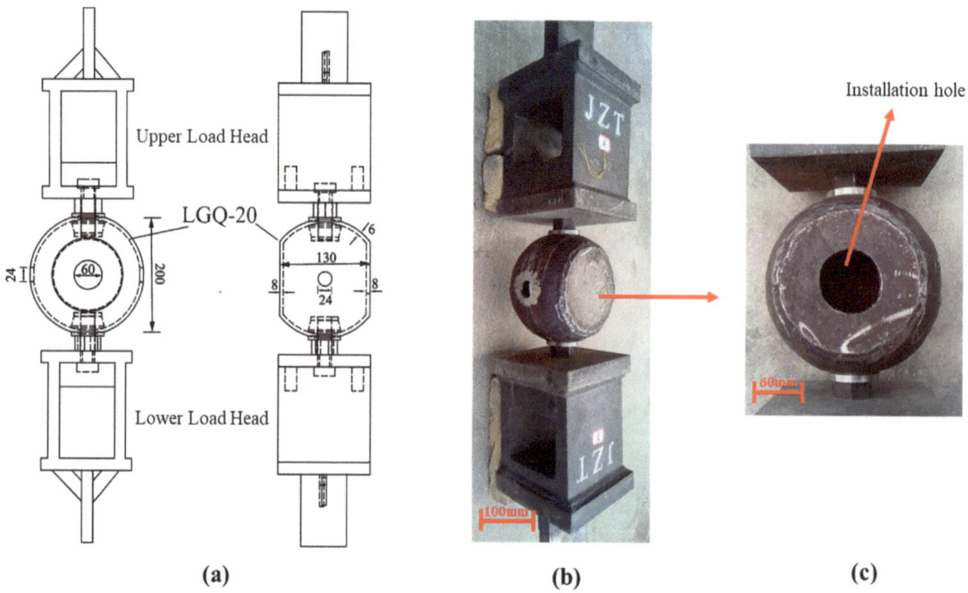

Figure 5. Schematic diagram of specimens: (**a**) dimensions and details of specimens (unit: mm), (**b**) picture of specimens, (**c**) installation hole.

The specimens for this experiment were assembled from high-strength bolts, curved washers, curved European nuts, and drum-shaped spherical shells. High-strength bolts were fastened through the installation holes on the drum surface. Three groups of specimens were selected for the static load axial compression test to ensure the experiment's reliability, labeled with the serial numbers LGQ-20-1, LGQ-20-2, and LGQ-20-3, respectively.

3.2. Measurement of Strain and Stress Distribution

Strain gauges were placed on the surface of the joint specimen to monitor and record strain data pertaining to the joint's stress distribution under axial compression. The specifications of the strain gauges included a resistance value of $120 \pm 0.1\ \Omega$, and a sensitivity coefficient of $2.0 \pm 1\%$. It was stipulated that the tangent direction of the arc line formed by the intersection of the plane perpendicular to the joint's stress direction and the drum body was called the circumferential direction, while the tangent direction of the arc line formed by the intersection of the plane parallel to the joint's stress direction (drum surface) and the drum body was called the axial direction. The arc surface formed between the two drum surfaces was referred to as the short arch, and the arc surface formed between two adjacent bolt holes was referred to as the long arch, as shown in Figure 6.

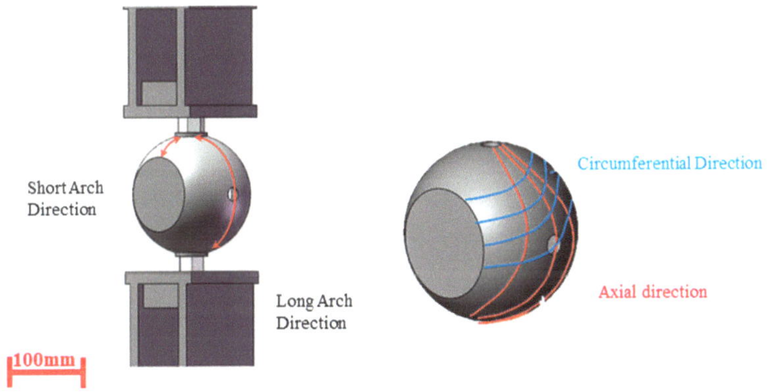

Figure 6. Schematic diagram of different directions of the joint.

Four strain rosettes were placed along the centerline in the long arch direction on both sides of the loaded bolt hole, and two in the short arch direction. Four strain gauges were set on the drum surface along the stress direction. Additionally, two strain gauges were placed on both sides of the fixture at the same displacement to ensure no eccentricity in the load application. Static specimens were labeled 1–8 from left to right along the long arch, and 9–12 from front to back along the short arch, as depicted in Figure 7.

3.3. Loading Equipment and Loading Scheme

Given that the upper chord members primarily bear axial compression, a uniaxial compressive force was applied along the centerline during the experiment. The loading equipment used was an MTS fully digitized electro-hydraulic servo testing machine, with a loading range of ±500 kN, as shown in Figure 8. Before each experiment, specimens were pre-loaded to check the functionality of the measuring instruments, ensuring the load applied during pre-loading did not exceed 10% of the bearing capacity. Both load and displacement loading methods were employed during formal loading to achieve more accurate measurement results and ensure sufficient structural deformation under the load. Before the specimen yielded, the load was applied in increments of 10 kN at a rate of 5 kN/min, with each level held for 3 min. As the yield status approached, displacement-controlled loading was employed at a 0.1 mm/min rate, with each level held for 2 min. Load and axial displacement data were automatically collected by the testing machine.

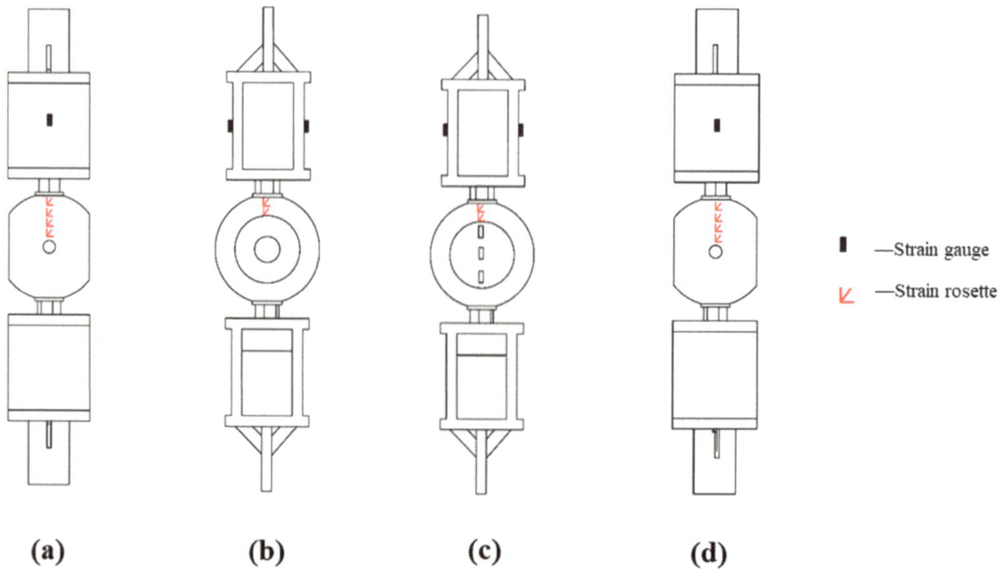

Figure 7. The layout of the strain measurement points: (**a**) left view; (**b**) front view; (**c**) rear view; (**d**) right view.

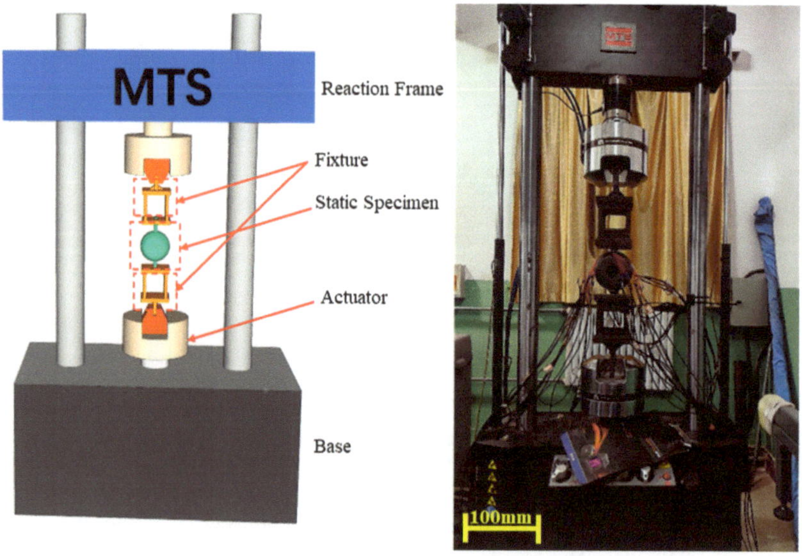

Figure 8. Diagram of the loading equipment for the static load test.

3.4. Criterion for the Ultimate Load and Bearing Capacity

According to the "Criteria for Evaluation of Grid Constructional Engineering Quality Inspection" (JGJ78-91) [26] and Yu et al. [27], the criteria for determining the ultimate load of the spherical joints in the experiment could be quantitatively obtained through summarizing the experiments and FEM analysis. It is stipulated that when any one of the following situations occurs during the experiment, the joint can be considered to have reached its ultimate bearing capacity in the failure mode:

(1) During tensile loading, when the deformation of the sphere reaches 1.2% of its diameter, the joint is considered to have reached its ultimate bearing capacity.
(2) When the ratio of the load increment to average displacement increment in the load–displacement curve of the experiment is lower than 1%, the joint is considered to have reached its ultimate bearing capacity, taking the load applied in the previous step as the ultimate load.

3.5. Experimental Results

Throughout the axial compression process, the specimens roughly experienced three phases: elastic phase, elastoplastic phase, and plastic phase. By continuously monitoring the load–displacement curve and the deformation of the joints, it was possible to determine whether the specimens entered the elastic or elastoplastic phases. In the early loading stage, the load and displacement increased linearly, with nearly the same deformation increment under the same load increment, exhibiting overall elastic deformation. As the load continued to increase, the curved washers gradually deformed and bent, leading to a change in the slope of the load–displacement curve. Subsequently, the loading was continued with controlled displacement, and the trend of the load–displacement curve gradually flattened, indicating that the plastic portion on the drum-shaped spherical shell expanded. At this point, the curved washers experienced significant indentation. Upon continued loading, it was observed that the load curve did not show a noticeable drop. The same static load axial compression tests were conducted on all three groups of specimens, revealing that the final failure modes were similar across all specimens, as shown in Figure 9.

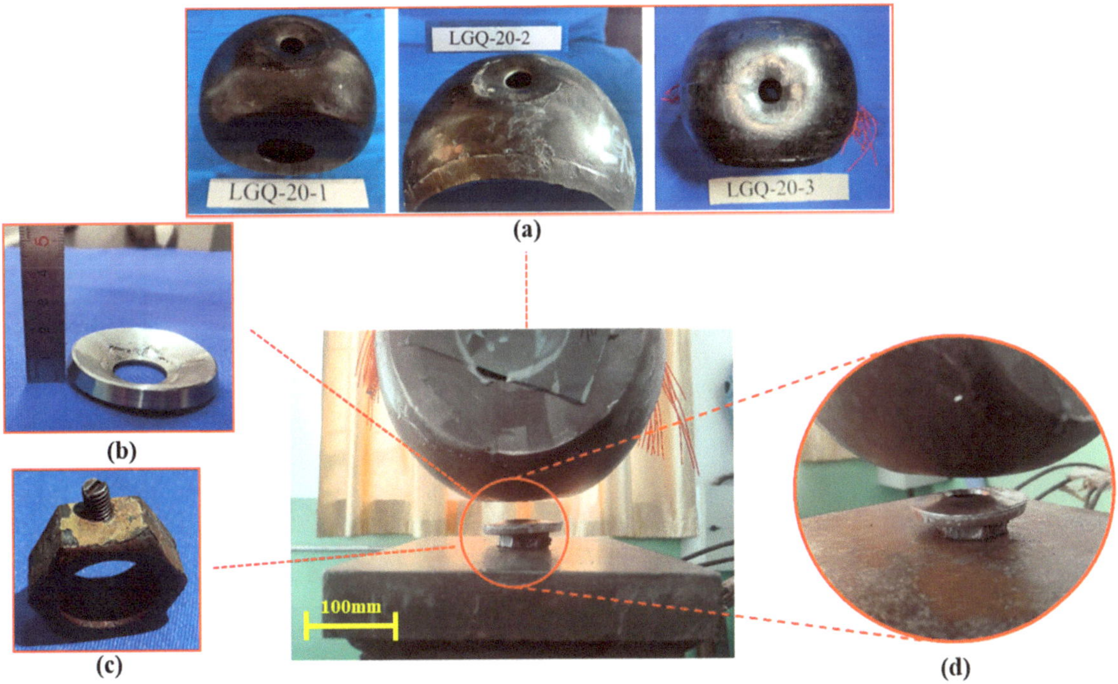

Figure 9. Final failure mode of the specimen: (**a**) hollow drum-shaped spherical shell deformation diagram, (**b**) curved washer deformation diagram, (**c**) sleeve deformation diagram, (**d**) failure modes.

Static load tests were conducted on the three groups of specimens according to the loading scheme, and the experimental data are presented in Table 1. The curves in the figure showed that as the displacement increased, the load gradually stabilized, without exhibiting a similar criterion for determining the failure mode as in welded spheres; this phenomenon has also been observed in other static load tests [28]. Throughout the test, the load did not decrease. Therefore, the maximum allowable deformation was used to determine the ultimate bearing capacity. The load sufficient to cause the local deformation of the drum body to reach 0.03 D (diameter of the drum-shaped spherical shell) was selected as the ultimate bearing capacity of the joint. The average measured diameter of the joint drum-shaped spherical shell was D = 202 mm, and the deformation under the ultimate bearing capacity was 6.06 mm, rounded to 6 mm.

Table 1. Experimental data for the load–displacement curve.

LGQ-20-1			LGQ-20-2			LGQ-20-3		
Load (kN)	Displacement (mm)	Displacement Increment (mm)	Load (kN)	Displacement (mm)	Displacement Increment (mm)	Load (kN)	Displacement (mm)	Displacement Increment (mm)
10	0.60		10	0.46		10	0.54	
20	1.15	0.55	20	0.74	0.28	20	0.83	0.29
30	1.55	0.4	30	0.99	0.25	30	1.14	0.31
40	1.86	0.31	40	1.18	0.19	40	1.49	0.35
50	2.10	0.24	50	1.39	0.21	50	1.9	0.41
70	2.71	0.31	70	1.85	0.24	70	2.48	0.3
80	3.00	0.31	80	2.17	0.32	80	2.84	0.36
90	3.44	0.44	90	2.46	0.29	90	3.27	0.43
100	4.01	0.57	100	2.87	0.41	100	4.02	0.75
110	4.88	0.87	110	3.59	0.72	110	5.74	1.68
120	7.40	2.52	120	5.25	1.66	120	11.37	5.63
125	13.52	6.12	125	6.76	1.51	125	16.80	5.43
130	21.4	7.88	134	19.99	13.23	128	19.99	3.19

4. Numerical Simulation

4.1. Finite Element Model

A three-dimensional FEM was established using the ABAQUS, a software for finite element analysis, with the model size identical to the actual specimens' size. The model was entirely constructed using C3D8R elements, ignoring welding defects and heterogeneities. The load transmission path under the axial compression of the joint is as follows: steel tube → cone/seal plate → sleeve → hollow drum-shaped spherical shell. Therefore, when simulating the compression of the node, only the curved washer, sleeve, and drum-shaped node need to be modeled. The drum-shaped hollow spherical shell was of specification Q345B, with the same material used for the sleeves and bolts. The curved washer and curved nut were modeled using quenched 45# steel. A reinforced elastoplastic bilinear model was used for Q345B, while an ideal elastoplastic bilinear model was used for 45# steel. The elastic modulus (E) was taken as 210 GPa, with a Poisson's ratio of 0.3. The yield strength for Q345 was 345 MPa, with an ultimate strength of 480 MPa; the yield strength for 45# steel was 420 MPa. The constitutive models for these materials are illustrated in Figure 10 below.

The joint was compressed uniformly downward through the loading fixture in the actual loading process. To simplify the calculation, reference points RP-1 and RP-2 were set on the sleeve surface in a coupled manner in the model. RP-1 was used as the loading point through displacement for load application, while RP-2 was the reference point for boundary constraints, subject to fixed constraints. In this model, the grid size for the wall thickness was divided into three layers, each about 2~3 mm. In this model with fewer component types, structured mesh division could be conducted for the curved washers and curved European nuts, while swept mesh division was performed on the drum-shaped spherical shell to minimize grid distortion. Friction contact was set between the sleeve's bottom surface and the curved washer's contact surface, and between the curved washer and the surface of the drum-shaped shell. The normal contact was set to be "hard", allowing separation after surface-to-surface contact; the tangential friction coefficient was set as 0.2, as shown in Figure 11. The finite element simulation load matches the experimental load.

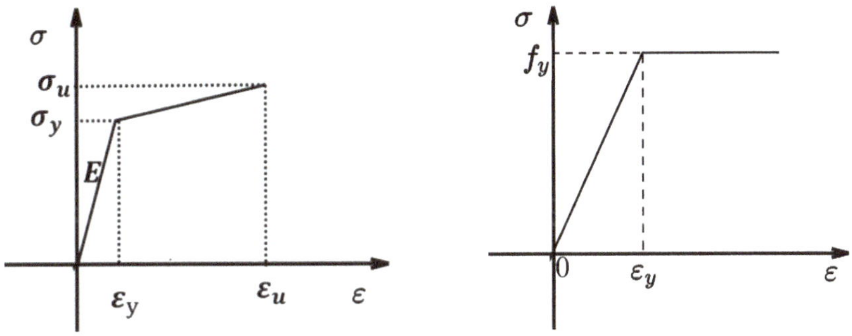

(a) (b)

Figure 10. The models for different materials: (**a**) Q345B; (**b**) 45# steel.

(a) (b)

Figure 11. The FEM: (**a**) the overall model, (**b**) the model with mesh division.

4.2. Simulation Results

Under the axial compressive load, the BDSSJs underwent certain plastic deformation, and were determined to have reached the ultimate bearing capacity due to excessive deformation. During the loading process, the joint underwent elastic deformation, elastoplastic development, and the local indentation of the drum body. The three main phases of the axial compressive failure mode are shown in Figure 12 below, and the load–displacement curve is given in Figure 13.

In the first stage, the load–displacement curve of the joint showed linear growth during the loading process. The stress at the bolt holes of the drum body gradually increased, pushing this part into the yield status first, while other sections were far below the material's yield strength, remaining in the elastic phase. The loaded bolt holes had no obvious deformation, as shown in Figure 12a.

In the second stage, the area entering the plastic phase near the loaded bolt holes gradually enlarged with the continuous load increase. The inner surface of the drum-shaped shell near the loaded area joined the yield status. As the load gradually increased, the area of the yield status on the inner and outer surfaces continually spread, initiating a stress redistribution process in the shell joint; the load–displacement curve showed nonlinear growth, as shown in Figure 12b.

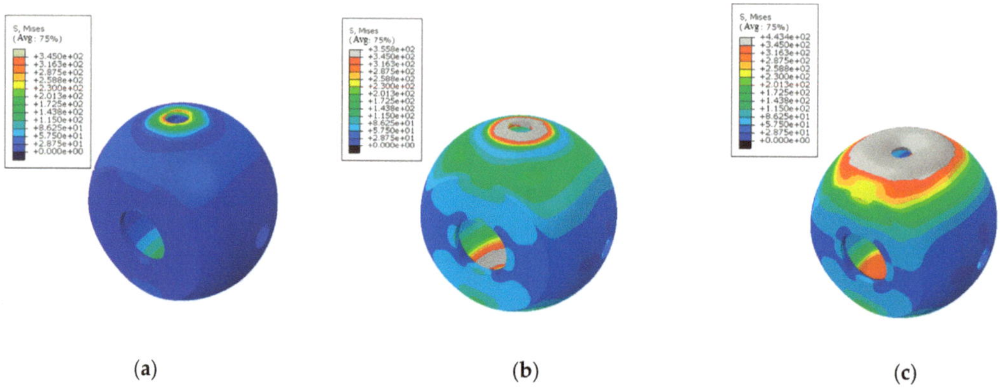

Figure 12. The phases of the BDSSJ's failure mode: (a) stress cloud diagram of elastic stage, (b) stress cloud diagram of plastic development stage, (c) stress cloud diagram of local depression stage.

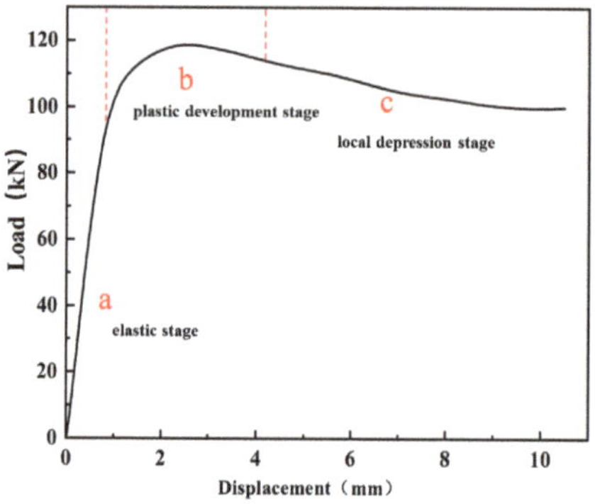

Figure 13. The load–displacement curve of the BDSSJ (a: elastic stage, b: plastic development stage, c: local depression stage).

In the third stage, an indentation occurred at the loaded bolt holes. During this phase, a wide range of plastic zones appeared in the curved washer and drum-shaped shell, leading to certain plastic deformation. The non-loaded bolt holes and the cover plate areas did not yield. The indented deformation intensified with the increase in loading displacement, as depicted in Figure 12c.

Moreover, throughout the loading process, the curved washer gradually underwent plastic deformation, and the bottom of the sleeve became warped, featuring plastic deformation. Upon reaching the ultimate load, a large area of the drum body at the loaded bolt holes yielded, locally forming a through-thickness area of plastic penetration. Upon further loading, the load drop was not sharp, but the excessive joint deformation led to the eventual failure of the joint.

5. Discussion

5.1. Analysis of the Ultimate Bearing Capacity

From Table 1, it can be deduced that before the load reached 90 kN, the displacement increment under every 10 kN load increment was around 0.3 mm. With the load exceeding 90 kN, the joint entered the elastoplastic stage, with uneven growth in the displacement increment. With 6mm as the deformation of the joint under the ultimate load, the ultimate bearing capacity of the static specimens is shown in Table 2 below:

Table 2. Bearing capacity results of BDSSJs.

Serial No.	Specimen ID	Deformation (mm)	Measured Ultimate Bearing Capacity (kN)	Average (kN)	Relative Deviation
1	LGQ-20-1	6	116.22		0.38%
2	LGQ-20-2	6	122.77	116.67	4.97%
3	LGQ-20-3	6	111.00		5.1%

Luo [29] has carried out numerical analyses on this type of joint and derived the corresponding formula for the joint's bearing capacity:

$$N_R^c = \left(250.66 \times Te + 18005.56\frac{T^2e^2}{D^2}\right) \tag{1}$$

Note: D is the outer diameter of the spherical shell; e represents the diameter of the curved washer; and T stands for the wall thickness of the spherical shell.

By inserting the parameters of the joint from the experiment (T = 6 mm, D = 200 mm, e = 60 mm) into the previously mentioned formula, a result of a theoretical bearing capacity was obtained (N_R^c = 148.57 kN). When compared with the average value obtained from the three experiments, it was found that the theoretical value was higher. A parameter η was introduced to align the theoretical bearing capacity formula more closely with the experimental results. By applying this parameter, a more accurate estimation of the compressive bearing capacity of the BDSSJ can be achieved. The modified formula is as follows, with the value of η being 0.78:

$$N_R^c = \eta\left(250.66 \times Te + 18005.56\frac{T^2e^2}{D^2}\right) \tag{2}$$

5.2. Analysis of the Strain and Stress Distribution

The stress distribution of the drum-shaped spherical shell could be provided by strain gauges attached to the shell. While stress can be deduced from strain during small deformations, the relationship became nonlinear as more significant deformations pushed the joint into the plastic phase. Hence, strain was used as the target parameter for the stress distribution analysis. The results are illustrated in Figure 14.

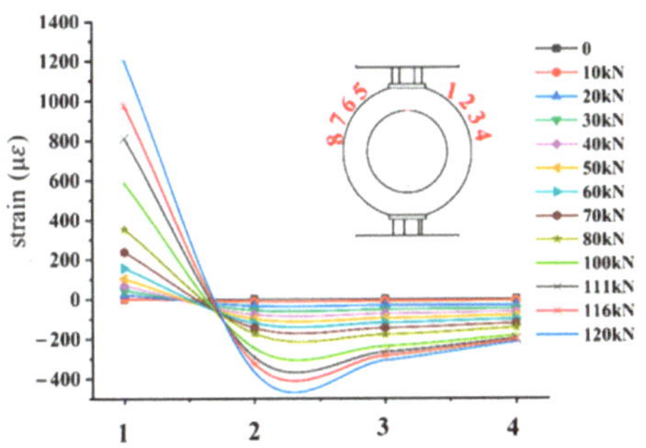

(a)

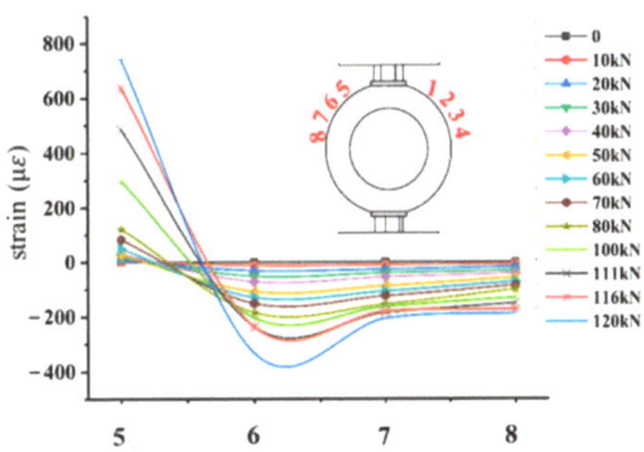

(b)

Figure 14. *Cont.*

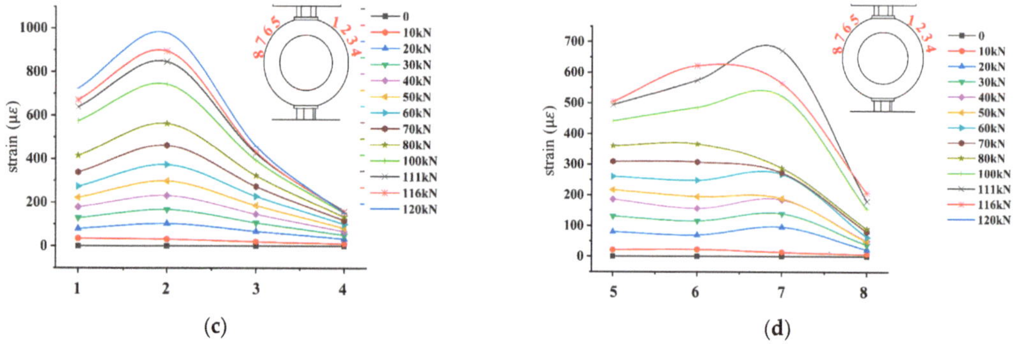

Figure 14. Strain diagram in the direction of the long arch of the drum body of LGQ-20-1: (**a**) axial strain of strain gauges No. 1–4; (**b**) axial strain of strain gauges No. 5–8; (**c**) circumferential strain of strain gauges No. 1–4; (**d**) circumferential strain of strain gauges No. 5–8.

The test results indicated that on the long arch of the drum body, the axial strain gradually decreased from the edge of the curved washer to the non-loaded bolt hole, and then changed reversely with a gradual increase. The maximum strain occurred in the influence zone of the curved washer edge, while the minimum strain occurred at the non-loaded bolt hole. During this period, the strain changed from tensile to compressive strain. On the long arch of the drum body, the circumferential strain at all eight points was tensile, indicating tension; the strain exhibited a trend of first increasing and then decreasing, with all the strains not exceeding the yield point. The similarity in the trend of the axial and circumferential strain variation curves on the long arches on both sides of the drum body, albeit with different peak values, demonstrated reliability in the experimental data. On the other hand, it suggested that there might be uneven stress distribution when the BDSSJ was under load. This unevenness may also be related to the uniformity of the wall thickness and welding performance during the specimen fabrication.

The analysis of the axial and circumferential strain data on the short arches of the drum body revealed that the axial strain at the edge of the curved washer increased with the growing load, entering the yield status with a load of around 100 kN. The axial strain across the short arches was all tensile. The circumferential strain near the curved washer edge gradually increased with the load, and the strain abruptly transitioned from compressive to tensile when the load exceeded 100 kN. The intersection of the drum body on the short arches with the cover plate experienced minimal stress, with all stress remaining within the elastic range, as illustrated in Figure 15

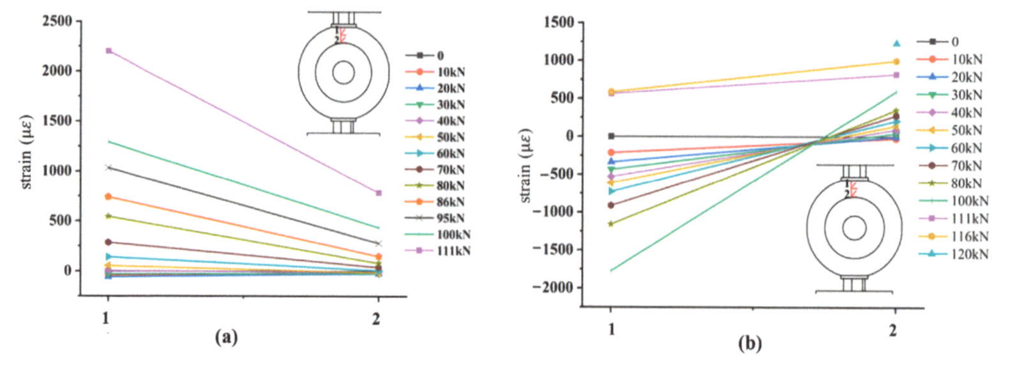

Figure 15. Circumferential and axial strain diagram in the direction of the short arch of the drum body of LGQ-20-1: (**a**) circumferential strain diagram; (**b**) axial strain diagram.

Measurements were also conducted on the stress changes in the cover plate under axial compression. Four strain gauges were evenly attached to the cover plate from top to bottom, and the results are shown in the above figure. It was found that the cover plate was under compressive stress during the loading process, and the stresses were relatively small, far below the material's yield strength. The ends of the cover plate experienced higher stress compared to the middle. As the load increased, the compressive strain on the cover plate slowly increased, indicating a significant reserve of strength, as illustrated in Figure 16.

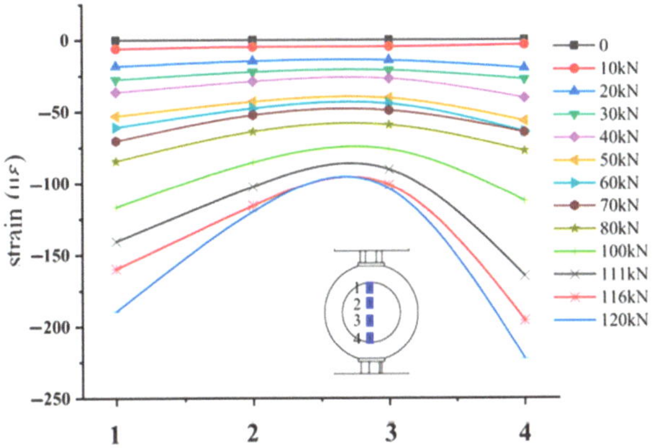

Figure 16. Strain diagram of the cover plate of LGQ-20-1.

6. Conclusions

Based on the axial compression tests and numerical research on the BDSSJs, the following conclusions could be drawn:

(1) No significant drop occurred in the load–displacement curve obtained from the experiment on the BDSSJs. According to the maximum allowable deformation, the joint's ultimate bearing capacity could be calculated as the load value that caused the local deformation value to reach 3% of the outer diameter of the drum-shaped spherical shell.

(2) Static load axial compression tests were carried out on the BDSSJ specimen LGQ-20, and the ultimate bearing capacity of the joint was found to be 116 kN.

(3) Under the axial compression load, both the inner and outer surfaces of the shell within the range of the curved washer were in a compressed state, with the stress being the lowest at the non-loaded bolt holes. The outer surface of the shell's long arch in the axial direction showed compression between the curved washer and non-loaded bolt holes area, while the inner surface exhibited tension. The cover plate was in a compressed state, and its stress level was significantly below the material's yield strength.

(4) The FEM analysis revealed that the failure process of the joint under unidirectional axial compression could be divided into three stages. The first stage was the elastic stage, during which a small area near the loaded bolt holes yielded, but the shell as a whole remained elastic. The second stage was the elastoplastic deformation stage, where the plastic area at the bolt holes enlarged and the inner surface of the shell yielded. In the third stage, significant indentation occurred at the loaded bolt holes, and the joint deformation tended towards a limiting value.

Author Contributions: Conceptualization, Y.Y.; methodology, Y.Y. and Z.Z.; software, Y.Y. and X.M.; validation, M.L.; formal analysis, Y.Y. and X.M.; investigation, M.L.; resources, M.L.; data curation, Y.Y. and Z.Z.; writing—original draft preparation, Y.Y. and Z.Z.; writing—review and editing, Y.Y., Z.Z., M.L. and X.M.; visualization, Y.Y. and Z.Z.; supervision, M.L. and Z.Z.; project administration,

Z.Z.; funding acquisition, Y.Y. and X.M. All authors have read and agreed to the published version of the manuscript.

Funding: This research was funded by the National Nature Science Foundation of China, grant number No. 51178286.

Institutional Review Board Statement: Not applicable.

Informed Consent Statement: Not applicable.

Data Availability Statement: All data generated or analyzed in this research were included in this published article. Additionally, readers can access all data used to support the conclusions of the current study from the corresponding author upon request.

Conflicts of Interest: M.L. was employed by Shanxi 6th Construction Group Co., Ltd. The remaining authors declare that the research was conducted in the absence of any commercial or financial relationships that could be construed as a potential conflict of interest.

References

1. Caglayan, O.; Yuksel, E. Experimental and Finite Element Investigations on the Collapse of a Mero Space Truss Roof Structure—A Case Study. *Eng. Fail. Anal.* **2008**, *15*, 458–470. [CrossRef]
2. Ebadi Jamkhaneh, M.; Davoodi, M.R.; Ebadi Jamkhaneh, J. Assessment of the Ball Joint Behavior under Combine Loading. *Modares Civ. Eng. J.* **2017**, *17*, 149–159.
3. Ghasemi, M.; Davoodi, M.R.; Mostafavia, S.A. Tensile Stiffness of MERO-Type Connector Regarding Bolt Tightness. *J. Appl. Sci.* **2010**, *10*, 724–730. [CrossRef]
4. Doaei, Y.; Aghakouchaki Hosseini, S.E.; Momenzadeh, A.; Harirchian, E. Investigating the Effect of Screw Size on the Stress Level in MERO Joint for Space Frame Structures. *Appl. Syst. Innov.* **2021**, *4*, 84. [CrossRef]
5. Xu, J.; Wang, C.; Li, H.; Zhang, C.; Hao, J.; Fan, S. Health Monitoring of Bolted Spherical Joint Connection Based on Active Sensing Technique Using Piezoceramic Transducers. *Sensors* **2018**, *18*, 1727. [CrossRef]
6. Fan, F.; Ma, H.; Chen, G.; Shen, S. Experimental Study of Semi-Rigid Joint Systems Subjected to Bending with and without Axial Force. *J. Constr. Steel Res.* **2012**, *68*, 126–137. [CrossRef]
7. Si, Q.; Tang, Y.; Zong, L.; Liu, H.; Kang, B. Experimental Study and Numerical Simulation of the Tensile Properties of Corroded Bolt-Sphere Joints. *Buildings* **2022**, *12*, 1989. [CrossRef]
8. Ding, B.; Zhao, Y.; Huang, Z.; Cai, L.; Wang, N. Tensile Bearing Capacity for Bolted Spherical Joints with Different Screwing Depths of High-Strength Bolts. *Eng. Struct.* **2020**, *225*, 111255. [CrossRef]
9. Yang, D.; Li, M.; Fu, F.; Wu, J. Experimental and Numerical Studies on a New Type of Bolt-Ball Joint for Spatial Grid Structures. *J. Constr. Steel Res.* **2022**, *188*, 107035. [CrossRef]
10. Huang, B.; Lu, M.; Fu, Y.; Yang, F. Experimental Investigation on Tensile Properties of the Bolt in Sphere Joints Under Fire. *Int. J. Steel Struct.* **2020**, *20*, 1355–1363. [CrossRef]
11. Tian, J.; Wang, X.; Li, H.; Wang, Z.; Pan, J. Ultra-Low Cycle Fatigue Performance of Grid Structure with Bolted Spherical Joints. *J. Constr. Steel Res.* **2023**, *201*, 107728. [CrossRef]
12. Yang, X.; Lei, H.; Chen, Y.F. Constant Amplitude Fatigue Test Research on M20 High-Strength Bolts in Grid Structure with Bolt–Sphere Joints. *Adv. Struct. Eng.* **2017**, *20*, 1466–1475. [CrossRef]
13. Qiu, B.; Yang, X.; Zhou, Z.; Lei, H. Experimental Study on Fatigue Performance of M30 High-Strength Bolts in Bolted Spherical Joints of Grid Structures. *Eng. Struct.* **2020**, *205*, 110123. [CrossRef]
14. Zhou, Z.; Lei, H.; Qiu, B.; Zhang, S.; Wang, G. Experimental Study on Fatigue Performance of M60 High-Strength Bolts with a Huge Diameter under Constant Amplitude Applied in Bolt–Sphere Joints of Grid Structures. *Appl. Sci.* **2022**, *12*, 8639. [CrossRef]
15. Fernández Landeta, J.; Fernández Valdivielso, A.; López De Lacalle, L.N.; Girot, F.; Pérez Pérez, J.M. Wear of Form Taps in Threading of Steel Cold Forged Parts. *J. Manuf. Sci. Eng.* **2015**, *137*, 031002. [CrossRef]
16. Bustillo, A.; Urbikain, G.; Perez, J.M.; Pereira, O.M.; Lopez de Lacalle, L.N. Smart Optimization of a Friction-Drilling Process Based on Boosting Ensembles. *J. Manuf. Syst.* **2018**, *48*, 108–121. [CrossRef]
17. Xing, J.; Qiu, C.; Wang, M.; Yang, N. Uniaxial Failure Mechanism and Design Strength of High-Strength Welded Hollow Spherical Joint. *Eng. Struct.* **2022**, *256*, 113897. [CrossRef]
18. Liu, H.; Ying, J.; Chen, Z.; Zhou, Y.; Yan, X. Ultimate Tensile and Compressive Performances of Welded Hollow Spherical Joints with H-Beam. *J. Constr. Steel Res.* **2018**, *150*, 195–208. [CrossRef]
19. Zhang, J.; Lei, H.; Jin, S. Experimental Study on Constant-Amplitude Fatigue Performance of Weld Toe in Steel Tube of Welded Hollow Spherical Joints in Grid Structures. *Adv. Mater. Sci. Eng.* **2019**, *2019*, 6204302. [CrossRef]
20. Duan, Y.; Lei, H.; Jin, S. Experimental Study on Fatigue Performance of Welded Hollow Spherical Joints Reinforced by CFRP. *Coatings* **2022**, *12*, 1585. [CrossRef]
21. Yan, R.; Yu, Z.; Wang, S.; Liu, J. Influence of Welding Residual Stress on Bending Resistance of Hollow Spherical Joints. *J. Constr. Steel Res.* **2023**, *208*, 108004. [CrossRef]

22. Zhao, Z.; Dai, B.; Xu, H.; Li, T. Bending Capacity of Corroded Welded Hollow Spherical Joints with Considering Interaction of Tension Force and Bending Moment. *Structures* **2021**, *34*, 2656–2664. [CrossRef]
23. Huang, B.; Lu, M.; Cao, Y.; Yang, F. Experimental Study on Residual Performance of Welded Hollow Spherical Joints Subjected to Axial Compression after a Fire. *Structures* **2021**, *30*, 996–1005. [CrossRef]
24. Shu, T.; Xu, X.; Pan, W.; Huang, W.; Luo, Y. Compressive Performance of Welded Hollow Spherical Joints with External Triangular Ribs. *Eng. Struct.* **2023**, *280*, 115717. [CrossRef]
25. *JG/T 10-2009*; Bolted Spherical Node of Space Grid Structures. China Architecture and Building Press: Beijing, China, 2009.
26. *JGJ78-91*; Criteria for Evaluation of Grid Constructional Engineering Quality Inspection. Architecture & Building Press: Beijing, China, 1991.
27. Yu, K.; Yu, J.; Tang, B. Experimental and finite element analysis of the ultimate bearing capacity of hollow sphericl joints with ribbd stiffener. *Ind. Constr.* **2011**, *41*, 85–90. [CrossRef]
28. Liu, Y.; Fan, J.; Liu, Z.; Jiao, J.; Lei, H. Finite element analysis of the influence of detailed structural measures on static performance of complex tubular joints. *J. Taiyuan Univ. Technol.* **2022**, *53*, 149–155. [CrossRef]
29. Luo, C. Design and Analysis of Round Steel Tube Supporting Assembly Nodes. Master's Thesis, Taiyuan University of Technology, Taiyuan, China, 2019.

coatings

MDPI

Article

Improving the Wear Resistance Properties of 7A04 Aluminum Alloy with Three Surface Modification Coatings

Jinmeng Hu [1], Cheng Zhang [1,2], Xiaodong Wang [1], Xiaobo Meng [1], Caihong Dou [1], Hua Yu [1,2], Changji Wang [1,2], Jun Xue [3], Ziping Qiao [3,*] and Tao Jiang [1,4,*]

[1] National Joint Engineering Research Center for Abrasion Control and Molding of Metal Materials, School of Materials Science and Engineering, Henan University of Science and Technology, Luoyang 471000, China; 15638265067@163.com (J.H.); zhangch06@126.com (C.Z.); nmxdwang@163.com (X.W.); mengxiao6o@163.com (X.M.); dch1805@163.com (C.D.); kedayuhua@126.com (H.Y.); wchj_1989@haust.edu.cn (C.W.)

[2] Longmen Laboratory, Luoyang 471003, China

[3] Science and Technology on Transit Impact Laboratory, No. 208 Research Institute of China Ordnance Industries, Beijing 102202, China; oil1999@163.com

[4] Luoyang Wanji Aluminum Processing Co., Ltd., Changjiang Avenue, Industrial Agglomeration Zone, Luoyang 471800, China

* Correspondence: maryqiao@163.com (Z.Q.); tedivy@163.com (T.J.)

Abstract: Multiple advantages, such as good formability, high specific strength, excellent thermal conductivity, and high corrosion resistance, enable aluminum alloy wide application in various fields; however, low surface hardness and poor wear resistance limit its further development. In this study, three surface modification coatings were successfully prepared on the surface of 7A04 aluminum alloy by microarc oxidation (MAO) and a combination of hard anodizing treatment (HA) and physical vapor deposition (PVD), named MAO, HA+W+DLC, and HA+Ti+ta-C, respectively. The microstructure, hardness, and tribological properties of the three coatings and the 7A04 aluminum alloy substrate were studied. The results show that the surface quality and hardness of the coated samples were higher than those of the 7A04 aluminum alloy and that the HA+Ti+ta-C coating possessed the highest hardness of 34.23 GPa. Moreover, the wear resistance of the two multilayer coatings was significantly improved during the ring-block wear tests under oil lubrication, exhibiting a wear rate of 1.51×10^{-7} $mm^3/N{\cdot}m$ for HA+W+DLC and 1.36×10^{-7} $mm^3/N{\cdot}m$ for HA+Ti+ta-C.

Keywords: aluminum alloy; MAO coating; HA+W+DLC multilayer coating; HA+Ti+ta-C multilayer coating; wear resistance

Citation: Hu, J.; Zhang, C.; Wang, X.; Meng, X.; Dou, C.; Yu, H.; Wang, C.; Xue, J.; Qiao, Z.; Jiang, T. Improving the Wear Resistance Properties of 7A04 Aluminum Alloy with Three Surface Modification Coatings. *Coatings* **2024**, *14*, 476. https://doi.org/10.3390/coatings14040476

Academic Editors: Elena Villa and Joachim Albrecht

Received: 7 March 2024
Revised: 7 April 2024
Accepted: 10 April 2024
Published: 13 April 2024

1. Introduction

Aluminum alloys are widely used in the aerospace industry, especially in commercial transport aircraft and military fighter aircraft, owing to their excellent properties, such as good formability, light weight, high specific strength, and low density (approximately 2.7 g/cm^3) [1–5]. However, the low surface hardness and poor wear resistance of aluminum alloys severely limit their application. 7A04 is a kind of superhard aluminum alloy, the dosage of which is second only to steel in modern industry, and surface modification can significantly improve its performance. In recent years, the surface modification of 7A04 aluminum alloy has become an active research subject owing to its potential to improve the wear resistance of aerospace aluminum alloy parts to ensure a long service life, offer good reliability and production quality, reduce production costs, and obtain good economic benefits [6–9].

Hard oxidation is known as hard anodizing (HA) treatment, in which a metal is placed in an electrolyte as the anode such that an oxide film of thickness ranging from tens to hundreds of microns is formed on the metal surface. The formation of such an oxide layer film endows the metal with corrosion and wear-resistant properties, but it also has the disadvantage of low hardness [10]. Microarc oxidation (MAO) is a surface treatment technology developed from anodic oxidation. It increases the electrode voltage of common anodic oxidation from the Faraday zone to the high voltage discharge zone, produces microarc plasma spark discharge, and in situ forms a ceramic coating on the surface of valve metals such as Al, Mg, Ti, and their alloys [11–13]. Diamond-like carbon coating is a substable amorphous carbon film combining sp^3-hybridized bonds (diamond structure) and sp^2-hybridized bonds (graphite structure), which is considered to be an ideal surface-protective coating to improve the performance and life of aluminum alloys due to its high hardness, excellent abrasion resistance, low coefficient of friction, high elastic modulus, and good chemical inertness [14–21]. Depending on its structure, it can be divided into hydrogen-containing DLC coatings and hydrogen-free tetrahedral amorphous carbon (ta-C) coatings. Many researchers have studied these three coating preparation methods and found that the hard anodic oxide layer has a significant influence on the wear behavior of aluminum alloy. Soffritti et al. [22] prepared anodic aluminum oxide coatings with different thicknesses by different hard oxidation methods and observed their microstructure and mechanical properties. Through the wear test, it was concluded that the anodic aluminum oxide coatings have a great influence on the wear performance of aluminum alloys. P Kwolek et al. [23] prepared hard oxide coatings on 5005 and 6061 aluminum alloys. The wear resistance of hard oxide coatings prepared on different aluminum alloys was studied by a scratch test and a wear test. The results show that the wear resistance of coated 6061 aluminum alloy is higher than that of 5005 aluminum alloy, which is mainly related to the lower porosity and higher hardness of hard oxide coatings prepared on 6061 aluminum alloy. In addition, the wear resistance of MAO coatings has also been studied. For example, J.J. Zhuang et al. [24] formed MAO coatings on aluminum alloys and studied the effects of oxide films formed at different oxidation times on their wear resistance. C. Yang et al. [25] prepared MAO coating on aluminum alloy by changing the concentration of phosphate, which increased the hardness of the coating and reduced the wear rate. Furthermore, some studies were carried out on the effects of DLC coatings on the surface of aluminum alloys [26,27]. It was found that the hardness was significantly higher than that of uncoated aluminum alloys, but there is a high residual stress and poor adhesion between the DLC film and the substrate. The surface properties can be optimized by doping elements. It has been reported that metal elements (such as Cr [28], Ti [29], W [30], Zr [31], Ni [32], and Cu [33]) can effectively release residual stress by changing the structure of the DLC film. Therefore, the above three surface modification coatings can effectively improve the hardness and wear resistance of aluminum alloy, but comparisons of these three types of coated aluminum alloy are scarce, so this aspect is worth studying.

In this study, three types of coatings, named MAO, HA+W+DLC, and HA+Ti+ta-C, were deposited on 7A04 aluminum alloy and compared with an aluminum alloy substrate without surface treatment. The microstructure, hardness, and wear behavior of these three coatings were investigated.

2. Test Materials and Methods

2.1. Test Material

In this study, a 7A04 aluminum alloy cuboid sample with a size of 19 mm × 12 mm × 12 mm was used as the test matrix material. A 25Cr3Mo3NiNbZr die steel ring with an inner diameter of 42 mm, an outer diameter of 50 mm, and a width of 13 mm was used as the grinding material. The nominal compositions of the materials are listed in Tables 1 and 2.

Table 1. Chemical composition of 7A04 aluminum alloy (wt.%).

Element	Si	Fe	Cu	Mn	Mg	Cr	Zn	Ti	Al
content	0.5	0.5	1.4–2.0	0.2–0.6	1.8–2.8	0.1–0.25	5.0–7.0	0.1	Balance

Table 2. Chemical composition of 25Cr3Mo3NiNbZr steel (wt.%).

Element	C	Si	Mn	Cr	Mo	Ni	Nb	Zr	V
content	0.28	<0.1	0.18	3.03	2.94	0.55	0.14	0.0012	Balance

2.2. Test Method

Prior to coating deposition, the 7A04 aluminum alloy substrates were polished with 2000 mesh sandpaper. Microarc oxidation was carried out using an electrolyte containing 30 g/L $(NaPO_3)_6$, 3 g/L KOH, 8 g/L $Na_2B_4O_7$, and 15 g/L glycerin for 35 min under a current density of 10 $A{\cdot}dm^{-2}$. The temperature of the electrolyte was maintained at 303 K, while the sample was placed as an anode, and the stainless steel was used as the cathode. When preparing multilayer diamond-like carbon coatings, the Cr/Ni-doped oxide film transition layer was firstly coated on the surface of aluminum alloy by hard oxidation, which was conducted using a 15 wt.% sulfuric acid electrolyte for 30 min under a current density of 50 $mA{\cdot}cm^{-2}$, and then conducted by repetition of dipping in 3 M Cr/Ni-containing electrolyte for 1 min and heat treatment in a 400 °C electric furnace for 30 min. Secondly, the W/Ti transition layer was prepared using a nonequilibrium magnetron sputtering system (UDP-650, Teer Coatings Co., Ltd., Droitwich, UK) with a W/Ti target separately, under the conditions of a constant current of 1 A and a negative bias voltage of −60 V for 5 min to enhance the bonding strength. Finally, the DLC layer was prepared by magnetron sputtering under the conditions of a constant current of 3.5 A and a negative bias voltage of −60 V for 150 min. The ta-C layer was prepared by arc ion plating with optimized parameters of a current of 60 A, a bias voltage of −120 V, and a depositing time of 60 min.

Friction and wear tests were carried out for 4000 cycles over 1200 s using a high-speed ring-block friction meter (MRH-3, Jinan Shunmao Corporation Ltd., Jinan, China) under oil lubrication with a load of 300 N and a speed of 200 rpm/min. The test was carried out at room temperature and atmospheric pressure. Bulk samples were prepared from uncoated and differently coated 7A04 aluminum alloys. The ring specimens were made of DLC-coated 25Cr3Mo3NiNbZr mold steel. After the test, the samples were firstly shaken ultrasonically for 10 min using petroleum ether, which acted as the organic solvent to remove the surface lubricating oil, and then were ultrasonic cleaned in anhydrous ethanol for 10 min and dried at 60 °C for 0.5 h. After drying, the samples were weighed immediately (five times for each measurement). The maximum and minimum values were removed, and the remaining average value was considered. The weighing method was the same as that used after the testing.

The wear surface morphologies were analyzed by a laser confocal microscope (OLS5100, Reco System Integration Ltd., Beijing, China). A field-emission scanning electron microscope (JSM-IT800, JEOL Companies, Tokyo, Japan) was used to observe the surface and cross-sectional morphology before and after wear and to analyze the characteristics of friction wear and wear behavior. It was combined with an SEM-supporting energy spectrometer to perform energy spectral analysis. The X-ray diffractometer (Brux-D8, Bruker AXS Companies, Karlsruhe, Germany) was used to analyze the phase composition of the worn specimen surface using Cu-Kα radiation at 40 kV and 40 mA. Cu was used as the target material for the tests, and the scanning range was set from 20° to 80° with a scanning speed of 6°/min. In these experiments, a nanoindentation tester (Keysight G200, Keysight Technologies, Santa Rosa, CA, USA) was chosen to measure the coating hardness using the continuous stiffness method and the 10% coating thickness method, with a point-to-point spacing of 20 μm, a load setting of 50 mN, and a Poisson's ratio of 0.26. The bonding states of carbon atoms in two multilayer coatings were measured by Raman spectroscopy (InVia, Renishaw, Lundon, UK).

3. Results and Discussion

3.1. Pre-Wear Microscopic Morphology Analysis

The surface morphologies of different coatings before wear are shown in Figure 1. As shown in Figure 1a,b, the MAO-coated specimens have an inhomogeneous surface and are endowed with a typical "crater" porous structure constituting several micropores, a few small particles, and microcracks. Micropores are formed by molten oxides and bubbles discharged from the microarc discharge channels, through which molten alumina flows out and solidifies rapidly. Small particles also solidify from these molten oxides. Simultaneously, a large amount of the anionic component PO_3^- and a small amount of $B_4O_7^{2-}$ in the electrolyte enter the channel, leaving a clear and distinct boundary. The cracks are mainly caused by the different coefficients of phase expansion in the coatings and thermal stresses [34]. As shown in Figure 1c–f, continuous dense diamond-like coatings are successfully deposited on the surface of the aluminum alloy, and both coating surfaces appear smooth and flat owing to the dense arrangement of small amorphous particles. The HA+W+DLC coating has more cracks than the HA+Ti+ta-C coating, which has a smoother surface topography. The surface morphologies of the substrate material shown in Figure 1g,h are not flat and have stray scratches and cracks related to the preparation of the specimens for machining. The cross-sectional morphologies of the investigated coatings and the corresponding elemental line scans are shown in Figure 2. The thickness of the as-deposited MAO coating, HA+W+DLC coating, and HA+Ti+ta-C coating is 37.1 μm, 9.8 μm, and 3.2 μm, respectively. It can be seen that the three coatings are dense with no pores and well bonded with the substrate, and the interfaces of each sublayer in the two multilayer coatings are clear. Additionally, the distribution of Al and O elements is relatively uniform in MAO, and the composition of HA+W+DLC and HA+Ti+ta-C conforms to the design, which is conducive to binding with the substrate.

3.2. XRD Patterns

The X-ray diffraction patterns of the coating and substrate after abrasion are shown in Figure 3. The 7A04 aluminum alloy substrate shows a strong aluminum peak, mainly composed of the aluminum phase, while the MAO coating also shows an aluminum phase, owing to the high porosity [35] and low thickness of the coating. The MAO coating is mainly composed of α-Al_2O_3 and γ-Al_2O_3. The diffraction peak of γ-Al_2O_3 is stronger than that of α-Al_2O_3, indicating that the oxide of the MAO coating is mainly composed of γ-Al_2O_3. The HA+W+DLC and HA+Ti+ta-C coatings are very thin, and the rays can penetrate the coating directly, showing the aluminum phase corresponding to the substrate.

3.3. Hardness Analysis

The hardness of the MAO, HA+W+DLC, HA+Ti+ta-C, and the substrate was measured using a nanoindentation instrument. Using the continuous stiffness and 10% coating thickness methods, five indents were created in each sample; the average value of the maximum and minimum values of the five indentation removals was considered to be the surface hardness of the specimen. Finally, the average hardness of each coating specimen was calculated, as shown in Figure 4. The average hardness of the 7A04 aluminum alloy substrate was 1.74 GPa, and the average hardness of the MAO coating was 7.33 GPa, which was four times higher than that of the substrate. The average hardness of the HA+W+DLC coating after hard anodizing treatment was 22.64 GPa, which was 13 times higher than that of the substrate. The average hardness of the HA+Ti+ta-C coating after hard anodizing treatment was 34.23 GPa, which was 19 times higher than that of the substrate. It was found that the substrate and MAO coating had a better bonding force, which was beneficial to the performance of coating hardness. The hardness of the HA+W+DLC and HA+Ti+ta-C coatings after hard anodizing treatment considerably improved owing to the typical diamond-like structure within the HA+W+DLC coating.

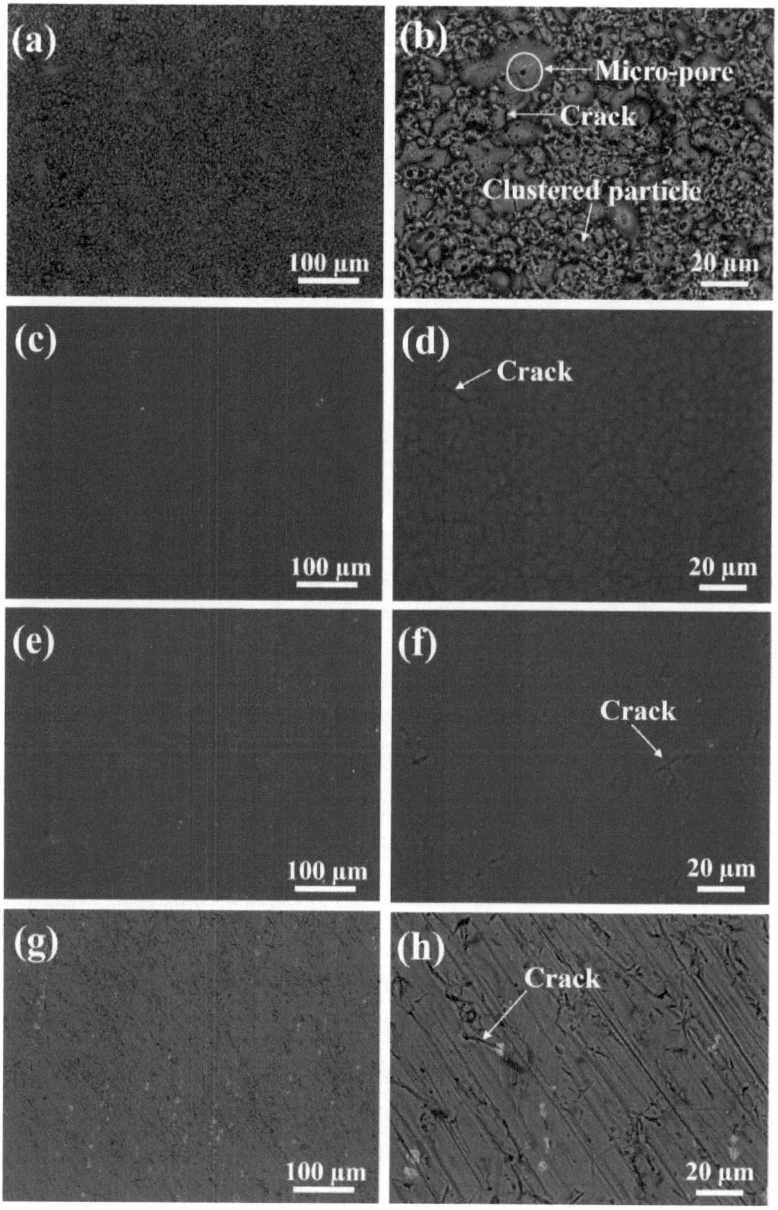

Figure 1. Surface morphologies of (**a**) MAO coating (×200 times), (**b**) MAO coating (×800 times), (**c**) HA+W+DLC coating (×200 times), (**d**) HA+W+DLC coating (×800 times), (**e**) HA+Ti+ta-C coating (×200 times), (**f**) HA+Ti+ta-C coating (×800 times), (**g**) 7A04 substrate (×200 times), and (**h**) 7A04 substrate (×800 times).

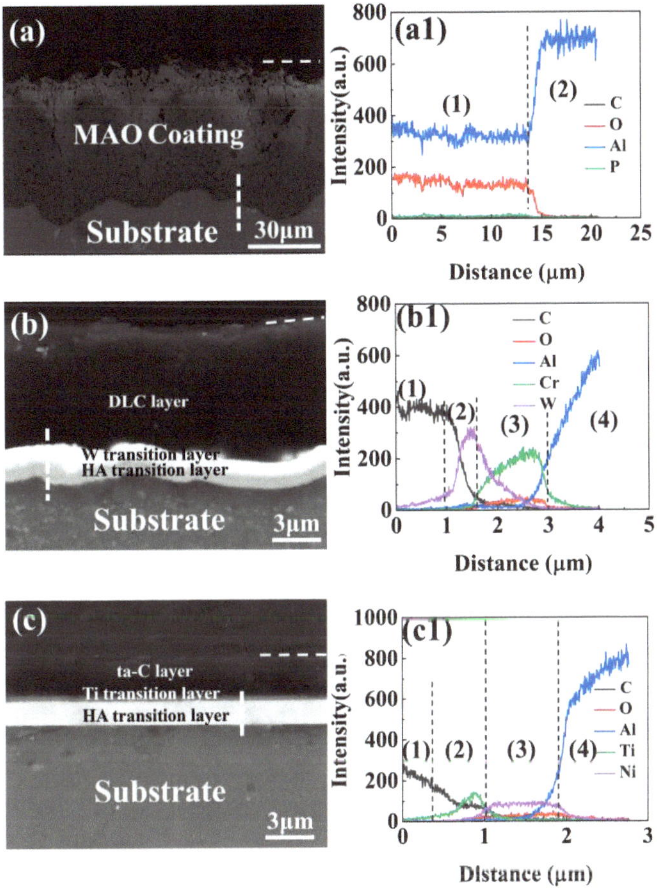

Figure 2. Cross-section morphologies and the corresponding elemental line scans of (**a**) MAO coating, (**b**) HA+W+DLC coating, and (**c**) HA+Ti+ta-C coating.

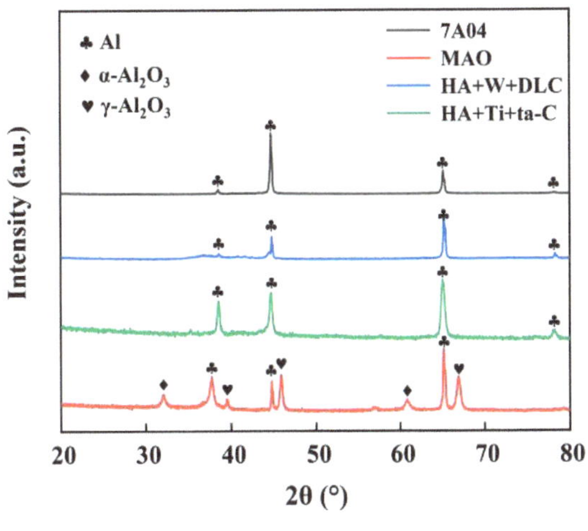

Figure 3. XRD patterns of MAO, HA+W+DLC, HA+Ti+ta-C, and 7A04 substrate.

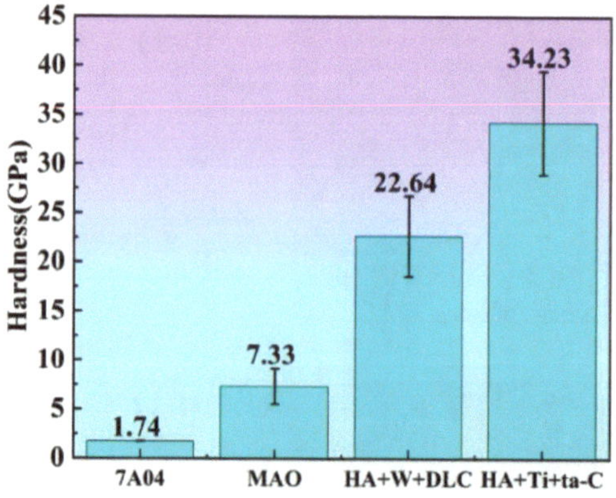

Figure 4. Hardness diagrams of MAO, HA+W+DLC, HA+Ti+ta-C, and 7A04 substrate.

3.4. Wear Test

3.4.1. Friction Coefficient and Wear Rate

The coefficient of friction of different coatings against the substrate as a function of the wear time is shown in Figure 5. The wear process is typically divided into three stages—initial, break-in, and stable wear [36]. As shown in Figure 5a, in the initial stage (0–10 s), due to its rough, porous structure, the tangential resistance is large during wear, and the friction coefficient gradually increases. It can be preliminarily determined that the coating begins to break, gradually peeling off. The loose, porous shape of the exterior surface stores the abrasive particles generated by friction. Subsequently, the abrasive particles form a relatively smooth protective coating on the surface, leading to a decrease in the coefficient of friction, which stabilizes at 0.03 [37]. As evident in Figure 5b,c, for the HA+W+DLC and HA+Ti+ta-C coatings, their average friction coefficients of 0.028 and 0.025, respectively, are very low after the break-in period. The average friction coefficients of the HA+Ti+ta-C coatings are slightly less than those of the HA+W+DLC coatings. It can be determined that the friction coefficient gradually fluctuates with an increase in the wear time, but there is no obvious "transition point" in the three coatings, indicating that the coating still exists [38,39]. In Figure 5d, the friction coefficient curves of the three coatings are compared with those of the 7A04 aluminum alloy substrate. The friction coefficient of the substrate is between 0.01 and 0.04, and it fluctuates considerably. In the initial stage of wear, the coefficient first increases rapidly and then decreases, possibly related to the accelerated oxidation of the frictional heat generated during the wear process. For the untreated substrate material, there is no protective coating on the surface, and the rapid fluctuation in the friction coefficient within 90 s indicates that the steel ring undergoes serious wear on the surface of the aluminum alloy. The combined extent of wear and wear rate are shown in Figure 5e,f. All three coatings reduce the mass loss and wear rate compared to the 7A04 substrate. It can be seen that all three coatings prepared on the 7A04 surface have a good anti-wear effect, resulting in an improved wear resistance of the substrate. The lower mass loss and wear rate of the HA+W+DLC coating and the HA+Ti+ta-C coating after hard anodizing treatment are 1.36×10^{-7} $mm^3/N{\cdot}m$ and 1.51×10^{-7} $mm^3/N{\cdot}m$, respectively. Mainly due to the typical amorphous material after hard anodizing treatment [40], HA+W+DLC and HA+Ti+ta-C have both sp^3- and sp^2-type bonding structures. The vibration mode of sp^2 represents the characteristics of graphite and hence has some characteristics of graphite, such as lubrication. Hence, the MAO, HA+W+DLC, and HA+Ti+ta-C coatings are slightly abraded, and the wear rate of HA+Ti+ta-C is lower than that of HA+W+DLC [41]. All coatings exhibit slight wear, with HA+Ti+ta-C showing a lower wear rate than HA+W+DLC, followed by MAO.

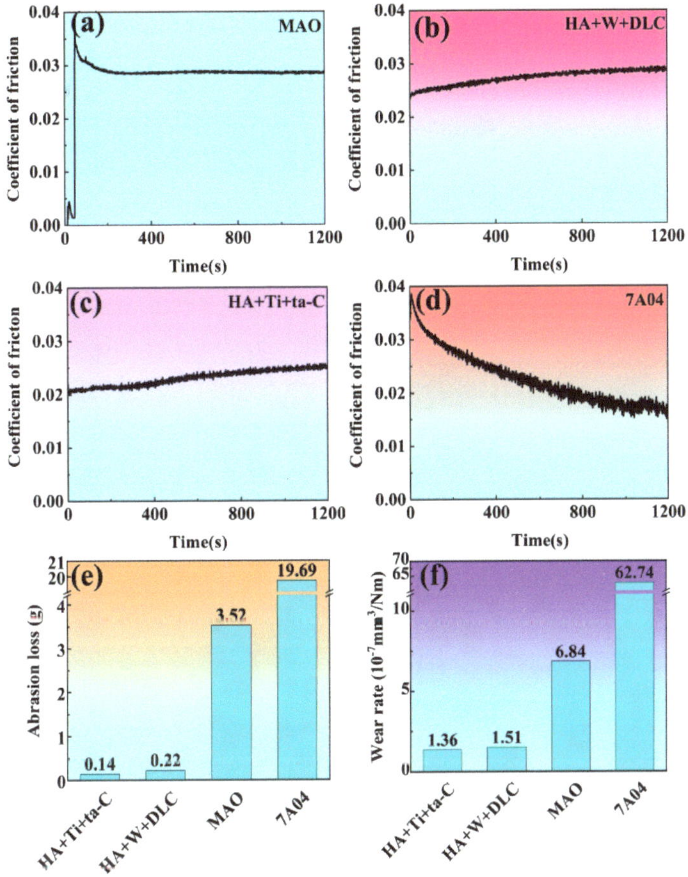

Figure 5. (**a–d**) Coefficient of friction of MAO coating, HA+W+DLC coating, HA+Ti+ta-C coating, and 7A04 substrate and corresponding (**e**) abrasion loss and (**f**) wear rate.

3.4.2. Three-Dimensional Morphology

The surface morphologies of the substrate and coating were observed using a three-dimensional profiler. The three-dimensional morphology of the three coatings and their surfaces are shown in Figure 6. It is evident that compared with the wear depth of 179.05 µm of the substrate material (Figure 6d,d1), the wear trace of the HA+Ti+ta-C coating is the shallowest, at 13.95 µm (Figure 6c,c1), followed by that of the HA+W+DLC coating after hard anodizing treatment (18.55 µm, as shown in Figure 6b,b1), and the wear depth of the MAO coating (35.33 µm, as shown in Figure 6a,a1). Combined with the adhesion between the coating and substrate, the wear of the HA+W+DLC and HA+Ti+ta-C coatings is the least, which is related to their having the lowest-roughness substrate, the preparation method, and the thickness of the coating. The results show that compared with the wear degree of the 7A04 aluminum alloy substrate, the wear resistance of the other three coated aluminum alloys is improved. Among them, the wear depth of HA+Ti+ta-C is the shallowest, which is consistent with the law of the friction coefficient wear rate, indicating better wear resistance.

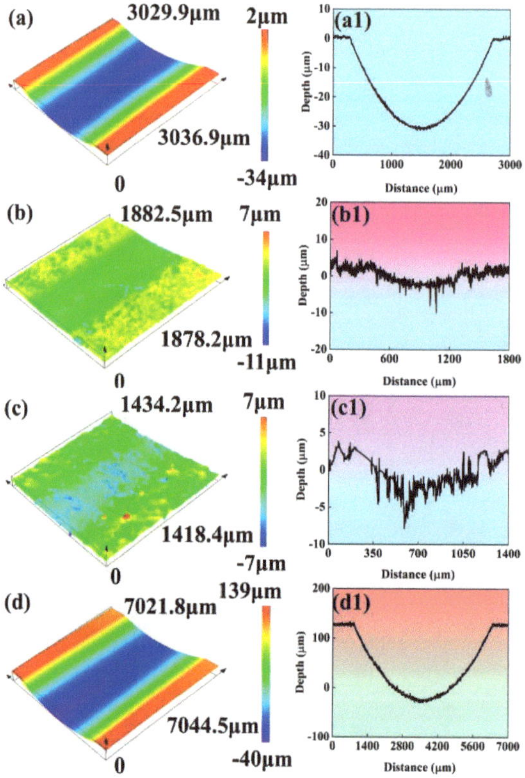

Figure 6. Three-dimensional topographies and height profiles of (**a**,**a1**) MAO coating, (**b**,**b1**) HA+W+DLC coating, (**c**,**c1**) HA+Ti+ta-C coating, and (**d**,**d1**) 7A04 substrate.

3.4.3. Wear Surface Morphology Analysis

The surface morphologies of the four kinds of samples with the worn part and the unworn part are shown in Figure 7. It can be seen from Figure 7a that the wear mechanism of the MAO coating is mainly abrasive wear. This is due to the friction between the MAO coating and the friction pair after wear. The friction pair first contacts the convex part of the MAO coating to form a nonuniform contact, and the actual contact occurs at the contact point. Under a certain load and speed, some microprotrusions are worn off, and wear debris is formed on the worn track. In the subsequent friction and wear test, microcutting occurs to accelerate the wear of the MAO coating, indicating that the wear mechanism is abrasive wear. After a period of wear, the surface of the microprotruding becomes flat, which increases the contact area between the MAO surface and the friction pair. When the load is greater than the strength of the microconvex part on the MAO coating, the convex part cracks to form a large abrasive particle. Under the cyclic action of the contact stress between the MAO and the friction pair, the crack propagates along the pore edge of the MAO coating, resulting in the separation of the MAO through abrasive wear. Combined with the EDS element distribution, it can be seen that the distribution of O, Al, and P elements is relatively uniform, and there is no significant difference. It can also be proved that the MAO coating is not completely destroyed. As shown in Figure 7b,c, the HA+W+DLC and HA+Ti+ta-C coatings have flake morphology. It can be seen that wear of the coating occurs on the surface of HA+W+DLC and HA+Ti+ta-C, accompanied by abrasive wear on the substrate due to debris generated after wear. According to the distribution of EDS elements, the surface of HA+W+DLC coating has Cr, W, C, O, and Al, and the surface of HA+Ti+ta-C coating has Ti, Ni, C, O, and Al. Among them, Cr, W, Ti, and Ni are doped elements, which are used to reduce problems regarding the high internal stress and poor thermal stability of the coating and substrate [42]. As shown in Figure 7d, it can be seen that there are deep parallel furrows in the aluminum alloy matrix after the wear test, and obvious abrasive wear can be seen.

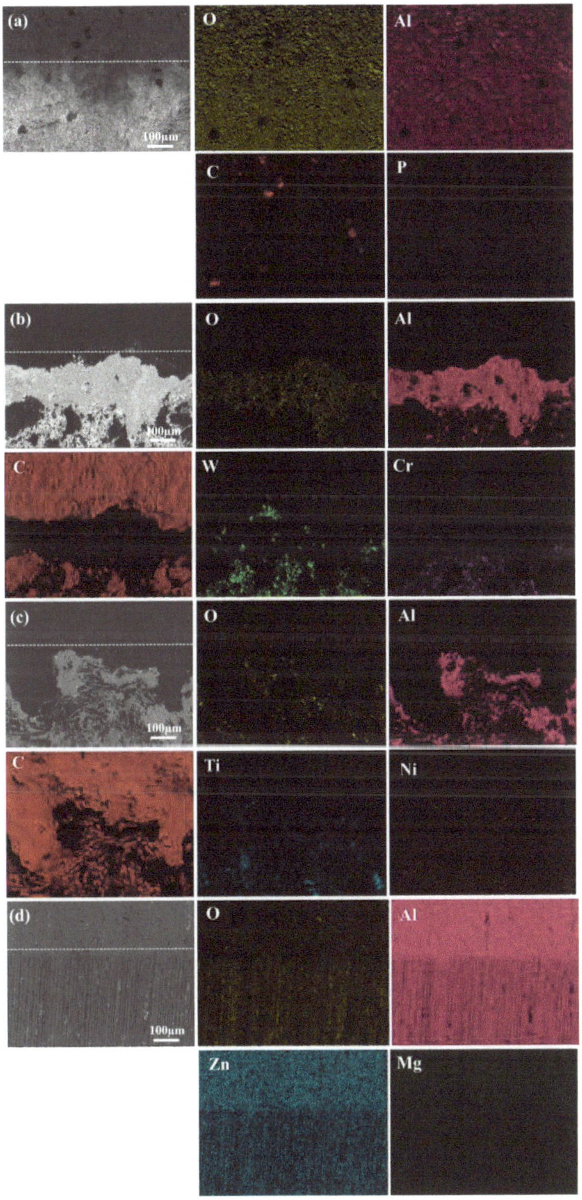

Figure 7. SEM surface morphologies and element distribution maps of (**a**) MAO coating, (**b**) HA+W+DLC coating, (**c**) HA+Ti+ta-C SEM, and (**d**) 7A04 substrate.

The SEM cross-sectional morphologies and element distributions of the different samples are shown in Figure 8. It can be seen from Figure 8a that the thickness of the MAO coating after wear is approximately 32 μm, and no connected pores are observed at the cross-section of the MAO coating. Furthermore, it does not extend to the substrate, which indicates that the bonding strength between the MAO coating and the substrate is high [24]. The endpoint of the diffusion of electrolyte to the substrate is between the MAO coating and substrate, which is the starting point of the microarc discharge channel [43]. The high temperature and pressure induced by the microarc discharge can promote the diffusion of O^{2-} and Al^{3+} in the region to form molten Al-O compounds, simultaneously transferring heat to the substrate to melt Al. The molten Al-O and Al undergo chemical microalloying reactions under the effects of thermochemistry, electrochemistry, and plasma chemistry. As shown in the line scan in Figure 8a1, P, O, and C diffuse from the electrolyte; Al and Mg diffuse from the matrix; the concentrations of Al and O are the highest; and the transition at the bonding interface is evident. This indicates that in the MAO reaction, aluminum mainly diffuses from the inside to the outside, and the electrolyte may also provide a small amount. This indirectly indicates that the MAO coating is generated in situ on the surface of the 7A04 aluminum alloy. The bonding at the interface is based on the metallurgical bonding of chemical bonding and diffusion, which prevent further separation of the coating layer and effectively reduce wear. From Figure 8b,b1,c,c1, it is evident that the thickness of the HA+W+DLC and HA+Ti+ta-C coatings after wear is approximately 9 μm and 3 μm, respectively. In the HA+W+DLC coating, Cr and W are the doping elements, W forms a transition layer, and Cr is hard-oxidized to oxide. Ti and Ni are the doping elements of the HA+Ti+ta-C coating, Ti forms a transition layer, and Ni is hard oxidized to oxide. The whole coating structure is dense and uniform without obvious pores or cracks, and there is a clear boundary between the substrate, the transition layer, and the coating.

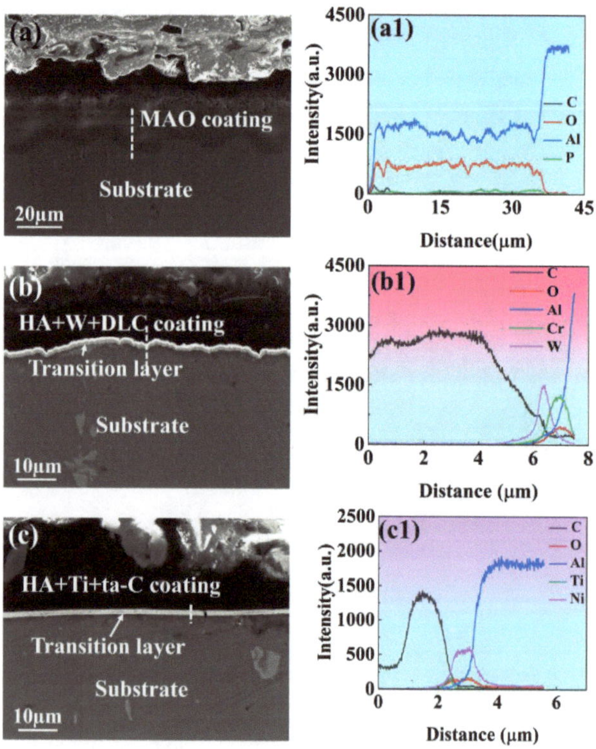

Figure 8. SEM cross-sectional morphologies and element distributions of (**a**,**a1**) MAO coating, (**b**,**b1**) HA+W+DLC coating, and (**c**,**c1**) HA+Ti+ta-C coating.

3.4.4. Raman Spectroscopy

Raman spectroscopy is a reliable method of analyzing the microstructure of a DLC coating and characterizing sp^2 and sp^3 C-C hetero-bonds [44,45]. The DLC coatings demonstrated a broad diffuse peak at 1200–1700 cm^{-1} and a weak shoulder peak at 1300–1400 cm^{-1}. Two Gauss peaks were obtained by fitting the 'D' peak near 1332 cm^{-1} (characterizing the sp^3 C-C hetero-bonds) and the 'G' peak near 1575 cm^{-1} (characterizing the sp^2 C-C hetero-bonds) with the Gauss function. These represent the characteristic Raman peaks of diamond and graphite, respectively.

The Raman spectra of the HA+W+DLC and HA+Ti+ta-C coatings are shown in Figure 9. The laser wavelength was 532 nm, and the wavelength range was 200–2000 nm^{-1}. Figure 9a,b show the Raman spectra of two typical DLC coatings fitted using the original software. The two DLC coatings prepared in this study exhibit typical Raman characteristics. The intensity ratio of the D peak to the G peak, ID/IG, was proportional to the number ratio of sp^2/sp^3 C-C bonds [46]. The ID/IG values of the HA+W+DLC and HA+Ti+ta-C coatings after hard anodizing treatment, calculated using Origin software(2022), were 1.11 and 0.89, respectively. Thus, the highest content of sp^3 C-C bonds was in the HA+Ti+ta-C coating after hard anodizing treatment, followed by the HA+W+DLC coating. It is known from the Raman spectrum that the D peak represents the mixed vibration mode of the sp^2, or sp^2 and sp^3, bond structure in the coating. The higher the content of the sp^3 bond structure in the system, the more prevalent the tetrahedral structure in the coating, the more the structure is biased toward the diamond structure, and the greater the hardness of the coating. A comparison of the Raman spectra of the two coatings shows that the HA+Ti+ta-C coatings had a higher sp^3 bond content, which is consistent with these results.

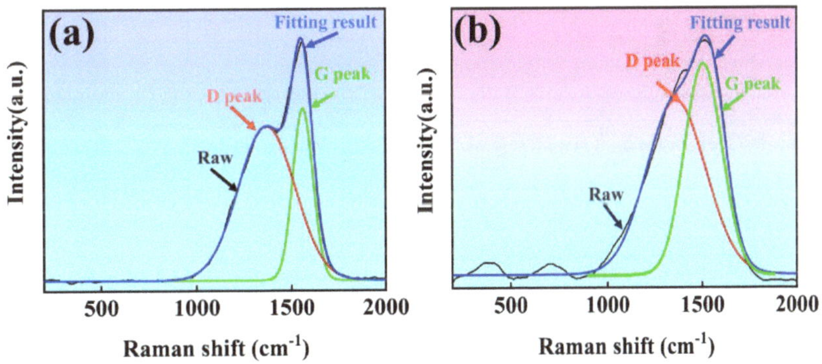

Figure 9. Raman spectra of (**a**) HA+W+DLC and (**b**) HA+Ti+ta-C.

4. Conclusions

(1) In this work, MAO, HA+W+DLC, and HA+Ti+ta-C coatings were successfully prepared on a 7A04 aluminum alloy substrate. The surface hardness was significantly improved after coating. The hardnesses of three coated samples were 7.33 GPa, 22.64 GPa, and 34.23 GPa, respectively. The highest hardness of the HA+Ti+ta-C coatings resulted from the high sp^3 C-C bond content.

(2) During the ring-block wear tests under oil lubrication, both multilayer coatings exhibited excellent wear resistance. The average coefficient of friction and wear rate of HA+W+DLC and HA+Ti+ta-C were, respectively, 0.028 and 1.51×10^{-7} $mm^3/N{\cdot}m$, and 0.025 and 1.36×10^{-7} $mm^3/N{\cdot}m$. The higher surface hardness of the HA+Ti+ta-C coating led to better wear resistance, which suggests that the coating can be applied in the surface protection of aluminum alloys.

Author Contributions: Conceptualization, Methodology, Investigation, and Data Curation, J.H. and C.Z.; Writing—Original Draft, J.H.; Software, X.W. and X.M.; Conceptualization and Writing—Review and Editing, C.D., H.Y. and C.W.; Methodology, Study Design, and Writing—Review and Editing, J.X., Z.Q. and T.J. All authors have read and agreed to the published version of the manuscript.

Funding: This project was supported by the State Key Lab of Advanced Metals and Materials (2022-Z17), Frontier Exploration Projects of Longmen Laboratory (NO. LMQYTSKT011), and Scientific and Technological Project of Henan Province (222102230033).

Institutional Review Board Statement: Not applicable.

Informed Consent Statement: Not applicable.

Data Availability Statement: Data are contained within the article.

Acknowledgments: We wish to take this opportunity to thank the Provincial and Ministerial Co-construction of Collaborative Innovation Center for Non-ferrous Metal New Materials and Advanced Processing Technology for their support.

Conflicts of Interest: Tao Jiang was employed by the company Luoyang Wanji Aluminum Processing Co., Ltd., Ziping Qiao and Jun Xue was employed by the company Science and Technology on Transit Impact Laboratory, No.208 Research Institute of China Ordnance Industries. The remaining authors declare that the research was conducted in the absence of any commercial or financial relationships that could be construed as a potential conflict of interest.

References

1. Heinz, A.; Haszler, A.; Keidel, C.; Moldenhauer, S.; Benedictus, R.; Miller, W.S. Recent development in aluminium alloys for aerospace applications. *Mater. Sci. Eng. A* **2000**, *280*, 102–107. [CrossRef]
2. Dursun, T.; Soutis, C. Recent developments in advanced aircraft aluminium alloys. *Mater. Des.* **2014**, *56*, 862–871. [CrossRef]
3. Tisza, M.; Czinege, I. Comparative study of the application of steels and aluminium in lightweight production of automotive parts. *Int. J. Lightweight Mater. Manuf.* **2018**, *1*, 229–238. [CrossRef]
4. Xu, L.; Yu, X.; Hui, L.; Zhou, S. Fatigue life prediction of aviation aluminium alloy based on quantitative pre-corrosion damage analysis. *Trans. Nonferrous Met. Soc. China* **2017**, *27*, 1353–1362. [CrossRef]
5. Vigneshwar, M.; Selvamani, S.T.; Hariprasath, P.; Palanikumar, K. Analysis of mechanical, metallurgical and fatigue behavior of friction welded AA6061-AA2024 dissimilar aluminum alloys in optimized condition. *Mater. Today Proc.* **2018**, *5*, 7853–7863. [CrossRef]
6. Xiao, F.; Chen, H.; Miao, J.; Du, J. Effect of oxidation time on the microstructure and properties of ceramic coatings prepared by microarc oxidation on 7A04 superhard aluminum alloy. *Int. J. Mod. Phys. B* **2017**, *31*, 1744026. [CrossRef]
7. Zhou, H.; Hu, B.; Yao, W.; Hong, X.; Song, S. Atmospheric Corrosion of Anodized Pure Al 1060, Al-alloys 2A12 and 7A04 Exposed to Polluted Atmospheric Environment at Jiangjin Region. *J. Chin. Soc. Corros. Prot.* **2017**, *37*, 273–278.
8. Su, Y.; Gui, X.; Xie, D.; Li, S.Y.; Sun, H.; Leng, Y.; Huang, N. The effect of a TiN interlayer on the tribological properties of diamond-like carbon films deposited on 7A04 aluminum alloy. *IEEE Trans. Plasma Sci.* **2011**, *39*, 3144–3148. [CrossRef]
9. Chen, X.W.; Song, H.; Pu, H.; Zheng, R.; Zhang, M.; Zhang, D.F. Study of wear and corrosion resistance of ATO nanoparticle–doped micro-arc oxide film layers. *Int. J. Appl. Ceram. Technol.* **2024**, *21*, 1078–1093. [CrossRef]
10. Cheng, C.; Ngan, A.H.W. Chemo-mechanical softening during in situ nanoindentation of anodic porous alumina with anodization processing. *J. Appl. Phys.* **2013**, *113*, 184903. [CrossRef]
11. Walsh, F.C.; Low, C.T.J.; Wood, R.J.K.; Stevens, K.T.; Archer, J.; Poeton, A.R.; Ryder, A. Plasma electrolytic oxidation (PEO) for production of anodised coatings on lightweight metal (Al, Mg, Ti) alloys. *Trans. IMF* **2009**, *87*, 122–135. [CrossRef]
12. Shchedrina, I.; Rakoch, A.G.; Henrion, G.; Martin, J. Non-destructive methods to control the properties of MAO coatings on the surface of 2024 aluminium alloy. *Surf. Coat. Technol.* **2014**, *238*, 27–44. [CrossRef]
13. Yerokhin, A.L.; Nie, X.; Leyland, A.; Matthews, A.; Dowey, S.J. Plasma electrolysis for surface engineering. *Surf. Coat. Technol.* **1999**, *122*, 73–93. [CrossRef]
14. Khadem, M.; Penkov, O.V.; Pukha, V.E.; Maleyev, M.V.; Kim, D.E. Ultra-thin nano-patterned wear-protective diamond-like carbon coatings deposited on glass using a C60 ion beam. *Carbon* **2014**, *80*, 534–543. [CrossRef]
15. Yokota, T.; Sawa, T.; Yokouchi, M.; Tozawa, K.; Anzai, M.; Aizawa, T. Frictional properties of diamond-like carbon coated tool in dry intermittent machining of aluminum alloy 5052. *Precis. Eng.* **2014**, *38*, 365–370. [CrossRef]
16. Bewilogua, K.; Hofmann, D. History of diamond-like carbon films—From first experiments to worldwide applications. *Surf. Coat. Technol.* **2014**, *242*, 214–225. [CrossRef]
17. Robertson, J. Diamond-like amorphous carbon. *Mater. Sci. Eng. R Rep.* **2002**, *37*, 129–281. [CrossRef]
18. Donnet, C.; Erdemir, A. Historical developments and new trends in tribological and solid lubricant coatings. *Surf. Coat. Technol.* **2004**, *180*, 76–84. [CrossRef]
19. Volosova, M.; Grigoriev, S.; Metel, A.; Shein, A. The role of thin-coating vacuum-plasma coatings and their influence on the efficiency of ceramic cutting inserts. *Coatings* **2018**, *8*, 287. [CrossRef]

20. Grigoriev, S.N.; Volosova, M.A.; Fedorov, S.V.; Mosyanov, M. Influence of DLC coatings deposited by PECVD technology on the wear resistance of carbide end mills and surface roughness of AlCuMg2 and 41Cr4 workpieces. *Coatings* **2020**, *10*, 1038. [CrossRef]
21. Hilbert, J.; Mangolini, F.; McClimon, J.B.; Lukes, J.R.; Carpick, R.W. Si doping enhances the thermal stability of diamond-like carbon through reductions in carbon-carbon bond length disorder. *Carbon* **2018**, *131*, 72–78. [CrossRef]
22. Soffritti, C.; Fortini, A.; Nastruzzi, A.; Sola, R.; Merlin, M.; Garagnani, G.L. Dry sliding behavior of an aluminum alloy after innovative hard anodizing treatments. *Materials* **2021**, *14*, 3281. [CrossRef]
23. Kwolek, P.; Obłój, A.; Kościelniak, B.; Buszta, R.; Tokarski, T.; Krupa, K.; Gradzik, A.; Nowak, W.J.; Wojnicki, M.; Motyka, M. Wear resistance of hard anodic coatings fabricated on 5005 and 6061 aluminum alloys. *Arch. Civ. Mech. Eng.* **2024**, *24*, 51. [CrossRef]
24. Zhuang, J.J.; Guo, Y.Q.; Xiang, N.; Lu, X.Y.; Hu, Q.; Song, R.G. Sliding wear behaviour and microstructure of PEO coatings formed on aluminium alloy. *Mater. Sci. Technol.* **2016**, *32*, 1559–1566. [CrossRef]
25. Yang, C.; Zhu, J.; Cui, S.; Chen, P.; Wu, Z.; Ma, Z.; Fu, R.K.Y.; Tian, X.; Chu, P.K.; Wu, Z. Wear and corrosion resistant coatings prepared on LY12 aluminum alloy by plasma electrolytic oxidation. *Surf. Coat. Technol.* **2021**, *409*, 126885. [CrossRef]
26. Zou, Y.S.; Zhou, K.; Wu, Y.F.; Yang, H.; Cang, K.; Song, G.H. Structure, mechanical and tribological properties of diamond-like carbon films on aluminum alloy by arc ion plating. *Vacuum* **2012**, *86*, 1141–1146. [CrossRef]
27. Srinivasan, N.; Bhaskar, L.K.; Kumar, R.; Baragetti, S. Residual stress gradient and relaxation upon fatigue deformation of diamond-like carbon coated aluminum alloy in air and methanol environments. *Mater. Des.* **2018**, *160*, 303–312. [CrossRef]
28. Sui, X.; Liu, J.; Zhang, S.; Yang, J.; Hao, J. Microstructure, mechanical and tribological characterization of CrN/DLC/Cr-DLC multilayer coating with improved adhesive wear resistance. *Appl. Surf. Sci.* **2018**, *439*, 24–32. [CrossRef]
29. Shen, Y.; Luo, J.; Liao, B.; Zhang, X.; Zhao, Y.; Zeng, X.; Chen, L.; Pang, P.; Bao, F. Tribocorrosion and tribological behavior of Ti-DLC coatings deposited by filtered cathodic vacuum arc. *Diam. Relat. Mater.* **2022**, *125*, 108985. [CrossRef]
30. Yi, M.; Wang, T.; Liu, Z.; Lei, J.; Qiu, J.; Xu, W. Tribological Performance of Steel/W-DLC and W-DLC/W-DLC in a Solid–Liquid Lubrication System Additivated with Ultrathin MoS2 Nanosheets. *Lubricants* **2023**, *11*, 433. [CrossRef]
31. Bobzin, K.; Brögelmann, T.; Kalscheuer, C.; Thiex, M. Formation of tribochemical reaction layers on a metal modified amorphous carbon coating a-C: H: Zr (ZrCg). *Tribol. Int.* **2019**, *135*, 152–160. [CrossRef]
32. Mohammadinia, E.; Elahi, S.M.; Shahidi, S. Structural and optical properties of Ni embedded hydrogenated diamond-like carbon (Ni-DLC) prepared by co-deposition of RF-Sputtering and RF-PECVD method. *Mater. Sci. Semicond. Process.* **2018**, *7*, 7–12. [CrossRef]
33. Sun, L.; Guo, P.; Ke, P.; Li, X.; Wang, A. Synergistic effect of Cu/Cr co-doping on the wettability and mechanical properties of diamond-like carbon films. *Diam. Relat. Mater.* **2016**, *68*, 1–9. [CrossRef]
34. Chen, Q.; Jiang, Z.; Tang, S.; Dong, W.; Tong, Q.; Li, W. Influence of graphene particles on the micro-arc oxidation behaviors of 6063 aluminum alloy and the coating properties. *Appl. Surf. Sci.* **2017**, *423*, 939–950. [CrossRef]
35. Wang, K.; Koo, B.-H.; Lee, C.-G.; Kim, Y.-J.; Lee, S.-H.; Byon, E. Effects of electrolytes variation on formation of oxide layers of 6061 Al alloys by plasma electrolytic oxidation. *Trans. Nonferrous Met. Soc. China* **2009**, *19*, 866–870. [CrossRef]
36. Wen, S.; Huang, P. *Principles of Tribology*; John Wiley & Sons: Hoboken, NJ, USA, 2012.
37. Li, X.J.; Zhang, M.; Wen, S.; Mao, X.; Huo, W.G.; Guo, Y.Y.; Wang, Y.X. Microstructure and wear resistance of micro-arc oxidation ceramic coatings prepared on 2A50 aluminum alloys. *Surf. Coat. Technol.* **2020**, *394*, 125853. [CrossRef]
38. Xie, H.; Cheng, Y.; Li, S.; Cao, J.; Cao, L. Wear and corrosion resistant coatings on surface of cast A356 aluminum alloy by plasma electrolytic oxidation in moderately concentrated aluminate electrolytes. *Trans. Nonferrous Met. Soc. China* **2017**, *27*, 336–351. [CrossRef]
39. Martini, C.; Ceschini, L.; Tarterini, F.; Paillard, J.M.; Curran, J.A. PEO layers obtained from mixed aluminate–phosphate baths on Ti–6Al–4V: Dry sliding behaviour and influence of a PTFE topcoat. *Wear* **2010**, *269*, 747–756. [CrossRef]
40. Zuo, Y.; Zhao, P.H.; Zhao, J.M. The influences of sealing methods on corrosion behavior of anodized aluminum alloys in NaCl solutions. *Surf. Coat. Technol.* **2003**, *166*, 237–242. [CrossRef]
41. Qin, L.; Zhao, W.; Hou, H.; Jin, Y.; Zeng, Z.; Wu, X.; Xue, Q. Achieving excellent anti-corrosion and tribological performance by tailoring the surface morphology and chemical composition of aluminum alloys. *RSC Adv.* **2014**, *4*, 60307–60315. [CrossRef]
42. Cao, H.; Ye, X.; Li, H.; Qi, F.; Wang, Q.; Ouyang, X.; Zhao, N.; Liao, B. Microstructure, mechanical and tribological properties of multilayer Ti-DLC thick films on Al alloys by filtered cathodic vacuum arc technology. *Mater. Des.* **2021**, *198*, 109320. [CrossRef]
43. Krishna, L.R.; Gupta, P.; Sundararajan, G. The influence of phase gradient within the micro arc oxidation (MAO) coatings on mechanical and tribological behaviors. *Surf. Coat. Technol.* **2015**, *269*, 54–63. [CrossRef]
44. Ferrari, A.C.; Robertson, J. Interpretation of Raman spectra of disordered and amorphous carbon. *Phys. Rev. B Condens. Matter* **2000**, *61*, 14095–14107. [CrossRef]
45. Ferrari, A.C.; Robertson, J. Resonant Raman spectroscopy of disordered amorphous and diamond like carbon. *Phys. Rev. B Condens. Matter* **2001**, *64*, 5414–5426. [CrossRef]
46. Yoshikawa, M.; Katagiri, G.; Ishida, H.; Ishitani, A.; Akamatsu, T. Raman spectra of diamondlike amorphous carbon films. *Solid State Commun.* **1988**, *66*, 1177–1180. [CrossRef]

coatings

MDPI

Article

Optimized Functionalization of Graphene Oxide for Enhanced Mechanical Properties in Epoxy Resin Composites

Xin Li [1], Bing Yu [2], Jie Chen [1], Dongxia Huo [1], Jun Liu [2,*] and Ding Nan [1,2,*]

[1] College of Chemistry and Chemical Engineering, Inner Mongolia University, Hohhot 010021, China
[2] Inner Mongolia Key Laboratory of Graphite and Graphene for Energy Storage and Coating, School of Materials Science and Engineering, Inner Mongolia University of Technology, Hohhot 010051, China
* Correspondence: clxylj@163.com (J.L.); nd@imu.edu.cn (D.N.)

Abstract: Epoxy resins have widespread applications across various industries, such as anticorrosive coatings, owing to their exceptional attributes. However, there is a constant demand for enhancements to their mechanical characteristics to cater to the requirements of unique and specialized conditions. In this work, graphene oxide modified by 4,4′-Oxydianiline (MGO) was prepared using a covalent grafting reaction. The MGO in epoxy resin composites exhibited a rougher morphology and thin layers with a superior tensile strength (38 MPa), elastic modulus (358 MPa), flexural strength (85 MPa), flexural modulus (957 MPa), and hardness (62 HD). The results indicated that the mechanical properties of epoxy resin are significantly improved by MGO and the improved mechanical properties of epoxy resin composites are due to the strong interfacial bonding between MGO and epoxy.

Keywords: graphene oxide; 4,4′-Oxydianiline; epoxy; mechanical properties

Citation: Li, X.; Yu, B.; Chen, J.; Huo, D.; Liu, J.; Nan, D. Optimized Functionalization of Graphene Oxide for Enhanced Mechanical Properties in Epoxy Resin Composites. *Coatings* **2024**, *14*, 609. https://doi.org/10.3390/coatings14050609

Academic Editor: Alina Pruna

Received: 22 April 2024
Revised: 7 May 2024
Accepted: 9 May 2024
Published: 11 May 2024

1. Introduction

Epoxy resin has been extensively used in aerospace [1], electronic instruments [2], engineering construction [3], and other fields [2,4] due to its superior excellent physical properties, chemical resistance, stability, and unique processability [5]. However, bisphenol A epoxy resins, the most commonly used epoxy resins, are limited by defects in their mechanical properties, such as high brittleness and poor tensile strength and flexural strength [6]. Thus, a large number of methods have been reported to improve the mechanical properties of epoxy resins, including changing the molecular structure of the matrix and adding compatible functional inorganic fillers [7,8]. Simultaneously, researchers have explored altering the structure of the curing agent, a pursuit that has garnered considerable interest. For instance, Alexey Orlov et al. [9] undertook the synthesis of curing agents featuring varying phosphazene contents. Their investigations revealed that the complex comprising 30% modifier significantly enhanced both the mechanical robustness and fire-retardant properties of the epoxy resin. However, the change of curing agent will make the operation complicated and reduce the efficiency. Among these methods, combination with the second phase of superior nanofillers has been developed as a simple and efficient method [10,11].

Due to its incredible mechanical, electrical, and thermal properties, graphene, a unique two-dimensional layered material, has generated great interest [12,13]. Cahit Orek et al. [14] solved the problem of gold(I) N-heterocyclic monocarbene-adsorbing graphene as a drug nanocarrier via calculations. Studies have shown that graphene and its derivatives can effectively improve the mechanical properties of polymers, Anastasia Konstantinova et al. [15] ascertained that the incorporation of a phosphazene-containing curing agent, coupled with thermally expanded graphite, markedly enhanced the fire resistance, as well as the physical and mechanical characteristics, of epoxy resin. However, it could only happen if graphene nanoplatelets were well dispersed in the polymer matrix [16–19]. In order to improve the dispersion of GO, Zhang et al. [20] employed atom transfers radical polymerization to link

GO and boron nitride, subsequently incorporating this composite as a filler into the epoxy resin matrix. Their investigation revealed commendable attributes, including noteworthy thermal conductivity, electrical insulation, and compression capabilities. In this regard, surface covalent functionalization has been conducted to enhance uniform dispersion and enhance compatibility with polymer matrices [21–23]. But there are still some deficiencies because of complex modification reactions and severe conditions, which will reduce efficiency and increase costs, even causing the degradation of the polymer matrix.

Due to its excellent physical properties and corrosion resistance, epoxy resin is widely used as a film-forming material in coatings. We have previously investigated the impact of covalent and non-covalent modifications to GO on the corrosion resistance and weathering resistance of epoxy resin coatings [24–26]. In this work, we report a facile strategy to synthesize 4,4′-Oxydianiline-modified graphene oxide (MGO) via surface covalent functionalization and modified graphene oxide/epoxy resin composites (MGO/EP) via two-step synthesis. The morphology and mechanical properties of MGO/EP were compared, and the effect of MGO on the mechanical properties was investigated.

2. Experiments

MGO was synthesized using a covalent grafting reaction. Firstly, 0.25 g 4,4′-Oxydianiline and 0.5 g GO were added into 50 mL and 100 mL N,N-Dimethylformamide (DMF), respectively, and stirred for 1 h. Subsequently, the mixture was stirred and refluxed for 4 h at 80 °C and then dried in a vacuum oven at 60 °C overnight. EP/MGO were synthesized using a two-step synthesis. Firstly, MGO and EP were added into ethanol under stirring for 1.5 h in turn and then dried at 80 °C overnight. Polyamide curing agents (650) were added into EP under stirring for 0.5 h, and the mixed solution was poured into a preheated Teflon mold and cured in a vacuum oven at 80 °C overnight. Figure 1 shows the synthesis mechanism and preparation method of MGO and EP/MGO.

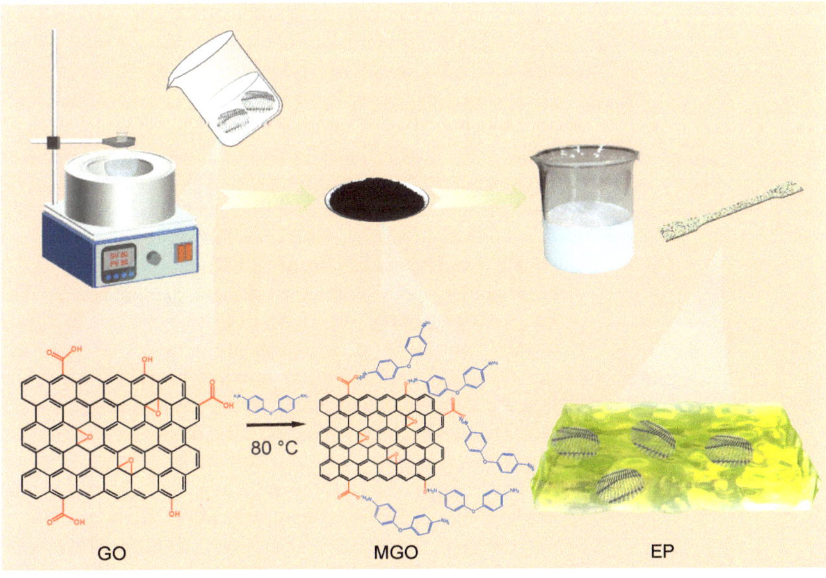

Figure 1. Schematic diagram of the preparation and the reaction between 4,4′-oxydianiline and GO.

The morphology of MGO was observed using scanning electron microscopy (SEM, FEI QUANTA 650) (FEI, Hillsboro, OR, USA) and transmission electron microscopy (TEM, JEOL 2100) (JEOL, Tokyo, Japan). The powder specimens were dispersed in ethanol and desiccated for SEM and TEM analysis. The modification of GO was evaluated using X-ray diffractometry (XRD), Fourier transform infrared (FTIR) spectrum, Raman spectrum, and

X-ray photoelectron spectrum (XPS) measurements. The GO or MGO powder specimens were promptly situated within the XRD sample chamber, and the XRD data were acquired within the angular span from 5° to 60°. FTIR was obtained using a Bruker TENSOR II spectrometer (Bruker, Karlsruhe, Germany), and the scanning range was 4000–400 cm^{-1}. The solid powder sample was mixed with a small amount of potassium bromide powder, and the mixed powder was fined and pressed into tablets for infrared spectroscopy. Raman spectra were obtained using a Horiba IHR320 spectrometer (Horiba, Kyoto, Japan) with a scanning range of 500–4000 cm^{-1} and an excitation wavelength of 532 nm. The solid powder was dispersed in ethanol and dried on a glass slide for Raman spectroscopy. The powder specimens were affixed directly onto the XPS sample platform for XPS examination. The tensile and bending properties of composite materials were tested using a microcomputer-controlled electronic universal testing machine, according to GB/T 13525-1992 [27] and GB/T 6569-2006 [28] standards.

3. Results and Discussion

Upon comparing the XRD (Figure S1), FTIR (Figure S2), and Raman (Figure S3) test results, the optimal modification reaction ratio between 4,4′-Oxydianiline and GO (1:2) was determined. MGO was synthesized using this optimal ratio for subsequent research endeavors. The XRD spectra of GO and MGO are shown in Figure 2a. The characteristic diffraction peaks at 12.445° of GO and 9.074° of MGO represent the 0 0 1 reflection peak, corresponding to an interlayer spacing of about 0.709 nm and 0.973 nm, respectively. In the FT-IR spectrum of GO, the peaks at 3397 cm^{-1}, 1732 cm^{-1}, 1612 cm^{-1}, 1236 cm^{-1}, and 1059 cm^{-1} are attributed to –OH, C=O, C=C, C–O–C, and C–O bands, respectively. This shift implies a reduction in the interlayer separation relative to that observed in GO. The computed interlayer spacing value provides compelling evidence indicating the covalent grafting of 4,4′-Oxydianiline molecules onto the surface of GO, thereby corroborating the presence of such molecules [26]. Meanwhile, in the FT-IR spectrum of MGO, the C=O band nearly disappears and new bands at 1496 cm^{-1} and 823 cm^{-1} arise, representing the C–N and N–H stretch vibrations of the methylene group (Figure 2b). The Raman spectra of GO and MGO displays two peaks at 1354 cm^{-1} (D band) and 1583 cm^{-1} (G band), and the I_D/I_G value increased from 1.89 of GO to 2.84 of MGO, indicating that covalent bonding occurs between 4,4′-Oxydianiline and GO without the significant destruction of the carbon lattice [29]. Furthermore, it is proven that, subsequent to covalent bonding, the degree of disorder within MGO tends to escalate when juxtaposed with GO. Figure 2d shows the XPS spectra of GO and MGO. Compared with GO, there is a characteristic peak at N 1*s* originating from 4,4′-Oxydianiline, and the appearance of C–N further confirms the covalent functionalization of GO. The high-resolution C 1*s* XPS spectrum of GO can be meticulously deconvoluted to discern distinct bands—namely, C–C (at 284.56 eV), C–O (at 286.87 eV), and C=O (at 289.01 eV)—as shown in Figure 2e. Meanwhile, the C–C peak of MGO undergoes attenuation, and a fresh C–N peak (at 285.69 eV) emerges, as shown in Figure 2f. The observed distinctions signal the alterations to the chemical bonding and functional groups stemming from the covalent grafting modification with 4,4′-Oxydianiline.

The morphology of GO and MGO was observed and is shown in Figure 3. The diagram clearly illustrates the pronounced layered structure inherent in both GO and MGO. As shown in Figure 3a,b, there are a few folds on the surface and edges of GO, which are stacked in layers. After the modification treatment, the surface of MGO becomes looser and rougher (Figure 3c,d). Thin layers of GO and MGO can be observed in Figure 3e,f, and the corresponding selected area electron diffraction (SAED) pattern (inset in Figure 3e,f) also suggests that MGO has a more obvious crystal structure than GO.

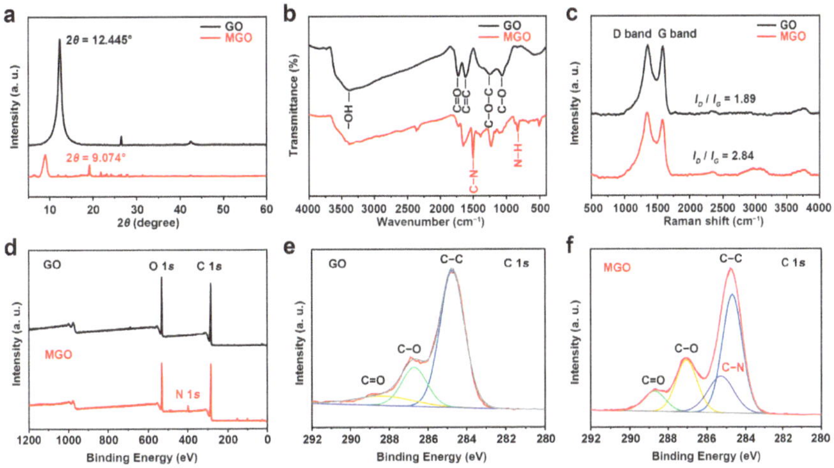

Figure 2. XRD spectra (**a**), FT-IR spectra (**b**), Raman spectra (**c**), and XPS spectra (**d**) of GO and MGO, and C 1s XPS spectra of (**e**) GO and (**f**) MGO.

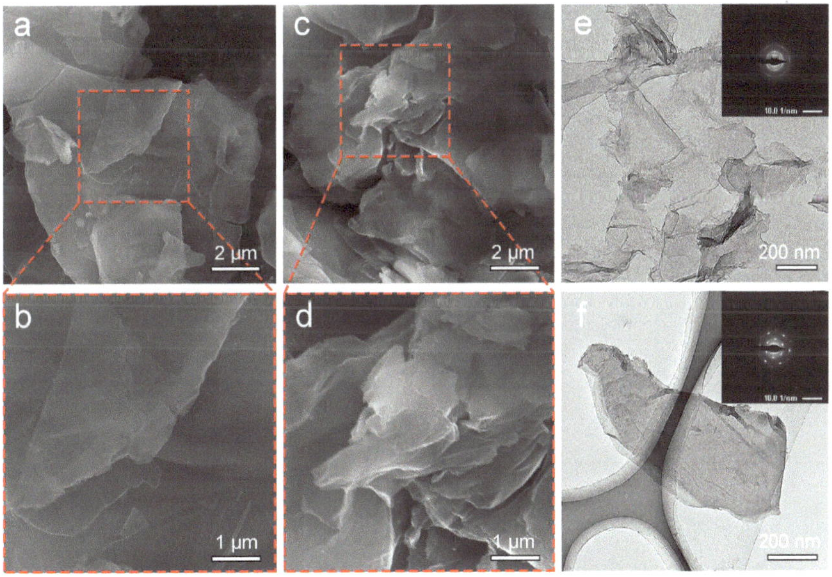

Figure 3. SEM images of GO (**a**,**b**) and MGO (**c**,**d**), and TEM images of GO (**e**) and MGO (**f**). The inset of (**e**,**f**) shows the corresponding SAED pattern.

In order to investigate the effect of MGO addition on the mechanical properties of EP/MGO, 0.1 wt.%, 0.3 wt.%, 0.5 wt.%, and 1.0 wt.% of MGO were added into epoxy resin, which was named EP/MGO01, EP/MGO03, EP/MGO05, and EP/MGO10, respectively. The tensile properties of EP and EP/MGO were tested and compared, as shown in Figure 4. The tensile strength of EP/MGO01 increased from 24 MPa to 38 MPa, and the elastic modulus increased from 248 MPa to 358 MPa. As the amount of MGO added increased, the tensile property and flexural property decreased in varying degrees, as shown in Figure 4b. Similarly, the elongation at break originally presented a significant increase and then decreased to varying degrees (Figure 4c).

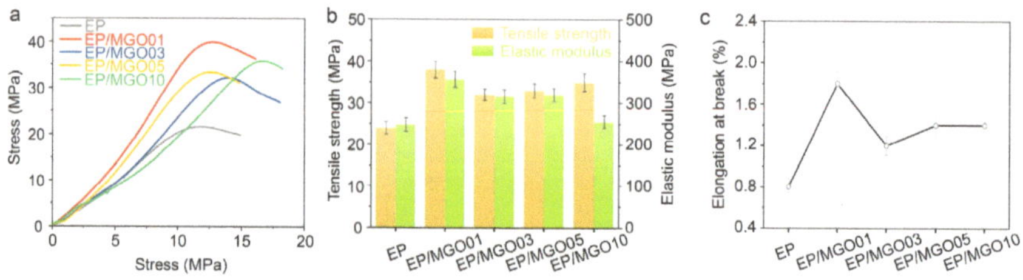

Figure 4. Tensile properties of epoxy and its composites with different MGO loadings: stress–strain curves (**a**), tensile strength and elastic modulus (**b**), and elongation at break (**c**).

The fractured surfaces of EP, EP/GO (the addition amount of GO is 0.1 wt.%), and EP/MGO01 were observed using SEM, as shown in Figure 5. There are small faceted features on the fracture surface of EP, indicating a brittle fracture with poor tensile strength. Likewise, the fracture surface of EP/GO retains small-faceted attributes akin to EP, suggestive of its brittle fracture nature. However, the fracture surface of EP/GO exhibits a heightened roughness compared to EP, indicative of GO incremental enhancement of EP mechanical resilience to a certain degree. Meanwhile, EP/MGO01 exhibits a relatively rough fracture surface with river-like patterns, which is a typical feature of ductile fracture. Moreover, no obvious pull-out trace can be observed, indicating the strong MGO-EP interfacial bonding as a result of the functional group of MGO, which has a significant influence on the mechanical properties of epoxy resin composites.

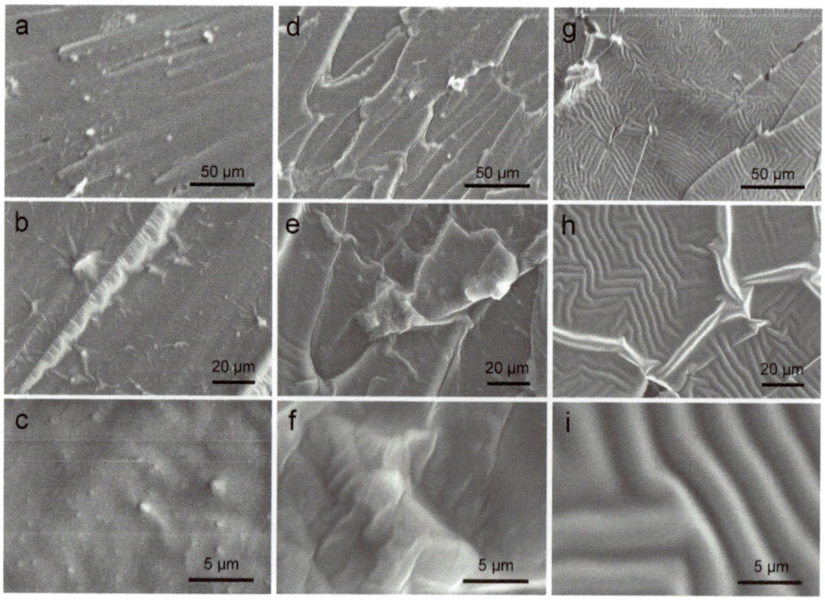

Figure 5. SEM images of the fractured surface of EP (**a**–**c**), EP/GO (GO content: 0.1 wt.%) (**d**–**f**), and EP/MGO01 (**g**–**i**).

The flexural properties and hardness of EP and EP/MGO were tested and compared, as shown in Figure 6. Figure 6a shows the typical load–displacement curves of EP and EP/MGO, and the smooth curves indicate that the cracks propagated in a controlled manner [30]. The flexural strength of EP/MGO01 increased from 48 MPa to 85 MPa, and the flexural modulus increased from 473 MPa to 957 MPa (Figure 6b). The impact strength of EP is 4.507 kJ/m^2, while the impact strength of EP/MGO01 is 8.462 kJ/m^2.

In comparison to EP, the impact strength of EP/MGO01 increased by 88% (Figure 6c). Moreover, the hardness of EP is significantly improved by MGO, as shown in Figure 6d, and the highest hardness of EP/MGO01 is 62 HD, which is 1.44 times that in EP (43 HD).

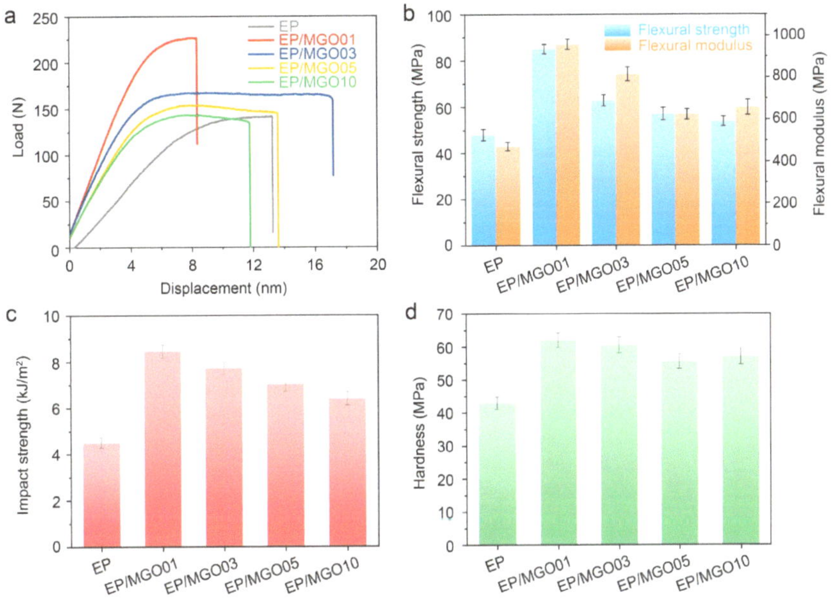

Figure 6. Load–displacement curves (**a**), flexural strength and flexural modulus (**b**), and impact strength (**c**) and hardness (**d**) of epoxy and its composites with different MGO loadings.

When MGO is evenly distributed within epoxy resin, the strong interfacial bonding between MGO and epoxy in composites offers several benefits for enhancing mechanical properties. The strong bond ensures efficient load transfer between the MGO filler and the epoxy matrix, leading to increased strength and stiffness in the composite material. And a well-bonded interface can contribute to lower wear and friction between the MGO and the epoxy matrix, resulting in improved tribological properties in the composite. Moreover, strong interfacial bonding minimizes the risk of interface delamination or debonding, thereby improving the long-term durability and reliability of the composite material. Therefore, the addition of MGO improves the mechanical properties of epoxy resin.

However, when there is an excessive amount of MGO added, the particles may agglomerate and clump together, leading to localized stress concentrations and interface breakdown, which ultimately reduce the overall performance of the epoxy matrix. And excessive amounts of MGO may weaken interface adhesion, as the reduced distance between filler particles hinders effective interactions between the epoxy resin and the filler, thereby diminishing the strength and toughness. Therefore, excessive amounts of MGO may lead to a decrease in the mechanical properties of epoxy resin, requiring appropriate control of the additive amount to achieve optimal performance. In this work, the optimal MGO addition amount is 0.1 wt.%.

4. Conclusions

MGO has been successfully fabricated via a covalent grafting reaction of GO and 4,4′-Oxydianiline. The looser and rougher surface and thin layers of MGO were confirmed using SEM and TEM. As feature-enhanced fillers, MGO in epoxy resins showed a significant increase in mechanical properties; the tensile strength of EP/MGO01 increased from 24 MPa to 38 MPa, the elastic modulus increased from 248 MPa to 358 MPa, the flexural strength increased from 48 MPa to 85 MPa, the flexural modulus increased from

473 MPa to 957 MPa, and the hardness increased from 43 HD to 62 HD. The aforementioned outcomes stem from the incorporation of MGO, which serves to ameliorate imperfections such as pores and microfractures within the EP matrix while concurrently augmenting its density. Furthermore, the introduction of MGO fortifies the cohesion at the MGO-EP interface. The results demonstrate that the MGO is very promising as an alternative effect filler candidates for high-mechanical-property composites.

Supplementary Materials: The following supporting information can be downloaded at: https://www.mdpi.com/article/10.3390/coatings14050609/s1, Figure S1: XRD spectra of GO and MGO samples; Figure S2: FT-IR spectra of GO and MGO samples; Figure S3: Raman spectra of GO and MGO samples; Figure S4: Tensile properties of epoxy and its composites with different MGO: stress-strain curves (a), tensile strength (b), elongation at break (c), and elastic modulus (d).

Author Contributions: X.L.: Conceptualization, Formal analysis, Writing—Original draft preparation; B.Y.: Data curation, Validation, Investigation; J.C.: Writing—Original draft preparation; D.H.: Writing—Original draft preparation; J.L.: Writing—Review and Editing; D.N.: Supervision, Writing—Review and Editing, Funding acquisition. All authors have read and agreed to the published version of the manuscript.

Funding: The financial supporters of this work are the Inner Mongolia Major Science and Technology Project (No. 2020ZD0024), Local Science and Technology Development Project of the Central Government (No. 2021ZY0006, 2022ZY0011), the Project of Innovation Research in Postgraduate in Inner Mongolia (B20231023Z), and Inner Mongolia Autonomous Region Key Research and Technological Achievements Transformation Plan Project (No. 2023YFHH0063).

Institutional Review Board Statement: Not applicable.

Informed Consent Statement: Not applicable.

Data Availability Statement: Data are contained within the article.

Conflicts of Interest: The authors declare no conflict of interest.

References

1. Paolillo, S.; Bose, R.K.; Santana, M.H.; Grande, A.M. Intrinsic Self-Healing Epoxies in Polymer Matrix Composites (PMCs) for Aerospace Applications. *Polymers* **2021**, *13*, 201. [CrossRef] [PubMed]
2. Salunke, D.R.; Gopalan, V. Thermal and Electrical behaviors of Boron Nitride/Epoxy reinforced polymer matrix composite—A review. *Polym. Compos.* **2021**, *42*, 1659–1669. [CrossRef]
3. Wang, Z.; Jia, Z.; Feng, X.; Zou, Y. Graphene nanoplatelets/epoxy composites with excellent shear properties for construction adhesives. *Compos. Part B Eng.* **2018**, *152*, 311–315. [CrossRef]
4. Jeong, S.-G.; Cha, J.; Kim, S.; Seo, J.; Lee, J.-H.; Kim, S. Preparation of thermal-enhanced epoxy resin adhesive with organic PCM for applying wood flooring. *J. Therm. Anal. Calorim.* **2014**, *117*, 1027–1034. [CrossRef]
5. Baig, M.M.A.; Samad, M.A. Epoxy\Epoxy Composite\Epoxy Hybrid Composite Coatings for Tribological Applications—A Review. *Polymers* **2021**, *13*, 179. [CrossRef] [PubMed]
6. Wei, J.; Ma, S.; Yue, H.; Wang, S.; Zhu, J. Comparison of Hydrogenated Bisphenol A and Bisphenol A Epoxies: Curing Behavior, Thermal and Mechanical Properties, Shape Memory Properties. *Macromol. Res.* **2018**, *26*, 529–538. [CrossRef]
7. Khan, M.; Khurram, A.A.; Li, T.; Zhao, T.; Subhani, T.; Gul, I.; Ali, Z.; Patel, V. Synergistic effect of organic and inorganic nano fillers on the dielectric and mechanical properties of epoxy composites. *J. Mater. Sci. Technol.* **2018**, *34*, 2424–2430. [CrossRef]
8. Sun, J.; Qian, L.; Li, J. Toughening and strengthening epoxy resin with flame retardant molecular structure based on tyrosine. *Polymer* **2021**, *230*, 124045. [CrossRef]
9. Orlov, A.; Konstantinova, A.; Korotkov, R.; Yudaev, P.; Mezhuev, Y.; Terekhov, I.; Gurevich, L.; Chistyakov, E. Epoxy Compositions with Reduced Flammability Based on DER-354 Resin and a Curing Agent Containing Aminophosphazenes Synthesized in Bulk Isophoronediamine. *Polymers* **2022**, *14*, 3592. [CrossRef]
10. Wan, Y.-J.; Gong, L.-X.; Tang, L.-C.; Wu, L.-B.; Jiang, J.-X. Mechanical properties of epoxy composites filled with silane-functionalized graphene oxide. *Compos. Part A Appl. Sci. Manuf.* **2014**, *64*, 79–89. [CrossRef]
11. Naebe, M.; Wang, J.; Amini, A.; Khayyam, H.; Hameed, N.; Li, L.H.; Chen, Y.; Fox, B. Mechanical Property and Structure of Covalent Functionalised Graphene/Epoxy Nanocomposites. *Sci. Rep.* **2014**, *4*, 4375. [CrossRef]
12. Sun, Y.W.; Papageorgiou, D.G.; Humphreys, C.J.; Dunstan, D.J.; Puech, P.; Proctor, J.E.; Bousige, C.; Machon, D.; San-Miguel, A. Mechanical properties of graphene. *Appl. Phys. Rev.* **2021**, *8*, 021310. [CrossRef]
13. Kovalev, S.; Hafez, H.A.; Tielrooij, K.-J.; Deinert, J.-C.; Ilyakov, I.; Awari, N.; Alcaraz, D.; Soundarapandian, K.; Saleta, D.; Germanskiy, S.; et al. Electrical tunability of terahertz nonlinearity in graphene. *Sci. Adv.* **2021**, *7*, eabf9809. [CrossRef] [PubMed]

14. Orek, C.; Bartolomei, M.; Coletti, C.; Bulut, N. Graphene as Nanocarrier for Gold(I)-Monocarbene Complexes: Strength and Nature of Physisorption. *Molecules* **2023**, *28*, 3941. [CrossRef] [PubMed]
15. Konstantinova, A.; Yudaev, P.; Orlov, A.; Loban, O.; Lukashov, N.; Chistyakov, E. Aryloxyphosphazene-Modified and Graphite-Filled Epoxy Compositions with Reduced Flammability and Electrically Conductive Properties. *J. Compos. Sci.* **2023**, *7*, 417. [CrossRef]
16. Papageorgiou, D.G.; Li, Z.; Liu, M.; Kinloch, I.A.; Young, R.J. Mechanisms of mechanical reinforcement by graphene and carbon nanotubes in polymer nanocomposites. *Nanoscale* **2020**, *12*, 2228–2267. [CrossRef] [PubMed]
17. Pang, Y.; Yang, J.; Curtis, T.E.; Luo, S.; Huang, D.; Feng, Z.; Morales-Ferreiro, J.O.; Sapkota, P.; Lei, F.; Zhang, J.; et al. Exfoliated Graphene Leads to Exceptional Mechanical Properties of Polymer Composite Films. *ACS Nano* **2019**, *13*, 1097–1106. [CrossRef] [PubMed]
18. Wang, J.; Song, F.; Ding, Y.; Shao, M. The incorporation of graphene to enhance mechanical properties of polypropylene self-reinforced polymer composites. *Mater. Des.* **2020**, *195*, 109073. [CrossRef]
19. Belyaeva, L.A.; van Deursen, P.M.G.; Barbetsea, K.I.; Schneider, G.F. Hydrophilicity of Graphene in Water through Transparency to Polar and Dispersive Interactions. *Adv. Mater.* **2018**, *30*, 1703274. [CrossRef]
20. Zhang, L.; Liu, H.; Wang, Z.; Sui, W.; Gong, Y.; Cui, J.; Ao, Y.; Shang, L. Functional boron nitride/graphene oxide three-dimensional skeleton co-heat transfer epoxy resin composite. *J. Alloys Compd.* **2024**, *985*, 173935. [CrossRef]
21. González, M.C.R.; Leonhardt, A.; Stadler, H.; Eyley, S.; Thielemans, W.; De Gendt, S.; Mali, K.S.; De Feyter, S. Multicomponent Covalent Chemical Patterning of Graphene. *ACS Nano* **2021**, *15*, 10618–10627. [CrossRef] [PubMed]
22. Assies, L.; Fu, C.; Kovaříček, P.; Bastl, Z.; Drogowska, K.A.; Lang, J.; Guerra, V.L.P.; Samorì, P.; Orgiu, E.; Perepichka, D.F.; et al. Dynamic covalent conjugated polymer epitaxy on graphene. *J. Mater. Chem. C* **2019**, *7*, 12240–12247. [CrossRef]
23. Valenta, L.; Kovaříček, P.; Valeš, V.; Bastl, Z.; Drogowska, K.A.; Verhagen, T.A.; Cibulka, R.; Kalbáč, M. Spatially Resolved Covalent Functionalization Patterns on Graphene. *Angew. Chem. Int. Ed.* **2018**, *58*, 1324–1328. [CrossRef] [PubMed]
24. Nan, D.; Li, X.; Liu, Q.; Wang, B.; Gao, X.; He, N.; Xu, Y.; Liu, J. Epoxy primer topcoat syncretic coating prepared by using modified graphene oxide incorporated with TiO_2 nanoparticles. *Mater. Today Sustain.* **2023**, *21*, 100282. [CrossRef]
25. Nan, D.; Li, X.; Li, D.; Liu, Q.; Wang, B.; Gao, X.; Ma, T.; He, N.; Xu, Y.; Dong, J. Preparation and Anticorrosive Performance of Waterborne Epoxy Resin Composite Coating with Amino-Modified Graphene Oxide. *Polymers* **2022**, *15*, 27. [CrossRef] [PubMed]
26. Li, X.; Li, D.; Chen, J.; Huo, D.; Gao, X.; Dong, J.; Yin, Y.; Liu, J.; Nan, D. Melamine-Modified Graphene Oxide as a Corrosion Resistance Enhancing Additive for Waterborne Epoxy Resin Coatings. *Coatings* **2024**, *14*, 488. [CrossRef]
27. *GB/T 13525-1992*; Test Method for Tensile-Impact Property of Plastics. State Technical Supervision Bureau: Beijing, China, 1992.
28. *GB/T 6569-2006*; Fine Ceramics (Advanced Ceramics, Advanced Technical Ceramics)—Test Method for Flexural Strength of Monolithic Ceramics at Room Temperature. Standardization Administration of China: Beijing, China, 2006.
29. Pourhashem, S.; Vaezi, M.R.; Rashidi, A.; Bagherzadeh, M.R. Exploring corrosion protection properties of solvent based epoxy-graphene oxide nanocomposite coatings on mild steel. *Corros. Sci.* **2017**, *115*, 78–92. [CrossRef]
30. Wan, Y.-J.; Tang, L.-C.; Gong, L.-X.; Yan, D.; Li, Y.-B.; Wu, L.-B.; Jiang, J.-X.; Lai, G.-Q. Grafting of epoxy chains onto graphene oxide for epoxy composites with improved mechanical and thermal properties. *Carbon* **2014**, *69*, 467–480. [CrossRef]

coatings

Article

Effective Promotion of Micro Damping of GO Hybrid PU–PF Copolymer Grinding Wheels on Precision Machining

Shaoling Xia [1,*], Hongying Zhang [1], Jixian Xu [2], Yingliang Liu [3], Cong Liu [1], Shengdong Guo [1], Xudong Song [1,*], Jin Peng [1], Yu Jia [1] and Jialu Li [1]

1 School of Materials Science and Engineering, Henan University of Technology, Zhengzhou 450001, China; m13253392659_1@163.com (H.Z.); 15294763611@163.com (C.L.); 17549213027@163.com (S.G.); jin_peng@haut.edu.cn (J.P.); jiayu@stu.haut.edu.cn (Y.J.); 221071000230@stu.haut.edu.cn (J.L.)
2 Langfang Shengsen Grinding Tools Co., Ltd., Langfang 065900, China; 13513012175@163.com
3 School of Materials Science and Engineering, Zhengzhou University, Zhengzhou 450052, China; liuylxn@zzu.edu.cn
* Correspondence: shaoling_xia@haut.edu.cn (S.X.); xudong_song@haut.edu.cn (X.S.); Tel./Fax: +86-371-6775-8721 (S.X.)

Abstract: The influence of damping and friction performance of grinding wheels on precision grinding was explored for the first time. GO hybrid PU-modified PF copolymers were prepared by in situ synthesis and adopted as a matrix for fabricating grinding wheels. FT-IR, DSC, TG, and mechanical property tests showed the optimal modification when PU content was 10 wt% and GO addition was 0.1 wt%. Damping properties were investigated by DMA, and tribological characteristics were measured by sliding friction and wear experiments. The worn surfaces and fracture morphologies of GO hybrid PU–PF copolymers were observed by SEM. Distribution of components on the worn surfaces was explored by Raman mapping and EDS. The research results revealed that the PU component tended to be dispersed around the edges of corundum abrasives acting as a buffer layer of abrasive particles, which could provide micro-damping characteristics for abrasives, making the grinding force more stable during precision machining and facilitating a smoother surface quality of the workpiece.

Keywords: damping; tribological characteristics; graphene oxide; precision machining

Citation: Xia, S.; Zhang, H.; Xu, J.; Liu, Y.; Liu, C.; Guo, S.; Song, X.; Peng, J.; Jia, Y.; Li, J. Effective Promotion of Micro Damping of GO Hybrid PU–PF Copolymer Grinding Wheels on Precision Machining. *Coatings* **2024**, *14*, 632. https://doi.org/10.3390/coatings14050632

Academic Editor: Alexander Tolstoguzov

Received: 17 April 2024
Revised: 13 May 2024
Accepted: 13 May 2024
Published: 16 May 2024

1. Introduction

Precision machining is an important link in the development of contemporary science [1] that plays a crucial role in contemporary high-tech fields and military and civilian industries, especially in the field of electrical automation, such as ultra-large-scale integrated circuits, high-precision disks, precision radars, missile fire control systems, precision instruments, optical lens, copier drums, etc. [2–7]. Its development trend has been from micrometer and submicron scales towards nanoscale processing technology [8–10]. In order to achieve satisfactory machining accuracy, machine tools, especially grinding wheels, need to have excellent dynamic performance and chatter avoided as much as possible [11,12]. Chatter in machining processes is strongly dependent on the dynamic compliance behavior of the machine tool and workpiece. The critical cutting depth where chatter occurs is in inverse proportion to the absolute value of the negative real part of the complex dynamic compliance response function of the machine tool. Therefore, while designing a machine tool, active damping systems should be designed to help avoid chatter of the machine tool [13]. Numerical structural models were established to analyze the contribution of vibration damping to machine tools [14]. Some scholars increased damping property by mounting the abrasive rim of the wheel via a flexible coupling or by modifying the flexibility of the grinding wheel shaft device [15–17]. These attempts have been successful, and the grinding accuracy and surface quality of the workpiece have been further improved. Since good damping characteristics have been proven to be beneficial for improving machining results [18–23], in theory, improving the damping performance of the grinding wheel itself could further improve the precision machining effect.

With excellent damping property, polyurethane (PU) has attracted much interest from researchers for use in damping and vibration isolation [24–28]. It was widely explored as a modifier to improve the damping performance of other polymer materials, such as epoxy [29–34], polystyrene [35], poly(tetramethylene glycol) [36], vinyl ester resin/ethyl acrylate [37], poly(vinylidene fluoride) [38], and so on. These studies achieved satisfactory results, but there have been few reports on enhancing damping properties of phenol formaldehyde resin (PF) by PU modification. As is well known, due to its excellent strength, rigidity, heat resistance, wear resistance, and low cost, PF is widely used as a matrix material for grinding tools. The drawback of PF grinding tools is the lack of damping, which is not conducive to obtaining fine surface quality of a workpiece in precision grinding. We attempted to improve the machining effect of the grinding wheel by increasing the damping property of PF through PU modification.

On the other hand, the formation of a self-lubricating layer by adding lubricants can reduce friction between the grinding wheel and workpiece, improving the surface quality of a workpiece [39]. As far as lubricants are concerned, graphene oxide (GO) cannot be ignored. The superior mechanical properties and high thermal conductivity of GO led to its excellent tribology applications in polymer nanocomposites [40,41]. The atomically smooth surface and weak van der Waals force between the GO layers eased interlayer sliding and contributed to its self-lubrication characteristics [42,43]. Furthermore, our previous work displayed that a 2D structure of GO with high specific surface area was also good for load transferring [44,45].

In this paper, in order to enhance damping and vibration properties, PU-modified PF copolymers were synthesized as the matrix of grinding wheels. GO was also introduced into the system to improve the tribology and grinding performance of materials. We conducted comprehensive characterization and tests of the GO hybrid PU–PF copolymers and conducted precision grinding experiments, which showed excellent machining results. This paper focuses on exploring the copolymers' damping characteristics, friction performance, and their impact mechanism on precision machining.

2. Experimental

2.1. Materials

Pre-polyurethane (PPU) with NCO content of 4% was supplied by Henan Institute of Chemistry. Phenol was obtained from Chengdu Kelong Chemical Co., Ltd. (Chengdu, China), formaldehyde (37%) and polyformaldehyde were supplied by Xilong Science Co., Ltd. (Chengdu, China), and hydrochloric acid (HCl) was purchased from Henan X.Z.C Reagent Ltd. (Zhengzhou, China). Dibutyltin dilaurate was produced by Ningxia Lingshi New Material Technology Co., Ltd. (Ningxia, China). All the materials were of analytical purity and used without purification.

2.2. In Situ Synthesis of GO Hybrid PU–PF Copolymers

PU–PF copolymer was firstly synthesized as shown in Figure 1A. PPU, phenol, and dibutyltin dilaurate catalysts were added to a flask and heated to 76 °C for 2 h. Then, the temperature was reduced to 60 °C, formaldehyde was added to the flask (the mole ratio of phenol to formaldehyde was 7:6), and a 25% HCl aqueous solution was added to the mixed solution four times. The interval each addition was 1 h, and the reaction temperatures were 75 °C, 80–85 °C and 95 °C, each temperature lasting for 2 h. After the reaction reached the end point, the product was put into a vacuum-drying oven and dried at 120 °C for 2 h. After cooling, a light-yellow translucent solid was obtained. This procedure has been proved to be effective in our previous research [46]. In that paper, a series of PU–PF copolymers with PU content of 0 wt%, 5 wt%, 10 wt%, 15 wt%, and 20 wt% were synthesized. Among them, 10 wt% content displayed the optimal mechanical properties, such as hardness, bending strength, and impact strength. As such, in this study, we selected 10 wt% PU content in the PU–PF copolymer as the matrix of GO-hybridized copolymer.

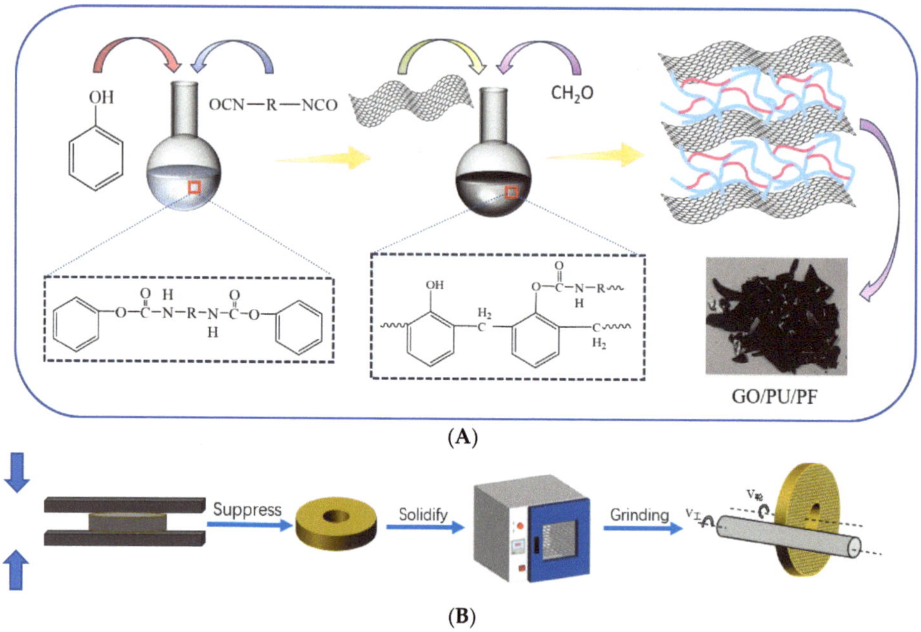

Figure 1. Scheme of synthesis of GO hybrid PU–PF copolymer (**A**) and fabrication of grinding wheels (**B**).

As for the preparation of GO hybrid PU–PF copolymers, GO was synthesized from graphite through a modified Hummers method according to our previous procedure [44,45] and kept in deionized water as a suspension form. The GO suspension was directly mixed with formaldehyde, and the other experimental procedures were the same as described above. A series of GO hybrid PU–PF copolymers with different GO content of 0.05 wt%, 0.1 wt%, 0.15 wt%, and 0.2 wt% were synthesized. The resultant GO hybrid PU–PF copolymer was black powder. Polyformaldehyde was added to the resultant powder as curing agent.

Grinding wheels, as well as samples for mechanical and DMA testing, were fabricated as shown in Figure 1B. GO hybrid PU–PF copolymers, corundum abrasive, and hollow ball filler were mixed into the mold at a volume ratio of 4:5:1, heated to 110 °C in flat vulcanizer, flattened and kept at 180 °C for 30 min, then demolded after natural cooling. Then, the samples were secondary cured at 120 °C, 140 °C, 160 °C and 180 °C for 2 h.

2.3. Testing Methods

FT-IR (Shimadzu, IR Prestige-21, Kyoto, Japan) was employed to study the chemical structure of GO hybrid PU–PF copolymer at a resolution of 0.5 cm^{-1}. TG (Peking Optical Instrument Factory, WCT-2, Beijing, China) and DSC (NETZSCH, 200F3, Selb, Germany) were tested with a heating rate of 10 °C/min in air. Dynamic thermomechanical analysis (DMA, NETZSCH, DMA242C, Selb, Germany) was used to investigate the dynamic mechanical properties using the compassing mode at a frequency of 45 Hz and at a heating rate of 3 °C/min. Tensile strength and elongation were tested by a tensile instrument (Jinan Faen Instrument Factory, WDW-S, Ji'nan, China) according to GB/T 528-1998 [47]. The bending strength was tested by 3-point support mode. Tribology characteristics was studied by a pin-on-disk friction wear testing machine (Lanzhou Zhongke Co., Ltd., Lanzhou, China,

QG-700) at room temperature in dry conditions. The counterpart pin is fabricated by steel with a diameter of 5 mm. The normal load and sliding speed were 0.4 MPa and 0.5 m/s, respectively. Experimental data were collected during 60 min of wearing for disk-like samples with a diameter of 60 mm and a thickness of 5 mm. Precision machining experiments were carried out using a CNC cylindrical grinder (MK1320C-500, Wuxi Yulin Machine Tool Co., Ltd., Wuxi, China). The worn surfaces and fracture morphologies of GO hybrid PU–PF copolymers were observed by SEM (Inspect F50, Thermo Fisher Scientific, Waltham, MA, USA). Distribution of components in grinding wheels was revealed by energy-dispersive spectroscopy (EDS). Chemical changes on the worn and unworn surfaces were explored via Raman mapping (Thermo Fisher Scientific DXRxi Raman imaging spectrometer) at a laser wavelength of 633 nm with laser energy of 4.3 MW. The wave-number range of Raman spectra was 0–3000 cm^{-1} with a spectral resolution of 1 cm^{-1}.

3. Results and Discussion

3.1. Fabrication of GO Hybrid PU–PF Copolymers

GO was synthesized by the same method as our previous research, and its characterization results are given in papers published by our research group [44,45]. The GO hybrid PU–PF copolymers prepared in this study were characterized by FI-IR, TG, and DSC. Mechanical properties were also tested. The results are shown in Figure 2.

The FT-IR spectra of pure PF, PPU, GO, and GO hybrid PU–PF copolymer are shown in Figure 2A. In the GO hybrid PU–PF copolymer spectra, strong C-H and -C=O stretching vibration peaks in PU chains appeared at 2943 cm^{-1}, 2860 cm^{-1} and 1708 cm^{-1}, respectively, but the stretching band of the isocyanate group (NCO) at 2274 cm^{-1} disappeared on the spectra of GO hybrid PU–PF copolymer, implying the -NCO groups in PU completely reacted with -OH groups in phenol.

The thermal stability of resulting nanocomposites was investigated by TG. The TG curves of PU–PF (a), GO hybrid PU–PF copolymers with different GO weight content 0.05 wt% (b), 0.1 wt% (c), 0.15 wt% (d), 0.2 wt% (e), and GO (f) are shown in Figure 2B. Data on characteristics are summarized in Table 1. It is obviously in Figure 2B that the curves moved to the right with the increase in GO content, which meant an increase in thermal stability. Both 20% and 50% weight-loss temperature were significantly increased with the increase GO content. The semi-disassembly temperatures of GO hybrid PU–PF copolymers with GO content of 0 wt%, 0.05 wt%, 0.1 wt%, 0.15 wt%, and 0.2 wt% were 528 °C, 543 °C, 565 °C, 580 °C, and 595 °C, respectively. This was attributed to strong intermolecular interactions, such as hydrogen bonds or van der Waals forces, between PF, PU, and GO, also acting as cross-linking points that could limit the movement of chain segments and consequently enhance the thermal stability.

Table 1. Temperatures of 20% and 50% weight loss in PU–PF (GO 0 wt%) and GO hybrid PU–PF copolymers with different GO content.

Weight-Loss Rate	GO wt%				
	0	**0.05**	**0.1**	**0.15**	**0.2**
20%	405	428	424	486	482
50%	528	543	565	580	595

The DSC curves of GO hybrid PU–PF copolymers with different GO content are shown in Figure 2C. Melting peaks can be seen below 100 °C. The peaks that appear between 140 °C and 170 °C were considered to be curing reaction peaks. Table 2 shows the temperature corresponding to the lowest point of the peak in Figure 2C, that is, the curing temperature. As can be seen from Figure 2C and Table 2, when GO content was below 0.1 wt%, the curing temperature increased with the increase in GO content, but they tended in the opposite direction when GO content were more than 0.1 wt%. The curing temperature reached the highest value of 159 °C when the GO content was 0.1 wt%. The reason was conjectured as following: when GO content were below than 0.1 wt%, GO pieces could be evenly intercalated between the PF and PU molecular chains, which resulted in the enlargement of the space between PF and PU molecular chains, and the distance between the reaction groups became further apart. What's more, the GO intercalation could also inhibit the movement of polymer molecular chains, so that the cross-linking reactions were more difficult to generate, resulting in the increment of curing temperature. When the GO content continued to increase, GO concentration became too high, resulting in agglomeration and accumulation, so that it could not be evenly intercalated between the polymer chains and begin to produce phase separation with the polymers, so the influence on the distance between the reaction groups and the hindrance of the molecular chains was weakened, and the cross-linking reaction was easier to occur, so the curing temperature began to decrease with further increase in GO content.

Table 2. Curing temperatures of PU–PF (GO 0 wt%) and GO hybrid PU–PF copolymers with different GO content.

GO Content	0 wt%	0.05 wt%	0.1 wt%	0.15 wt%	0.2 wt%
Curing temperatures	148 °C	150 °C	159 °C	155 °C	154 °C

Mechanical properties test also displayed the same tendency. Figure 2D showed bending strength and impact strength of GO hybrid PU–PF copolymers with different GO content. Both bending strength and impact strength reached the maximum when the GO content was 0.1 wt%. Compared to the PU–PF copolymer without the GO hybrid, bending strength was enhanced by 14.8% and impact strength was enhanced by 18.8%. This phenomenon confirms our previous suspicions. We believe that 0.1 wt% GO content was optimal for the even intercalation of GO pieces in the PU–PF copolymer chains. The unique layer structure of GO and the molecular interaction between GO and copolymer chains effectively transferred the stress, reduced distortion and fracture, and thus improved the mechanical properties, but too much GO hybrid caused agglomeration, resulting in stress concentration, which led to the reduction in mechanical properties.

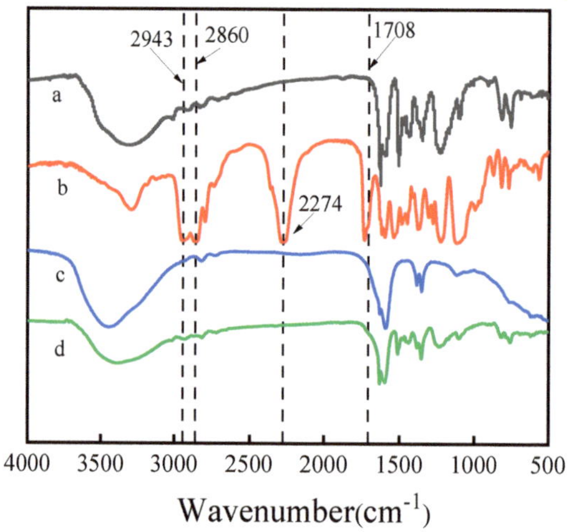

(A)

Figure 2. *Cont.*

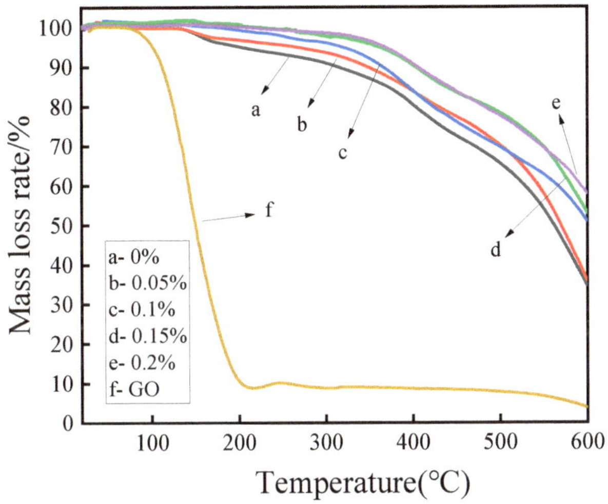

(B)

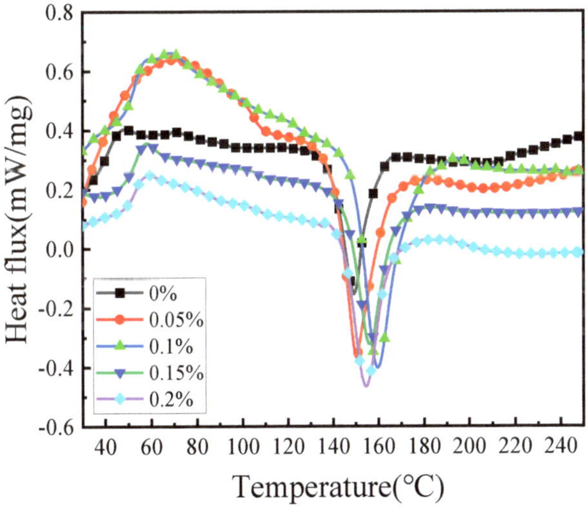

(C)

Figure 2. *Cont.*

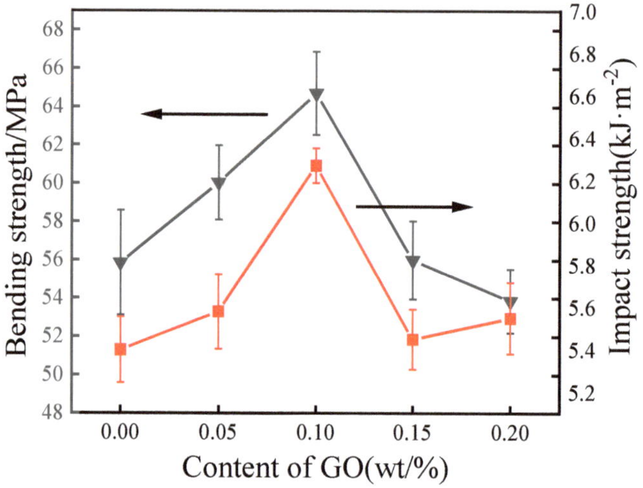

(D)

Figure 2. FT–IR spectra (**A**) of pure PF (a), PPU (b), GO (c) and GO hybrid PU–PF copolymer (d), TG (**B**), DSC (**C**) and bending/impact strength (**D**) of resultant materials.

3.2. Damping Property

DMA was detected at a frequency of 45 Hz, which corresponded to 35 m/s rotational speed of the grinding wheel during precision machining. As shown in Figure 3A, the rectangular shaded area lay in the range of the precision machining temperature. It can be seen in this area that there is a huge peak on the curve of PU (a), which is completely nonexistent on the curve of PF (c). But on the curve of PU–PF copolymer (b) that contained 10 wt% PU, a small peak appears. This peak resulted from the glass transition of PU, which meant some segments on PU were able to move gradually. This result indicates that the PU ingredients endowed the PU–PF copolymer with micro damping properties.

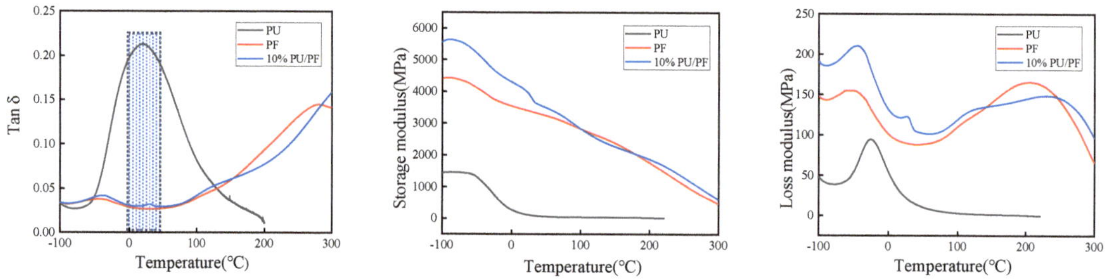

(A) (B) (C)

Figure 3. Tan δ (**A**), energy storage modulus (**B**), and loss modulus (**C**)—temperature curves of PU, PF, and PU–PF copolymer.

Figures 3B and 3C respectively, display the energy storage and loss modulus–temperature curves of PU, PF, and PU–PF copolymer. The two graphs show the same phenomenon in that both storage modulus and loss modulus of the copolymer were greater than that of each homopolymer at room temperature. This meant that both elasticity and viscosity of the copolymer had been significantly improved.

For different GO-content samples of GO hybrid PU–PF copolymers, it can be seen in Figure 4A that glass transition peaks almost appear at room temperature, but no glass transition peak appeared on the curve of 0.2 wt% GO-content PU–PF copolymer. To explore the reason, we checked their energy storage and loss modulus–temperature curves (Figure 4B,C). It was found that an energy storage modulus of the 0.2 wt% GO-content sample was the maximum in all the GO content samples tested, but its loss moduli were lower than that of 0.1 wt% and 0.15 wt% GO-content samples. Since Tanδ is a quotient of loss modulus and storage modulus, the Tanδ of 0.2 wt% GO content sample was reduced. It was presumed that a large number of GO pieces intercalated in the copolymer chains were able to transfer and store energy well, but GO sheets that were too rigid stacked the links of the polymer and hindered the movement of the segments.

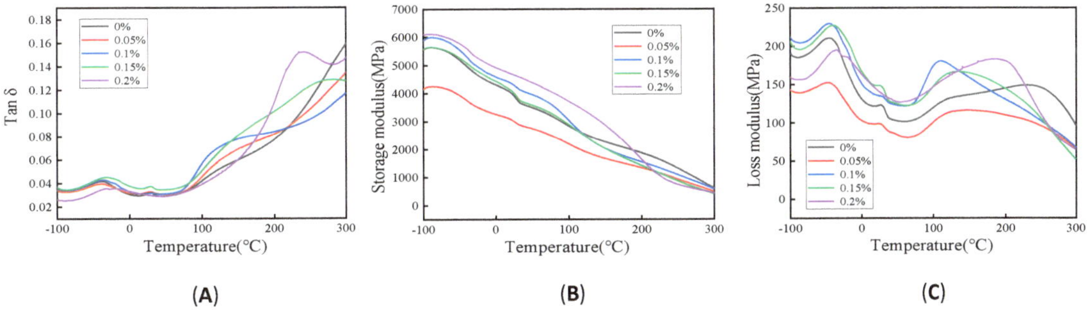

Figure 4. Tan δ (**A**), storage modulus (**B**), and loss modulus (**C**)—temperature curves of GO hybrid PU–PF copolymers with different GO content.

In Figure 4A, the difference in Tanδ is very small at room temperature, but was significantly enlarged in the 100–150 °C zone. The maximum belonged to the sample of 0.1 wt% GO content, and the Tanδ values of samples of 0.15 wt%, 0.05 wt%, 0 wt%, and 0.2 wt% decreased sequentially. In fact, at this temperature range, the copolymers began to soften and partial areas to melt, so some segments and molecules obtained enough energy to move. Figure 4C indicates that the loss modulus of the 0.1 wt% GO-content sample increased more significantly than other samples in 100–150 °C range, due to the homogeneous dispersion of the GO pieces between the molecular chains creating a strong intermolecular force that hindered the thermal movement of the molecular chains. By comparison, we found the change trend in Tanδ of different GO content in the range of 100–150 °C was similar to that of DSC and mechanical properties. The experimental results were mutually confirmed.

3.3. Tribological Character

Dry sliding behavior of PU, PF, and PU–PF with 10 wt% PU content and the GO (0.1 wt%) hybrid PU (10 wt%)–PF copolymer during friction are displayed in Figure 5. It can be seen that the PU curve exhibits great noise, The curve of the PU–PF copolymer is very smooth, and the curves of PF and GO hybrid PU–PF copolymers are a little rough. Average friction coefficients (COFs) were calculated and are listed in Table 3. The COF of the PU–PF copolymer was 0.58, which lay in an intermediate position between pure PF (0.3) and PU (1.03). With the aim of understanding the tribology mechanism, the worn surfaces of sliding friction samples of PU, PF, PU–PF, and GO hybrid PU–PF copolymers were detected by SEM, as shown in Figure 6. Figure 6a shows the morphology of the worn surface of pure PU. It can be seen that the PU surface was damaged homogeneously. Only some slim ridges were left when the counterpart was moved away. Signs of melting can be seen on the edges of ridges. Evidently, the most of material on the surface has been worn off and melt under frictional heat. The friction mechanism of PU was classified as adhesion wear, since the surface was mainly damaged by viscous tearing. The friction force was very unstable during the friction process, which could explain the great noise in the PU sliding curve, while there were lots of fish scale-like protrusions on the worn surface of pure PF in Figure 6b, with slight peeling of the epidermis. Owing to outstanding thermal stability, stiffness, and mechanical strength, the surface of PF resin was not easily destroyed. Only on some microscopic protrusions was frictional resistance increased, leading to some surface layers being scraped off. The fluctuation in frictional resistance was deemed to be the reason that the sliding curve of PF in Figure 5 was relatively rough. Above all, the friction mechanism of PU belongs to abrasive and fatigue wear mechanisms. On the worn surface of the PU–PF copolymer in Figure 6c, more severe surface detachment has occurred, and some relatively deep furrow traces appear in the detachment area. This phenomenon was caused by the soft–hard phase-combination structure of PU–PF on the surface, where the PF phase formed the hard region and the PU phase formed the soft region. When the surface was exposed to the counterpart pressure, the PU soft region was depressed via elastic deformation, while the PF hard region could not deform, so the hard region was ploughed into the soft region by the counterpart pin. The situation was changed by the addition of GO. As displayed in Figure 6d, the worn surface of the GO hybrid PU–PF copolymer appeared relatively intact, with only some shallow scratches appearing. Upon investigation, we believed that since the hybrid of GO enhanced the hardness and strength of the whole PU–PF copolymer, the PU soft region was also reinforced, so its surface resistance to friction damage was stronger. There were also lots of fish scale-like protrusions on it, which induced uneven frictional resistance resulting in a relatively rough sliding curve of the GO hybrid PU–PF copolymer in Figure 5.

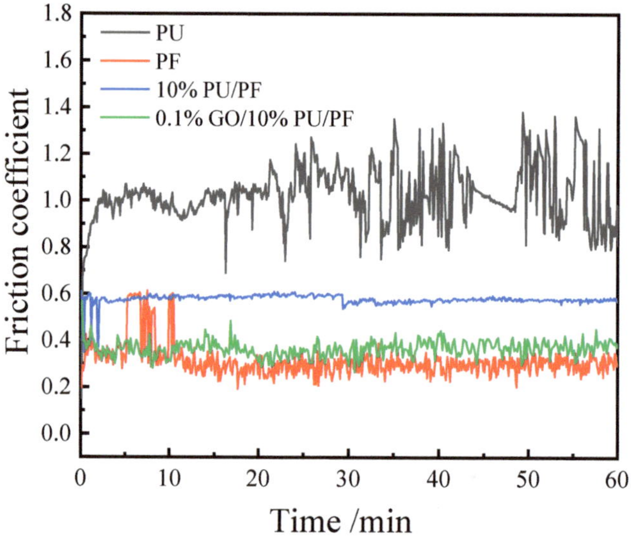

Figure 5. Sliding behavior curves of PU, PF, and PU–PF copolymers during wearing process.

Table 3. COFs of PU, PF, and PU–PF and GO hybrid PU–PF copolymers.

Resin Type	PU	PF	10%PU–PF	0.05%GO + 10%PU–PF	0.1%GO + 10%PU–PF	0.15%GO + 10%PU–PF	0.2%GO + 10%PU–PF
COFs	1.03	0.3	0.58	0.42	0.37	0.4	0.39

The dry sliding behavior of GO hybrid PU–PF copolymers with different GO content is illustrated in Figure 7, with the COF curve in the inset. This indicates that COFs of the copolymers were reduced by GO hybridization. When GO addition was 0.1 wt%, the COF was at its minimum of 0.37. The specific wear rate (listed in Table 4) revealed the same tendency with COFs. Likewise, the 0.1 wt% GO-content sample showed the lowest value

(1.16×10^{-4} mm^3/Nm), which was reduced by 73.5% compared with that of the PU–PF copolymer without GO hybrid (4.38×10^{-4} mm^3/Nm).

Table 4. Specific wear rate of PU, PF, and PU–PF and GO hybrid PU–PF copolymers.

Resin Type	PU	PF	10%PU–PF	0.05%GO + 10%PU–PF	0.1%GO + 10%PU–PF	0.15%GO + 10%PU–PF	0.2%GO + 10%PU–PF
Specific wear rate (10^{-4} mm^3/Nm)	53.44	3.13	4.38	2.77	1.16	1.7	1.89

An interesting phenomenon can be observed in Figure 7 where all the sliding curves of GO hybrid PU–PF copolymers converge together independently of GO content, indicating almost the same COF. This result was consistent with our previous research, and the reasons for this are analyzed in detail in our previous published papers [44,45]. The most important observation was that the identical steady friction coefficients of GO hybrid PU–PF copolymers with different GO contents were derived from the same friction bodies in the three-body friction model, which were composed of the GO-strengthened PU–PF copolymer surface, the metallic counterpart, and the GO-wrapping polymer particles as the wear debris in the transfer film.

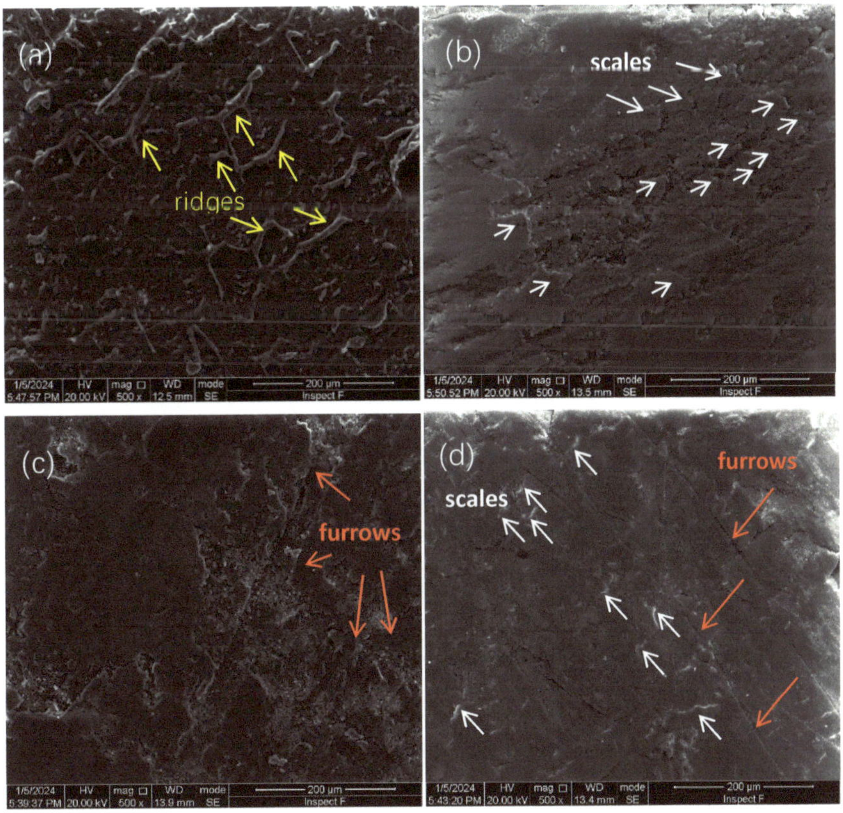

Figure 6. SEM pictures of worn surfaces of PU (**a**), PF (**b**), and PU–PF (**c**) and GO hybrid PU–PF copolymers (**d**). (yellow arrows for ridges, white arrows for scales, red arrows for furrows).

To verify the observations above, the substance distribution on the worn surface of the PU–PF copolymers with and without GO hybrid samples was tested by Raman mapping, as shown in Figure 8. Both tests were focused on the interface between friction and non-friction areas. The two images present two completely different phenomena. There is a uniform picture on the surface of PU–PF copolymers without GO hybrid in Figure 8A, with no difference between the friction surface and non-friction surface. A small number of PU components (blue dots) are evenly distributed in the PF matrix (red area). However, it is completely different on the surface of PU–PF copolymers with GO hybrid in Figure 8B: the components on the friction and non-friction surfaces are clearly different. The friction area shows a large amount of GO component coverage. This result was consistent with our previous research findings on GO hybrid PU–EP IPNs [44]. In the GO hybrid complex, the GO components gathered on the friction surface to form a GO transfer film after a period of friction, which meant the friction behavior changed from dry friction to boundary friction. That could be another explanation for why the COFs of composites with different GO content were very close.

By comparing the frictional performance of the GO hybrid PU–PF materials with those of materials (GO hybrid PU–EP) in the literature [44,45], similar results were obtained, which has beneficial theoretical value for the promotion and application of GO hybrid composite materials in the field of friction and wear. Due to its superior strength and heat resistance, the GO hybrid PU–PF material was more suitable for precision grinding of difficult-to-machine metal materials such as bearing steel than the GO hybrid PU–EP materials, so in this study, we explored the precision machining property of GO hybrid PU–PF materials.

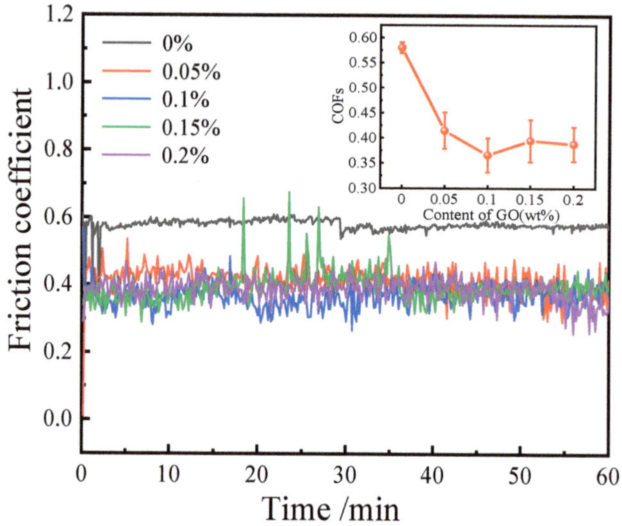

Figure 7. Sliding behavior curves (inset: COFs) of GO hybrid PU–PF copolymers with different GO content during wearing process.

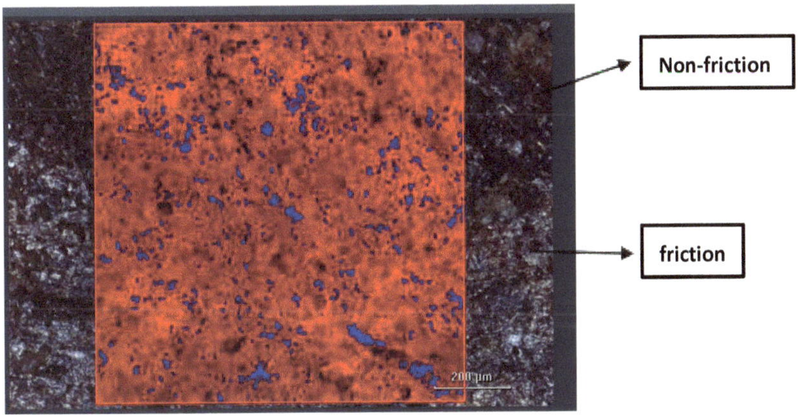

(A)

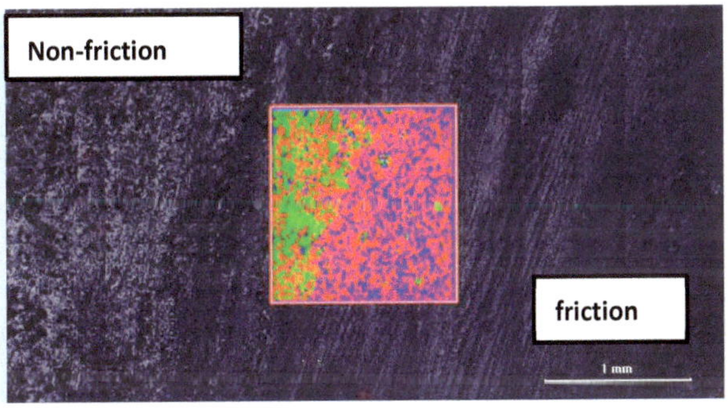

(B)

Figure 8. Raman mapping images of PU–PF copolymers without (**A**) and with (**B**) GO hybrid sample surfaces before and after friction.

3.4. Precision Machining

Three types of grinding wheels were manufactured using pure PU, PU–PF, and GO hybrid PU–PF as matrix material. Corundum was applied as abrasive, and for each matrix type, three corundum particle sizes of w14, w10, and w5 were selected to manufacture grinding wheels, respectively. Therefore, a total of nine grinding wheel samples were prepared and machining tests were conducted on high-carbon chromium bearing steel GCr15 workpieces. The surface roughness (Ra) of the workpieces before and after machining was measured using white-light interferometry, as shown in Figure 9. The resultant values of surface roughness (Ra) are shown in Figure 10A. It was found that no matter which particle-size grinding wheel was used for machining, the Ra of the workpiece was significantly reduced. The finer the particle size of corundum abrasive, the lower the surface roughness of the workpiece. For grinding wheels of the same particle size, GO hybrid PU–PF grinding wheels had the lowest Ra, followed by PU-modified PF. This was consistent with our prediction. PU modification improved the damping and vibration reduction performance of the grinding wheel matrix material PF. During the machining process, the friction force between the grinding wheel and the workpiece was more uniform and stable, resulting in uniform removal of micro-roughness on the workpiece surface, resulting in lower surface roughness of the workpiece. The hybridization of GO further improved the damping characteristics of the grinding wheel substrate material, and formed a transfer film between the tool and the workpiece during the machining process, which further reduced the grinding force and was more conducive to obtaining high-quality surfaces.

After grinding with the GO hybrid PU–PF grinding wheel with a corundum particle size of w5, the Ra value of the high-carbon chromium bearing steel GCr15 workpiece was reduced to 29 nm, which was 35% lower than that of pure PF grinding wheels.

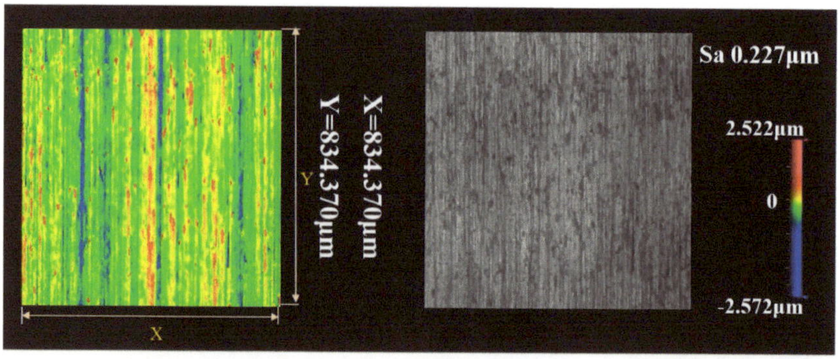

(A)

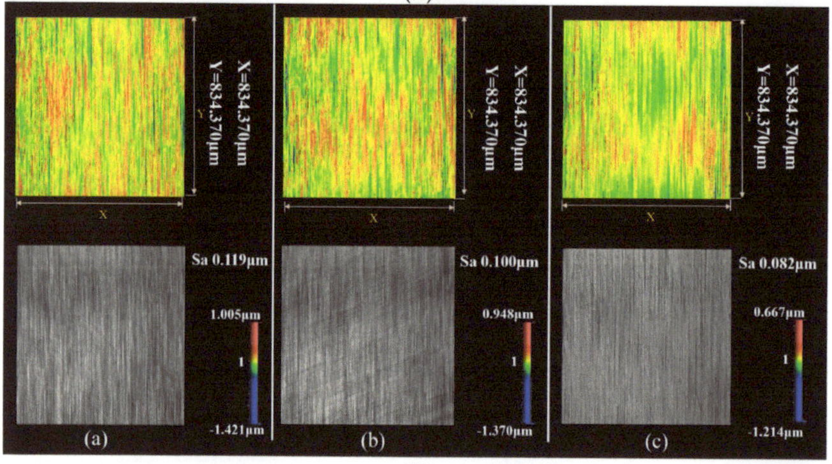

(B)

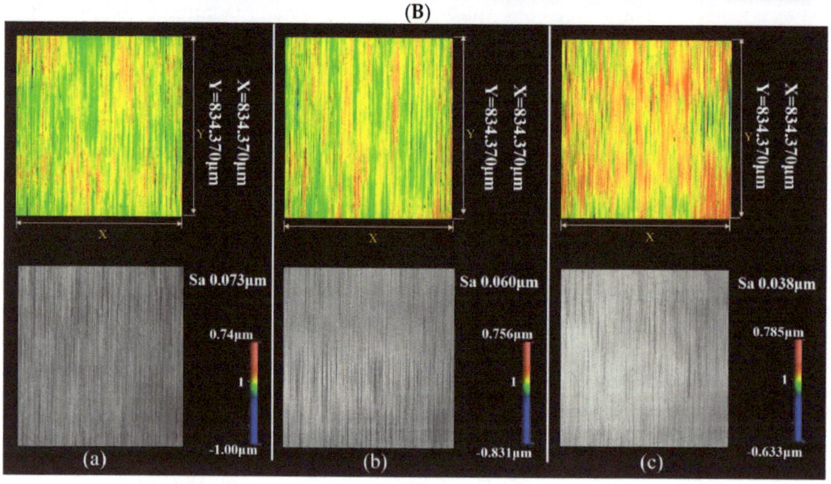

(C)

Figure 9. *Cont.*

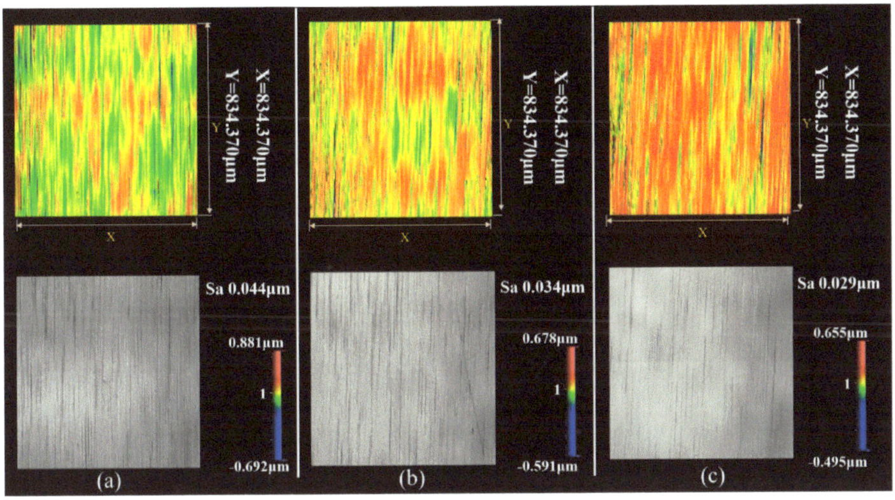

(**D**)

Figure 9. White-light interference pattern on the surface of workpieces before (**A**) and after machining by grinding wheels made of pure PF (**a**), PU–PF (**b**), and GO hybrid PU–PF (**c**) with corundum particle sizes of w14 (**B**), w10 (**C**), and w5 (**D**).

In Figure 10B, the grinding sharpness of grinding wheels made of pure PF, PU–PF and GO hybrid PU–PF copolymers with different corundum particle sizes (w14, w10, and w5) are also evaluated. During the machining process, the reduction in diameter of axial workpieces per unit time was used to measure the sharpness of the grinding wheel. The results showed that the sharpness of both grinding wheels made of PU–PF and GO hybrid PU–PF were much higher than that of pure PF, and the sharpness of the grinding wheel made of PU–PF was the maximum.

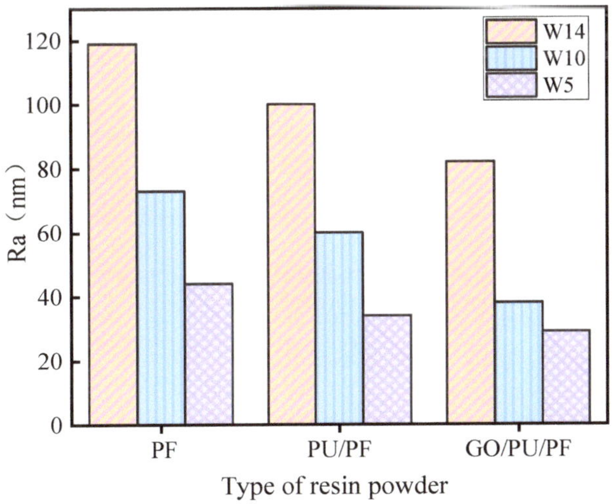

(**A**)

Figure 10. *Cont.*

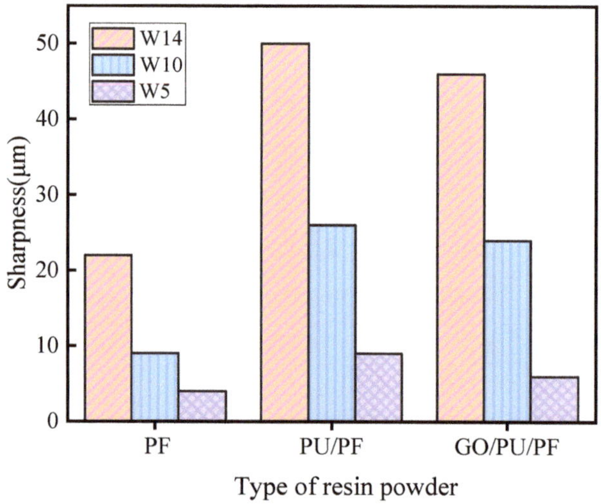

(B)

Figure 10. Surface roughness (Ra) of workpieces (**A**) and grinding sharpness (**B**) machined by grinding wheels made of pure PF, PU–PF, and GO hybrid PU–PF copolymers with different corundum particle sizes (w14, w10, and w5).

In order to explore the mechanism for improving the machining effect, SEM and EDS mapping was conducted on the grinding surface of the grinding wheel with corundum particle sizes of w14, as shown in Figure 11. By comparing with the images on the left, it can be observed that there are significant differences in the grinding surfaces of the three types of grinding wheels. For the grinding wheel made of pure PF, the grinding surfaces look relatively flat and smooth. The grinding wheel of PU–PF appears somewhat uneven, with many protrusions of corundum particles on the surface looking like mountain ridges. For that of GO hybrid PU–PF, it became rougher with many micro-gaps, and the EDS mapping graph (Figure 11(Cb)) shows that the corundum abrasive was more evenly wrapped by the resin. The formation cause could be very complex. It may be related to the viscosity and interactions between components. We will carry out further research specifically on this issue. However, the morphology described above could well explain the difference in sharpness of the grinding wheels. The mountain ridges formed by corundum particle edges just like blades cut the workpiece sharply, allowing the substance on the surface of the workpiece to be quickly ground off. As for the slightly lower sharpness of GO hybrid PU–PF grinding wheels compared to that of PU–PF, the reason was speculated to be that the lubricating effect of GO reduced the grinding force to a certain extent, which resulted in the reduction in sharpness.

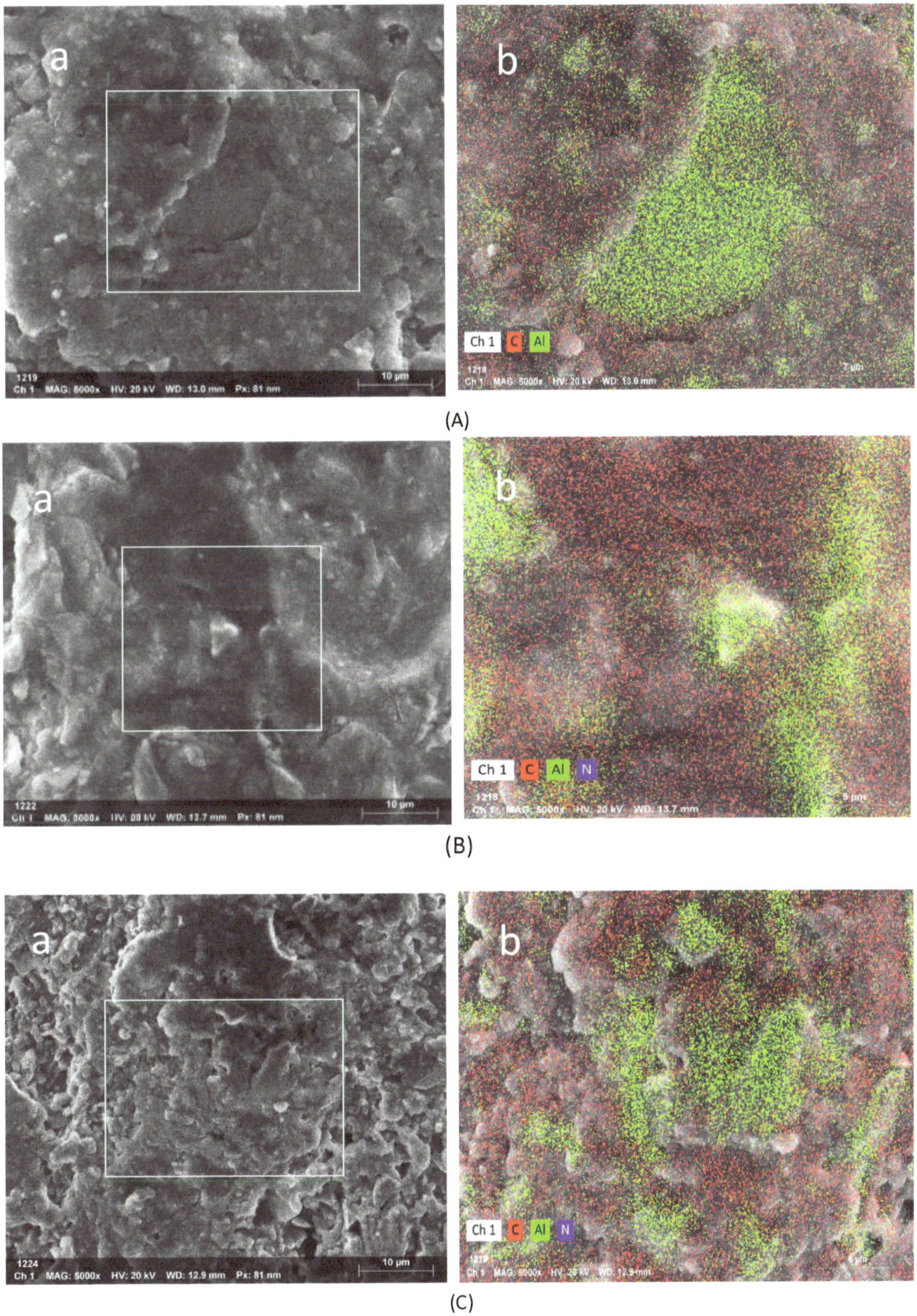

Figure 11. EDS pictures of grinding wheels made of pure PF (**A**), PU–PF (**B**), and GO hybrid PU–PF (**C**). (graph **b** is the energy mapping of the white box area in graph **a**).

Figure 12 displays Al and N element distribution maps of grinding wheels made of PU–PF and GO hybrid PU–PF. The N element came from -NCO groups in PU chains, and Al from the corundum abrasive (Al_2O_3). For the N element distribution maps in Figure 12(Ab,Bb), one can see upon careful observation that both images show that the N atoms are not uniformly dispersed outside the corundum region, but arranged in a linear pattern. Also, the contour profile matches the edge position of the corundum abrasive to a large extent. After analysis, we speculated that this was caused by the strong intermolecular forces generated from carbonyl and amino groups in the PU molecular chain and the Al-O bond on the surface of the corundum abrasive, since all of them possess strong polarity. As such, the PU component tended to be dispersed around the edges of the corundum abrasive, acting as a buffer layer of abrasive particles, as depicted in Figure 13. This distribution provided micro damping characteristics for the abrasive, which could play a role in buffering the grinding force on the abrasive, making the grinding force on the workpiece more stable during precision machining and thus achieving smoother surface quality of the workpiece. This discovery has great significance and provides a scientific basis for the design and fabrication of precision machining tools.

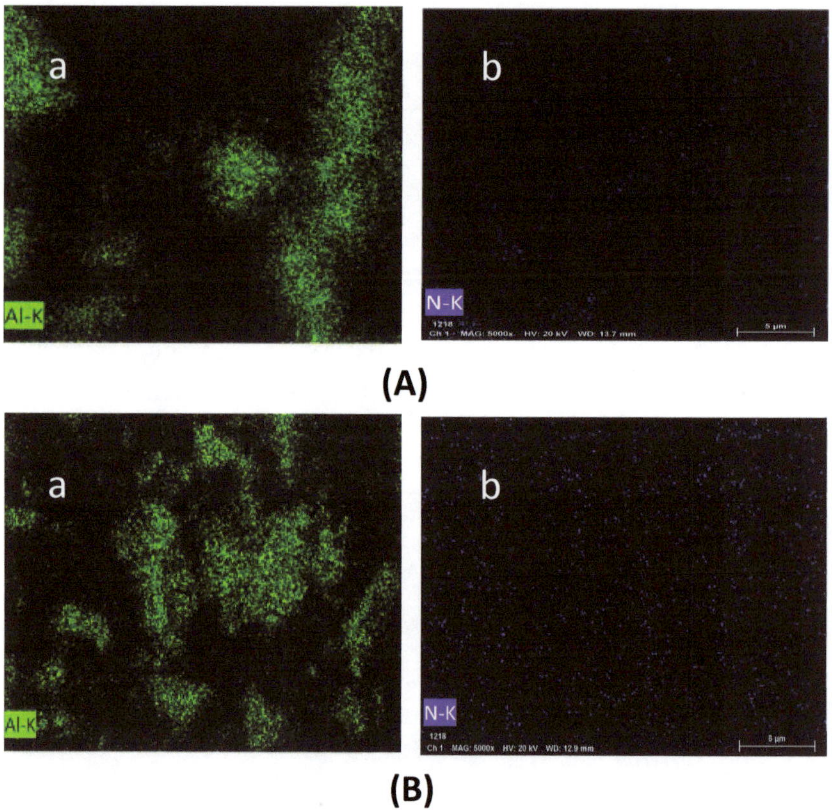

Figure 12. Al (**a**) and N (**b**) element distribution maps of grinding wheels made of PU–PF (**A**) and GO hybrid PU–PF (**B**).

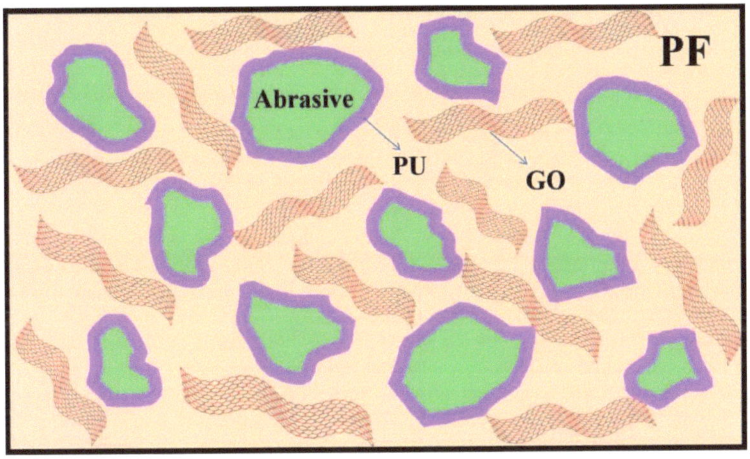

Figure 13. Component distribution diagram of grinding wheels made of GO hybrid PU–PF.

4. Conclusions

To understand the influence of damping and friction performance of grinding wheels on precision grinding, GO hybrid PU-modified PF copolymers were prepared by in situ synthesis. The optimal performance of the copolymer was achieved when the PU addition was 10 wt% and the GO addition was 0.1 wt%. With this ratio, the bending strength of 0.1% GO hybrid PU–PF was increased by 14.8% and the impact strength was increased by 18.8% compared to the PU–PF copolymer without GO hybrid. DMA results displayed small glass transition peaks on the curve of PU–PF and GO hybrid PU–PF copolymers, which meant the PU ingredients endowed the modified PF micro damping properties.

The sliding behavior of the GO hybrid PU–PF copolymers was displayed when GO addition was 0.1 wt% and the COF was at its minimum of 0.37. COFs of the copolymers with different GO content were very similar. Likewise, the 0.1 wt% GO-content sample obtained the lowest value, 1.16×10^{-4} mm^3/Nm, which was 73.5% less than that of the PU–PF copolymer without GO hybrid (4.38×10^{-4} mm^3/Nm).

Grinding wheels were manufactured using pure PU, PU–PF, and GO hybrid PU–PF as matrix material. Machining tests were conducted on high-carbon chromium bearing steel GCr15 workpieces. GO hybrid PU–PF grinding wheels significantly reduced the surface roughness of the workpiece. The sharpness of both grinding wheels made of PU–PF and GO hybrid PU–PF was much higher than that of pure PF. Raman mapping images provided favorable evidence for the formation of GO transfer film during the friction process. EDS mapping revealed the PU component tended to be dispersed around the edges of the corundum abrasive, acting as a buffer layer of abrasive particles. It was supposed to provide micro damping characteristics for abrasives, making the grinding force more stable during precision machining and facilitating smoother surface quality of the workpiece. This discovery provides a scientific basis for the design and preparation of precision grinding tools.

Author Contributions: Conceptualization, S.X.; Methodology, Y.L.; Software, S.G.; Validation, Y.J. and J.L.; Formal analysis, X.S.; Investigation, H.Z.; Data curation, J.P.; Writing—review & editing, C.L.; Funding acquisition, J.X. All authors have read and agreed to the published version of the manuscript.

Funding: This research was funded by Henan Province Science and Technology Research Projects (242102230179), Henan Province Undergraduate Innovation and Entrepreneurship Training Program Project (221071000210), the Key R & D projects of Henan Province (241111233100) and Langfang Shengsen Grinding Tools Co., Ltd.

Institutional Review Board Statement: Not applicable.

Informed Consent Statement: Not applicable.

Data Availability Statement: Data are contained within the article.

Conflicts of Interest: Author Jixian Xu was employed by the company Langfang Shengsen Grinding Tools Co., Ltd. The remaining authors declare that the research was conducted in the absence of any commercial or financial relationships that could be construed as a potential conflict of interest.

References

1. Brinksmeier, E.; Mutlugünes, Y.; Ohmori, H. Ultra-precision grinding. *CIRP Ann.* **2010**, *59*, 652–671. [CrossRef]
2. Lee, E.S.; Baek, S.Y. A study on optimum grinding factors for aspheric convex surface micro-lens using design of experiments. *Int. J. Mach. Tool. Manuf.* **2007**, *47*, 509–520. [CrossRef]
3. Yan, Y.; Zhang, Z.; Liu, J. Study on the algorithm of three-dimensional surface residual material height of nano-ZrO_2 ceramics under ultra-precision grinding. *Micromachines* **2021**, *12*, 1363. [CrossRef] [PubMed]
4. Xu, J.; He, Q.; Zhang, X. Euler's equation grinding track. microgrinding mechanism, and process research of space optical crystal materials. *Int. J. Adv. Manuf. Technol.* **2023**, *128*, 5209–5221. [CrossRef]
5. Jahnel, K.; Michels, R.; Wilhelm, D.P. Investigation of surface integrity induced by ultra-precision grinding and scratching of glassy carbon. *Micromachines* **2023**, *14*, 2240. [CrossRef]
6. Wang, S.; Zhao, Q.; Wu, T. An investigation of monitoring the damage mechanism in ultra-precision grinding of monocrystalline silicon based on AE signals processing. *J. Manuf. Process.* **2022**, *81*, 945–961. [CrossRef]
7. Kakinuma, Y.; Konuma, Y.; Fukuta, M. Ultra-precision grinding of optical glass lenses with La-doped CeO_2 slurry. *CIRP Ann.* **2019**, *68*, 345–348. [CrossRef]
8. Jiang, C.; Li, H.; Mai, Y. Material removal monitoring in precision cylindrical plunge grinding using acoustic emission signal. *Proc. Inst. Mech. Eng. Part C J. Mech. Eng. Sci.* **2014**, *228*, 715–722. [CrossRef]
9. Sun, Z.; Guan, C.; Dai, Y. Ultra-precision time-controlled grinding for flat mechanical parts with weak stiffness. *J. Manuf. Process.* **2023**, *99*, 105–120. [CrossRef]
10. Monier, A.; Guo, B.; Zhao, Q. Effects of the grinding conditions on geometry of microstructured surfaces fabricated via designed precision grinding. *Proc. Inst. Mech. Eng. Part B J. Eng. Manuf.* **2023**, *237*, 573–587. [CrossRef]
11. Xiao, G.; Huang, Y. Adaptive belt precision grinding for the weak rigidity deformation of blisk leading and trailing edge. *Adv. Mech. Eng.* **2017**, *9*, 1687814017731705. [CrossRef]
12. Cai, R.; Wan, N.; Mo, R. Prediction of un-uniform grinding wheel wear based on instantaneous engagement of multi-axis grinding. *Int. J. Adv. Manuf. Technol.* **2022**, *119*, 3407–3425. [CrossRef]
13. Brecher, C.; Baumler, S.; Brockmann, B. Avoiding chatter by means of active damping systems for machine tools. *J. Mach. Eng.* **2013**, *13*, 117–128.
14. Bianchi, G.; Cagna, S.; Cau, N. Analysis of vibration damping in machine tools. *Procedia CIRP* **2014**, *21*, 367–372. [CrossRef]
15. Ghorbani, S.; Polushin, N.I. Effect of composite material on damping capacity improvement of cutting tool in machining operation using taguchi approach. *Eng. Technol. Int.* **2015**, *9*, 1339–1349.
16. Devin, L.N.; Osadchii, A.A. Improving performance of CBN cutting tools by increasing their damping properties. *J. Superhard Mater.* **2012**, *34*, 328–335. [CrossRef]
17. Sexton, J.S.; Howes, T.D.; Stone, B.J. The use of increased wheel flexibility to improve chatter performance in grinding. *Proc. Inst. Mech. Eng.* **1982**, *196*, 291–300. [CrossRef]
18. Shen, R.; Qian, X.; Zhou, J. Characteristics of passive vibration control for exponential non-viscous damping system: Vibration isolator and absorber. *J. Vib. Control.* **2023**, *29*, 5078–5089. [CrossRef]
19. Zhu, L.; Hao, J.; Lu, L. Research on influence of damping on the vibration noise of transformer. *IEEE Access* **2022**, *10*, 92128–92136. [CrossRef]
20. Shum, K.M. Tuned vibration absorbers with nonlinear viscous damping for damped structures under random load. *J. Sound. Vib.* **2015**, *346*, 70–80. [CrossRef]
21. Chen, S.; Yang, Z.; Ying, M. Parallel load-bearing and damping system design and test for satellite vibration suppression. *Appl. Sci.* **2020**, *10*, 1548. [CrossRef]
22. Luo, H.; Guo, S.; Yu, C. Vibration suppression analysis and experimental test of additional constrained damping layer in space science experiment cabinet. *Compos. Adv. Mater.* **2021**, *30*, 2633366X20978659. [CrossRef]
23. Gou, Z.; Li, B.; Fei, J. Frictional characteristics and vibration behavior of silicone rubber powders modified resin-based friction materials. *J. Appl. Polym. Sci.* **2024**, *141*, e55344. [CrossRef]
24. Babkina, N.; Antonenko, O.; Kosyanchuk, L. Effect of polyurethane material design on damping ability. *Polym. Adv. Technol.* **2023**, *34*, 3426–3437. [CrossRef]
25. Sain, T.; Yeom, B.; Waas, A.M. Effect of soft segment and clay volume fraction on rate dependent damping of polyurethane and polyurethane-clay nanocomposites. *J. Reinf. Plast. Comp.* **2014**, *33*, 2129–2135. [CrossRef]

26. Khanouki, M.A.; Ohadi, A. Improved acoustic damping in polyurethane foams bythe inclusion of silicon dioxide nanoparticles. *Adv. Polym. Technol.* **2018**, *37*, 2799–2810. [CrossRef]
27. Li, L.; Tian, B.; Li, L. Preparation and characterization of silicone oil modified polyurethane damping materials. *J. Appl. Polym. Sci.* **2019**, *136*, 47579. [CrossRef]
28. Chen, S.; Wang, Q.; Wang, T. Preparation, tensile, damping and thermal properties of polyurethanes based on various structural polymer polyols: Effects of composition and isocyanate index. *Polym. Res.* **2012**, *19*, 9994. [CrossRef]
29. Shi, M.; Zheng, J.; Huang, Z. Synthesis of polyurethane prepolymers and damping property of polyurethane/epoxy composites. *Adv. Sci. Lett.* **2014**, *4*, 740–744. [CrossRef]
30. Su, Y.; Li, T.; Liu, Y. Mechanical and damping properties of graphene-modified polyurethane-epoxy composites for structures. *Polym-Korea* **2021**, *45*, 483–490. Available online: https://10.7317/pk.2022.45.4.483 (accessed on 26 July 2021). [CrossRef]
31. Zhang, C.; Chen, Y.; Li, H. Facile fabrication of polyurethane/epoxy IPNs filled graphene aerogel with improved damping, thermal and mechanical properties. *RSC Adv.* **2018**, *8*, 27390–27399. [CrossRef]
32. Feng, Q.; Shen, M.; Zhu, J. Realization of polyurethane/epoxy interpenetrating polymer networks with a broad high-damping temperature range using β-cyclodextrins as chain extenders. *Mater. Des.* **2021**, *212*, 110208. [CrossRef]
33. Yu, W.; Zhang, D.; Du, M. Role of graded length side chains up to 18 carbons in length on the damping behavior of polyurethane/epoxy interpenetrating polymer networks. *Eur. Polym. J.* **2013**, *49*, 1731–1741. [CrossRef]
34. Chen, S.; Wang, Q.; Pei, X. Dynamic mechanical properties of castor oil-based polyurethane/epoxy graft interpenetrating polymer network composites. *J. Appl. Polym. Sci.* **2010**, *118*, 1144–1151. [CrossRef]
35. Chen, S.; Wang, T.; Wang, Q. Studies on tribological, damping and thermal properties of polyurethane/polystyrene interpenetrating polymer networks. *J. Polym. Mater.* **2016**, *33*, 503–512.
36. Wang, Q.; Chen, S.; Wang, T. Damping, thermal, and mechanical properties of polyurethane based on poly(tetramethylene glycol)/epoxy interpenetrating polymer networks: Effects of composition and isocyanate index. *Appl. Phys. A-Mater.* **2011**, *104*, 375–382. [CrossRef]
37. Qin, C.; Jin, Z.; Bai, X. Compatibility of polyurethane/(vinyl ester resin)(ethyl acrylate) interpenetrating polymer network. *Polym. J.* **2007**, 1365–1372. [CrossRef]
38. Statharas, E.C.; Yao, K.; Rahimabady, M. Polyurethane/poly(vinylidene fluoride)/MWCNT composite foam for broadband airborne sound absorption. *J. Appl. Polym. Sci.* **2019**, *136*, 47868. [CrossRef]
39. Xu, X.; Huang, S.; Guan, J. Grinding performance and self–lubrication mechanism of phenolic resin-bonded grinding wheel filled with inclusion complex of β-cyclodextrin and dialkyl pentasulfide. *J. Mater. Process Technol.* **2015**, *221*, 163–171. [CrossRef]
40. Bashandeh, K.; Lan, P.; Meyer, J.L. Tribological performance of graphene and PTFE solid lubricants for polymer coatings at elevated temperatures. *Tribol. Lett.* **2019**, *67*, 99. [CrossRef]
41. Aliyu, I.K.; Mohammed, A.S.; Al-Qutub, A. Tribological performance of ultra high molecular weight polyethylene nanocomposites reinforced with graphene nanoplatelets. *Polym. Compos.* **2019**, *40*, E1301–E1311. [CrossRef]
42. Aliyu, I.K.; Mohammed, A.S.; Al-Qutub, A. Tribological performance of UHMWPE/GNPs nanocomposite coatings for solid lubrication in bearing applications. *Tribol. Lett.* **2018**, *66*, 144. [CrossRef]
43. Ma, C.; Shan, Y.; Wang, N. High-damping polyurethane-based composites modified with amino-functionalized graphene. *High. Perform. Polym.* **2023**, *35*, 795–811. [CrossRef]
44. Xia, S.; Liu, Y.; Pei, F. Identical steady tribological performance of graphene-oxide-strengthened polyurethane/epoxy interpenetrating polymer networks derived from graphene nanosheet. *Polymer* **2015**, *64*, 62–68. [CrossRef]
45. Yue, R.; Liu, Y.; Xia, S. Raman imaging evidence for mechanical/tribological quasi-steady state in GO strengthening polyurethane/epoxy interpenetrating polymer network. *Macromol. Res.* **2022**, *30*, 477–485. [CrossRef]
46. Zhang, H.Y.; Xia, S.L.; Liu, C. In-situ closed synthesis of phenolic resin from PPU and its damping properties. *Comp. Sci. Eng.* **2024**, 1–9. Available online: http://kns.cnki.net/kcms/detail/10.1683.TU.20240227.1807.006.html (accessed on 23 November 2023). [CrossRef]
47. *GB/T 528-1998*; Rubber, Vulcanized or Thermoplastic—Determination of Tensile Stress-Strain Properties [Including MODIFICATION 1]. Ministry of Chemical Industry of the People Republic of China: Beijing, China, 1998.

coatings

MDPI

Article

Mechanical and Microstructural Characterization of AISI 316L Stainless Steel Superficially Modified by Solid Nitriding Technique

Rene Guardian-Tapia [1], Isai Rosales-Cadena [1,*], Jose Luis Roman-Zubillaga [1] and Sergio Ruben Gonzaga-Segura [2]

[1] Center of Research in Engineering and Applied Science, Autonomous University of Morelos State, Av. Univ. 1001 Col. Chamilpa, Cuernavaca 62209, Mexico; rene.guardian@uaem.edu.mx (R.G.-T.); jlroman@uaem.mx (J.L.R.-Z.)

[2] Institute of Physical Sciences, National University Autonomous of Mexico, Av. Univ. 1001 Col. Chamilpa, Cuernavaca 62210, Mexico; sergio.gonzaga@icf.unam.mx

* Correspondence: faye12@uaem.mx; Tel./Fax: +(52)-777-329-7984

Abstract: AISI 316L austenitic stainless steel superficially modified by the solid nitriding technique was investigated at different nitriding times (2, 4, 6, 8, 12 and 24 h) and at 450 °C. The microstructural characterization was conducted using scanning electron microscopy (SEM) analysis and X-ray diffraction analyses, finding the presence of $Fe_{2–3}N$, Fe_4N and Cr_2N, among others. The mechanical behavior of the modified surfaces was carried out by developing hardness profiles and relating it with the nitride layer thickness evaluated using scanning electron microscopy (SEM), obtaining layers up to 70 μm wide. The nitrogen diffusion produced species above and below the surface sample with a transformation from the austenitic phase to an expanded austenite (γ_N) phase, which is responsible for producing an increase in hardness of up to 1200 HV in the samples treated at 24 h, which is four times higher than the untreated steels. Wear evaluations of the obtained layers were performed using a pin-on-disk system under zero lubrication, indicating that the samples with 12 and 24 h of treatment present the best wear resistance promoted by an oxidative–adhesive mechanism. The obtained results are positively comparable with those of the ion nitriding technique but with a lower implementation cost.

Keywords: surface modification; stainless steel; hardness; mechanical properties; wear

Citation: Guardian-Tapia, R.; Rosales-Cadena, I.; Roman-Zubillaga, J.L.; Gonzaga-Segura, S.R. Mechanical and Microstructural Characterization of AISI 316L Stainless Steel Superficially Modified by Solid Nitriding Technique. *Coatings* **2024**, *14*, 1167. https://doi.org/10.3390/coatings14091167

Academic Editors: Pinghu Chen, Yun Zhang and Ruiqing Li

Received: 9 August 2024
Revised: 5 September 2024
Accepted: 6 September 2024
Published: 10 September 2024

1. Introduction

Hardened surfaces of components exposed to a relative friction is an important issue that for many decades has been treated with the goal of minimizing the effect of wear in different areas of the productive sector due to the great economic losses derived from this phenomenon. Therefore, implementing alternatives to harden only the surface of the components that are in contact is important, mainly to reduce their repositioning time, which is one of the imperative goals of this surface engineering [1,2]. On the other hand, obtaining hardened surfaces and a core with a lower hardness is very important, since, in this way, the component toughness may increase, reducing crack nucleation in the hardened zone and, therefore, minimizing the component brittleness [2,3]. Several studies have been conducted to establish that there are many surface treatments available to obtain hardened surfaces. Lin [4] established the importance of the surface texture after nitriding to improve the tribological performance of an AISI 316L. However, there are few methods that involve the combination of two hardening mechanisms such as precipitation and phase transformation [5,6]. Within these methods, nitriding is one of the most important, due to multiple benefits such as uniform surface appearance, very low dimensional variation, controlled depth of the layer and low energy consumption for heating [7–9]. Regarding the ion nitriding technique, Christiansen et al. suggests a pretreatment to prevent a surface re-passivation [1]. Surface modification of stainless steels has been investigated for many years, providing important information about the nitrided layer formation using the ion nitriding technique [10–14]. It is well known that the main cause of hardness increment in nitrided austenitic steels is produced due to the (γ_N) phase formation [15] and the possible carbonitrides precipitated during the process [16,17] on the surface and in the subsurface due to the nitrogen atom diffusion and the subsequent species formation. However, this technique is expensive due to the high cost of infrastructure; therefore, the implementation of an effective and economical technique for nitride parts is attractive. This is the case in solid nitriding, which consists of diffusing nitrogen atoms into the steel using Fe_4KCN, with the samples being surrounded by this compound and confined inside of a sealed container heated to a certain temperature [18]. Until now, the use of the abovementioned solid nitriding technique remains with limited information [5,18]. Hence, the aim of this present investigation is to generate knowledge of the nitrogen diffusion mechanism and also the mechanisms that improve the hardness and wear behavior of the modified surfaces by solid nitriding of AISI 316L stainless steel. This investigation is carried out at relatively low temperature because if an austenitic stainless steel, such as AISI 316L, is treated at temperatures higher than 450 °C, it can suffer a significant decrease in its corrosion resistance due to the excessive formation of chromium nitride and therefore diminishing the chromium in its solid solution [19]. This work provides an understanding that solid nitriding, through the formation of nitrides of the alloying elements of the steel, can be an effective, reproducible and economical surface treatment to obtain hardened layers in components for industrial applications, which is highly comparable to the ion nitriding technique.

2. Materials and Methods

2.1. Materials Synthesis

Several specimens of AISI 316L-type stainless steel were obtained by cutting coupons with dimensions of 1 × 1 × 1 cm^3 (see Table 1 for chemical compositions). Before the nitriding treatments, the samples were superficially pretreated with a based HNO_3, HCl and HF solution for 3 min. For solid nitriding, the samples were surrounded (20 mm thickness) with a potassium ferrocyanide (Fe_4KCN) compound inside a hermetic stainless steel container (to keep the produced gases inside of the container) internally coated with ammonia nitrate solution (NH_4NO_3) in order to enhance the nitrogen diffusion. After that, the container was placed inside a chamber using a Felisa 301 Series furnace (Guadalajara, Mexico) and heated at 450 °C for 2, 4, 6, 8, 12 and 24 h. Once the nitriding time was over, the containers were removed from the furnace for cooling to avoid the annealing of the samples.

Table 1. Chemical composition for stainless steel AISI 316L in [wt%].

C	Si	Mn	Cr	Ni	Mo	Fe
0.08	1.0	2.0	18.0	11.50	2.30	Bal.

2.2. Metallographic Evaluation

Specimens were sanded using SiC sanding paper up to 600 grade, and afterward they were polished using alumina powder with a particle size of 0.03 μm. The samples were electro-etched with a reactant of 10 g oxalic acid in 90 mL deionized water for 20 s using 0.5 volts in DC for 120 s. Microstructural and surface analyses were carried out in an LEO-1450 VP Scanning Electron Microscope (Southampton, UK), whereas chemical microanalysis was performed using the energy disperse system (EDS) attached to the equipment. X-ray diffraction analyses were carried out using a Bruker D2 PHASER X-ray diffractometer equipment (Billerica, MA, USA), with a radiation of Cu-Kα: λ = 0.15406 nm and a 3°/min scanning speed.

2.3. Mechanical Testing

Hardness Vickers measurements were carried out on the sample surface (previously polished with diamond paste with an average size 0.03 μm), using an LECO, LM300AT Microhardness Tester (St. Joseph, MI, USA), with a 0.2 kg load for 15 s. For this, a series of 10 indentations in different zones of the surface sample were taken to obtain a better precision in the average and standard deviation of the hardness value, which is in agreement with ASTM E384-22. Five samples of each time of treatment were used. A conventional pin-on-disk TRB^3 Anton Paar Tribometer (Graz, Austria) wear system was used to evaluate the wear behavior of the pin samples with dimensions of 6 mm diameter by 8 mm length and a surface roughness with alumina 0.5 μm, under a load of 5 N, with a constant disk rotation of 100 rpm and zero lubrication. AISI-4140 steel (oil quenched) was used as a counterpart with a hardness of 51 ± 3 HRC in agreement with ASTM-G99. Worn surfaces were observed and analyzed using a SEM (scanning electron microscopy, Leo-1450 VP, Southampton, UK) and a total of three samples were used to calculate wear factor, weight loss and worn surfaces analyses. A Mitutoyo Surftest SJ-210 JP Profilometer Roughness Tester JP (Kawasaki, Japan) surface profilometer was used to determine the surface roughness before and after solid nitriding. The surface roughness was characterized by measuring the average roughness, Ra, according to ISO 21920-2:2021. The measurement length was chosen as 0.5 mm. For each specimen, the surface roughness was determined developing 10 measurements, and the results are shown in Table 2.

Table 2. Roughness values for the different nitriding times before and after treatment.

Treatment Time [h]	Surface Roughness Before Solid Nitriding [μm]	Surface Roughness After Solid Nitriding [μm]
2	0.030	0.113
4	0.032	0.127
6	0.027	0.145
8	0.031	0.157
12	0.027	0.194
24	0.026	0.216

3. Results and Discussion

3.1. SEM Characterization

Nitrogen Diffusion and Line Scan Characterization

Diffusion in steel is considered as a mass transport process in a non-steady state where the atoms that diffuse into the steel do so as a function of time, concentration and temperature in the system [15,16]. The formation of a primary and secondary layer is produced due to the initial nitrogen diffusion reaching a saturation point, and then, a secondary layer starts to be created with the nitrogen atoms that diffuse from the primary layer [17,19–21]. Analyses were carried out on the cross section of the treated samples, with the acronym used in the figures being as follows: P for the primary layer, S for the secondary layer and DL for the diffusion layer. Figure 1A shows the image of the sample treated for 2 h. The presence of a primary layer of approximately 6 μm wide can be observed, with a secondary layer of approximately 2 μm, with the total depth of the diffused layer being 35 μm with a nitrogen surface concentration of 5.28 at.%. Similar behavior was observed in Figure 1B for the sample after 4 h, which presents a nitrogen surface value of 7.8 at.%. For the treatment for 6 and 8 h, shown as a primary layer of approximately 17 μm with a secondary layer of 6 μm, as seen in Figure 1C,D, the layer is observed as homogeneous. Although, it is interesting to observe the apparition of several precipitates along the nitrided layer segregated mainly to the interphase between the primary and the secondary layer across the modified layer, which is produced due to the reaction between the nitrogen with the constituent elements of the steel [22–26], with the total depth of the layer being 50 μm. For the treatment for 12 and 24 h in Figure 1E and 1F, respectively, it is observed that the primary layer formation of 19 μm with a secondary layer of 8 μm approximately has few precipitates with an average size of 0.3 μm. These are the largest diffusion layers obtained in these experiments with a maximum width of 70 μm on average and a surface nitrogen content of 9.40 at.%.

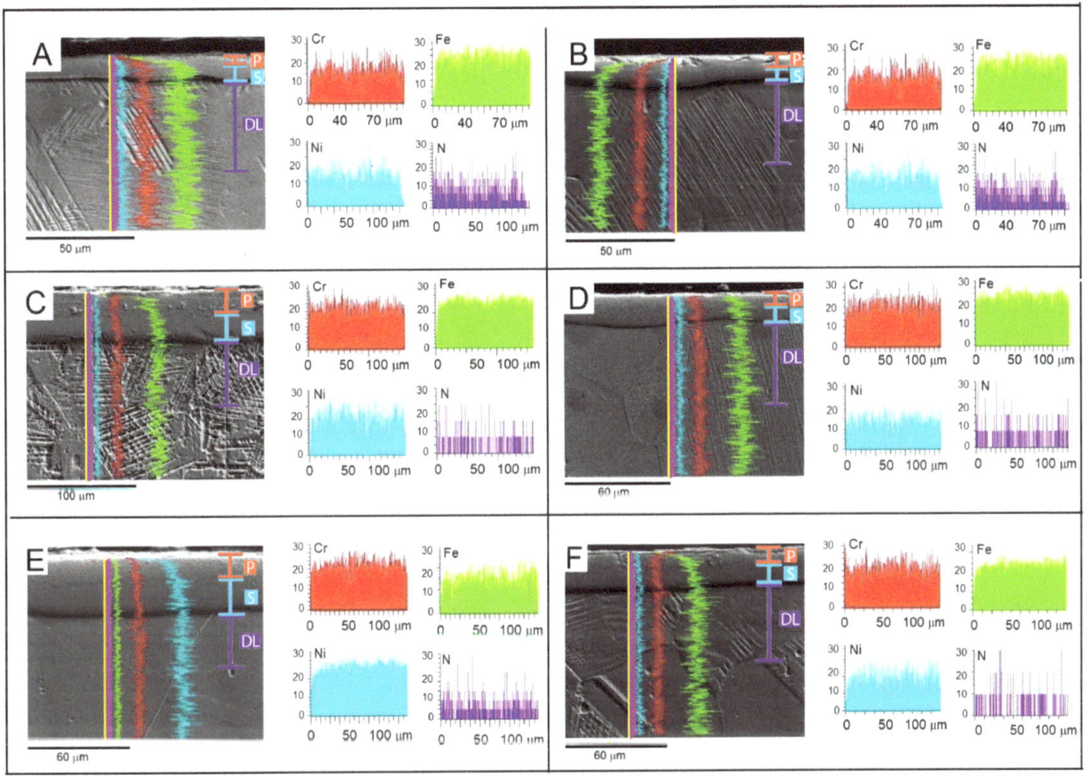

Figure 1. Nitriding layers morphologies at different treatment times: (**A**) 2 h, (**B**) 4 h, (**C**) 6 h, (**D**) 8 h, (**E**) 12 h and (**F**) 24 h. The line scan profiles show the nitrogen diffusion toward the matrix sample. Total diffusion layers cannot be clearly observed in the images of Figure 1 due to the over-etching of the samples.

Nitrogen chemical analyses are observed in the nitrogen diffusion profiles of Figure 2, which were obtained by performing punctual chemical analyzes along the nitrided layer from the surface to the center of the treated sample, supporting the results observed in the line scan analyses. It can be observed that the samples with 2 and 4 h of treatment present 34 and 39 μm of diffusion, respectively, while the samples with 6, 8, 12 and 24 h of treatment show a nitrogen diffusion of 55, 60, 66 and 70 μm, respectively. It is interesting to observe that there is an interval between the sample with 4 h and the sample with 6 h of treatment of approximately 15 μm of nitrogen diffusion. This effect can be attributed due to the strong nitrogen concentration on the surface sample with 6 h of treatment that again promotes the continuous nitrogen diffusion. In other words, nitrogen atoms can move from one position in the lattice to another, through an activated process in which the ion moves over an energy barrier [5].

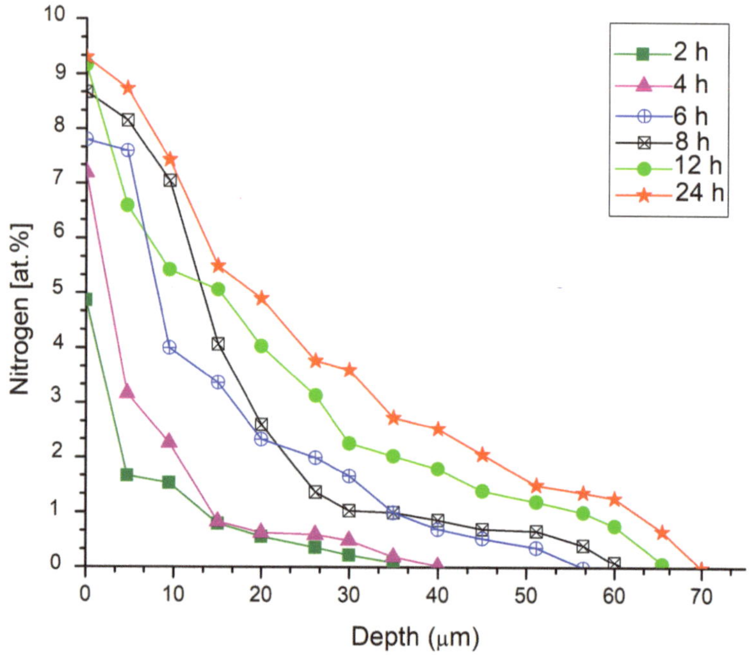

Figure 2. Nitrogen diffusion profiles obtained after the solid nitriding process with treatments of 2, 4, 6, 8, 12 and 24 h.

3.2. X-ray Diffraction Analyses

Figure 3 shows the XRD patterns of the solid nitriding samples treated with different exposure times. Due to the atomic radii of nitrogen being 0.5427 Å, the diffusion mechanism is clearly interstitial [27], where nitrogen diffuses into the sample and reacts with the γ-phase (matrix) and the other elements of the steel (Fe, Cr, Ni) [28–32]. Therefore, the formation of the γ_N phase is indicative that the nitriding process has been obtained [17]. For the sample treated for 2 h, the XRD spectrum mainly shows the presence of γ_N and γ'-Fe_4N formed during the treatment. The presence of Fe-α and Cr-α was detected in several samples with different times of the solid nitriding treatment. In the sample treated for 4 h, it mainly shows the presence of γ'-Fe_4N, while the CrN phase does not appear in this diffraction pattern (and neither in the others patterns), which is indicative that chromium does not react completely with nitrogen, creating chromium nitrides, which is one of the main objectives of not affecting the corrosion resistance of the steel [19]. For the treatments for 6 and 8 h, the formation of Cr_2N, ε-Fe_2N and ε-Fe_3N was detected, together with the α-Fe presence, while in the diffraction pattern for the samples treated at 12 and 24 h, the presence of Cr_2N and Fe_4N can be observed. The formation of these nitrides for both nitriding times may indicate that the nitrogen, which comes from the potassium ferrocyanide, has been depleted throughout the treatment due to temperature, i.e., the nitrogen concentration has decreased at very low levels, reducing the possibility of a continuing diffusion. The ICDD files used to identify the peaks were Fe_4N-(111), ICDD: 98-0030-05401; $Fe_{2\text{–}3}$ N-(110), ICDD: 98-009-8404; Fe-α-(110), ICDD: 98-007-1819; Cr_2N-(110), ICDD: 98-002-5566; and Cr-α-(200), ICDD: 01-085-1336.

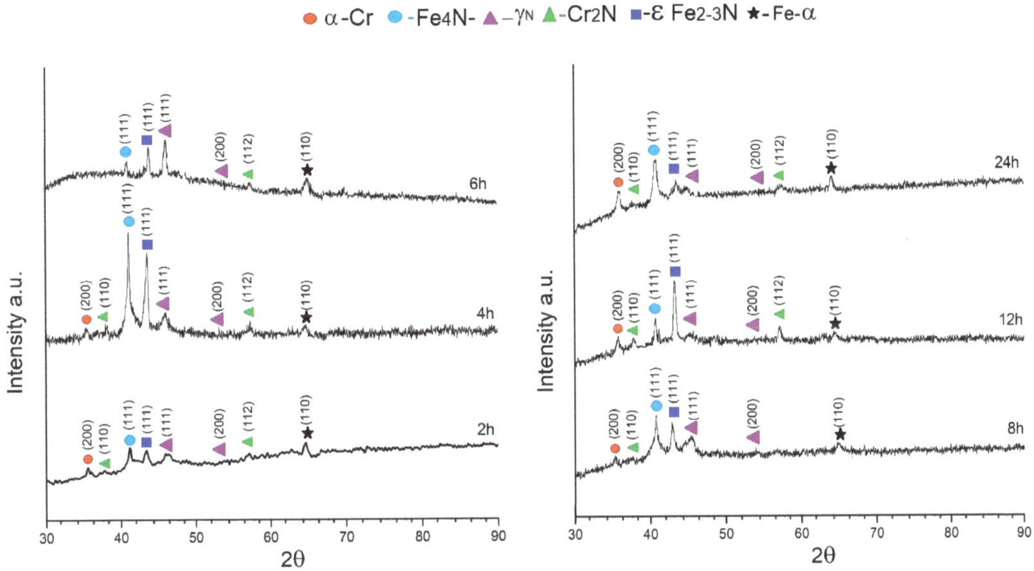

Figure 3. X-ray diffraction patterns of AISI 316L stainless steel with solid nitriding treatment with times of 2, 4, 6, 8, 12 and 24 h.

3.3. Mechanical Properties

3.3.1. Microhardness Evaluation

Hardening after the nitriding process is mainly attributed to the (γ_N) formation [15,16] in addition to the formation of different types of nitrides such as Fe_2N, Fe_4N and Cr_2N [18]. Figure 4 shows the hardness values for the solid nitriding samples, and also shows the hardness value for the untreated sample of 280 HV as a reference. Hardness values were obtained along the cross section of the treated samples from the surface sample to the final sample of the modified layer. For the treatment time of 2 h, an average hardness of 1010 HV with a depth of 48 μm occurs, showing a sudden decrease in hardness after 5 μm, perhaps after the primary layer finishes due to the compounds formation, which is basically due to the γ_N phase formation [16]. On the other hand, it is important to mention that hardening in the modified layer cannot be exclusively attributed to the formation of the nitriding compounds, because it is also due to the lattice distortion produced by the diffusion of nitrogen within the austenitic structure via the interstitial mechanism [18,31]. The sample with 4 h of treatment had a hardness value of 1104 HV with a hardened layer of 45 μm, also showing a drop in hardness of approximately 6 μm. For the sample treated for 6 h, a hardness of 1116 HV was obtained over the surface with a hardened layer of 80 μm. In this case, the drop in hardness occurs at 25 μm below the surface. These hardness drops observed in the plots can be correlated with the change in layer morphologies observed in Figure 1, from the primary layer to the secondary layer, i.e., the zone where there is a transition of compound formation from high saturation to a constant nitrogen diffusion [9,27]. That is, if the concentration of nitrogen atoms is lower than the diffusion rate of nitrogen driven by the atoms gradient at greater depths, a layered structure of Fe-N phases (interface) can develop at different layer depths, which is a function of the decrease in the chemical potential of nitrogen as the distance from the surface increases. Hence if the compositions on both sides of this interface are equal to the compositions prevailing under equilibrium conditions, then a local equilibrium occurs. However, if there is an accumulation or decrease in nitrogen atoms on either side of the interface, then the local equilibrium is affected because the transport of nitrogen atoms across the interface is not restricted, generating a rapid mobility of nitrogen atoms through this interface and, therefore, a real equilibrium can be reached again, according to the applied nitriding temperatures [27,31]. For the 8 h treatment, a hardness of 1090 HV was obtained with a depth of 60 μm. For the 12 h sample, a hardness of 1106 HV was obtained with a depth of 90 μm. Finally, in the case of the sample treated for 24 h, a hardness of 1200 HV was obtained with a depth of 65 μm. Thus, the hardness increment reached by this thermochemical treatment was approximately five times higher in comparison with the untreated sample, attributing this hardening behavior from the microstructural point of view to the creation of defects such as precipitates and dislocations due to lattice distortion [32] in the structure that inhibits the plastic flow [7,9,23]. The hardness reached by this treatment is comparable with the results in hardness reported by Lin et al. [4] of 1050 HV in plasma-nitrided AISI-316 stainless steel and the results reported by Gokcekaya et al. [15] in a range of 900 to 1400 HV in ion-nitrided AISI-316. The standard deviation in the hardness measurements was of 0.5% on average, which indicates a variation of 5.5 HVN.

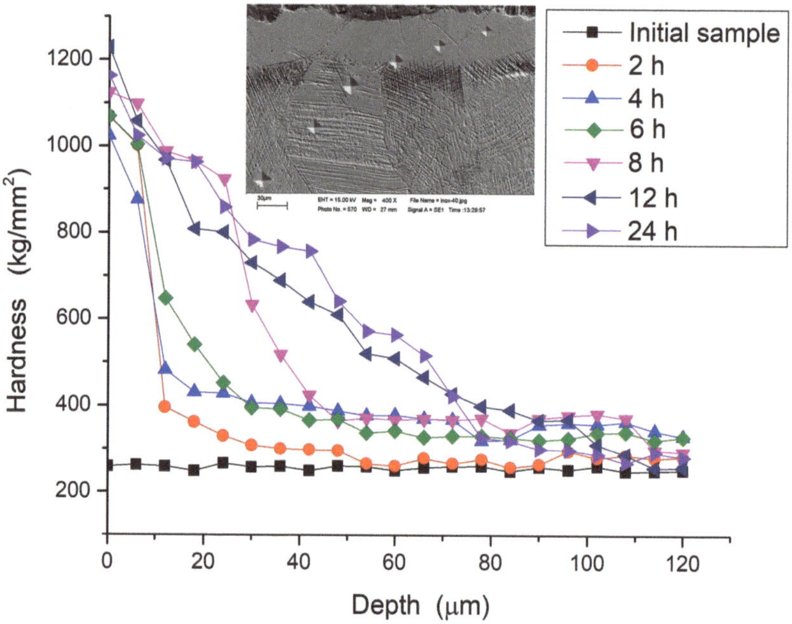

Figure 4. Hardness values for the samples with different diffusion treatments compared with the untreated sample. The inset image shows the dimensions and distribution of the indentations in the 24 h sample.

3.3.2. Wear Behavior

Weight Loss Analyses

The weight loss curves as a function of sliding distance are shown in Figure 5. In this figure, the wear behavior of the samples is observed with solid nitriding treatments for 2, 4, 6, 8, 12 and 24 h compared with the sample without treatment (initial sample). It can be observed that the initial wear behavior of the sample after 2 h treatment in a range from 0 to 450 m increases slightly. After this distance, the weight losses present a gradual increment until the end of the test. This sudden increment is due to the coupling between the two surfaces during the wear process. In the case of the samples treated with 4, 6 and 8 h, the initial wear remains without increasing until 700 m. After this distance, it increases up to 1000 m to form a plateau until 2000 m, and finally there is a slight increase in the weight losses until the end of the test. This behavior can be attributed to the intrinsic superficial hardness of the samples reached by the formation of the γ_N phase [4,15,27]. For the samples treated with 12 and 24 h solid nitriding treatment, a remarkable wear resistance is observed from the initial to the end of the test at 4000 m without any apparent weight loss in these samples. The moderated changes in the curves indicate the formation of a stable oxide interface adhered to the worn surfaces. This oxide layer is created by the relative frictional heating that produces a strong bond between the formed layers due to the high hardness of the steel surface [4].

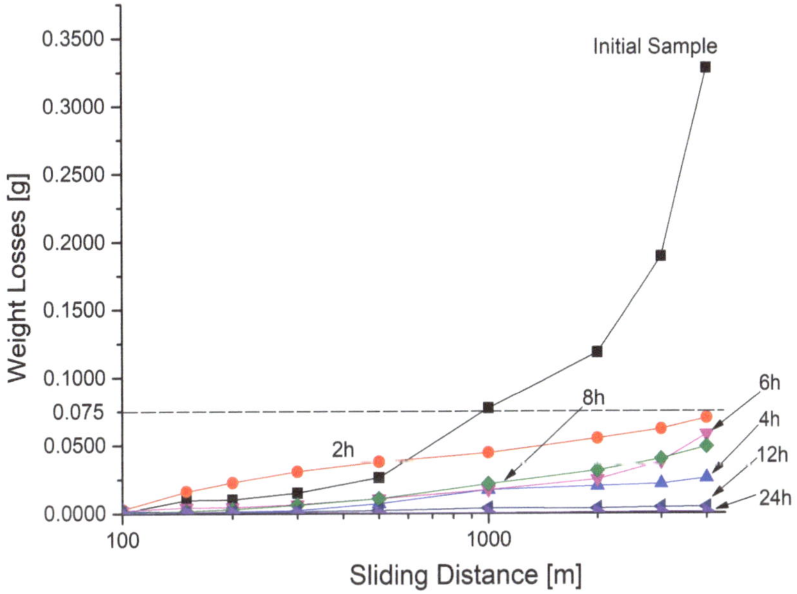

Figure 5. Weight losses against sliding distance of the samples with solid nitriding treatment: initial sample, 2, 4, 6, 8, 12 and 24 h.

Worn Surfaces Analyses

The analysis of the worn surface is presented in Figure 6, where it can be observed that for the initial sample (untreated sample) in Figure 6A, abrasion grooves are observed in the sliding direction with small zones with particles adhesion, which is characteristic of the abrasion mechanism [3,13]. The elemental analysis obtained shows the presence of oxygen in small quantities in addition to Cr, Fe and Ni in a greater proportion, suggesting that there is no significant oxide formation with the elements present in the steel. For the treatment for 2 h in Figure 6B, deep grooves over the wear area are mainly observed, where it clearly shows the sliding direction and presents the accumulated material with small adhered particles. The chemical analysis in this area shows the presence of oxygen in a greater proportion as well as a high percentage of Cr, Fe, Ni. These elements are part of the worn material, although this finding does not exclude the possibility that Fe comes from the counterpart [29]. In Figures 6C, 6D and 6E, which correspond to the worn surfaces of the samples with 4, 6 and 8 h of heat treatment, respectively, a pattern of few grooves is observed on the worn surface with the presence of oxide accumulation, which implies that there is an oxygen reaction with the elements present in the surface sample as is shown in the chemical analyses of the worn surfaces where the presence of oxygen is high. On the other hand, the images of the worn surfaces corresponding to the samples with 12 and 24 h of treatment are presented in Figure 6F,G where few zones with oxide agglomeration are observed (indicated by an arrow), although grooves are observed in most of the area, but with a low level of depth. In the chemical analyses of these samples, a relatively low concentration of oxygen was detected along with Ni, Cr, Fe and Mo. These results broadly support the observations of the wear curves, thus establishing that the operating wear mechanism is abrasive with a mixed oxidative–adhesive mode [3,13,29]. On the other hand,

Mo was not detected in several samples due to the presence of dense oxide layers formed on the surface sample after wear, which mask the molybdenum presence and therefore inhibit the Mo detection in these specific zones.

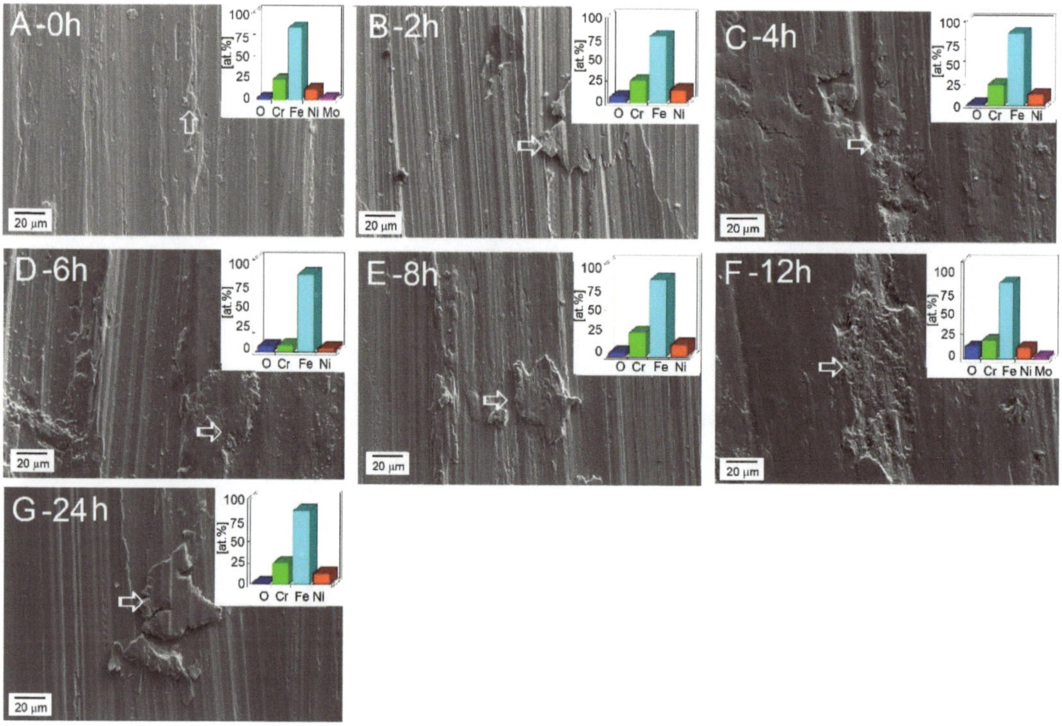

Figure 6. Worn surfaces images of the samples with solid nitriding treatment: (**A**) initial sample, (**B**) 2 h, (**C**) 4 h, (**D**) 6 h, (**E**) 8 h, (**F**) 12 h and (**G**) 24 h.

Wear Factor Analyses

Because a wear phenomenon involves the weight loss as a function of time and the sliding distance, then it is important to know the wear index between these parameters, hence the results of this indicator is presented in Figure 7 where the graph of the wear factor obtained from the wear curves of the initial sample and 2, 4, 6, 8, 12 and 24 h of treatment is presented. In the graph, a noticeable value of the wear factor for the initial sample is observed, which is approximately 80% higher compared with the treated samples. This phenomenon is generated by the difference in hardness between the pin sample and the counterpart [29] and is notoriously influenced by the amount of Cr and Fe nitrides generated on the worn surface [13,33] and the oxides of the elements present in the steel. For the worn samples treated during 2, 4, 6 and 8 h, a resultant wear factor average of 1.2×10^{-5} g/m was obtained while for the worn samples treated for 12 and 24 h, a wear factor of 5×10^{-6} g/m was obtained. These results clearly indicate the advantage of implementing a treatment for 12 and 24 h, where a minimum value in the wear factor is achieved and consequently an increase in the useful life period of the steel [13,29]. The measured error in the experiments was of 1.8%, which falls within the experimental error.

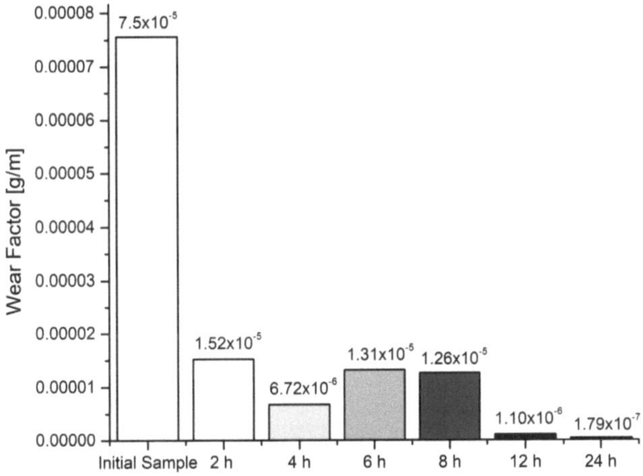

Figure 7. Wear factor of the samples with solid nitriding treatment compared with the untreated sample. Treatment times of 2, 4, 6, 8, 12 and 24 h.

4. Conclusions

In this research, the scientific and technological information related to the microstructural and mechanical behavior of an AISI 316L stainless steel surface modified by the solid nitriding technique at 450 °C and different nitriding times is presented. The following conclusions were obtained:

1. The layer thickness consisting of different nitrides increases as a function of the nitriding time.
2. The γ_N, γ'-Fe_4N, Cr_2N and $Fe_{2-3}N$ phases were produced in the diffusion layer in most of the different nitriding times. Although at 12 and 24 h, the γ_N phase decreases due to the reduction in nitrogen concentration.
3. The samples treated for 24 h exhibited the highest surface hardness values. This hardness increment is attributed to the presence of the γ_N phase, which is one of the main factors responsible for the hardness increase, followed by the lattice distortion of an interstitial mechanism.
4. Wear evaluation indicates an increase in wear resistance in the samples treated at 12 and 24 h exposure times, clearly evidenced and supported by the wear factor results, with the different nitrided phases being responsible for this increase in wear resistance, and the produced oxide layers due to intrinsic frictional heating. Hence, they suffered a small depletion due to the low residual stresses in the nitrided layer.
5. A disadvantage of applying this method is that the pieces obtained have a rough surface finish after the solid nitriding treatment. However, with a polishing treatment, an acceptable surface finish is obtained without affecting the modified layer.

The nitrided sample at 24 h exhibited a thick nitrided layer with minimal or zero cracks and/or delamination generation, with the best wear and hardness performance. This sample is suitable for use in wear and corrosion applications. Therefore, by applying this method under these conditions, results similar to those of ion nitriding can be obtained.

Author Contributions: All authors participated in samples preparation. Experiments, data collection analysis results; R.G.-T., J.L.R.-Z. and S.R.G.-S. The first draft of the manuscript was written by I.R.-C., R.G.-T., J.L.R.-Z. and S.R.G.-S. commented on previous version of the manuscript, developing a complete discussion of results. The final version of the manuscript was performed by I.R.-C. All authors have read and agreed to the published version of the manuscript.

Funding: This research received no external funding.

Institutional Review Board Statement: Not applicable.

Informed Consent Statement: Not applicable.

Data Availability Statement: Data are contained within the article.

Acknowledgments: Technical support from J. Macedonio, A. Aguilar is recognized.

Conflicts of Interest: The authors declare no conflicts of interest.

References

1. Christiansen, T.; Somers, M.A.J. Characterization of low temperature surface hardened stainless steel. *Struers J. Metallogr.* **2006**, *9*, 2–17.
2. Munz, W.D.; Hofmann, D.; Harting, K. A high rate sputtering process for the formation of hard-reducing TiN coatings of tools. *Thin Solid Films* **1982**, *96*, 79–86. [CrossRef]
3. Devaraju, A.; Elayaperumal, A.; Alphonsa, J.; Kailas, S.V.; Venugopal, S. Microstructure and dry sliding wear resistance evaluation of plasma nitrided austenitic stainless steel type AISI 316LN against different sliders. *Surf. Coat. Technol.* **2012**, *207*, 406–412. [CrossRef]
4. Lin, N.; Liu, Q.; Zou, J.; Guo, J.; Li, D.; Yuan, S.; Ma, Y.; Wang, Z.; Wang, Z.; Tang, B. Surface Texturing-Plasma Nitriding Duplex Treatment for Improving Tribological Performance of AISI 316 Stainless Steel. *Materials* **2016**, *9*, 875. [CrossRef]
5. De Almeida, E.A.D.S.; Da Costa, C.E.; Milan, J.C.G. Study of the nitrided layer obtained by different nitriding methods. *Matéria* **2015**, *20*, 460–465. [CrossRef]
6. Günzel, R.; Betzl, M.; Alphonsa, I.; Ganguly, B.; John, P.; Mukherjee, S. Plasma-source ion implantation compared with glow-discharge plasma nitriding of stainless steel. *Surf. Coat. Technol.* **1999**, *112*, 307–309. [CrossRef]
7. Liang, W.; Juncai, S.; Xiaolei, X. Low pressure plasma arc source ion nitriding compared with glow-discharge plasma nitriding of stainless steel. *Surf. Coat. Technol.* **2001**, *145*, 31–37. [CrossRef]
8. Qayyum, A.; Naveed, M.A.; Zeb, S.; Murtaza, G.; Zakaullah, M. Glow Discharge Plasma Nitriding of AISI 304 Stainless Steel. *Plasma Sci. Technol.* **2007**, *9*, 463–468. [CrossRef]
9. Grigoriev, S.; Metel, A.; Volosova, M.; Melnik, Y.; Ney, H.A.; Mustafaev, E. Surface Hardening of Massive Steel Products in the Low-pressure Glow Discharge Plasma. *Technologies* **2019**, *7*, 62. [CrossRef]
10. Ashrafizadeh, F. Influence of plasma and gas nitriding on fatigue resistance of plain carbon (Ck45) steel. *Surf. Coat. Technol.* **2003**, *173–174*, 1196–1200. [CrossRef]
11. Foadian, F.; Feyzi, F.; Haghighat, S.; Aghajani, H. Nitrogen diffusion in different microstructures of plasma nitrided CK45 steel. *Mater. Sci. Technol.* **2013**, *30*, 86–90. [CrossRef]
12. Yamada, Y.; Hirohito, E.; Takahashi, K. Influence of Crystal Structure of Nitride Compound Layer on Torsion Fatigue Strength of Alloy Steel. *Metals* **2019**, *9*, 1352. [CrossRef]
13. Allen, C.; Li, C.X.; Bell, Y.; Sun, Y. The effect of fretting on the fatigue behaviour of plasma nitrided stainless steels. *Wear* **2003**, *254*, 1106–1112. [CrossRef]
14. Yıldız, F.; Yetim, A.; Alsaran, A.; Çelik, A.; Kaymaz, İ. Fretting fatigue properties of plasma nitrided AISI 316L stainless steel: Experiments and finite element analysis. *Tribol. Int.* **2011**, *44*, 1979–1986. [CrossRef]
15. Gokcekaya, O.; Ergun, C.; Gulmez, T.; Nakano, T.; Yilmaz, S. Structural Characterization of Ion Nitrided 316L Austenitic Stainless Steel: Influence of Treatment Temperature and Time. *Metals* **2022**, *12*, 306. [CrossRef]
16. Oliveira, R.M.; Ueda, M.; Silva, L.L.G.; Reuther, H.; Lepienski, C.M. Characteristics of austenitic stainless steel nitrided in a hybrid glow discharge plasma. *Braz. J. Phys.* **2009**, *39*, 554–558. [CrossRef]
17. Li, G.Y.; Lei, M.K. Microstructure and Properties of Plasma Source Nitrided AISI 316 Austenitic Stainless Steel. *J. Mater. Eng Perform* **2017**, *26*, 418–423. [CrossRef]
18. Dos Santos de Almeida, E.A.; Giubilei Milan, J.C.; Edil da Costa, C. Acquired Properties Comparison of Solid Nitriding, Gas Nitriding and Plasma Nitriding in Tool Steels. *Mater. Res.* **2015**, *18*, 27–35. [CrossRef]
19. Zhang, Z.L.; Bell, T. Structure and corrosion resistance of plasma nitrided stainless steel. *Surf. Eng.* **1985**, *1*, 131–136. [CrossRef]
20. Baranowska, J. Characteristic of the nitrided layers on the stainless steel at low temperature. *Surf. Coat. Technol.* **2004**, *180–181*, 145–149. [CrossRef]
21. Drouet, M.; Stinville, J.C.; Villechaise, P.; Rivière, J.P.; Templier, C. Surface evolution during low temperature plasma assisted nitriding of austenitic stainless steel. *Eur. Phys. J. Appl. Phys.* **2008**, *43*, 349–351. [CrossRef]
22. Wei, R.; Vajo, J.J.; Matossian, J.N.; Wilbur, P.J.; Davis, J.A.; Williamson, D.L.; Collins, G.A. A comparative study of beam ion implantation, plasma ion implantation and nitriding of AISI 304 stainless steel. *Surf. Coat. Technol.* **1996**, *83*, 235–242. [CrossRef]
23. Liang, W. Surface modification of AISI 304 austenitic stainless Steel by plasma nitriding. *Appl. Surf. Sci.* **2003**, *211*, 308–314. [CrossRef]
24. Krawczyńska, A.T.; Zdunek, J.; Sitek, R.; Lewandowska, M. Formation of the Nitrided Layers on an Austenitic Stainless Steel with Different Grain Structures. *Adv. Eng. Mater.* **2017**, *20*, 1701049. [CrossRef]
25. Li, Y.; Wang, Z.; Wang, L. Surface properties of nitrided layer on AISI 316L austenitic stainless steel produced by high temperature plasma nitriding in short time. *Appl. Surf. Sci.* **2014**, *298*, 243–250. [CrossRef]

26. Díaz-Guillén, J.C.; Granda-Gutiérrez, E.E.; Vargas-Gutiérrez, G.; Díaz-Guillén, M.R.; Aguilar-Martínez, J.A.; Álvarez-Contreras, L. Effect of Nitriding Current Density on the Surface Properties and Crystallite Size of Pulsed Plasma-Nitrided AISI 316L. *J. Mater. Sci. Chem. Eng.* **2015**, *03*, 45–51. [CrossRef]
27. Kochmański, P.; Długozima, M.; Baranowska, J. Structure and Properties of Gas-Nitrided, Precipitation-Hardened Martensitic Stainless Steel. *Materials* **2022**, *15*, 907. [CrossRef]
28. Gorbachev, I.; Popov, V.V.; Pasynkov, A.Y. Thermodynamic Simulation of the Formation of Carbonitrides in Steels with Nb and Ti. *Phys. Met. Metallogr.* **2012**, *113–117*, 687–695. [CrossRef]
29. Binder, C.; Bendo, T.; Hammes, G.; Klein, A.N.; de Mello, J.D.B. Effect of nature of nitrided phases on sliding wear of plasma nitrided sintered iron. *Wear* **2015**, *332–333*, 995–1005. [CrossRef]
30. Wang, L.; Ji, S.; Sun, J. Effect of nitriding time on the nitrided layer of AISI 304 austenitic stainless steel. *Surf. Coat. Technol.* **2006**, *200*, 5067–5070. [CrossRef]
31. Xu, X.; Yu, Z.; Wang, L.; Qiang, J.; Hei, Z. Phase depth distribution characteristics of the plasma nitrided layer on AISI 304 stainless steel. *Surf. Coat. Technol.* **2003**, *162*, 242–247. [CrossRef]
32. Mittemeijer, E.J.; Vogels, A.B.P.; Van Der Schaaf, P.J. Morphology and lattice distortions of nitrided iron and iron-chromium alloys and steels. *J. Mater. Sci.* **1980**, *15*, 3129–3140. [CrossRef]
33. Tang, D.; Zhang, C.; Zhan, H.; Huang, W.; Ding, Z.; Chen, D.; Cui, G. High-Efficient Gas Nitridation of AISI 316L Austenitic Stainless Steel by a Novel Critical Temperature Nitriding Process. *Coatings* **2023**, *13*, 1708. [CrossRef]

coatings

MDPI

Article

Effect of Coating Treatment on the Properties of Extruded Mg-1.0Zn-0.3Zr-1.0Y-2.0Sn Alloys

Junguang He [1,*], Qinglei Gao [1], Jiuba Wen [1,2], Yuan Gong [1], Wuyun Feng [1], Zhenfei Cheng [1] and Xuyang Li [1]

[1] School of Materials Science and Engineering, Henan University of Science and Technology, Luoyang 471023, China; 15236470608@163.com (Q.G.); wenjiuba12@163.com (J.W.); gy199802121234@163.com (Y.G.); fengwuyun2022@163.com (W.F.); ly1849974632@163.com (Z.C.); lixuyang_0301@163.com (X.L.)

[2] Collaborative Innovation Center of Nonferrous Metals, Luoyang 471023, China

[*] Correspondence: he.ellen@163.com

Abstract: The impact of fluoride-based coatings on the microstructure and mechanical integrity of extruded Mg-1.0Zn-0.3Zr-1.0Y-2.0Sn alloys was assessed utilizing optical microscopy (OM), scanning electron microscopy (SEM), X-ray diffraction (XRD), immersion testing, electrochemical analysis, and tensile testing. It was observed that the magnesium alloys could be immersed in hydrofluoric acid (HF) for varying durations to achieve coatings of distinct thicknesses, with the coating thickness stabilizing at approximately 8 μm after a 48 h immersion period. The application of the fluoride coating significantly enhanced the corrosion resistance of the alloys, with a corrosion rate (CR_H) of 0.13 ± 0.012 mm/y. Upon a 20-day immersion in simulated body fluid (SBF), the degradation rates of the yield strength (YS), tensile strength (UTS), and elongation (EL) for the cast alloys were recorded as 62%, 59%, and 64%, respectively. For the extruded alloys, these rates escalated to 77%, 76%, and 95%. In contrast, the fluorine-coated alloys exhibited significantly lower degradation rates of 28%, 23%, and 39% after a 25-day immersion in SBF. Upon extrusion, the specimens exhibit a diminished corrosion resistance and a more substantial decline in mechanical properties compared to their as-cast state. Upon the application of the coating, there is a discernible reduction in the rate of mechanical property degradation observed in the specimens. This indicates that the fluorinated coating can mitigate the corrosion rate and enhance the corrosion resistance of magnesium alloys.

Keywords: Mg-1.0Zn-0.3Zr-1.0Y-2.0Sn; fluoride coating; corrosion resistance

Citation: He, J.; Gao, Q.; Wen, J.; Gong, Y.; Feng, W.; Cheng, Z.; Li, X. Effect of Coating Treatment on the Properties of Extruded Mg-1.0Zn-0.3Zr-1.0Y-2.0Sn Alloys. *Coatings* **2024**, *14*, 1350. https://doi.org/10.3390/coatings14111350

Academic Editor: Michał Kulka

Received: 27 September 2024
Revised: 13 October 2024
Accepted: 18 October 2024
Published: 24 October 2024

1. Introduction

The aging of the global population represents an inescapable social challenge, with an associated escalation in the prevalence of bone and cardiovascular diseases. Consequently, there is an escalating demand for biomaterials. Bio-magnesium alloys, characterized by their biodegradability and biocompatibility, hold promising potential in cardiovascular scaffolding and orthopedic repair applications, positioning them as candidates for the next generation of degradable biomaterials [1–5]. Despite their promise, Mg alloys encounter several challenges: 1. Qi et al. implanted extruded ZK60 samples into the right humerus of mice and found that the biocompatibility of magnesium alloy needs to be further improved [6]. 2. Xie et al., through the study of the development of magnesium alloys in recent years, put forward the problem of coordinating the improvement of magnesium alloy strength and corrosion resistance [7]. 3. The low standard electrode potential of Mg and the porous nature of its surface oxide layer predispose Mg alloys to an increased susceptibility to corrosion [8]. Hence, the alignment of the mechanical properties and corrosion resistance in Mg alloys is a critical issue that demands immediate resolution to facilitate the advancement and utilization of bio-magnesium alloys in the biomedical sector.

Current strategies to enhance the mechanical properties and corrosion resistance of Mg alloys encompass alloying, heat treatment, hot deformation, and surface modification, with alloying and hot extrusion being particularly prevalent. Regarding the alloying elements, Zn serves as a prevalent alloying element, which significantly ameliorates both the mechanical integrity and corrosion resistance of the alloys [9]. Zr is known to refine the grain structure of Mg alloys, thereby diminishing the brittleness and concurrently augmenting the alloys' strength [10,11]. The addition of Y also confers improvements in the mechanical properties and corrosion resistance of Mg alloys [12,13]. Furthermore, the incorporation of Sn into Mg alloys facilitates the formation of the Mg_2Sn phase, which, due to its high thermal stability and fine-grained microstructure, contributes to the enhancement of the alloys' mechanical characteristics [14–16]. These alloying elements and processes are pivotal in the ongoing development of Mg alloys for biomedical applications, where both strength and durability are paramount.

The mechanical characteristics of the Mg-1.0Zn-0.3Zr-1.0Y-2.0Sn alloy are substantially enhanced through hot extrusion and deformation processes; however, this treatment concurrently diminishes its corrosion resistance, rendering it suboptimal for biomedical applications. Surface modification techniques offer a promising solution to this issue. These methods can produce a variety of coatings, including inorganic, polymeric, and composite layers, which have been extensively documented in the literature [17–20]. A specific surface modification approach involves immersing the magnesium alloy in an HF solution, which results in the formation of a fluorinated layer on the alloy's surface. This layer serves as a protective barrier for the alloy matrix and effectively mitigates the corrosion rate [21–23]. Additionally, Weber et al. [24] conducted an investigation employing a sinus $MgNd_2$ implant coated with MgF_2, which exhibited a superior histocompatibility and controlled degradation over a six-month observation period. The study revealed that the controlled release of fluoride ions during the degradation of this surface coating is considered non-toxic and biocompatible.

In this study, we have utilized an extruded Mg-1.0Zn-0.3Zr-1.0Y-2.0Sn alloy as a substrate to investigate the impact of varying durations of HF immersion on the microstructure, topography, and chemical composition of the resulting fluoride coating. The objective is to enhance the alloy's mechanical integrity and corrosion resistance, thereby developing a more robust biodegradable alloy.

2. Experimental Procedure

2.1. Material Preparation

Mg (99.95%), Zn (99.99%), and Mg-30%Zr, Mg-20%Y, and Mg-20%Sn were employed to fabricate the as-cast Mg-1.0Zn-0.3Zr-1.0Y-xSn (x = 0, 0.5, 1, 1.5, 2, 2.5, 3) alloy series. The alloy preparation was conducted in a vacuum induction melting furnace (ZGJL0.01-40-4) under a protective atmosphere consisting of CO_2 (99% by volume) and SF_6 (1% by volume). The melting temperature was maintained at 750 °C, while the casting temperature was set at 720 °C. Prior to casting, the mold was preheated to 200 °C to ensure uniform heat distribution. Subsequently, the ingots were sectioned into billets with dimensions of Φ49 × 36 mm. Hot extrusion of these billets was performed at various temperatures, specifically, 340 °C, 360 °C, 380 °C, and 400 °C, employing an extrusion ratio of 10:1 and an extrusion velocity of 5 mm/min. The outcome of the hot extrusion process resulted in billets with dimensions of Φ16 mm × 260 mm.

Specimens for fluoride-coating treatment were selected from the extrusion bar produced at an extrusion temperature of 360 °C with an extrusion ratio (λ) of 10. The central portion of the extrusion bar was wire-cut to obtain specimens with dimensions of Φ11.3 mm × 8 mm, ensuring that the test surface of each specimen was aligned parallel to the extrusion direction. The specimens underwent a sequential polishing process using water-saturated silicon carbide abrasive papers with grit sizes of 240#, 600#, 800#, 1500#, and 2000#. The polished specimens were ultrasonically cleaned in alcohol for 5 min, then ultrasonically cleaned in acetone solution for 5 min, and then ultrasonically cleaned in alcohol for 5 min, and, finally, the specimens were taken out and placed in a drying oven for 10 min. The final preparation step involved immersing the specimens in a 40% HF solution. The specimens were soaked in the HF solution for durations of 12, 24, 48, and 72 h to ensure adequate and uniform fluoride-coating penetration.

2.2. Performance Testing

The microstructures of the as-cast alloy and the extruded variants, aligned with the parallel extrusion axis, were examined via optical microscopy. The etching solution employed was a picric-acid-based mixture, comprising 4.5 g of picric acid, 100 mL of anhydrous ethanol, 9 mL of deionized water, and 5 mL of glacial acetic acid. Grain size measurements of the alloys under scrutiny were conducted utilizing the Nano-Measurer software (1.2.5). The morphologies and composition of the second phase of the investigated alloys were characterized by SEM (JSM-5610LV, Tokyo, Japan, JEOL Ltd.) and EDS.

The tensile properties of specimens oriented perpendicular to the extrusion direction were evaluated at ambient temperature using an electronic universal testing machine (DNS100, China, SINOTEST). The specimens adhered to the GB/T 228.1-2010 standard, with dimensions of 2 mm in thickness, 3.5 mm in width, and 15 mm in gauge length [25]. The tensile tests were conducted at a strain rate of 1 mm/min, employing five replicate specimens for each experimental trial.

2.3. Immersion Corrosion Tests

Immersion tests were conducted on specimens of size Φ18 mm × 5 mm, submerged in SBF at a temperature of 37 ± 0.5 °C. The ratio of SBF volume to specimen surface area was maintained at 30 mL/cm^2. The specimens were immersed for a duration of 120 h, with the SBF solution being refreshed every 24 h. Hydrogen evolution was quantified daily using a gas collection apparatus. Post-immersion, the corrosion byproducts were eliminated using a boiling chromic acid solution composed of 3 g $AgNO_3$, 60 g CrO_3, and 300 mL deionized water. The mass loss was ascertained with a precision electronic analytical balance, accurate to 0.1 mg. The chemical constituents of the SBF are detailed in Table 1. The corrosion rate, derived from the mass loss, was calculated according to the formula referenced in [26]

$$CR_{\mathrm{w}} = (K_1 \times W)/(A \times T \times D) \tag{1}$$

$$CR_{\mathrm{H}} = K_2 \times V_H \tag{2}$$

where CR_{w} is the corrosion rate (mm/y), CR_{H} is the hydrogen evolution corrosion rate (mm/y), K_1 = 8.76 × 10^4, K_2 = 2.088, W is the weight loss value (g), A is the specimen surface area, T is the immersion time (h), D is the density of material (g/cm^3), and V_H is the total amount of hydrogen change (mL/cm^2/d).

Table 1. Chemical composition of the SBF (g/L).

$NaCl$	$CaCl_2$	KCl	$NaHCO_3$	$MgCl_2 \cdot 6H_2O$	$C_6H_{12}O_6$	$Na_2HPO_4 \cdot 12\,H_2O$	KH_2PO_4	$MgSO_4 \cdot 7H_2O$
8.00	0.14	0.40	0.35	0.10	1.00	0.06	0.06	0.06

Electrochemical assessments were conducted utilizing an Autolab electrochemical workstation (model AUT84580) within a three-electrode cell configuration. The working electrodes consisted of specimens with an exposed surface area of 1 cm^2. A graphite sheet served as the counter electrode, while a saturated calomel electrode (SCE) functioned as the reference electrode. The reported potentials are referenced to the SCE. Electrochemical impedance spectroscopy (EIS) measurements were initiated once the specimens, after being immersed in SBF for 1 h, had reached a stable open-circuit potential (OCP). The EIS was executed with an AC perturbation amplitude of 5 mV, spanning a frequency spectrum from 10 kHz to 0.1 Hz. Subsequently, the polarization curve was derived at a uniform potential scan rate of 5 mV/s, ranging from −1.9 V to −1.1 V.

3. Results and Discussion

3.1. Coating Morphology and Structure

The cross-sectional morphology of the fluorinated coating, following a 48 h immersion period, and the corresponding EDS analysis are depicted in Figure 1,the table in the figure is the red cross composition. Figure 2 illustrates the line-sweep morphologies of the cross-sections of the extruded Mg-1.0Zn-0.3Zr-1.0Y-2.0Sn alloy coatings after immersion in HF for intervals of 12, 24, 48, and 72 h. The EDS analysis (Figure 1) reveals that the coating is predominantly composed of Mg, F, and O elements, with an atomic ratio of Mg to F of approximately 1:2 and Mg to O of approximately 1:1, suggesting the presence of MgF_2 and MgO as the primary constituents. The line scan analysis (Figure 2) indicates an initial increase followed by a stabilization in the coating thickness with extended soaking times. The thickness measurements were approximately 2, 4, 8, and 8 μm at 12, 24, 48, and 72 h of immersion, respectively. Notably, at 48 and 72 h, the coating thicknesses were comparable, suggesting that the maximum thickness is achieved at 48 h. Collectively, the EDS and line-scan data indicate a progressive increase in coating thickness that plateaus with prolonged immersion. Elemental distribution patterns from the cross-sectional EDS and line-scan analyses confirm the coating's composition as primarily MgF_2 and MgO, corroborating the findings of Shi et al. [27].

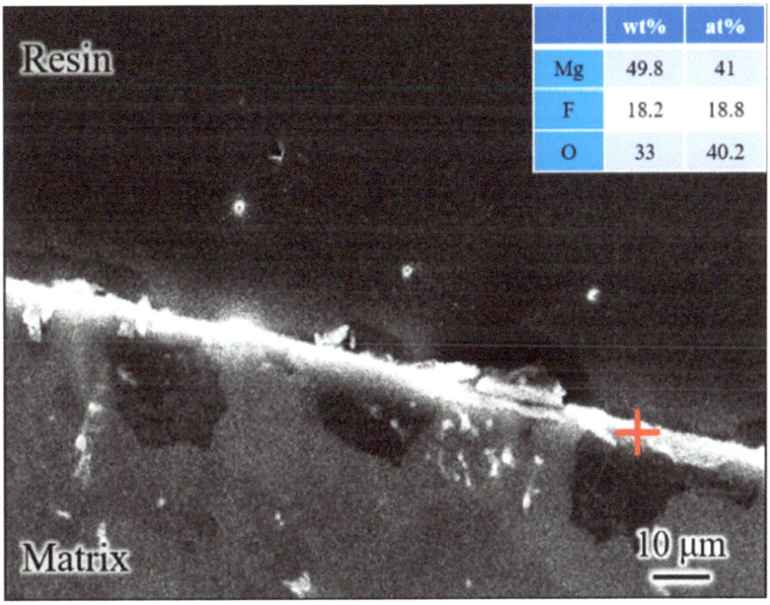

Figure 1. Cross-section morphology and EDS of fluorinated-coating (immersion time 48 h) alloy.

The analysis of the surface morphology and elemental composition of the fluorinated coating, following a 48 h immersion period, is presented in Figure 3. The surface scan data indicate that the coating's surface is predominantly composed of Mg, F, and O, with the highest concentrations observed for these elements. In contrast, the concentrations of Sn, Y, Zn, and Zr are minimal and effectively negligible. By integrating the findings from the EDS, line scan, and surface scan analyses as depicted in Figures 1–3, respectively, it is evident that the coating is primarily constituted by MgF_2 and MgO.

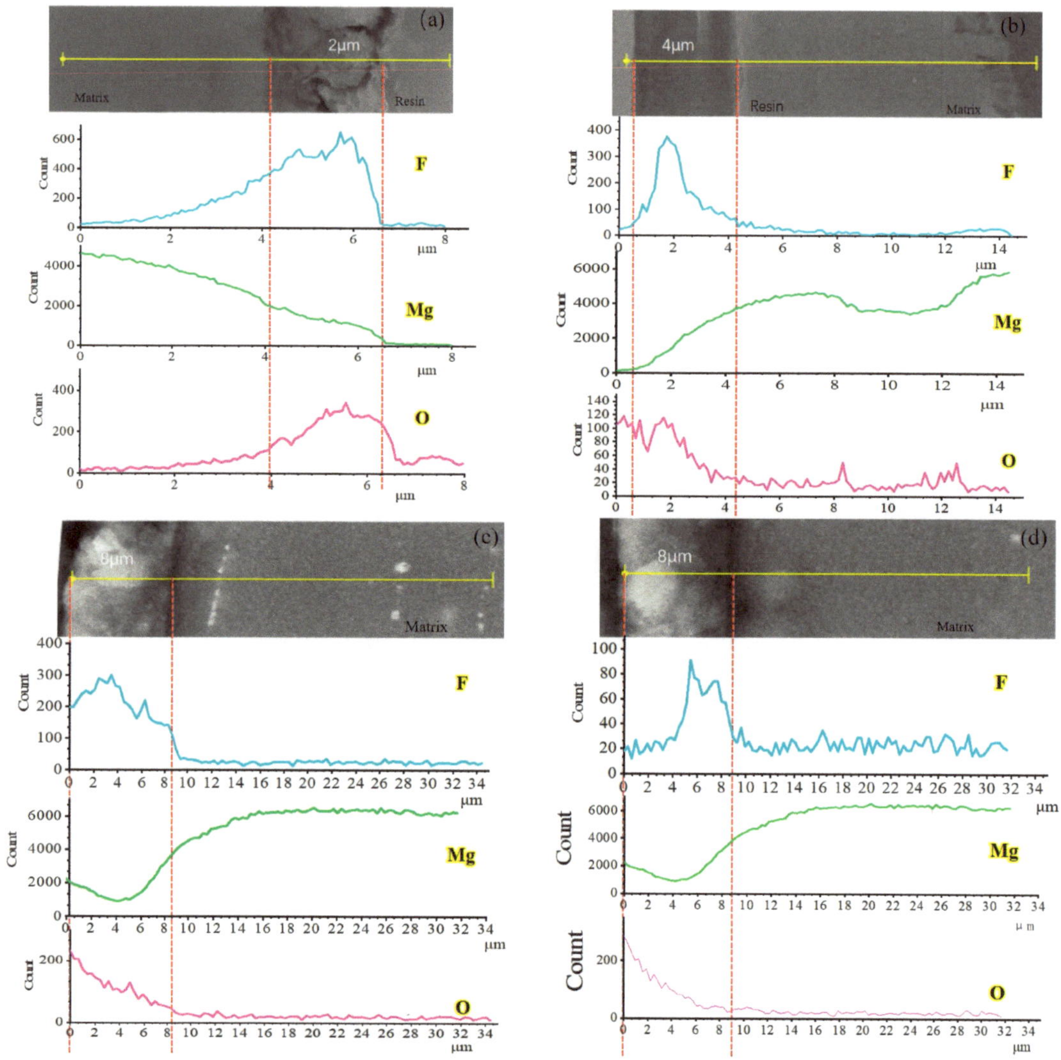

Figure 2. Cross-section morphology of extruded Mg-1.0Zn-0.3Zr-1.0Y-2.0Sn alloy at different soaking times in HF: (**a**)12 h; (**b**) 24 h; (**c**) 48 h; and (**d**) 72 h.

3.2. *Physical Composition of Coatings*

To elucidate the phase composition of the fluorinated coating, XRD was conducted on the specimens, with the findings presented in Figure 4. The XRD analysis revealed that the surface of the coating is predominantly composed of phases of MgF_2, MgO, ZnO, and ZrF_4. These findings are in closer alignment with the XRD results, which indicated atomic ratios of approximately 1:2 for Mg to F and 1:1 for Mg to O, as depicted in Figure 3. Consequently, it is postulated that the coating is primarily constituted by MgF_2 and MgO.

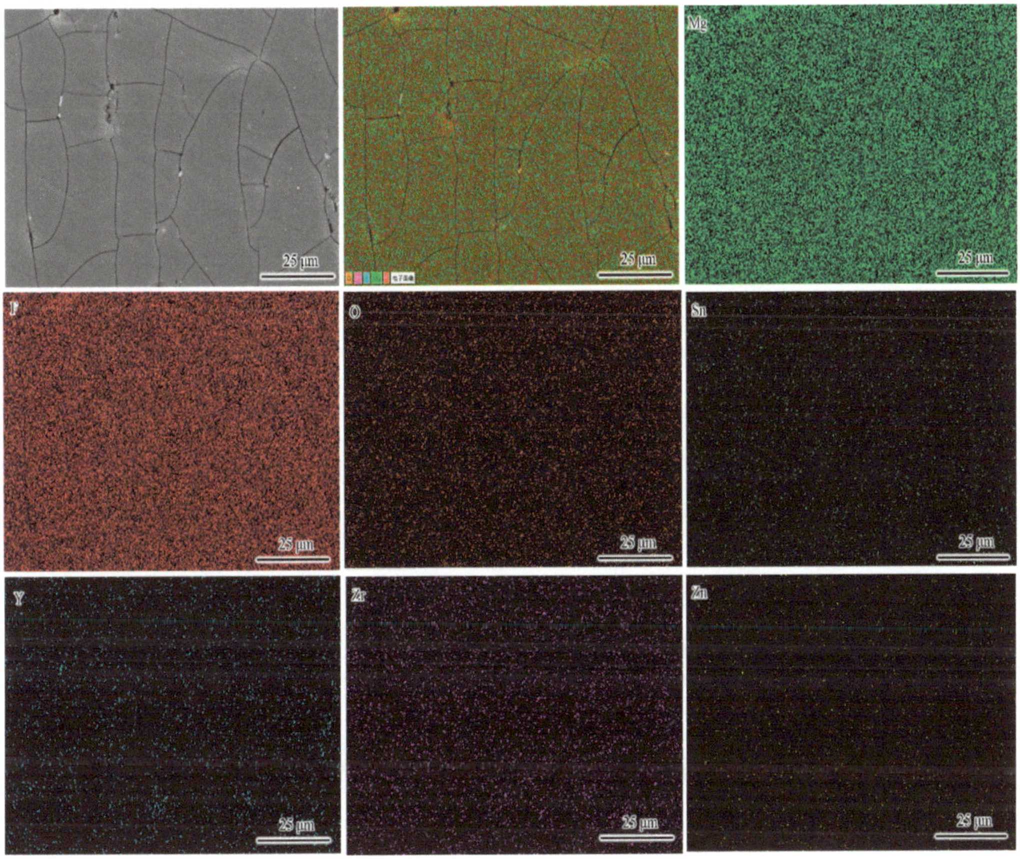

Figure 3. Element distribution of fluorinated-coating (immersion time of 48 h) alloy.

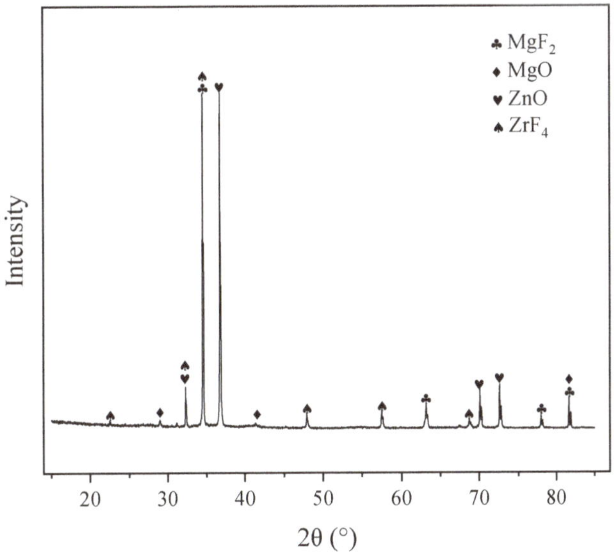

Figure 4. XRD of fluorinated-coated (48 h immersion) alloys.

3.3. Effect of Fluorinated Coatings on Mechanical Properties of Alloys in the Extruded

Figure 5 shows the macroscopic corrosion morphology of the Mg-1.0Zn-0.3Zr-1.0Y-2.0Sn alloy in cast, extruded, and fluoride-coated tensile specimens before corrosion in SBF (0 days), 10 days, and 20 days of corrosion. The corrosion products on the surface of the alloys in different states gradually increased with the increase in immersion time. It is noteworthy that the corrosion on the surface of the uncoated specimen is more severe compared to the surface of the fluorinated-coated specimen. When the cast specimens were immersed in SBF for 10 days, a small amount of corrosion products appeared on the surface of the specimens, and continued to be immersed, and, at 20 days, the surface of the specimens was basically covered with corrosion products. For extruded specimens in SBF immersed for 10 days, most of the specimen surface was covered with corrosion products, with the extension of the immersion time; in 20 days, the specimen surface has been completely covered with corrosion products. For fluorinated-coating specimens in SBF immersed in 10 days, the surface of the coating is basically intact, with almost no corrosion products; with the extension of the immersion time, the production of H_2 led to the specimen surface coating cracking; in 20 days, the specimen surface is covered with a small amount of irregularly distributed corrosion products. The comparison revealed that the alloys corrode differently in different states, with the extruded-state specimens corroding the most severely and the fluorine-coated specimens the least. This indicates that the presence of fluoride coating on the surface can effectively improve the corrosion resistance of the alloy.

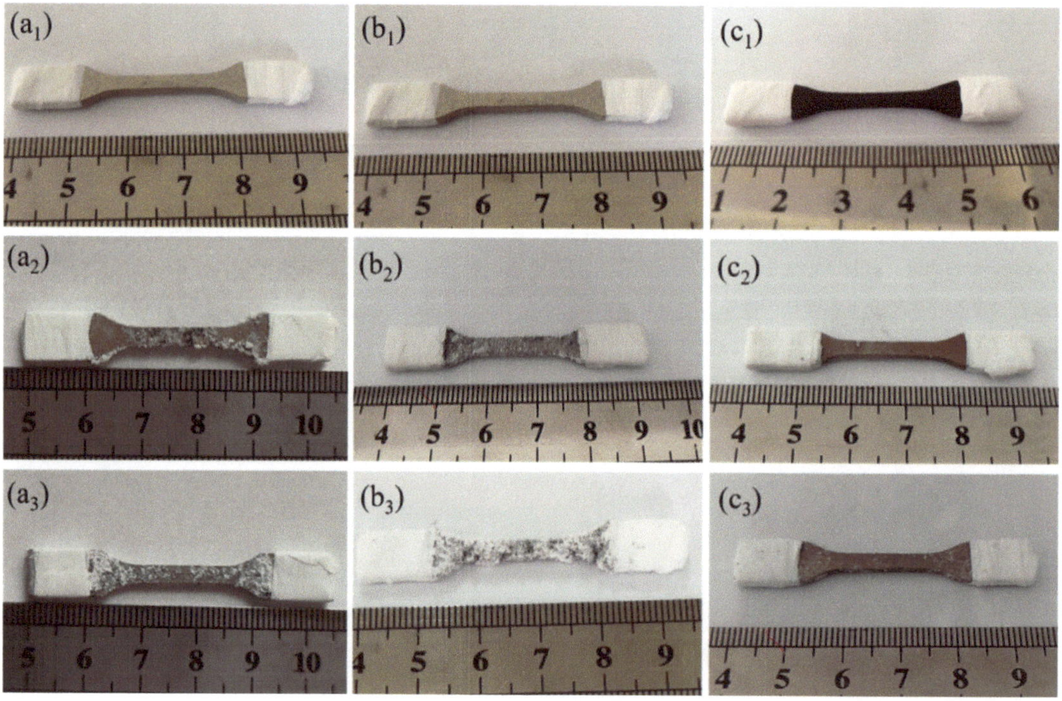

Figure 5. Macroscopic corrosion morphology of Mg-1.0Zn-0.3Zr-1.0Y-2.0Sn alloy extruded and fluorinated-coating tensile specimens soaked in SBF for different times: as-cast: (**a$_1$**) 0; (**a$_2$**) 10 d; and (**a$_3$**) 20 d; extruded alloy: (**b$_1$**) 0; (**b$_2$**) 10 d; and (**b$_3$**) 20 d; and fluorinated-coated: (**c$_1$**) 0; (**c$_2$**) 10 d; and (**c$_3$**) 20 d.

Figure 6 delineates the variation curves of YS, UTS, and EL as a function of the immersion time in SBF for Mg-1.0Zn-0.3Zr-1.0Y-2.0Sn alloys in their as-cast, extruded, and fluoride-coated conditions (data in Tables 2–4). Prior to immersion, the mechanical properties of the extruded and fluoride-coated specimens were found to be similar, suggesting that the application of the fluoride coating had a negligible impact on the alloy's microstructure. The analysis of the data presented in the figure indicates that the YS, UTS, and EL of the as-cast, extruded, and fluoride-coated Mg-1.0Zn-0.3Zr-1.0Y-2.0Sn alloy specimens exhibit a progressive decline with extended immersion in SBF. Nevertheless, the degradation of the mechanical properties in the fluoride-coated specimens was considerably less pronounced compared to the as-cast and extruded counterparts. The YS, UTS, and EL of the cast specimens decreased from 154 ± 10 MPa, 211 ± 10 MPa, and 19.1 ± 0.5% to 58 ± 10 MPa, 86 ± 10 MPa, and 6.8 ± 0.3% after 20 days of immersion in SBF, resulting in attenuation rates of 62%, 59%, and 64%, respectively. After 25 days of immersion, the YS, UTS, and EL further declined to 31 ± 9 MPa, 46 ± 9 MPa, and 2.2 ± 0.4%, with decay rates of 80%, 78%, and 88%, respectively. Consistent with the observations for the as-cast specimens, the extruded Mg-1.0Zn-0.3Zr-1.0Y-2.0Sn alloy specimens exhibited a significant reduction in YS, UTS, and EL after a 20-day immersion in SBF. Specifically, the YS, UTS, and EL values diminished from 221 ± 9 MPa, 277 ± 9 MPa, and 22.3 ± 0.5% to 52 ± 10 MPa, 68 ± 10 MPa, and 1.2 ± 0.5%, respectively, corresponding to decay rates of 77%, 76%, and 95%. After 25 days, the specimens were too severely corroded to facilitate the accurate measurement of their mechanical properties. Fluoride-coated specimens, however, displayed a more moderate decline in mechanical properties. After 20 days of immersion in SBF, the YS, UTS, and EL values decreased from 219 ± 9 MPa, 272 ± 9 MPa, and 22.1 ± 0.4% to 172 ± 8 MPa, 229 ± 10 MPa, and 15.7 ± 0.6%, with respective attenuation rates of 21%, 15%, and 29%. By the 25th day, these values further reduced to 158 ± 9 MPa, 210 ± 8 MPa, and 13.3 ± 0.4%, with decay rates of 28%, 23%, and 39%, respectively. The comparison of mechanical property attenuation between the extruded and fluoride-coated specimens indicates that the fluoride-coated specimens experienced the lowest degree of degradation. This finding underscores the protective role of the fluoride coating in reducing the corrosion rate and mitigating the degradation of the alloy's mechanical properties.

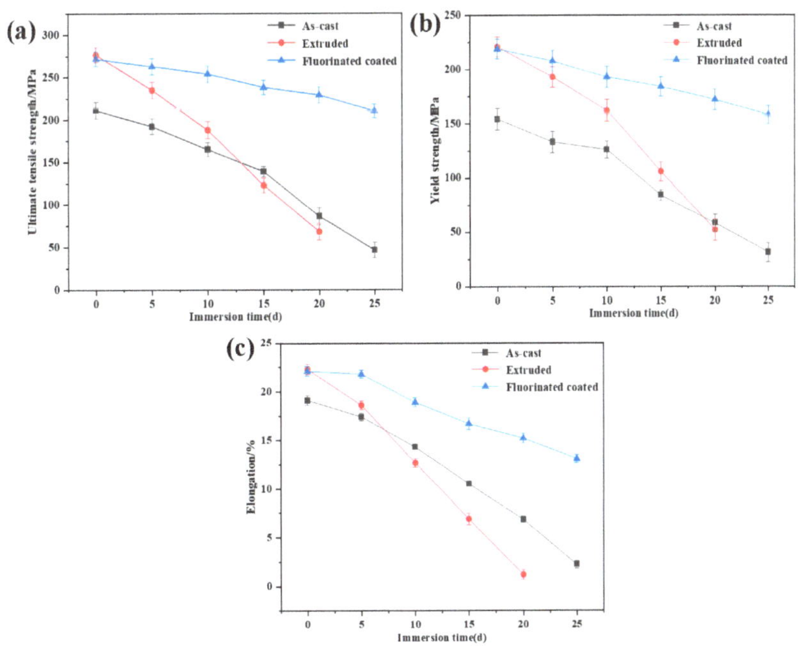

Figure 6. Mechanical properties of Mg-1.0Zn-0.3Zr-1.0Y-2.0Sn alloy as-cast, extruded, and fluorinated-coated tensile specimens soaked in SBF for different times: (**a**) tensile strength; (**b**) yield strength; and (**c**) elongation rate.

Table 2. Mechanical properties of as-cast Mg-1.0Zn-0.3Zr-1.0Y-2.0Sn alloy in Figure 6.

As-Cast	0 d	5 d	10 d	15 d	20 d	25 d
UTS (MPa)	211 ± 10	192 ± 9	165 ± 8	139 ± 6	86 ± 10	46 ± 9
YS (MPa)	154 ± 10	133 ± 10	126 ± 8	84 ± 5	58 ± 8	31 ± 9
EL (%)	19.1 ± 0.5	17.4 ± 0.4	14.3 ± 0.2	10.5 ± 0.2	6.8 ± 0.3	2.2 ± 0.4

Table 3. Mechanical properties of extruded Mg-1.0Zn-0.3Zr-1.0Y-2.0Sn alloy in Figure 6.

Extruded	0 d	5 d	10 d	15 d	20 d	25 d
UTS (MPa)	277 ± 9	235 ± 10	188 ± 10	123 ± 9	68 ± 10	-
YS (MPa)	221 ± 9	193 ± 10	162 ± 10	106 ± 9	52 ± 10	-
EL (%)	22.3 ± 0.5	18.6 ± 0.4	12.7 ± 0.5	6.9 ± 0.6	1.2 ± 0.5	-

Table 4. Mechanical properties of extruded Mg-1.0Zn-0.3Zr-1.0Y-2.0Sn alloy with fluorinated coating in Figure 6.

Coated	0 d	5 d	10 d	15 d	20 d	25 d
UTS (MPa)	272 ± 8	263 ± 10	254 ± 10	238 ± 8	229 ± 10	210 ± 8
YS (MPa)	219 ± 9	208 ± 9	193 ± 10	184 ± 9	172 ± 8	158 ± 9
EL (%)	22.1 ± 0.4	21.8 ± 0.5	18.9 ± 0.4	16.9 ± 0.4	15.7 ± 0.6	13.3 ± 0.4

Figure 7 presents the SEM fracture morphologies of the Mg-1.0Zn-0.3Zr-1.0Y-2.0Sn alloy specimens in the as-cast and extruded states, as well as those with fluorinated coatings, before corrosion (0 days) and after 10 and 20 days of corrosion in SBF, parallel to the extrusion direction. The surface corrosion of the specimens intensified with increasing immersion time; however, the fluoride-coated specimens exhibited significantly fewer and less severe corrosion pits compared to the as-cast and extruded specimens. Corroborated by the data in Table 2, a positive correlation is observed between the severity of surface corrosion and the degree of mechanical property attenuation in the alloy. In the as-cast specimens, Figure 7a–c reveals the presence of numerous pits of varying sizes on the fracture surface, attributed to the abundant Cl- in the SBF, which facilitate the formation of galvanic corrosion cells between the Mg matrix and secondary phases. Figure 7d–f depicts a similar, yet more severe, corrosion pattern on the extruded specimens, ascribed to the increased susceptibility to galvanic corrosion due to a higher concentration of secondary phases post-extrusion. As the immersion time extends, the number and depth of pitting corrosion sites escalate, leading to preferential fracture initiation at these severely corroded areas. The fluoride-coated specimens, as shown in Figure 7g–i, exhibit a fish-scale-like surface morphology, most pronounced after 0 days of SBF immersion. This morphology, also visible in Figure 3, is indicative of a mixture of MgF_2 and MgO. The fish-scale pattern arises from the higher elastic modulus of the coating compared to the magnesium matrix, resulting in a lack of cohesive plastic deformation during tensile stress. This disparity in elastic properties leads to a continuous rupture and the formation of fish-scale-like blocky features. Prolonged immersion reduces the coating's coverage area, suggesting gradual delamination. This may be attributed to the accumulation of corrosion products and the release of H_2, which exerts internal pressure at the substrate–coating interface, ultimately compromising the coating integrity [28].

An examination of the tensile fracture surfaces of the cast and extruded Mg-1.0Zn-0.3Zr-1.0Y-2.0Sn alloy specimens after various durations of immersion in SBF reveals that fracture initiation predominantly occurs at sites of severe corrosion, namely, the corrosion pits. These pits possess a more negative potential relative to the surrounding specimen surface, thereby acting as anodes in the corrosion cell. The larger surface area of the specimen in comparison to the pit area establishes a large-cathode-to-small-anode configuration, which intensifies the corrosion process. This configuration promotes the progression of pitting corrosion along the depth, culminating in a precipitous decline in the specimen's mechanical properties. Post-coating, while the specimens exhibited reduced corrosion, fluoride-coated specimens, nonetheless, accumulated corrosion products at the tensile fracture surface following SBF immersion. An escalation in the quantity of corrosion products was observed with prolonged immersion periods, signifying a gradual diminishment in the coating's protective efficacy. As the frequency and severity of pitting corrosion on the specimen surface increases, so does the degradation of the specimen's mechanical properties. This progressive deterioration predisposes the specimen to premature failure under tensile stress.

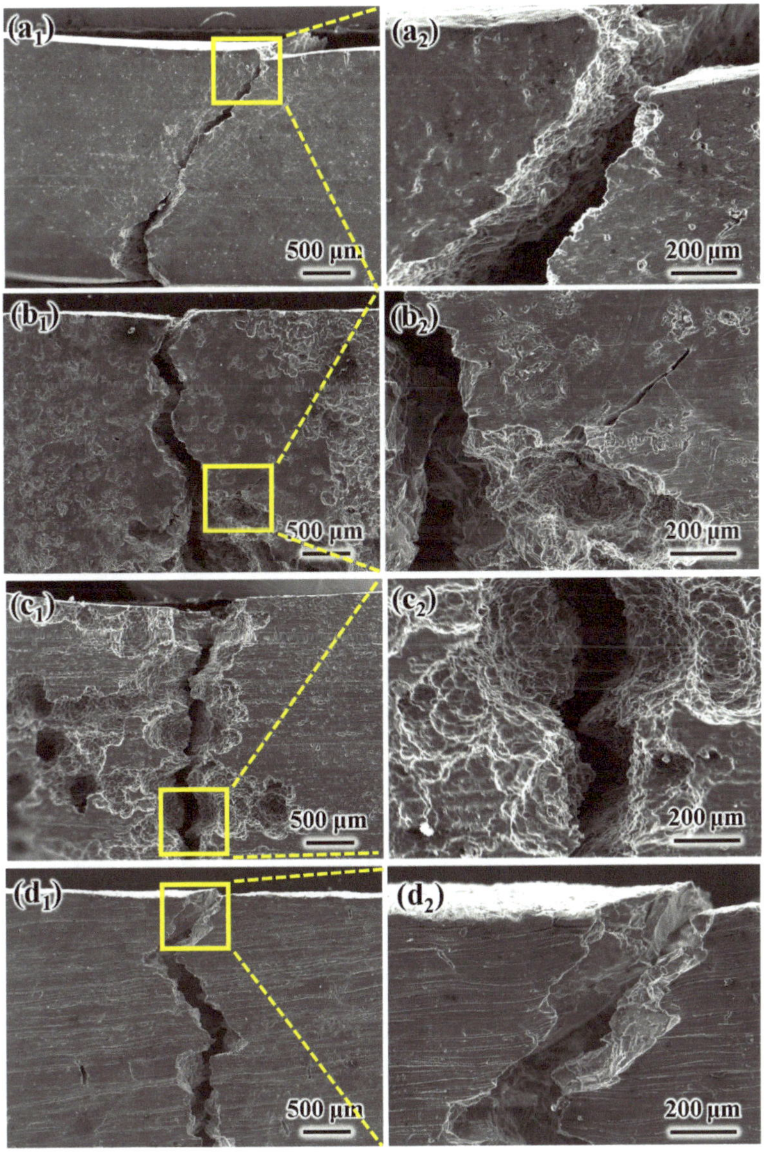

Figure 7. *Cont.*

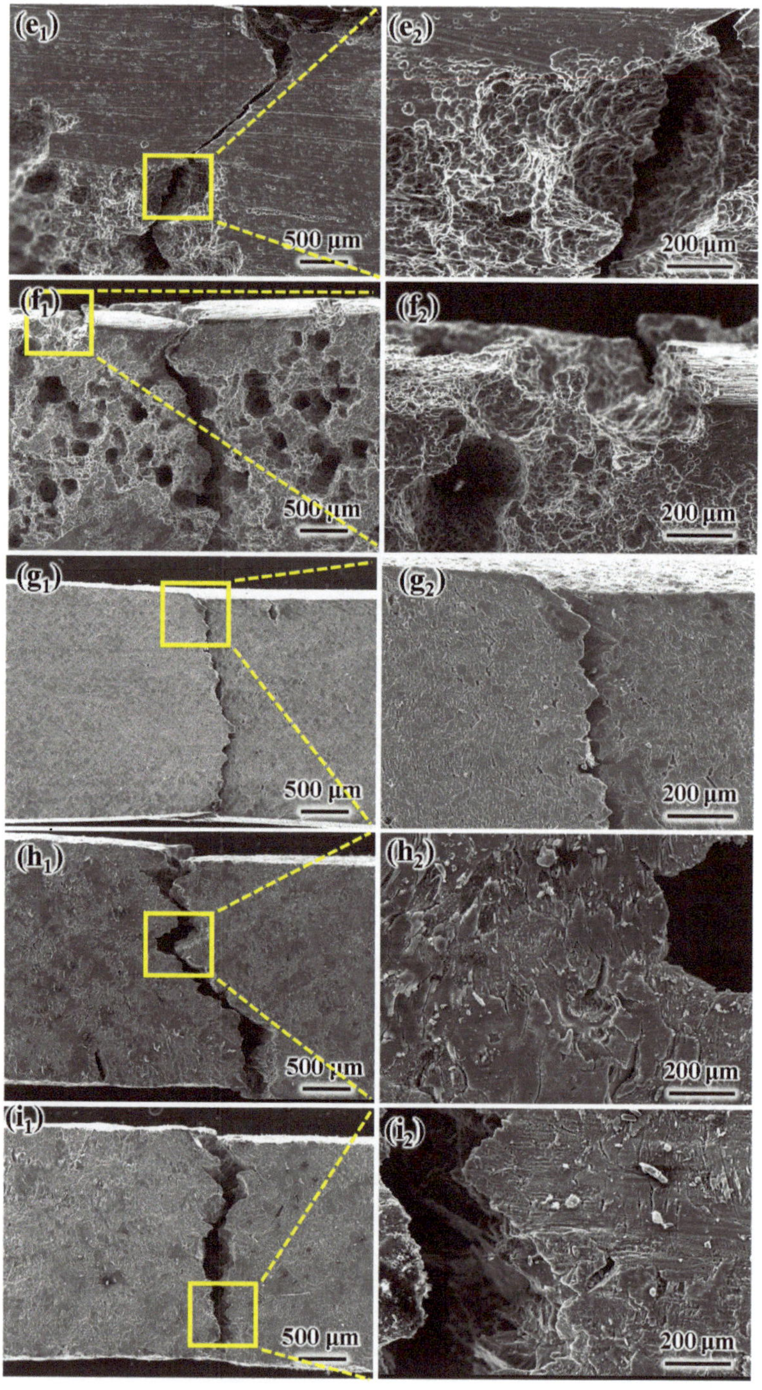

Figure 7. SEM fracture images of Mg-1.0Zn-0.3Zr-1.0Y-2.0Sn alloy in different states corroded in SBF at different times parallel to the extrusion direction: as-cast alloy: ($\mathbf{a_1}$,$\mathbf{a_2}$) 0 d; ($\mathbf{b_1}$,$\mathbf{b_2}$) 10 d; and ($\mathbf{c_1}$,$\mathbf{c_2}$) 20 d; extruded alloy: ($\mathbf{d_1}$,$\mathbf{d_2}$) 0 d; ($\mathbf{e_1}$,$\mathbf{e_2}$) 10 d; and ($\mathbf{f_1}$,$\mathbf{f_2}$) 20 d; and fluorinated coating: ($\mathbf{g_1}$,$\mathbf{g_2}$) 0 d; ($\mathbf{h_1}$,$\mathbf{h_2}$) 10 d; and ($\mathbf{i_1}$,$\mathbf{i_2}$) 20 d.

3.4. Effect of Fluorinated Coatings on the Corrosion Properties of Extruded-State Alloys

Figure 8 shows the microscopic corrosion morphology of the Mg-1.0Zn-0.3Zr-1.0Y-2.0Sn alloys in the cast state, extruded state, and with fluorinated coatings with unremoved corrosion products after 15 days of immersion in SBF. The surface corrosion products of the alloys in the extruded state were significantly greater compared to the cast alloys (Figure 8a,b). With the addition of the coating (Figure 8c), the surface corrosion products of the alloy were again significantly reduced, which indicates that the fluorinated coating can effectively improve the corrosion resistance of the alloy. After the cast-state alloy was immersed in SBF for 15 days (Figure 8a), the surface was covered with small clusters of corrosion products, and the corrosion areas showed an uneven distribution. The extruded alloy (Figure 8b) after 15 days of immersion in SBF showed more severe corrosion than the cast alloy, and, at the same time, the corrosion occurred in a more uneven distribution of the area, and presented an uneven shape, and the corrosion products on the surface were mainly in the form of large clusters. The fluoride-coated alloy (Figure 8c) was immersed in SBF for 15 days, and the surface coating was retained intact, with no large areas of detachment and only a small amount of corrosion products. Cracks appeared on the surface of the corrosion layer of the alloy in the extruded state as well as the fluoride-coated treated alloy, which was due to the dehydration of the corrosion products on the surface of the alloy [29]. Figure 9 shows the amount of hydrogen precipitation and pH change of the Mg-1.0Zn-0.3Zr-1.0Y-2.0Sn alloy after 10 days of immersion in SBF in the as-cast, as-extruded, and as-fluoride-coated state. Hydrogen precipitation corrosion mainly occurs during immersion in SBF, and the reaction formula is as follows: $Mg + 2H^+ = Mg^{2+} + H_2$. Therefore, the corrosion rate of the alloys can be reflected by the change in the amount of hydrogen precipitation in SBF, and the change in the pH value can reflect the degree of the corrosion reaction of the alloys [30]. In Figure 9a, the amount of H_2 analyzed in SBF of different alloys in different states shows a tendency of increasing and then leveling off with time. The amount of hydrogen precipitation was significantly reduced for the surface-covered coating specimen. The hydrogen precipitation rate of the as-cast alloy was 2.02 ± 0.06 mL/cm^2.

Figure 8. SEM images of Mg-1.0Zn-0.3Zr-1.0Y-2.0Sn alloy, as-cast, extruded, and with fluorinated-coating soaked in SBF for 10 days without removing corrosion products: (**a**) as-cast; (**b**) extruded; and (**c**) fluorinated coating.

Equation (1) delineates the corrosion rate (CR_H) of the alloy, which was determined to be 0.53 ± 0.05 mm/y. Comparatively, the as-cast alloy exhibited a reduced hydrogen precipitation rate compared to the extruded variant. Furthermore, the precipitation rate of hydrogen in the alloy declined significantly post-application of the fluoride coating, with the rate plummeting to 0.62 ± 0.012 mL/cm^2, and the CR_H was 0.13 ± 0.012 mm/y in combination with Equation (1). These findings suggest that the MgF_2 coating substantially mitigates the corrosion rate, thereby augmenting the alloy's corrosion resistance. As depicted in Figure 9b, the daily hydrogen precipitation for all alloy states surged precipitously on the initial day, followed by a progressive stabilization. This pattern is attributed to the rapid formation of a corrosion layer on the alloy surface, which subsequently inhibits further corrosion progression, leading to a plateau in hydrogen precipitation levels. Figure 9c illustrates a deceleration in the pH values of all alloys after a 24 h immersion in SBF. This observation signifies a gradual deceleration of the corrosion reaction, indicative of the alloy's adaptive response to the corrosive environment [31].

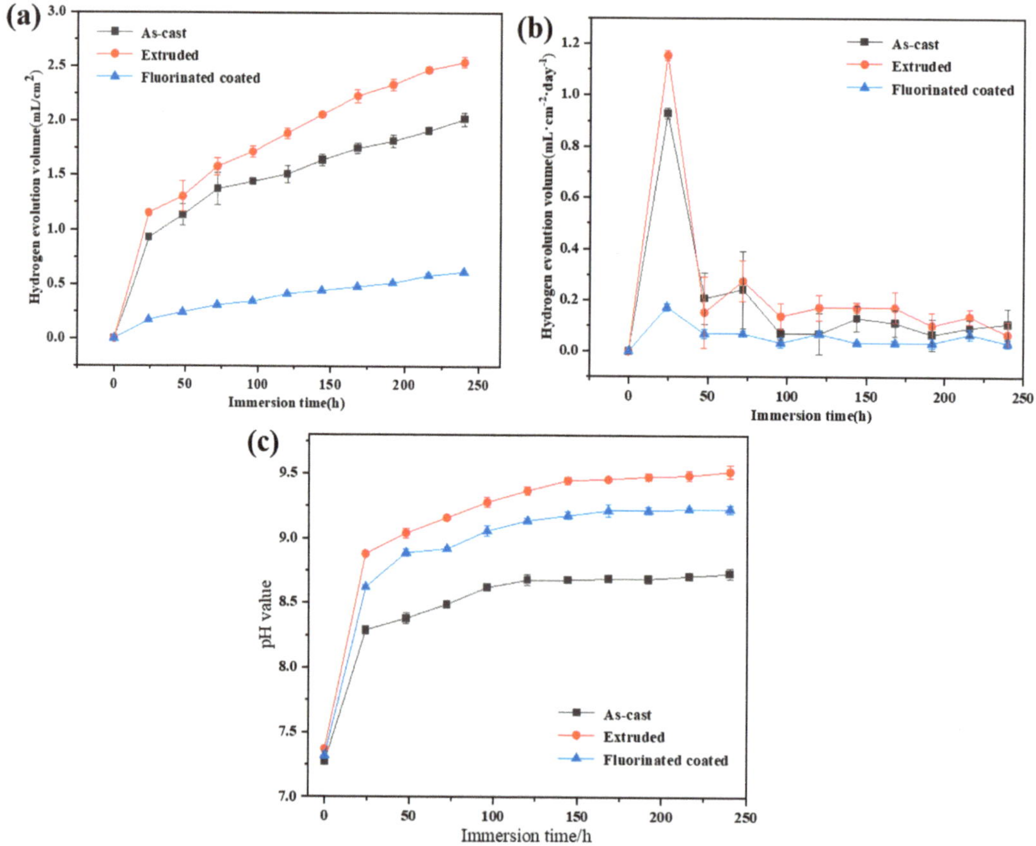

Figure 9. Changes in hydrogen evolution and pH value of Mg-1.0Zn-0.3Zr-1.0Y-2.0Sn alloy, extruded and with fluorinated coating, after immersion in SBF for 10 days: (**a**) total amount of hydrogen evolution change; (**b**) daily variation of hydrogen evolution; and (**c**) daily variation of pH value.

Figure 10 presents the EIS and corresponding equivalent circuit diagrams for the Mg-1.0Zn-0.3Zr-1.0Y-2.0Sn alloy in its as-cast, extruded, and fluoride-coated states following a 3600 s exposure to SBF. The Nyquist plots reveal minimal variance in the capacitive resistance arcs among the three alloy states, suggesting a comparable corrosion mechanism across these conditions [32]. The impedance magnitudes, denoted by the capacitive arcs and impedance Z in Figure 10a,b, indicate that the as-cast alloy exhibits higher impedance values than the extruded alloy. This observation implies a diminished corrosion resistance following the extrusion process. Conversely, the fluoride-coated alloy demonstrates the highest impedance values, signifying superior corrosion resistance. This enhancement is attributed to the protective fluoride coating, which shields the alloy surface, elevates the barrier to corrosion reactions, and bolsters the overall corrosion resistance of the alloy. By integrating the Bode and Nyquist diagrams (Figure 10a–c), an equivalent circuit model (Figure 10d) is employed to elucidate the alloy's corrosion behavior. In this model, CPE_1 and CPE_2 represent the constant phase elements associated with the surface corrosion layer and the bilayer, respectively. Rs, R1, and R2 correspond to the solution resistance, the resistance of the corrosion layer, and the Faradaic impedance (charge transfer resistance at the bilayer), respectively. The electrochemical parameters, extracted from the fitting of the equivalent circuit, are tabulated in Table 5. Notably, the fluoride-coated alloy exhibits the highest combined R1 and R2 values within the equivalent circuit. This finding indicates that the corrosion layer on the fluoride-coated alloy provides the most significant impediment to the charge transfer during the corrosion process, thereby conferring enhanced corrosion resistance in SBF.

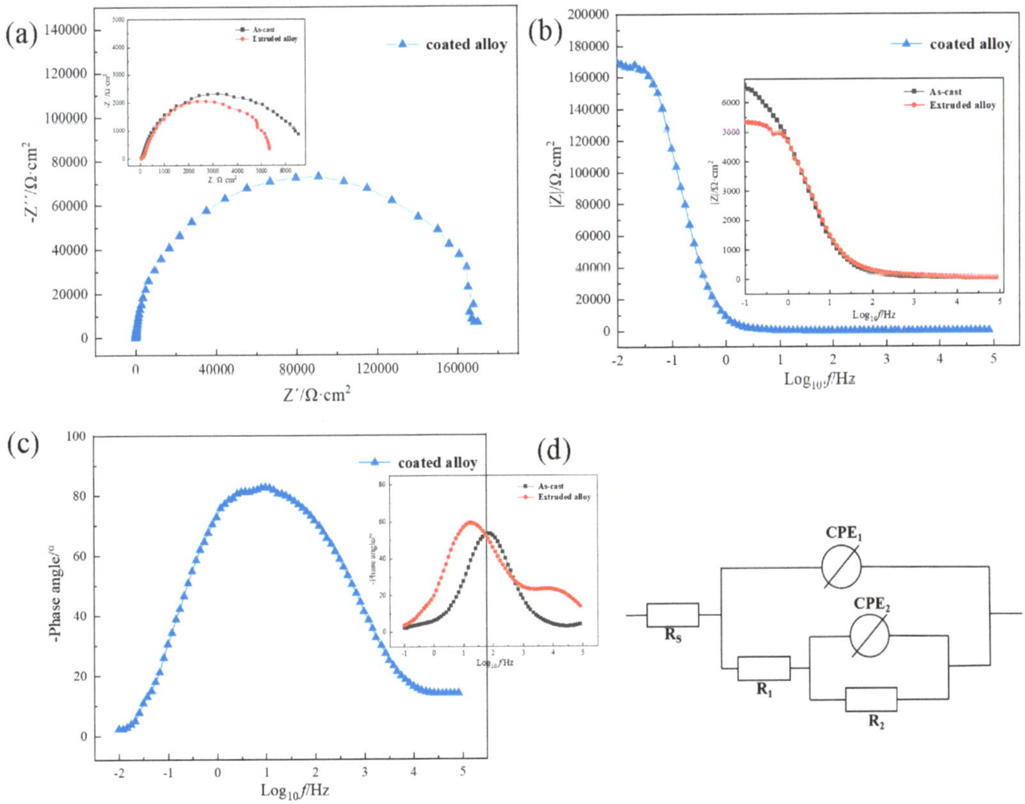

Figure 10. EIS diagram and equivalent circuit of Mg-1.0Zn-0.3Zr-1.0Y-2.0Sn alloy, as-cast, extruded, and with fluorinated coating: (**a**) Nyquist diagram; (**b**,**c**) Bode diagrams; and (**d**) equivalent circuit.

Table 5. EIS fitting parameter values for Mg-1.0Zn-0.3Zr-1.0Y-2.0Sn alloy, as-cast, extruded, and with fluorinated coating.

Sample	R_s ($\Omega \cdot cm^2$)	CPE_1 Y_{01} ($\Omega^{-1} \cdot cm^{-2} \cdot s^{-n}$)	CPE_1 n_1	R_1 ($\Omega \cdot cm^2$)	CPE_2 Y_{02} ($\Omega^{-1} \cdot cm^{-2} \cdot s^{-n}$)	CPE_2 n_2	R_2 ($\Omega \cdot cm^2$)
As-cast	36.47	1.41×10^{-5}	0.71	64.63	1.08×10^{-5}	0.89	6.58×10^{3}
Extruded alloy	38.78	8.23×10^{-6}	0.70	140.4	1.28×10^{-5}	0.87	5.44×10^{3}
Coated alloy	41.35	1.29×10^{-6}	0.91	44.01	4.48×10^{-6}	0.91	1.72×10^{5}

3.5. Corrosion Mechanism

The corrosion characteristics of the Mg-1.0Zn-0.3Zr-1.0Y-2.0Sn alloy exhibited notable variations when assessed in its cast, extruded, and fluorinated-coated forms within SBF. The discrepancies were further elucidated by integrating the outcomes from weightlessness and electrochemical tests conducted on the alloy in these distinct conditions. Figure 11 delineates the proposed corrosion mechanisms for the alloy in its as-cast, extruded, and fluorinated-coated states within SBF. As depicted in Figure 11a,b, Cl- initiates the corrosion process by targeting the vulnerable regions, specifically the dislocations within the oxide film [33]. The presence of a galvanic cell established due to the potential gradient between the second-phase particles and the magnesium matrix, results in the second phase functioning as the cathode and the magnesium substrate as the anode. This arrangement facilitates the onset of localized galvanic corrosion.

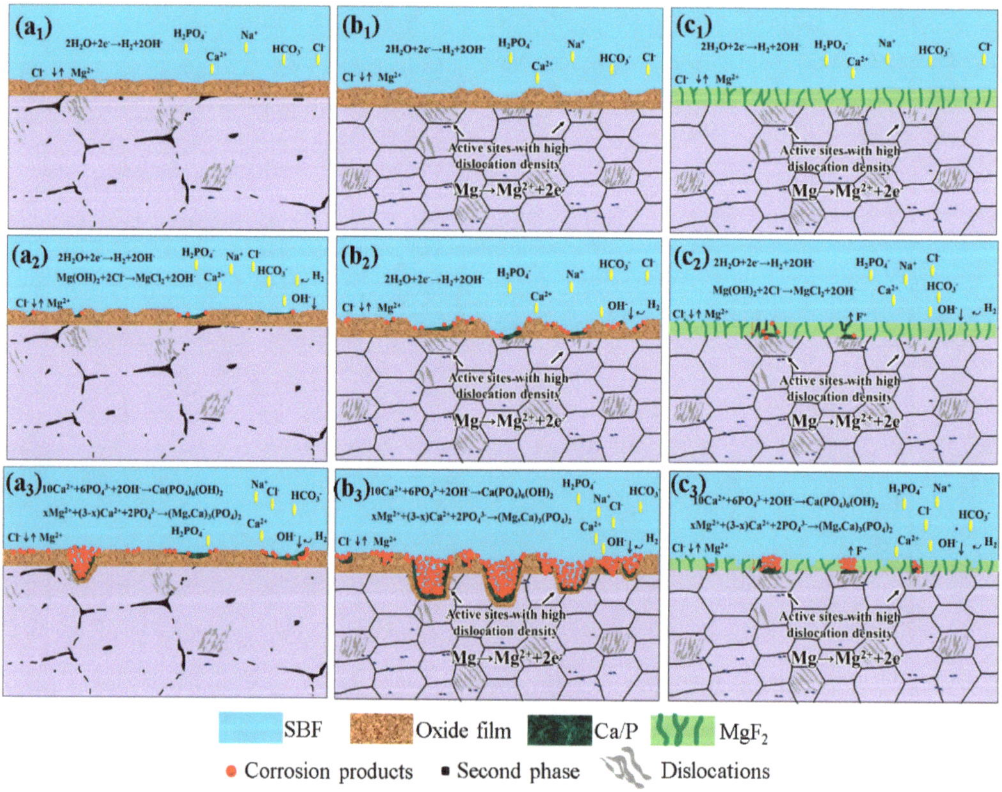

Figure 11. Corrosion mechanism model of Mg-1.0Zn-0.3Zr-1.0Y-2.0Sn alloy, as-cast, extruded, and with fluorinated coating in SBF: (**a_1,a_2,a_3**) as-cast; (**b_1,b_2,b_3**) extruded alloy; and (**c_1,c_2,c_3**) fluorinated coating.

The progression of corrosion within the Mg-1.0Zn-0.3Zr-1.0Y-2.0Sn alloy is characterized by an initial penetration along grain boundaries, followed by an internal expansion into the grains. The accumulation of H_2, a byproduct of the corrosion process, leads to the fracturing of corrosion products, thereby exacerbating the corrosion. Post-extrusion, the alloy's corrosion resistance is compromised due to the fragmentation of the second phase and an increase in dislocations, which facilitates the detachment of nanoscale second-phase particles, exposing fresh surfaces and further reducing corrosion resistance [34]. Additionally, the extrusion process refines the grain structure, increasing the number of grain boundaries and, thus, the susceptibility to intergranular corrosion. Consequently, the corrosion resistance of the extruded alloy is diminished compared to its as-cast state. The application of a protective coating, as illustrated in Figure 11c, introduces a layer of MgO that shields the underlying MgF_2 from direct exposure to SBF. However, corrosive ions can still infiltrate the coating, reaching the MgF_2 layer and initiating a reaction that transforms it into $Mg(OH)_2$. Simultaneously, smaller ions from the SBF permeate the MgF_2 coating along its columnar grain boundaries, engaging in reactions with the alloy matrix. The accumulation of corrosion products within the microchannels at the grain boundaries of MgF_2 creates a barrier that impedes the access of SBF to the substrate, thereby mitigating further corrosion reactions and decelerating the overall corrosion process. As the immersion duration progresses, the accumulation of corrosion products and the release of H_2 escalate, exerting internal pressure at the interface between the substrate and the coating. This pressure can induce stress within the coating. Upon reaching a critical threshold, the pressure may cause the MgF_2 coating to fracture, thereby enlarging the area susceptible to corrosive attack. Furthermore, the dissolution of the MgF_2 coating results in a less compact structure compared to its original dense state, allowing corrosive ions to permeate the now porous coating and initiate reactions with the underlying substrate [35–37]. Despite these challenges, the presence of the coating is generally beneficial in retarding the rate of corrosion. The protective layer serves to mitigate the corrosive effects, even as it undergoes degradation over time

4. Conclusions

In this study, we examined the impact of a fluorinated coating on the microstructure and characteristics of extruded Mg-1.0Zn-0.3Zr-1.0Y-2.0Sn alloys. The findings can be encapsulated as follows:

(1) The fluorinated coating's surface is predominantly constituted by MgO and MgF_2. As the duration of immersion in the HF solution extends, the coating's thickness on the extruded alloy increases initially, and then plateaus. After a 48 h immersion, the thickness stabilizes at approximately 8 μm.

(2) Post-coating treatment, the corrosion resistance of the extruded alloy is markedly enhanced, with a corrosion rate CR_H of 0.13 ± 0.012 mm/year. Following a 20-day immersion in SBF, the YS, UTS, and EL of the extruded alloy diminished by 77%, 76%, and 95%, respectively. In contrast, the mechanical properties of the fluoride-coated samples exhibit a more modest decline in YS, UTS, and EL by 21%, 15%, and 29%, respectively, after the same period.

(3) The extruded Mg-1.0Zn-0.3Zr-1.0Y-2.0Sn alloy exhibits a higher dislocation density and residual stress levels than its as-cast counterpart, which results in an elevated internal dislocation rate and a subsequent reduction in corrosion resistance. However, the application of a fluorinated coating significantly augments the alloy's corrosion resistance. This enhancement is attributed to the presence of MgO in the outer layer of the coating, which acts as a barrier to prevent direct contact between MgF_2 and SBF. Upon prolonged immersion, corrosive ions infiltrate the coating's vulnerable regions and initiate a reaction with MgF_2. Concurrently, the smaller ions present in the SBF solution engage in reactions with the alloy matrix, particularly along the columnar grain boundaries of MgF_2.

Author Contributions: Methodology, Z.C.; Validation, J.W. and W.F.; Investigation, Y.G.; Writing—original draft, J.H.; Writing—review & editing, Q.G. and X.L. All authors have read and agreed to the published version of the manuscript.

Funding: This work was supported by the Natural Science Foundation of Henan Province (242300420018), the Provincial and Ministerial Co-construction of Collaborative Innovation Center for Non-ferrous Metal new Materials and Advanced Processing Technology, and the Henan Key Laboratory of Non-ferrous Materials Science and Processing Technology.

Institutional Review Board Statement: Not applicable.

Informed Consent Statement: Not applicable.

Data Availability Statement: All data that support the findings of this study are included within the article.

Conflicts of Interest: The authors declare that there are no conflicts of interest.

References

1. Guo, X.; Liu, G.; Sang, S.; Lin, Q.; Qiao, Y. Characteristics and Surface Serviceability for Cryogenic Milling Mg-1.6Ca-2.0Zn Medical Magnesium Alloy. *J. Funct. Biomater.* **2022**, *13*, 179. [CrossRef] [PubMed]
2. Gong, Y.; He, J.G.; Wen, J.B.; Feng, W.Y.; Zheng, X.Y.; Li, H. Effect of Hot Extrusion on Microstructure and Properties of As-cast Mg-1Zn-0.3Zr-1Y-2Sn Magnesium Alloy. *Rare Met. Mater. Eng.* **2023**, *52*, 508–516.
3. Mani, G.; Feldman, M.D.; Patel, D.; Agrawal, C.M. Coronary stents: A materials perspective. *Biomaterials* **2007**, *28*, 1689–1710. [CrossRef] [PubMed]
4. Gong, Y.; He, J.G.; Wen, J.B.; Feng, W.Y.; Zheng, X.Y.; Li, H.A. Microstructure and Properties of As-Cast Mg–2.0Sn–1.0Zn–1.0Y–0.3Zr Alloys at Different Extrusion Ratios. *Adv. Eng. Mater.* **2023**, *25*, 2201715. [CrossRef]
5. Jiang, W.; Wu, D.M.; Dong, W.; Ding, J.J.; Ye, Z.C.; Zeng, P.; Gao, Y.Z. Design and Validation of a Nonparasitic 2R1T Parallel Hand-Held Prostate Biopsy Robot with Remote Center of Motion. *J. Mech. Robot.* **2024**, *16*, 051009. [CrossRef]
6. Qi, Z.R.; Zhang, Q.; Tan, L.L.; Lin, X.; Wang, Y. Comparison of degradation behavior and the associated bone response of ZK60 and PLLA in vivo. *J. Biomed. Mater. Res. Part A* **2014**, *102*, 125–1263. [CrossRef]
7. Xie, J.S.; Zhang, J.H.; You, Z.H.; Liu, S.J.; Guang, K.; Wu, R.Z.; Wang, J.; Feng, J. Towards developing Mg alloys with simultaneously improved strength and corrosion resistance via RE alloying. *J. Magnes. Alloys* **2021**, *9*, 41–56. [CrossRef]
8. Imus, P.H.; Blackford, A.L.; Bettinotti, M.; Iglehart, B.; Dietrich, A.; Tucker, N.; Bolaños-Meade, J. Major histocompatibility mismatch and donor choice for second allogeneic bone marrow transplantation. *Biol. Blood Marrow Transpl.* **2017**, *23*, 1887–1894. [CrossRef]
9. Doernberg, E.; Kozlov, A.; Schmid-Fetzer, R. Experimental Investigation and Thermodynamic Calculation of Mg-Al-Sn Phase Equilibria and Solidification Microstructures. *J. Phase Equilibria Diffus.* **2007**, *6*, 523–535. [CrossRef]
10. Munir, K.; Lin, J.; Wen, C.; Wright, P.F.A.; Li, Y. Mechanical, Corrosion, and Biocompatibility Properties of Mg-Zr-Sr-Sc Alloys for Biodegradable Implant Applications. *Acta Biomater.* **2020**, *102*, 493–507. [CrossRef]
11. Cai, C.; Song, R.; Wang, L.; Li, J. Surface Corrosion Behavior and Reaction Product Film Deposition Mechanism of Mg-Zn-Zr-Nd Alloys during Degradation Process in Hank's Solution. *Surf. Coat. Technol.* **2018**, *342*, 57–68. [CrossRef]
12. Qin, H.; Zhao, Y.; An, Z.; Cheng, M.; Wang, Q.; Cheng, T.; Wang, Q.; Wang, J.; Jiang, Y.; Zhang, X.; et al. Enhanced Antibacterial Properties, Biocompatibility, and Corrosion Resistance of Degradable Mg-Nd-Zn-Zr Alloy. *Biomaterials* **2015**, *53*, 211–220. [CrossRef] [PubMed]
13. Yu, X.; Jiang, B.; Yang, H.; Yang, Q.; Xia, X.; Pan, F. High Temperature Oxidation Behavior of Mg-Y-Sn, Mg-Y, Mg-Sn Alloys and Its Effect on Corrosion Property. *Appl. Surf. Sci.* **2015**, *353*, 1013–1022. [CrossRef]
14. Luo, D.; Wang, H.Y.; Chen, L.; Liu, G.J.; Wang, J.G.; Jiang, Q.C. Strong Strain Hardening Ability in an As-Cast Mg-3Sn-1Zn Alloy. *Mater. Lett.* **2013**, *94*, 51–54. [CrossRef]
15. Sasaki, T.T.; Yamamoto, K.; Honma, T.; Kamado, S.; Hono, K. A High-Strength Mg-Sn-Zn-Al Alloy Extruded at Low Temperature. *Scr. Mater.* **2008**, *59*, 1111–1114. [CrossRef]
16. Sasaki, T.T.; Oh-ishi, K.; Ohkubo, T.; Hono, K. Enhanced Age Hardening Response by the Addition of Zn in Mg–Sn Alloys. *Scr. Mater.* **2006**, *55*, 251–254. [CrossRef]
17. Wang, P.; Xiong, P.; Liu, J.; Gao, S.; Xi, T.F.; Cheng, Y. A silk-based coating containing GREDVY peptide and heparin on Mg–Zn–Y–Nd alloy: Improved corrosion resistance, hemocompatibility and endothelialization. *J. Mater. Chem. B* **2018**, *6*, 966–978. [CrossRef]
18. Poinern, G.E.J.; Brundavanam, S.; Fawcett, D. The mechanical properties of a porous ceramic derived from a 30 nm sized particle based powder of hydroxyapatite for potential hard tissue engineering applications. *Am. J. Biomed. Eng.* **2012**, *2*, 278–286. [CrossRef]
19. Shadanbaz, S.; Dias, G.J. Calcium phosphate coatings on magnesium alloys for biomedical applications: A review. *Acta. Biomater.* **2012**, *8*, 20–30. [CrossRef]

20. Hornberger, H.; Virtanen, S.; Boccaccini, A.R. Biomedical coatings on magnesium alloys—A review. *Acta. Biomater.* **2012**, *8*, 2442–2455. [CrossRef]
21. Li, H.; Wen, J.B.; Jin, J.; He, J.G. Roles of the micro-arc oxidation coating on the corrosion resistance and mechanical properties of extruded Mg-2Zn-0.5 Zr-1.5 Dy (mass%) alloy. *Mater. Corros.* **2022**, *73*, 414–426. [CrossRef]
22. Bakhsheshi, H.R.; Idris, M.H.; Kadir, M.R.A.; Daroonparvar, M. Effect of fluoride treatment on corrosion behavior of Mg-Ca binary alloy for implant application. *Trans. Nonferr. Metal. Soc.* **2013**, *23*, 699–710. [CrossRef]
23. Yan, T.; Tan, L.; Zhang, B.; Yang, K. Fluoride Conversion Coating on Biodegradable AZ31B Magnesium Alloy. *J. Optoelectron Adv. M.* **2014**, *30*, 666–674. [CrossRef]
24. Weber, C.M.; Eifler, R.; Seitz, J.M.; Maier, H.J.; Reifenrath, J.; Lenarz, T. Biocompatibility of MgF2-coated MgNd2 specimens in contact with mucosa of the nasal sinus—A long term study. *Acta Biomater.* **2015**, *18*, 249–261. [CrossRef]
25. *GB/T 228.1-2010*; Metallic Materials Tensile Testing Part 1: Method of Test at Room Temperature. The China National Standardization Management Committee: Beijing, China, 2011.
26. Qiao, Z.X.; Shi, Z.M.; Hort, N.; Abidin, N.I.; Atrens, A. Corrosion behaviour of a nominally high purity Mg ingot produced by permanent mould direct chill casting. *Corros. Sci.* **2012**, *61*, 185–207. [CrossRef]
27. Shi, H. *Microstructure and Properties of Mg-2Zn-0.5Zr-1.5Dy Magnesium Alloy*; Henan University of Science and Technology: Luoyang, China, 2021; Volume 5.
28. Li, S.; Yi, L.; Zhu, X.; Liu, T. Ultrasonic Treatment Induced Fluoride Conversion Coating without Pores for High Corrosion Resistance of Mg Alloy. *Coatings* **2020**, *10*, 996. [CrossRef]
29. Lou, G.; Xu, S.M.; Teng, X.Y.; Ye, Z.J.; Jia, P.; Wu, H.; Leng, J.F.; Zuo, M. Effects of Extrusion on Mechanical and Corrosion Resistance Properties of Biomedical Mg-Zn-Nd-xCa Alloys. *Materials* **2019**, *12*, 1049–1061. [CrossRef]
30. Zhao, M.; Liu, M.; Song, G.; Atrens, A. Influence of pH and chloride ion concentration on the corrosion of Mg alloy ZE41. *Corros. Sci.* **2008**, *50*, 3168–3178. [CrossRef]
31. Liu, Y.; Wen, J.; He, J.; Li, H. Enhanced mechanical properties and corrosion resistance of biodegradable Mg-Zn-Zr-Gd alloy by Y microalloying. *J. Mater. Sci.* **2020**, *55*, 1813–1825. [CrossRef]
32. Xu, D.; He, J.; Wen, J.; Shi, H. Effect of immersion time on corrosion behavior of Mg-2Nd-0.5Zn-0.4Zr-1Y alloy in simulated body fluid. Mater. *Res. Express* **2019**, *6*, 125414. [CrossRef]
33. Dai, C.; Gao, X.; Zhai, C.; Jia, Q.; Zhao, B.; Shi, H.; Gao, Q.; Cai, H.; Lee, E.S.; Jiang, H. Corrosion Evaluation of Pure Mg Coated by Fluorination in 0.1 M Fluoride Electrolyte. *Scanning* **2021**, *2021*, 5574946. [CrossRef] [PubMed]
34. Fu, W.; Wang, R.; Zhang, J.; Wu, K.; Liu, G.; Sun, J. The effect of precipitates on voiding, twinning, and fracture behaviors in Mg alloys. *Mat. Sci. Eng. A-struct.* **2018**, *720*, 98–109. [CrossRef]
35. Gollapudi, S. Grain size distribution effects on the corrosion behaviour of materials. *Corros. Sci.* **2012**, *62*, 90–94. [CrossRef]
36. Da Conceicao, T.F.; Scharnagl, N.; Blawert, C.; Dietzel, W.; Kainer, K.U. Surface modification of magnesium alloy AZ31 by hydrofluoric acid treatment and its effect on the corrosion behaviour. *Thin. Solid. Films* **2010**, *518*, 5209–5218. [CrossRef]
37. Makkar, P.; Kang, H.J.; Padalhin, A.R.; Park, I.; Moon, B.G.; Lee, B.T. Development and properties of duplex MgF2/PCL coatings on biodegradable magnesium alloy for biomedical applications. *PLoS ONE* **2018**, *13*, e0193927. [CrossRef]

coatings

MDPI

Article

Effect of Layer Thickness on the Practical Adhesion of Borided Monel 400 Alloy

Francisco Javier Alfonso-Reyes [1], José Martínez-Trinidad [1,*], Luis Alfonso Moreno-Pacheco [1], Osvaldo Quintana-Hernández [1], Wilbert Wong-Ángel [2] and Ricardo Andrés García-León [3,*]

1 Grupo Ingeniería de Superficies, Instituto Politécnico Nacional, SEPI-ESIME Zacatenco, México City 07738, Mexico
2 SEDENA, D.G.E.M., Universidad del Ejército y Fuerza Aérea, Escuela Militar de Ingenieros, México City 53960, Mexico
3 Programa de ASST, Universidad del Magdalena, Santa Marta 470004, Colombia
* Correspondence: jomartinezt@ipn.mx (J.M.-T.); ragarcial@ufpso.edu.co (R.A.G.-L.)

Abstract: This study presents new results on the practical adhesion behavior of a boride layer formed on Monel 400 alloy, developed using the powder-pack boriding (PPBP) at 1223 K for 2, 4, and 6 h of exposure times, obtaining layer thicknesses from approximately 7.9 to 23.8 μm. The nickel boride layers were characterized using optical microscopy, Berkovich nanoindentation, X-ray diffraction (XRD), and scanning electron microscopy (SEM) to determine microstructure, hardness distribution, and failure mechanisms over the worn tracks. Scratch tests were conducted on the borided Monel 400 alloy according to the ASTM C-1624 standard, applying a progressively increasing normal load from 1 to 85 N using a Rockwell-C diamond indenter, revealing that critical loads (L_{C1}, L_{C2}, and L_{C3}) increased with layer thickness. The tests monitored the coefficient of friction and residual stress in real time. Critical loads were determined based on the correlation between the normal force and visual inspection of the worn surface, identifying cracks (cohesive failure) or detachment (adhesive failure). The results exposed those cohesive failures that appeared as Hertzian cracks, while adhesive failures were chipping and delamination, with critical loads reaching up to 49.0 N for the 6 h borided samples. Also, the results indicated that critical loads increased with greater layer thickness. The boride layer hardness was approximately 12 ± 0.3 GPa, ~4.0 times greater than the substrate, and Young's modulus reached 268 ± 15 GPa. These findings underscore that PPBP significantly enhances surface mechanical properties, demonstrating the potential for applications demanding high wear resistance and strong layer adhesion.

Keywords: borided Monel 400; adhesion failure; layer thickness; scratch test; nickel borided

Citation: Alfonso-Reyes, F.J.; Martínez-Trinidad, J.; Moreno-Pacheco, L.A.; Quintana-Hernández, O.; Wong-Ángel, W.; García-León, R.A. Effect of Layer Thickness on the Practical Adhesion of Borided Monel 400 Alloy. *Coatings* **2024**, *14*, 1414. https://doi.org/10.3390/coatings14111414

Academic Editors: Pinghu Chen, Yun Zhang and Ruiqing Li

Received: 9 October 2024
Revised: 1 November 2024
Accepted: 5 November 2024
Published: 7 November 2024

1. Introduction

Nickel-based alloys are widely used in industrial applications due to their high resistance to corrosion and oxidation processes. In particular, Monel 400 is known for its excellent corrosion resistance in environments such as seawater, saline solutions, and aggressive acids, including hydrofluoric acid and phosphoric acid [1]. Based on these characteristics, its main applications include marine engineering, chemical processing, valves, pumps, tanks, electrical and electronic components, equipment for crude oil refining, food processing, gasoline manufacturing, power generation, and heat exchangers [2]. However, one of the main disadvantages of the Monel 400 alloy is its somewhat lower hardness compared to other nickel alloys. Since products made from this material are often exposed to high-flow conditions, they may be susceptible to failures caused by erosion, cavitation, or wear [3].

Different thermal, deposition, or thermochemical processes have been employed to improve the surface properties of nickel-based alloys. In this way, the powder-pack boriding process (PPBP) thermochemical treatment is widely used in the industry due to its relative simplicity and low costs, as well as its contribution to enhancing mechanical and chemical properties due to the high hardness achieved at the surface, leading to excellent performance in industrial applications [4]. PPBP involves the diffusion of active boron atoms (from a boron-rich source) into the material's surface to form a hard, resistant boride layer [1]. On the other hand, practical adhesion is defined in [5] as the force or work required to remove or detach a film or coating from a substrate, regardless of where the failure occurs. As mentioned in [4], it is one of the most critical issues in the industry concerning hard coatings, as adhesion problems often lead to failures.

Krelling et al. (2019) examined the microstructure and properties of borided Monel 400 alloy, reporting a fivefold increase in surface hardness due to the formation of a silicide/boride layer, which led to significant improvements in wear resistance [6]. Similarly, Kukliński et al. (2020) investigated the effects of chemical composition and microstructure on wear resistance and microhardness in Monel 400 treated by laser boriding. Their findings showed that wear resistance improved by a factor of 4 to 10, depending on the parameters used in the laser boriding process [7].

In 2022, Günen et al. proposed a kinetic growth model for boride layers, establishing different experimental conditions. The authors observed that the Ni_2B layer thickness ranged from 35 to 290 μm, with hardness values between 1002 and 1476 $HV_{0.0025}$. Additionally, an activation energy of 33.7 kJ/mol for Ni_2B was calculated [1].

Campos-Silva et al. (2018) conducted a practical adhesion study on borided Inconel 718 alloy, examining the relationship between the adhesion of the coating-substrate system and indentation properties (residual stress, elastic modulus, and hardness), as well as the critical loads that generated tear channels. The results indicated that critical loads increased to 76 N with increasing layer thickness. Cohesive failure manifested as Hertzian cracks, preceded adhesive failure in chipping and delamination [8].

This study aims to evaluate the practical adhesion behavior of the borided Monel 400 alloy surface, taking into account that not enough research has been reported on this material in the open literature. Initially, Monel 400 alloy samples were treated using PPBP to achieve nickel boride layer thicknesses ranging from approximately 7.9 to 23.8 μm. This thermochemical treatment not only reduces energy consumption and costs compared to other treatment methods but also enhances the overall performance of the alloy for industrial applications. Subsequently, the adhesion properties of the borided Monel 400 samples were assessed by determining the critical normal loads (L_{CN}). Additionally, the study presents a detailed analysis of the failure modes observed in the scratch tracks, contributing to a deeper understanding of the material's durability and performance under operational conditions.

2. Experimental Details

2.1. PPBP Thermochemical Treatment

Rectangular specimens of Monel 400, with nominal dimensions of 20 × 20 × 5 mm, were utilized for the study. The chemical composition of the alloy included 0.30 wt% C, 2.0 wt% Mn, 0.024 wt% S, 0.5 wt% Si, 63.0 wt% Ni, 28.0–34.0 wt% Cu, 2.5 wt% Fe, and 0.05 wt% Al. The specimens were metallographically prepared and packaged in a sealed cylindrical AISI 304 steel container, containing a powder mixture comprising 90 wt% B_4C and 10 wt% KBF_4 [4]. The samples were subjected to boriding at a temperature of 1223 K for durations of 2, 4, and 6 h, following the growth kinetics model proposed by Günen et al. (2022) [1]. Upon completing the powder-pack boriding (PPB) process, the container was removed from the furnace and allowed to cool slowly to room temperature. This process yielded nickel boride layer thicknesses in the range of approximately 7.9 to 23.8 μm, consisting of a mixture of nickel borides (Ni_2B, Ni_4B_3, NiB) and copper precipitates.

2.2. Chemical and Mechanical Characterization

The borided samples were analyzed using X-ray diffraction (XRD) with Cu-Kα radiation over a 2θ range of 30° to 100°, employing a PANalytical X'Pert/Pro diffractometer.

The resulting diffraction patterns were evaluated using the commercial software High Score Plus Version 4.5 to confirm the presence of nickel borides. Additionally, the samples obtained from the process were metallographically prepared in bakelite to observe the morphology of the cross-section and measure the layer thickness using optical microscopy (Olympus/GX-51, Olympus, Shinjuku City, Japan).

The nickel boride layers produced by PPBP were indented across the cross-section using a Berkovich nanoindentation technique with depth sensing, applying a constant load of 50 mN. Indentations were spaced approximately 5.0 μm apart to estimate the hardness (H) distribution and Young's modulus (E) using a TTX-NHT/CSM instrument and CETR software Version 1.8, following standard procedures of the ISO-14577-4, (2016) [9].

2.3. Scratch Test

The scratch test is widely employed to evaluate practical adhesion resistance in layer/substrate systems resulting from various thermal, thermochemical, or deposition processes, following the standard test method ASTM C1624-05 [5]. This test is commonly utilized to determine the mechanical strength of coatings with thicknesses up to 30 μm and subsequently calculate the critical loads required to induce layer failure. During the test, the coated sample is secured in the testing apparatus, where a diamond-tipped indenter moves at a constant speed. The normal load force (NF) is progressively increased, creating a wear track that exhibits various failure mechanisms on the layer's surface. These worn trajectories are analyzed in terms of the critical scratch load (L_{CN}) to estimate the characteristic failure mechanisms, which are correlated with visual inspections via microscopy, the friction coefficient (COF), and the residual depth (Rd) [10].

Scratch tests were performed on the polished surface of the boride layers obtained from the entire experimental set, as depicted in the schematic representation of the test in Figure 1. A commercial scratch testing machine (Revetest Xpress+, CSM Instruments, Graz, Austria) equipped with a Rockwell C diamond indenter with a 200 μm radius was utilized, according to the testing parameters outlined in Table 1. Multiple repetitions were conducted for each load, as stipulated by ASTM C1624-05, wherein NF, COF, and Rd were analyzed using a personal computer with Scratch-V6.2.12 software. After conducting the scratch tests, the channels on the surface of the borided Monel 400 were examined using optical microscopy and scanning electron microscopy with energy dispersive spectroscopy (SEM-EDS), using a JEOL/JSM-7800F instrument (JEOL, Tokyo, Japan) at 20 kV.

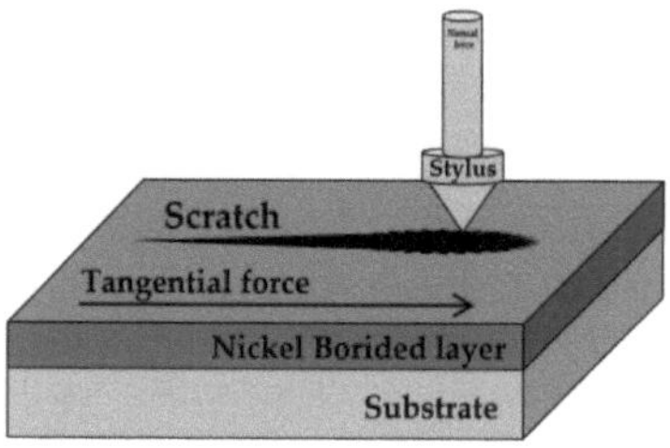

Figure 1. Schematic representation of the scratch test.

Table 1. Experimental parameters employed during scratch tests.

Parameter	Description
Range load	1 to 85 N
Speed rate	1.0 mm/min
Load rate	12 N/min
Scratch length	7.0 mm

Note: These parameters were selected to observe and identify cohesive failure mechanisms in detail, as described in the ASTM C1624-05 standard procedure.

The critical scratch loads (L_{CN}) on the worn tracks generated by the scratch tests were estimated by critical normal loads to the different damage levels of interest and the length using Equation (1).

$$L_{CN} = \left[L_{rate} \times \left(\frac{l_n}{X_{rate}}\right)\right] + L_{initial} \quad (1)$$

where L_{CN} is the specific critical scratch load, L_{rate} is the loading rate of force application in the specific scratch test, l_n is the scratch length, X_{rate} is the horizontal displacement rate, and $L_{initial}$ is the initial load established at the start of the scratch test [5].

3. Results and Discussion

3.1. Nickel Borides Microstructure

The morphology of the borided Monel 400 alloy layer exposed to 1223 K for 2, 4, and 6 h is presented in Figure 2. The cross-section reveals a thickness range of approximately 7.9 to 23.8 μm. X-ray diffraction (XRD) patterns were obtained under different conditions to verify the formation of nickel borides. According to the Ni-B phase diagram, various intermetallic compounds (NiB, Ni_2B, Ni_4B_3, Ni_3B) can form on the material's surface [11,12], as was observed in its characteristic degrees (see Figure 3). The formation of these compounds is linked to the composition of the powder mixture surrounding the sample, including factors such as the type of boron source and the proportions of the various ingredients. Additionally, the temperature and exposure time during the boriding process play a crucial role in this phenomenon.

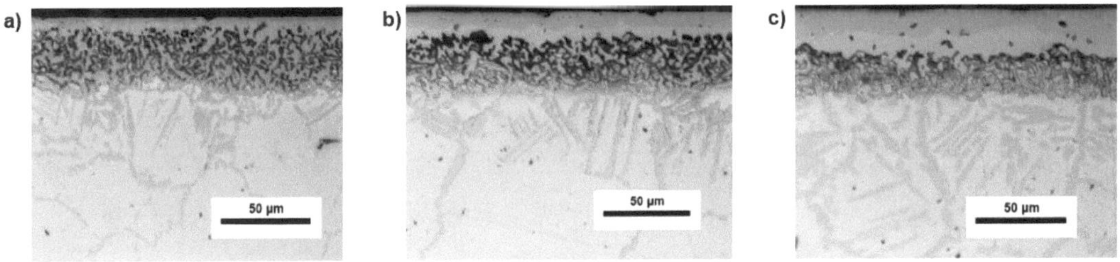

Figure 2. Cross-sectional views of the borided Monel 400 obtained at 1223 K with exposure times of (**a**) 2 h, (**b**) 4 h, and (**c**) 6 h.

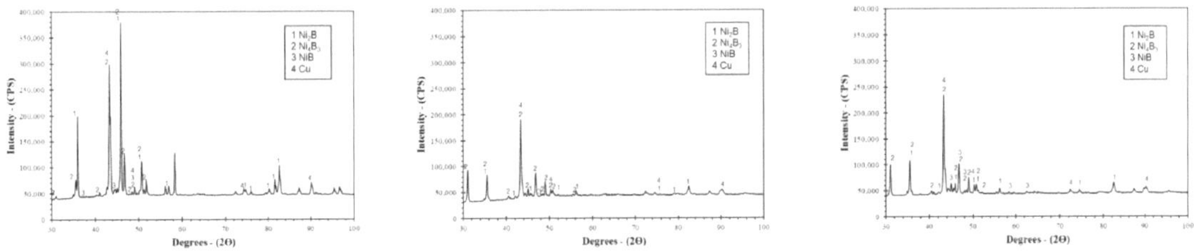

(**a**) (**b**) (**c**)

Figure 3. XRD patterns obtained at the surface of borided Monel 400 using scan normal procedure. The boriding condition was 1223 K: (**a**) 2 h, (**b**) 4 h, and (**c**) 6 h.

The X-ray diffraction (XRD) patterns of each sample, presented in Figure 3, are in good agreement with similar studies [1,2,6]. The XRD analysis predominantly reveals the presence of Ni_4B_3, with a preferential crystallographic plane of [0 2 2] at a 2θ angle of ~44°. Peaks corresponding to Ni_2B with a crystallographic plane of [2 1 1], NiB with a plane of [0 4 0], and Cu with a plane of [1 1 1] are also observed, appearing at angles ranging from 35° to 90°, with varying intensities [1].

This observation suggests that the boriding process conditions directly influence the crystallographic orientation and the stability of the phases formed on the Monel 400 alloy. The dominance of the Ni_4B_3 phase, alongside other borides like Ni_2B and NiB, indicates a well-developed boride layer. The presence of the Cu peaks, derived from the base alloy, remains consistent across the samples, unaffected by the boriding treatment. The variations in peak intensity across the samples further suggest differences in phase concentration, which may be attributed to factors such as boriding time, temperature, and the specific composition of the boriding environment [13]. The identification of these crystallographic planes is critical, as it provides insight into the material's microstructure and potential mechanical properties. The preferential orientation of the boride phases could play a significant role in enhancing surface hardness, wear resistance, and other mechanical attributes, making these findings valuable for applications requiring durable surface layers [14].

3.2. Nickel Borided Mechanical Properties

Figure 4 shows the hardness (H) and elastic modulus (E) profiles, indicating that the hardness value for the nickel boride layer is approximately ~12 ± 0.3 GPa, gradually decreasing across the layer until it reaches the substrate's hardness of ~3.3 GPa [2]. This implies that the hardness ratio between the nickel boride layer (PPBP) and the substrate is roughly four times greater (See Figure 5). Regarding the Young's modulus (E) of the borided Monel 400, the maximum value in the layer is about ~268 ± 15 GPa [6]. It is important to note that the values below ~4.5 GPa do not significantly affect the scratch behavior, which aligns with the findings reported by Cimenoglu et al. [15]. Generally, the H/E ratio values suggest that the PPBP samples possess good resistance to plastic deformation within the layer. This characteristic is associated with improved behavior to prevent ceramic layer adhesion failures and ensure high wear resistance [16].

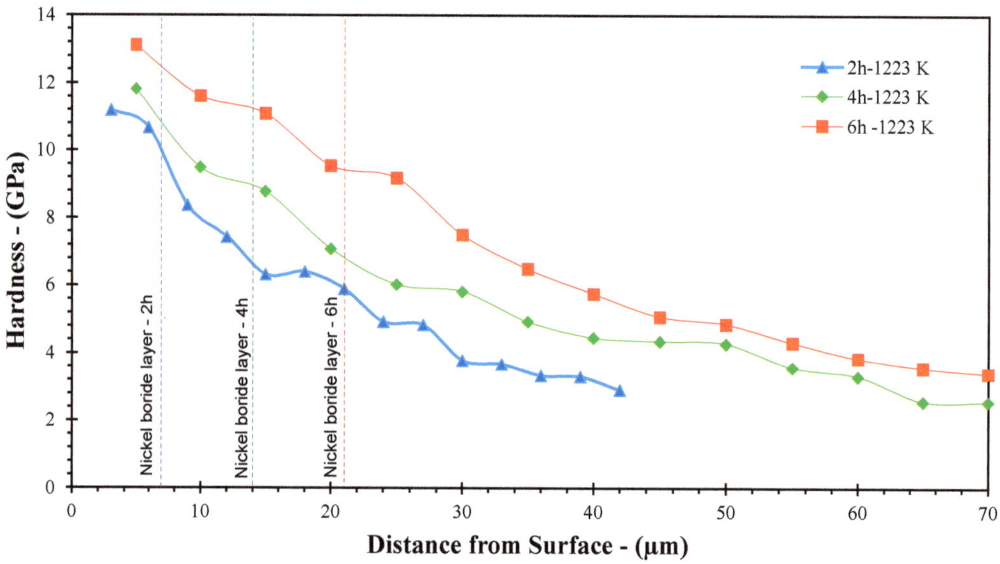

(**a**)

Figure 4. *Cont.*

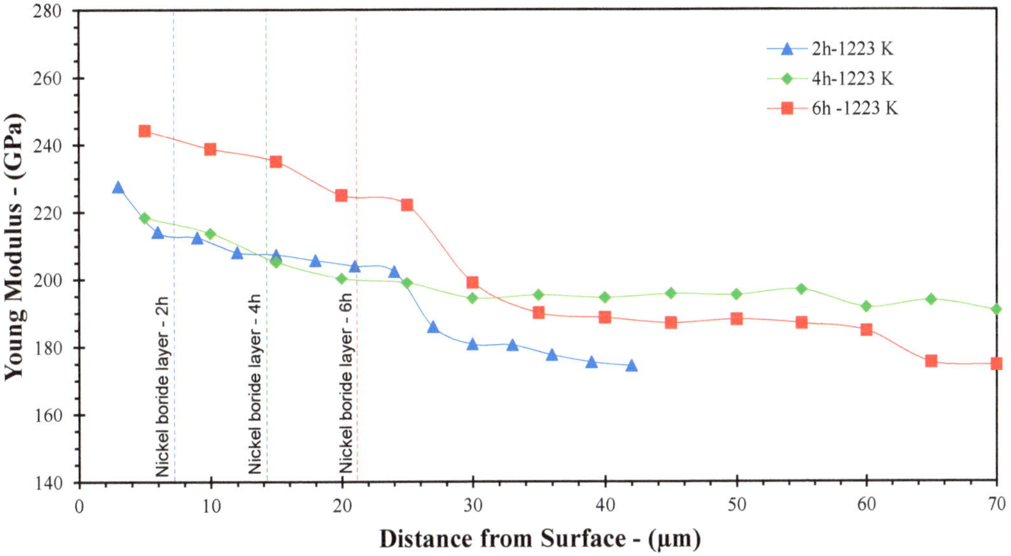

(**b**)

Figure 4. Profiles obtained on the cross-section of the nickel boride layer–substrate system: (**a**) Hardness (H) and (**b**) Young's modulus (E).

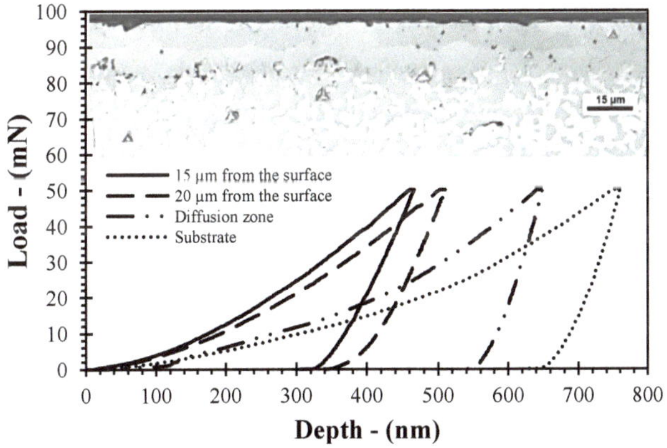

Figure 5. Load–unload curves obtained by nanoindentation at 6 h with 1223 K.

In Figure 5, the load-unload curves clearly illustrate the variation in mechanical properties as a function of depth from the material's surface to the substrate. According to Oliver and Pharr's theory, the region close to the surface (15 μm) exhibits the highest hardness and modulus of elasticity, with a slight decrease observed in the 20 μm region. The diffusion zone serves as a transitional area, displaying intermediate properties between the hardened surface and the substrate. This zone also shows increased ductility, reaching greater indentation depths under lower loads. Notice that the same behavior was observed in the layers exposed to 2 h and 4 h.

The residual stress distribution as a function of the distance from the surface for samples subjected to heat treatments at 1223 K for varying durations (2 h, 4 h, and 6 h) is shown in Figure 6, which was calculated using Chen et al. [17] expression. The results indicate a consistent pattern of compressive residual stress (σr) near the surface for all treatment durations, which gradually decreases toward zero as the distance from the surface increases. This trend is in line with typical residual stress profiles observed in surface-hardened materials, where the compressive stress generated near the surface plays a significant role in improving the material mechanical performance; some observations are described below:

- **Treatment (2 h):** The shortest treatment time exhibited a relatively lower magnitude of compressive stress at the surface, with significant variations in stress levels as the depth increased. This suggests that the 2 h treatment might lead to insufficient boron diffusion into the material, resulting in a less uniform stress distribution;
- **Treatment (4 h):** The intermediate treatment duration generates a more stable and gradual reduction in compressive stress from the surface towards the material's interior. This indicates more effective stress relief compared to the shorter treatment, suggesting improved boron diffusion and nickel boride layer formation;
- **Treatment (6 h):** The longest treatment duration resulted in the highest compressive stress near the surface, with a smoother and more consistent transition to lower stress values deeper within the material. This outcome aligns with expectations that extended treatment times facilitate greater boron diffusion, leading to a thicker nickel boride layer and enhanced surface hardening.

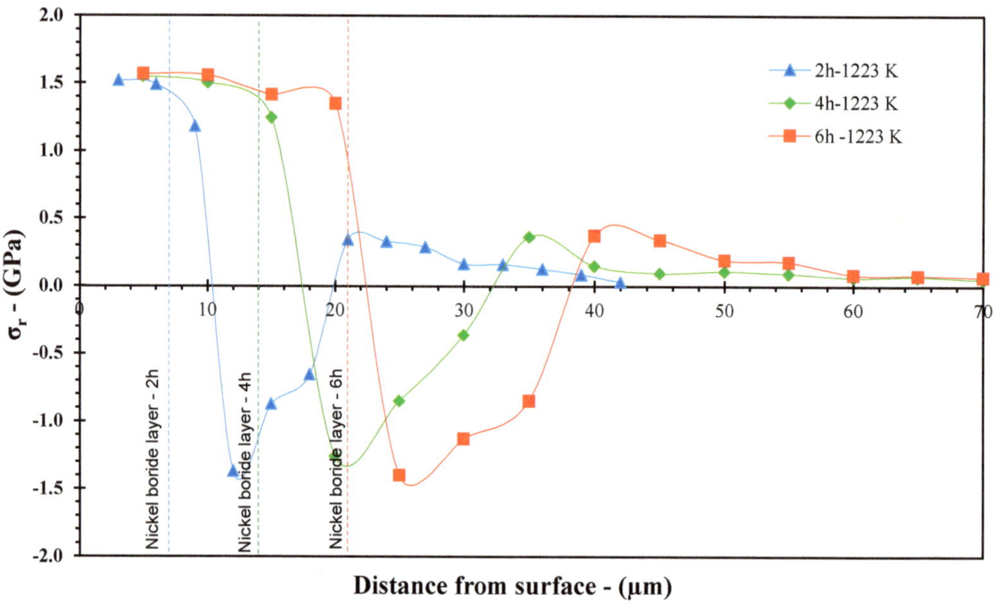

Figure 6. Residual stress distribution across the length of the nickel boride layer–substrate system.

On the other hand, the dashed vertical lines in Figure 6 represent the boundaries of the nickel boride layer for each treatment duration. As the treatment time increases, the nickel depth layer also increases, indicating that longer exposure to the boriding process enhances boron penetration into the substrate. This progression in layer thickness corroborates the trend of increasing surface compressive stress, which is known to improve surface durability and fatigue resistance significantly.

The observed decrease in compressive residual stress on the surface with longer treatment durations is consistent with the theoretical understanding of thermochemical treatment processes (less than ~2.0 GPa). Compressive residual stress is beneficial for mitigating crack propagation and enhancing the mechanical resilience of the material, which is critical for components subjected to cyclic loading conditions. Extended boriding times (notably the 6 h treatment) produce deeper boron diffusion, resulting in a thicker hardened layer and a more substantial increase in compressive stress [18]. This characteristic is particularly advantageous for applications requiring high fatigue, strength and wear resistance.

3.3. *Scratch Test Results*

3.3.1. Estimation of Critical Loads in Nickel Boride Systems

Figures 7–9 present the behavior of the friction coefficient (COF) and the normal force (NF) measurements in relation to the scratch track. These graphs illustrate the increase in the COF signal as damage is observed along the scratch track. Specific levels and types of damage were evaluated for the scratch tracks, corresponding to the critical loads (L_{C1}, L_{C2}, and L_{C3}, respectively), revealing critical loads with cohesive and adhesive failure modes [19].

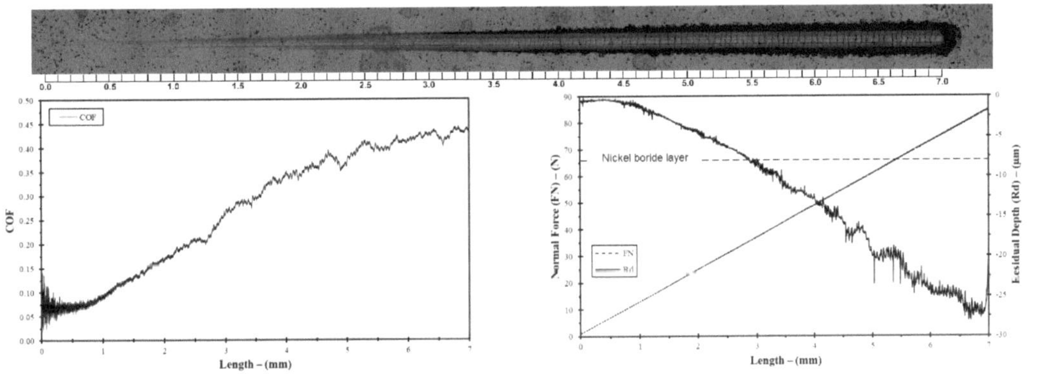

(a) (b)

Figure 7. Scratch track length of the Monel 400 borided obtained at 2 h. (**a**) COF behavior, and (**b**) normal force (NF) and the residual depth (Rd).

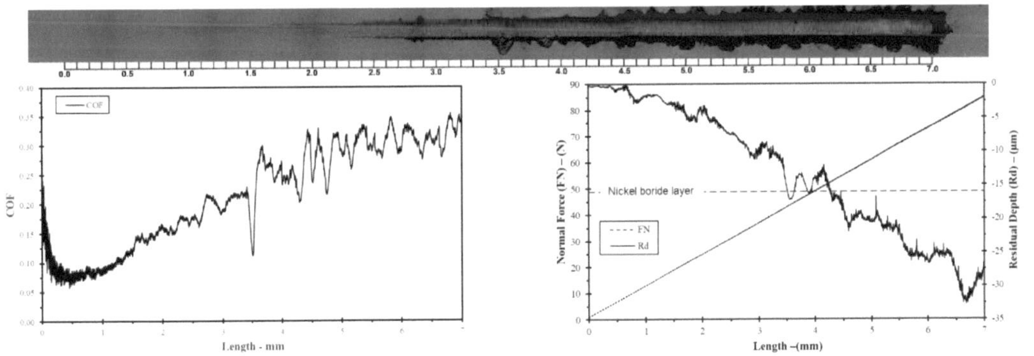

(a) (b)

Figure 8. Scratch track length of the Monel 400 borided obtained at 4 h. (**a**) COF behavior, and (**b**) normal force (NF) and the residual depth (Rd).

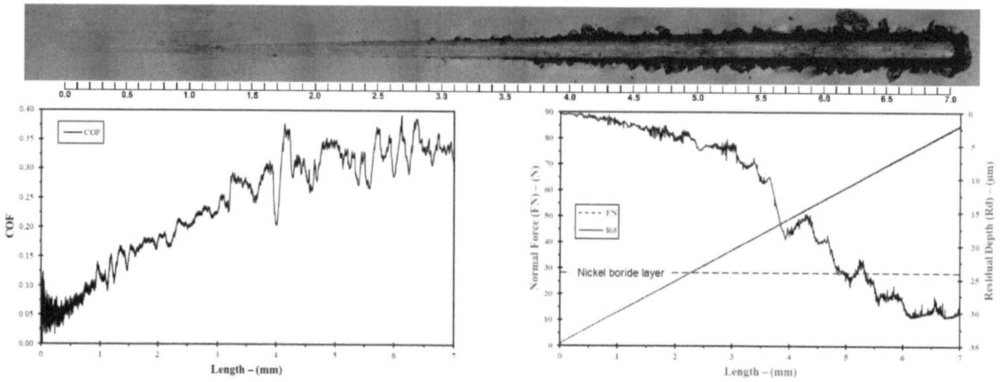

(a) (b)

Figure 9. Scratch track length of the Monel 400 borided obtained at 6 h. (**a**) COF behavior, and (**b**) normal force (NF) and the residual depth (Rd).

Initially, the indenter moves across the surface of the boronized Monel, generating plastic deformation without detecting cracks or only identifying minor ones that increase with the rise in normal force. It is important to note that COF values reach their peak value, approximately 0.45.

Between COF behavior at 2 h, 4 h, and 6 h of treatment, it was possible to observe that the material's frictional properties evolve over time. For the case of the 2 h Sample (Figure 7a), the COF increases quickly, reaching a maximum value of 0.45 as cohesive damage mechanisms appear. This behavior is related to low residual stress (A compressive state of −0.8 GPa was observed), indicating limited boron diffusion, reduced mechanical resistance, and coating stability under scratch loads evaluated. For the 4 h Sample, the COF shows a more gradual rise, reaching a peak around 0.40 (as was observed in Figure 8a). The boride layer, with a higher and more uniform stress profile (A compressive state of −1.2 GPa was observed), allows better resistance to damage mechanisms like chipping, withstanding higher loads before a significant COF increase is observed. This increase in stress correlates with greater resistance to damage. Finally, for the 6 h Sample at critical loads, the COF in Figure 9a increases to maximum values of 0.42–0.45, where failure mechanisms such as spalling and delamination are observed. These high COF values and elevated residual stresses suggest that while the layer resists initial deformation well, the internal stresses induced by prolonged boriding times may lead to more severe failure under high loads. As was observed, the highest recorded compressive stress was around −1.5 GPa; this sample demonstrates the greatest initial resistance to scratching. However, these residual stress and COF values highlight how optimized boriding times enhance coating durability (failure mechanisms), though extended treatments may introduce internal stresses that increase susceptibility to severe failure modes at higher loads [8].

3.3.2. Adhesion Failure Mechanisms

The failure mechanisms observed in the scratch tests can be classified into two general categories, depending on whether the material layer behaves in a brittle or ductile manner. Different failure modes have been observed and documented in the context of scratch tests on different coated materials [8]. In the case of the nickel boride/substrate system, several types of cracks were identified, including arc tensile cracks, delamination/chipping, and spalling [20].

Critical loads, failure mechanisms, and residual depth estimated over the surface of the worn tracks developed on the nickel boride layer–substrate system are summarized in Table 2. Notice that critical loads were calculated using Equation (1). Also, based on the data the following considerations based on the observations were performed:

- The thickness of the nickel boride layer significantly increases with longer exposure times, reaching values of 7.9 ± 0.7 μm, 15.8 ± 1.2 μm, and 23.8 ± 1.6 μm for exposure times of 2, 4 h, and 6 h, respectively. This increase in layer thickness indicates enhanced scratch resistance with longer exposure, which is crucial for improving the mechanical properties and durability of the treated system;
- **L_{C1} (Arc Tensile Cracks):** This failure mechanism appears consistently across all exposure times. However, the critical load for initiating arc tensile cracks decreases as the exposure time increases (19.0 N for 2 h, 16.6 N for 4 h, and 14.2 N for 6 h). This downward trend in critical loads suggests that although the layer becomes thicker, it may also accumulate internal stresses that make it more susceptible to this type of cracking under scratch forces;
- **L_{C2} (Chipping):** Chipping occurs at higher critical loads compared to arc tensile cracks, with values of 32.2 N for 4 h and 35.8 N for 6 h. This indicates that the mechanical integrity of the layer is improving with longer exposure times, as it can withstand greater forces before chipping occurs;
- **L_{C3} (Chipping):** Chipping is observed at the highest critical loads, reaching 43.0 N for 4 h and 49.0 N for 6 h. The increase in critical load with extended exposure time suggests that the cohesion between the nickel boride layer and the substrate strengthens over time, making it more resistant to severe damage under mechanical stress;
- The residual depth of the scratch tracks generally increases as the failure mechanisms progress from arc tensile cracks to spalling. For example, at 6 h of exposure, the residual depth increases from 2.3 ± 0.01 μm (L_{C1}) to 18.5 ± 0.55 μm (L_{C3}). This trend indicates that more severe failure mechanisms lead to deeper damage on the surface of the layer, highlighting the importance of optimizing the exposure time to balance mechanical properties and durability.

Table 2. Critical loads, failure mechanisms, and residual depth estimated over the surface of the worn tracks developed on the nickel borided layer-substrate system.

Exposure Time (h)	Nickel Boride Layer (μm)	Mechanism	Failure Mechanism	Critical Load—L_{CN} (N)	Track Length (mm)	Residual Depth (μm)
2	7.9 ± 0.7	L_{C1}	Arc tensile cracks	19.0 ± 0.10	1.5 ± 0.01	2.8 ± 0.02
4	15.8.0 ± 1.2	L_{C1}	Arc tensile cracks	16.6 ± 0.11	1.3 ± 0.01	2.5 ± 0.02
		L_{C2}	Chipping	32.2 ± 0.27	2.6 ± 0.03	7.9 ± 0.11
		L_{C3}	Spalling	43.0 ± 0.33	3.5 ± 0.02	16.0 ± 0.94
6	23.8.0 ± 1.6	L_{C1}	Arc tensile cracks	14.2 ± 0.09	1.1 ± 0.01	2.3 ± 0.01
		L_{C2}	Chipping	35.8 ± 0.22	2.9 ± 0.02	7.2 ± 0.34
		L_{C3}	Spalling	49.0 ± 0.35	4.0 ± 0.03	18.5 ± 0.55

In all repetitions of the experimental set conducted under the critical load (L_{C3}), the friction coefficient of the nickel boride/substrate system remained within the range of 0.15 to 0.45.

The classification of failure mechanisms can generally be approached in two ways, considering whether the layer behaves in a brittle or ductile manner during the scratch test (see Figure 10). The first category involves situations where the layer could be completely removed (brittle failure), while the second category implies that the layer experiences gradual wear [21]. Based on the experimental results obtained, Bull [19] proposed various failure modes as a consequence of scratch testing in coated materials. In the case of the nickel boride/substrate system, the presence of arc tensile cracks was observed at the initial critical loads (up to 35.8 N); these cracks develop as a result of unloading, similar to those reported by [22]. Subsequently, chipping and delamination appear due to the compressive stresses generated by the movement of the indenter during the last two critical loads, as illustrated in Figures 11 and 12.

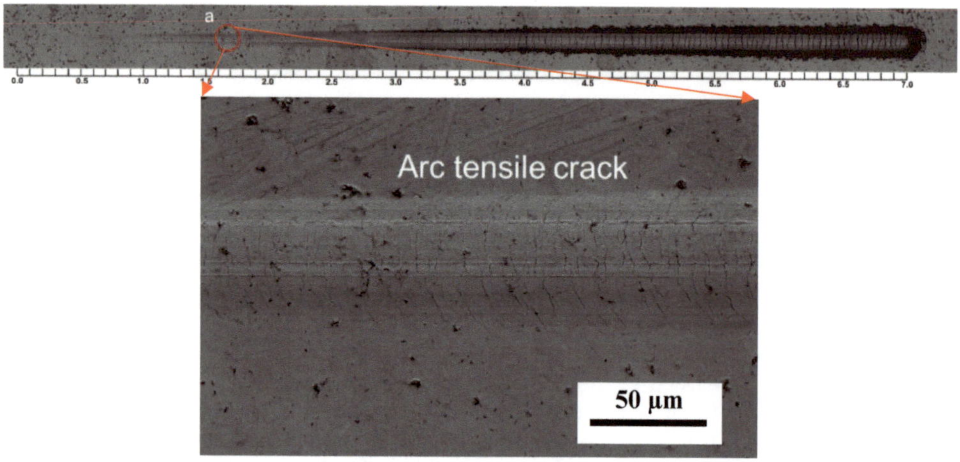

Figure 10. Failure mechanisms obtained over the worn surface for nickel boride layer–substrate system obtained at 1223 K with 2 h.

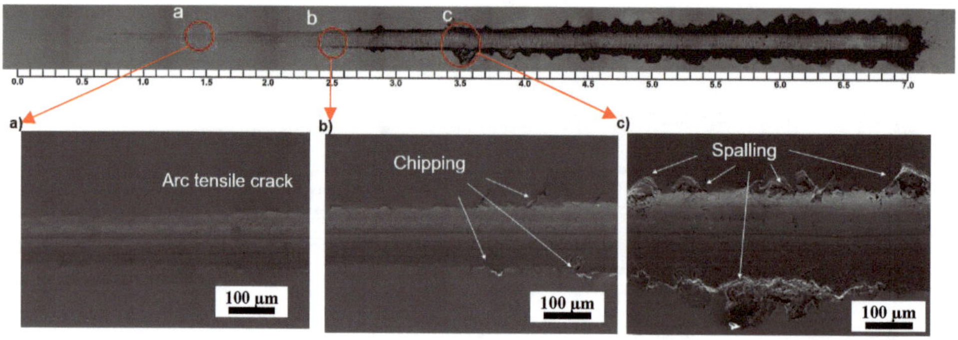

Figure 11. Failure mechanisms obtained over the worn surface for nickel boride layer–substrate system obtained at 1223 K with 4 h.

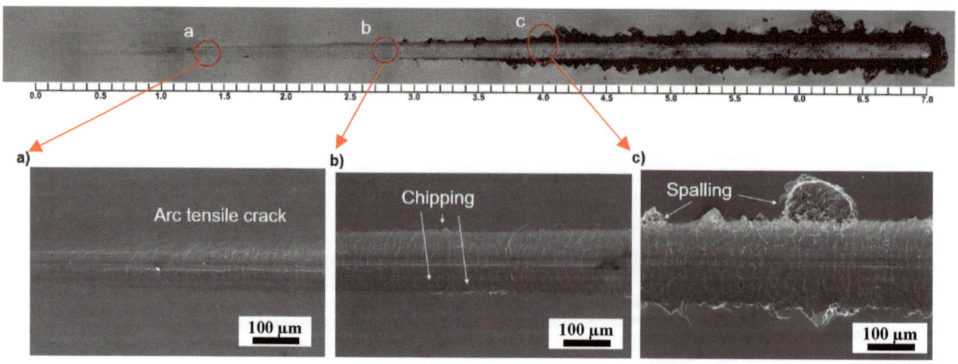

Figure 12. Failure mechanisms obtained over the worn surface for nickel boride layer–substrate system obtained at 1223 K with 6 h.

The critical loads (L_{CN}) were mainly classified into three categories, based on the results obtained from Figures 10–12, as well as Table 2:

I. In all cases, at the beginning of the scratch tracks for low loads, a series of nested microcracks formed within the groove (see Figure 10a, Figure 11a, and Figure 12a). This type of failure, known as arc tensile cracks (cohesive failure), occurs as a result of the generation of tensile stresses behind the tip of the indenter, where several of these arcs form and propagate from the surface to the substrate. The variation in the COF signal is consistent with the presence of tensile cracks detected on the surface of the layer through SEM and recorded as L_{C1};

II. With a further increase in load, reaching L_{C2}, the COF increases as layer spallation occurs (removal of chips from the layer outside the edge of the track without exposing the substrate). This type of failure (Figure 10, Figure 11b, and Figure 12b) is characterized by the separation of part of the nickel boride layer to minimize the amount of elastic energy stored by the compressive stress generated ahead of the indenter during testing [23];

III. Delamination of the layer (removal of chips from the layer outside the edge of the track with substrate exposure—adhesive failure) was observed, defining the critical load L_{C3}. Based on Figure 10, Figure 11c, and Figure 12c, delamination developed due to compressive stresses in front of the indenter at the point where the depth of the indenter is almost equal to the thickness of the layer. The compressive stresses are followed by tensile stresses resulting from the deformation of the nickel boride layer, which is independent of the boride layer thickness. The size (or cracking area) of both chipping and delamination was much greater than that of arc tensile cracks, resulting in more pronounced COF values at the point of failure. Finally, the layer was completely removed from the scratch track.

4. Conclusions

The practical adhesion behavior of the boride layer on Monel 400 was thoroughly investigated in this study. The main conclusions are summarized as follows:

1. The powder-pack boriding (PPBP) treatments produced layer thicknesses ranging from approximately 7.9 to 23.8 μm, depending on the exposure time (2, 4, and 6 h) at 1223 K. XRD analysis confirmed the presence of nickel borides, primarily Ni_2B, Ni_4B_3, and NiB, along with minor Cu inclusions. These phases indicate effective boron diffusion into the Monel 400 substrate, resulting in a well-structured and reinforced surface layer with enhanced mechanical properties;
2. Prolonged boriding times, particularly the treatment at 1223 K with 6 h, were most effective in generating a robust compressive residual stress profile, significantly improving the surface hardness (~12 ± 0.3 GPa) and structural integrity of the boride layer. The 2 h treatment induced a limited residual stress effect, while the 4 h treatment achieved a balanced stress profile and optimal layer thickness (~15.8 μm). These results underscore the need to adjust treatment durations based on specific application requirements to maximize compressive stress benefits while maintaining material stability;
3. The nickel boride layer's scratch resistance revealed that extended exposure times increased the layer's resilience against severe failure mechanisms, with critical loads for failure (L_{C1} for arc tensile cracks, L_{C2} for chipping, and L_{C3} for spalling) reaching up to 49 N in the 6 h sample. However, a decrease in critical load for arc tensile cracks suggests a trade-off between layer thickness and the accumulation of internal stresses. The coefficient of friction (COF) stability under critical load conditions indicates reliable mechanical performance, essential for high-stress applications;
4. During scratch tests, the COF and residual penetration depth were strongly correlated with the type and severity of damage in the boride layer. Consistent COF values (0.15 to 0.45) throughout the scratch test suggest that the enhanced surface properties contribute to the resistance against crack propagation and layer detachment under increasing load conditions;

5. Across the experimental conditions, adhesive failures (e.g., chipping and delamination) were observed as scratch loads increased, with the boride layer detaching from the substrate. This highlights the influence of compressive and tensile stresses on adhesion behavior. Balancing layer adhesion and toughness is critical to reducing these failure modes and extending the coated system's operational lifespan in demanding environments;
6. This study confirms the PPBP's effectiveness in enhancing surface hardness, adhesion, and wear resistance. Future research could further optimize layer properties by tailoring characteristics such as thickness, uniformity, and bonding strength. Additionally, finite element analysis (FEA) focusing on the stress distribution within the layer-substrate system would offer valuable insights for applications in diverse operational environments, enhancing the boride layer's functional performance.

Author Contributions: F.J.A.-R., investigation, formal analysis. J.M.-T., methodology, supervision, project administration, formal analysis, funding acquisition. L.A.M.-P., O.Q.-H. and W.W.-Á., conceptualization, methodology, other contributions. R.A.G.-L., conceptualization, methodology, formal analysis, writing—original draft, writing—review and editing. All authors have read and agreed to the published version of the manuscript.

Funding: This work was supported by the research grants 20241361 and 20232797 of the Instituto Politécnico Nacional of Mexico.

Institutional Review Board Statement: No applicable.

Informed Consent Statement: No applicable.

Data Availability Statement: No data were used for the research described in the article.

Acknowledgments: Thanks to the GIS group of the SEPI-ESIME Zacatenco.

Conflicts of Interest: The authors declare that they have no conflicts of interest for this work.

References

1. Günen, A.; Keddam, M.; Erdoğan, A.; Karakaş, M.S. Pack-Boriding of Monel 400: Microstructural Characterization and Boriding Kinetics. *Met. Mater. Int.* **2022**, *28*, 1851–1863. [CrossRef]
2. Küçük, Y.; Döleker, K.M.; Gök, M.S.; Dal, S.; Altınay, Y.; Erdoğan, A. Microstructure, Hardness and High Temperature Wear Characteristics of Boronized Monel 400. *Surf. Coatings Technol.* **2022**, *436*, 128277. [CrossRef]
3. Range, M.; Temperature, C. Monel Alloy 400. *Alloy Dig.* **1964**, *13*, Ni-93. [CrossRef]
4. García-León, R.A.; Martínez-Trinidad, J.; Campos-Silva, I. Historical Review on the Boriding Process Using Bibliometric Analysis. *Trans. Indian Inst. Met.* **2021**, *74*, 541–557. [CrossRef]
5. *ASTM C1624-05*; Standard Test Method for Adhesion Strength and Mechanical Failure Modes of Ceramic Coatings by Quantitative Single Point Scratch Testing. ASTM: West Conshohocken, PA, USA, 2012; pp. 1–29.
6. Krelling, A.P.; Melo, F.S.; Almeida, E.A.S.; Da Costa, C.E.; Milan, J.C.G. Microstructure and Properties of Borided Monel 400 Alloy. *Mater. Res. Express* **2019**, *6*, 106410. [CrossRef]
7. Kukliński, M.; Bartkowska, A.; Przestacki, D.; Kinal, G. Influence of Microstructure and Chemical Composition on Microhardness and Wear Properties of Laser Borided Monel 400. *Materials* **2020**, *13*, 5757. [CrossRef] [PubMed]
8. Campos-Silva, I.; Contla-Pacheco, A.D.; Ruiz-Rios, A.; Martínez-Trinidad, J.; Rodríguez-Castro, G.; Meneses-Amador, A.; Wong-Angel, W.D. Effects of Scratch Tests on the Adhesive and Cohesive Properties of Borided Inconel 718 Superalloy. *Surf. Coat. Technol.* **2018**, *349*, 917–927. [CrossRef]
9. *ISO-14577-4*; Metallic Materials-Instrumented Indentation Test for Hardness and Materials Parameters. ISO: Geneva, Switzerland, 2016; pp. 1–10.
10. Rodríguez-Castro, G.A.; Vega-Morón, R.C.; Meneses-Amador, A.; Jiménez-Díaz, H.W.; Andraca-Adame, J.A.; Campos-Silva, I.E.; Pardavé, M.E.P. Multi-Pass Scratch Test Behavior of AISI 316L Borided Steel. *Surf. Coat. Technol.* **2016**, *307*, 491–499. [CrossRef]
11. Chu, B.; Zhao, Y. Structural, Mechanical and Electronic Properties of Binary Ni–B Compounds under Pressure. *Solid State Commun.* **2021**, *340*, 114524. [CrossRef]
12. Kanca, Y. Microstructural Characterization and Dry Sliding Wear Behavior of Boride Layers Grown on Invar-36 Superalloy. *Surf. Coatings Technol.* **2022**, *449*, 128973. [CrossRef]
13. Kanca, Y.; Uçgun, M.C.; Günen, A. Microstructural and Tribological Behavior of Pack-Borided Ni-Based Hastelloy C-276 Superalloy. *Metall. Mater. Trans. A* **2023**, *54*, 671–687. [CrossRef]
14. Kulka, M. *Current Trends in Boriding*; Springer International Publishing: Cham, Switzerland, 2019; Volume 1, ISBN 9780873353410.

15. Cimenoglu, H.; Atar, E.; Motallebzadeh, A. High Temperature Tribological Behaviour of Borided Surfaces Based on the Phase Structure of the Boride Layer. *Wear* **2014**, *309*, 152–158. [CrossRef]
16. Taktak, S. Some Mechanical Properties of Borided AISI H13 and 304 Steels. *Mater. Des.* **2007**, *28*, 1836–1843. [CrossRef]
17. Chen, K.-S.; Chen, T.-C.; Ou, K.-S. Development of Semi-Empirical Formulation for Extracting Materials Properties from Nanoindentation Measurements: Residual Stresses, Substrate Effect, and Creep. *Thin Solid Films* **2008**, *516*, 1931–1940. [CrossRef]
18. Campos-Silva, I.E.; Rodríguez-Castro, G.A. 18-Boriding to Improve the Mechanical Properties and Corrosion Resistance of Steels. In *Thermochemical Surface Engineering of Steels*; Woodhead Publishing: Oxford, UK, 2015; pp. 651–702, ISBN 978-0-85709-592-3.
19. Bull, S.J. Failure Modes in Scratch Adhesion Testing. *Surf. Coat. Technol.* **1991**, *50*, 25–32. [CrossRef]
20. Mitjan, K. Friction and Wear of Ceramics. In *ASM Handbook, Friction, Lubrication, and Wear Technology*; Totten, G.E., Ed.; ASM International: West Conshohocken, PA, USA, 2017; Volume 18, pp. 542–550.
21. Márquez-Cortés, R.; Martínez-Trinidad, J.; García-León, R.A. Adhesion Behaviour of Borided AISI 4140 Steel. *Surf. Eng.* **2022**, *38*, 977–984. [CrossRef]
22. Du, S.; Hamdi, M.; Sue, H.-J. Experimental and FEM Analysis of Mar Behavior on Amorphous Polymers. *Wear* **2020**, *444–445*, 203155. [CrossRef]
23. Huff, M. Review Paper: Residual Stresses in Deposited Thin-Film Material Layers for Micro- and Nano-Systems Manufacturing. *Micromachines* **2022**, *13*, 2084. [CrossRef] [PubMed]

coatings

Article

Research on Ni-WC Coating and a Carbide Solidification Simulation Mechanism of PTAW on the Descaling Roll Surface

Chang Li [1,*], Fanhong Kong [1], Lei Feng [1], Han Sun [1], Xing Han [1] and Fenghua Luo [2]

1 College of Mechanical Engineering and Automation, University of Science and Technology Liaoning, Anshan 114051, China
2 State Key Laboratory of Powder Metallurgy, Central South University, Changsha 410083, China
* Correspondence: lichang2323-23@163.com

Abstract: The descaling roll is a critical component in a hot-rolling production line. The operating conditions are significantly impacted by water with high-pressure and dynamic shocks caused by high-temperature steel slab descaling. Roll surfaces often experience wear and corrosion failures. This is attributed to a combination of high temperatures, intense wear, and repeated thermal, mechanical, and fluid stresses. Production costs and efficiency are significantly affected by the replacement of descaling rolls. Practice shows that the use of plasma cladding technology forms high-performance coatings. Conventional metal surface properties can be significantly improved. In this study, a Ni-WC composite coating was prepared on the descaling roll surface by plasma-transferred arc welding (PTAW) technology. The microstructure and phase composition of the welding overlay were analyzed by scanning electron microscopy (SEM) and X-ray diffraction (XRD). Results show that the WC hard phase added to the molten pool dissolves, and subsequently M_7C_3 and W_2C phases are formed. To further explore the morphological evolution mechanism of the hard phase, numerical simulations were performed using a phase-field method to model M_7C_3 phase precipitation. The evolution from nucleation, rod-like growth, to eutectic structure formation was revealed. Experimental and simulation results show high consistency, validating the established phase-field model. In this study, a theoretical foundation for designing and preparing high-performance coatings is provided.

Keywords: PTAW; Ni-WC composite coating; M_7C_3 carbides; phase-field method; microstructure

Citation: Li, C.; Kong, F.; Feng, L.; Sun, H.; Han, X.; Luo, F. Research on Ni-WC Coating and a Carbide Solidification Simulation Mechanism of PTAW on the Descaling Roll Surface. *Coatings* **2024**, *14*, 1490. https://doi.org/10.3390/coatings14121490

Academic Editor: Victor J. Rico

Received: 2 November 2024
Revised: 20 November 2024
Accepted: 25 November 2024
Published: 27 November 2024

1. Introduction

During service, the descaling roll is faced with multiple harsh conditions such as high-temperature, high-pressure, corrosion, impact, and wear. Under such harsh conditions, descaling rolls made from conventional materials exhibit a short service life and require frequent replacement. The production efficiency is significantly reduced and steel manufacturing costs are increased [1,2]. Therefore, enhancing the surface performance of descaling rolls has become a critical technical challenge to be solved urgently [3]. As a current research focus, surface modification technology primarily involves methods such as plasma-transferred arc welding, laser cladding, high-velocity oxygen fuel spraying, and vapor deposition [4]. In recent years, significant progress has been made in applying plasma-transferred arc welding technology for descaling roll repair. The economic losses due to roller failure and frequent replacement have been substantially reduced by the technology. The service life of descaling rolls has been effectively extended and the economic efficiency of the steel industry has been enhanced.

Plasma-transferred arc welding uses a plasma arc as a heat source to melt alloy powders or welding wire, forming a molten pool that fuses evenly with the substrate surface. Upon cooling, a welding overlay with high wear resistance, low dilution rate, and excellent mechanical properties is formed. As a surface modification technique, plasma-transferred arc welding offers several advantages, including high deposition speed, superior quality, broad powder adaptability, low dilution of the substrate, and a small heat-affected zone. This technology has been widely applied in industries requiring wear resistance, corrosion resistance, and anti-friction properties, such as metallurgical machinery, mining machinery, and petrochemicals [5]. Particularly notable progress has been made in descaling roll repair, wear-resistant overlays for various valve sealing surfaces, and the repair of worn equipment in oil extraction scenarios.

Nickel-based alloys are preferred materials in surface coating technology due to their excellent wettability, oxidation resistance, self-fusion properties, and moderate cost [6,7]. However, their long-term application in harsh conditions is restricted by mechanical performance limitations. In contrast, WC is considered an ideal reinforcing phase for enhancing coating performance [8]. This is owing to its high strength, high melting point, low thermal expansion coefficient, and excellent high-temperature wettability [9,10].

As a continuum approach, the phase-field method allows the simulation of complex microstructure evolution without explicitly tracking the movement of interfaces [11]. This method describes the microstructure evolution of the study system by introducing field variables. Then, numerical solutions are used to determine the microstructural evolution [12]. The phase-field model is grounded in diffuse interface theory. The notable characteristic is the ability to distinguish different states of matter through field variables when state variables such as temperature, concentration, and external stress remain consistent [13]. Therefore, phase-field variables are considered independent thermodynamic state variables. Based on the theory, the concept of order parameters and their thermodynamic gradient functions are expanded by Ginzburg [14] and Landau [15]. Then, a model is established to describe superconductivity. Subsequently, Hillert [16] proposes the spinodal decomposition model, which describes phase transformations and solid liquid transitions in binary alloys. The model divides alloys into immiscible regions controlled by phase-field variables, each with a unique composition and thermodynamic state. In addition, the model also considers the phase transition process under non-isothermal conditions and determines the system evolution by temperature, alloy composition, and chemical potential. In the late 1950s to early 1960s, Cahn and Hilliard [17] introduced the continuous phase-field model, initially aimed at describing solid-state transformations and microstructural evolution in alloys. The core idea is to treat alloy composition variables as a continuously distributed "phase field" in space. Since the 21st century, the phase-field model has been increasingly used in microstructural simulations, eventually becoming one of the primary tools for material simulation.

In surface modification technology, material hardness, wear resistance, and microstructure are significantly influenced by the precipitation of hard phases. In recent years, numerous researchers have conducted in-depth studies on the precipitation behavior of hard phases under different process conditions. Guozhi X et al. [18] used plasma-transferred arc welding technology to prepare a Ni-based WC coating on AISI 1045 steel. The study shows that the coating forms a metallurgical bond with the substrate, comprising γ-Ni phase, carbides like M23C6, M7C3, and borides. Yang G et al. [19] used vacuum cladding technology to prepare multilayer Ni-WC alloy coatings. The study shows that the coating exhibits a dense and homogeneous microstructure with γ-Ni as the substrate phase, uniformly dispersed Cr7C3 and Cr23C6 carbides, and the grain boundaries are filled with WC particles, significantly improving macro-hardness. Jiang J et al. [20] used laser cladding on 42CrMo steel rollers to prepare Fe-based alloy coatings. They observe that with variations in process parameters, the microstructure evolved from planar crystals to cellular or dendritic structures, eventually forming equiaxed crystals. The eutectic phases in the coating consisted of network-like M_2B, granular $M_{23}(C, B)_6$, and $M_7(C, B)_3$ phases, resulting in hardness three times that of the substrate. Buytoz S et al. [21] used TIG technology to prepare Ni-Cr-B-Si/WC coatings with varying compositions. The study shows that under high-energy input conditions, WC dissolves and resolidifies in the substrate, promoting the formation of γ-Ni primary dendrites and dendritic Cr_7C_3 carbides. Luo K et al. [22] investigated the effect of welding current (70A–85A) on the Ni-WC composite coatings prepared by plasma transfer arc (PTA). The result shows that with increasing welding current, WC particle dissolution intensifies, particularly in the upper layer of the coating. At the same time, the volume fractions of M_6C and $M_7(B, C)_3$ precipitates also significantly increase.

The above studies have shown that the carbides precipitated during plasma-transferred arc welding play a significant role in the properties of coatings. However, systematic studies on the carbide evolution mechanism in the molten pool, especially using advanced simulation tools such as the phase-field method, are still limited. Therefore, it is of great significance to reveal the carbide evolution mechanism in the molten pool using the phase-field method. In this study, Ni-WC composite coatings are prepared on the surface of descaling roller by plasma-transferred arc welding technology, and their microstructure and phase composition are characterized by SEM and XRD. In order to deeply understand the morphology evolution mechanism of the hard phase, a phase field model of M7C3 phase precipitation was established, and the evolution mechanism of WC hard phase dissolution and complex secondary phase precipitation was explored, confirming the reliability of the phase field model. The research results provide a theoretical basis for the organization regulation and performance optimization of high-performance coatings.

2. Plasma-Transferred Arc Welding Experiment

2.1. Experimental Materials

In this experiment, the 42CrMo descaling roller provided by China Zhengfa Company is used as the base material, and the chemical composition is shown in Table 1.

Table 1. Composition and content of 42CrMo (wt%).

Element	Ni	Mo	Cr	Mn	Si	C	P	S	Fe
Content (wt%)	0.38	0.25	0.96	0.75	0.17	0.38	0.04	0.03	Bal.

The self-prepared 40%NiCrBSi + 60%WC self-melting nickel-based alloy composite powder was used to the surfacing powder material. Both nickel-based alloy powder and WC powder were produced by China Zhengfa Company. The chemical composition was identified by inductively coupled plasma. The elemental content of the nickel-based alloy powder is shown in Table 2 and the elemental content of the WC powder is shown in Table 3. The required Ni-based alloy powder and WC powder were weighed accurately with an electronic balance to obtain the required weight. Then, the weighed powder was added to a ball mill and mixed uniformly at 300 rpm for 10 h. Finally, the powder was packed in a clean container and placed in a vacuum drying oven and dried at 120 °C for 2 h.

Table 2. Nickel-based elemental composition.

Element	C	Si	B	Cr	Fe	Ni
Content (wt%)	0.3	3.2	1.8	9	2.4	Bal.

Table 3. WC elemental composition.

Element	C	W
Content (wt%)	6.0	Bal.

2.2. Experimental Results of Plasma-Transferred Arc Welding

The welding overlays were prepared by a plasma-transferred arc welding machine model DH4-LC401-B12 (Duomu Industrial Co., LTD, Shanghai, China) in this study. The experimental block size is 100 mm × 100 mm × 20 mm. Before welding, the substrate surface was polished with sandpaper to remove burrs and rust. During preheating of the experimental blocks, the furnace temperature was set to 400 °C, with a holding time of 3–5 h. The mixed powder was filled into the powder feeder with argon gas as a feeding and shielding gas. The performance of the welding overlay depends on the plasma-transferred arc welding process and optimizing process parameters is critical to enhancing its properties. The schematic diagram of plasma-transferred arc welding is shown in Figure 1. After multiple experiments, the optimal process parameters were determined, as shown in Table 4.

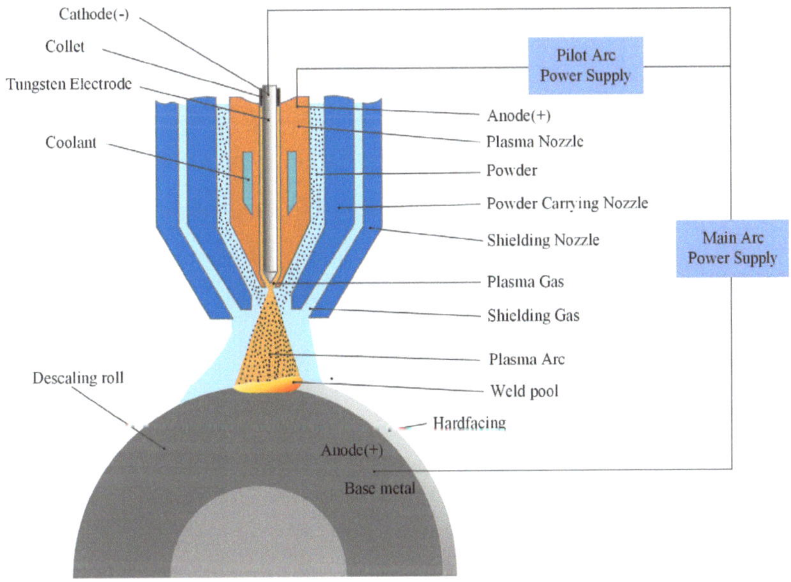

Figure 1. Schematic diagram of plasma-transferred arc welding.

Table 4. Process parameters of the plasma-transferred arc welding machine.

Welding Parameter	Value
Welding voltage/U	29–31
Welding current/A	135–145
Welding speed/(mm/s)	75
Shielding gas velocity/(L/min)	9–10
Powder feed rate/(g/min)	18–20

After welding, the experimental blocks were held for 1–2 h and then air-cooled to room temperature. Pre-welding preheating and post-welding holding are performed to reduce stress differences after welding. The welded specimens were cut by EDM wire and transversely sectioned along the welding overlay into several test pieces. The sections were polished step by step with 200# to 2000# sandpaper until there were no obvious scratches. Then, the test block was polished with w2.5 and w1.5 polishing agents combined with a P-1 polishing machine. For microstructural observation, the polished samples were treated with aqua regia for 7–8 s, immediately rinsed with anhydrous ethanol, and dried by a blower. Finally, XRD and SEM analyzes were conducted.

2.3. Phase Analysis of the Surfacing Layer

The microstructure and properties of the welding overlay are significantly influenced by the phase composition. In the experiment, the phase composition of the surfacing layer was analyzed by an X'Pert Powder X-ray diffractometer made by Panaco. The diffractometer employs a Kα ray source and a Cu target. The scanning angle ranges from 20° to 90°, with a scanning time of 2 min. The experimental results were imported into data processing software to generate diffraction curves. The XRD results of the Ni-WC welding overlay, as shown in Figure 2. Phases γ-Fe/Ni, Fe-Ni, M_7C_3, and W_2C are observed at the diffraction angle of 40~50 degrees. At the diffraction angle of 60–70 degrees, Fe-Ni, M_7C_3, and WC phases are observed. At the diffraction angle of 80–90 degrees, Fe-Ni and NiCrFe are observed. It can be seen that the peaks where M_7C_3 is located are of high intensity and large in number, indicating that it occupies an important position in the coating phase composition. Thermal transformation phenomenon occurs in the WC hard phase under the action of plasma arc heat. The dissolved C elements diffuse and react with the alloying elements in the nickel-based phase to produce M_7C_3 carbides.

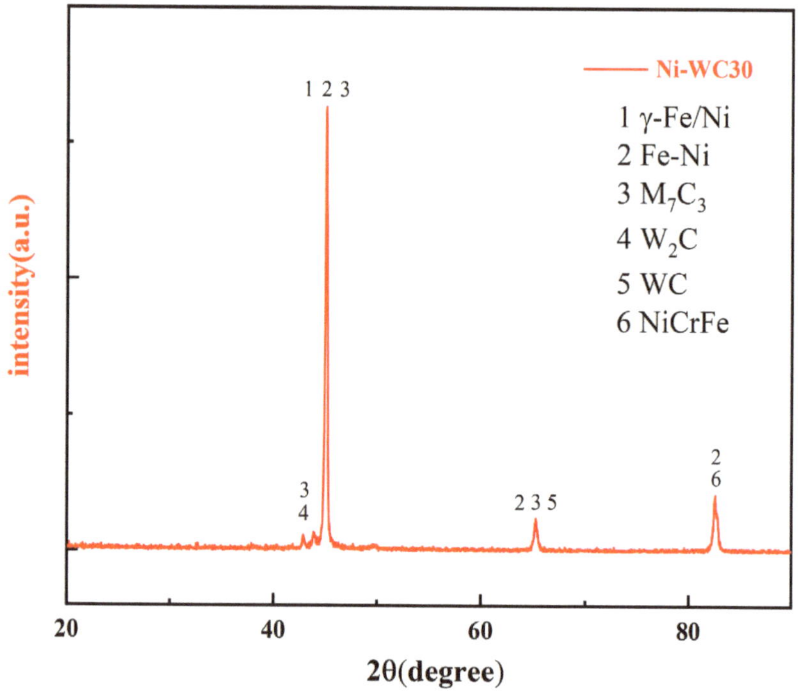

Figure 2. XRD test results of surfacing layer at different diffraction angles.

3. Phase Field Method

3.1. Dynamic Equation

The phase field model employs conservative field variable c (to characterize the concentration field) and non-conservative field variable η (to characterize the phase field) [23]. These variables are used to describe the heterogeneity of the system. The evolution of these field variables follows the Cahn-Hilliard equation and the Allen-Cahn equation, respectively [24]. They are expressed as:

$$\frac{\partial c_i(r,t)}{\partial t} = \nabla \cdot \sum_j M_{ij} \cdot \nabla \frac{\delta F}{\delta c_i(r,t)} + \xi_{ci}(r,t), i = 2,3 \tag{1}$$

$$\frac{\partial \eta(r,t)}{\partial t} = -M_\eta \frac{\delta F}{\delta \eta} + \xi_\eta(r,t) \tag{2}$$

where $c_i(r,t)(i = 2, 3$ is Cr, C atom, respectively) is the concentration distribution of the atom i at a specific time t and position r. $\xi_{ci}(r,t)$ and $\xi_\eta(r,t)$ are Gaussian Langevin random noise.

The total free energy F of the system includes both chemical free energy and interfacial energy components [17]. The phase-field variable $\eta(r,t)$ is the spatial distribution of the FCC phase and the M_7C_3 phase. M_η is the mobility coefficient, which is set as a constant to simplify the calculation, taking a value of 0.05 m s^{-1} J^{-1}. The total chemical free energy F is expressed as:

$$F = \int_V \left(f_{ch}(\{c_i(r,t),\eta\}) + \frac{1}{2}\sum \kappa_c(\nabla c_i)^2 + \frac{1}{2}\kappa_\eta(\nabla\eta)^2 \right) dV \tag{3}$$

where κ_c and κ_η are the gradient energy coefficients for the composition and order parameters, respectively, using the same parameters as those reported in the literature [25]. V is the molar volume

$$f_{ch}(c_i(r,t),\eta) = [1-h(\eta)]G_m^{fcc}\{c_i(r,t),T\} + h(\eta)G_m^{m_7c_3}\{c_i(r,t),T\} + Wg(\eta) \tag{4}$$

where $G^{fcc}\{c_i(r,t),T\}$ and $G^{m_7c_3}\{c_i(r,t),T\}$ are the Gibbs free energies of the FCC phase and the M_7C_3 phase, respectively. The height of the potential well W is the resistance that must be overcome during the phase transformation, with a value of 5000 J·mol^{-1}.To achieve a smooth transition of the precipitated phase from the FCC phase to the M_7C_3 phase, continuous interpolation functions $h(\eta)$ and $g(\eta)$ with values ranging from [0, 1] are introduced. The function $g(\eta)$ takes on a double-well form to characterize the energy barrier during the phase transformation process [26]. $h(\eta)$ and $g(\eta)$ are expressed as:

$$h(\eta) = \eta^2(3-2\eta),\ g(\eta) = \eta^2(1-\eta)^2 \tag{5}$$

where M_{ij} is the dependent diffusion mobility [27], expressed as:

$$M_{ij} = \sum_{n=1}^{3}(\delta_{in}-c_i)(\delta_{jn}-c_j)c_n M_n \tag{6}$$

$$M_n = \frac{D_n^\eta}{RT},\ D_n^\eta = D_n^{\eta,0}\exp\left(\frac{-Q_n^{0,\eta}}{RT}\right) \tag{7}$$

where δ_{in} is the Kronecker delta function ($i = n = 1, I \neq n - 0$). M_n is the atomic mobility of component n (n = 1, 2, 3 corresponds to Ni, Cr, and C, respectively). R is the gas constant, with a value of 8.31448 J mol^{-1} K^{-1}. D_n^η is the self-diffusion coefficient of component n in the η phase. $Q_n^{0,\eta}$ is the thermal diffusion activation energy of component n in the η phase.

To provide a more detailed description of the microstructural evolution process of the system, a sublattice model was introduced based on the phase-field model in this study. In the FCC solid solution, the substitutional sublattice is composed of metal elements (Cr, Ni), while the interstitial sublattice consists of carbon atoms (C) and vacancies (V). This interstitial solid solution can be represented as $(Cr, Ni)_a(C, V)_c$. For the FCC structure, take $a = c = 1$, while in the BCC structure, take $a = 1$ and $c = 3$. The solid solution can be characterized by four pseudo-compounds: Cr_aC_c, Cr_aV_c, Ni_aC_c, and Ni_aV_c. Based on this, the molar Gibbs free energy of the four-component system can be expressed as:

$$\begin{aligned} G_m^{f'cc} &= y_4G_{2:4}^{fcc} + y_2y_3G_{2:3}^{fcc} + y_1y_4G_{1:4}^{fcc} + y_1y_3G_{1:3}^{fcc} + RT \\ &[a(y_2\ln y_2 + y_1\ln y_1) + c(y_3\ln y_3 + y_4\ln y_4)] + G_m^{ex} \end{aligned} \tag{8}$$

$$G_m^{ex} = y_2y_1\left(y_3L_{2,1:3}^{fcc} + y_4L_{2,1:4}^{fcc}\right) + y_4y_3\left(y_2L_{2:4,3}^{fcc} + y_1L_{1:4,3}^{fcc}\right) \tag{9}$$

where 1, 2, 3, and 4 are Ni, Cr, C, and V, respectively. The overall composition of the solid solution is determined by the molar fractions of each component within the substitutional and interstitial sublattices, which can be expressed as:

$$y_1 = \frac{x_1}{1 - x_3}; \; y_1 + y_2 = 1 \tag{10}$$

$$y_3 = \frac{x_3}{1 - x_3}; \; y_3 + y_4 = 1, \tag{11}$$

The M_7C_3 carbide is described by a sublattice model (Ni, Cr, V)$_a$C$_c$, where $a = 7$ and $c = 3$. By adjusting the values of a and c in the two sublattices, the stoichiometric ratio of the carbide can be determined. The molar Gibbs free energy of this carbide can be expressed as:

$$G_m^{m_7c_3} = y_1 G_{1:3}^{m_7c_3} + y_2 G_{2:3}^{m_7c_3} + aRT(y_1 \ln y_1 + y_2 \ln y_2) + y_2 y_2 L_{1,2:3}^{m_7c_3} \tag{12}$$

where $G_{2:3}^{m_7c_3}$ is the Gibbs free energy of pure binary carbide.

The Gibbs free energies for the FCC and M_7C_3 phases are available in the literature, and the thermodynamic parameters for the Ni-Cr-C system are listed in the Table 5.

Table 5. Thermodynamic parameters of Ni-Cr-C system [28,29].

Phase	Parameter, J/mol
M_7C_3 phase	$G_{2:3}^{M_7C_3} = H^{ser} - 20975 + 980.29T - 170.5T \ln T - 0.690921T^2$
	$G_{1:3}^{M_7C_3} = 193000 + 7G_1^{fcc} + 3G_3^{gra}$
	$L_{1:2;3}^{M_7C_6} = 100000$
FCC phase	$G_{2:4}^{fcc} = G_2^{fcc} + 7284 + 0.163T$
	$G_{2:3}^{fcc} = G_2^{fcc} + G_3^{gra} + 25000 G_{1:3}^{fcc} = H_1^{ser} + H_3^{ser} + 62000 - 7.6T + GHSERNI + GHSERCC$
	$G_{1:4}^{fcc} = H_1^{ser} + GHSERNI$
	$G_1^{fcc} = H^{ser} - 5179.159 + 117.854T - 22.096T \ln T - 0.00048407T^2$
	$G_2^{bcc} = H^{ser} - 8851.93 + 157.48T - 26.908T \ln T + 0.00189435T^2$ $-1.47721 \times 10^{-6} T^3 + 139250T^{-1}$
	$G_3^{gra} - H^{ser} = -17369 + 170.73T - 24.3T \ln T - 4.723 \times 10^{-4} T \ln T$ $-4.723 \times 10^{-4} T \ln T^2 + 2562600T^{-1} - 2.643 \times 10^8 T^{-2} + 1.2 \times 10^{10} T^{-3}$
	$L_{2:3,4}^{fcc} = -29686 - 18T$
	$L_{2,1:4}^{0fcc} = 8347 - 12.1038T$
	$L_{2,1:4}^{1fcc} = 29895 - 16.3838T$
	$L_{2,1:3}^{fcc} = -81265 + 81.8T$
	$L_{1:4,3}^{fcc} = -14902 + 7.5T$
Symbols	$GHSERCC = -17368.441 + 170.73T - 24.3T \ln T - 4.723 \times 10^{-4} T^2$ $+2562600T^{-1} - 2.643 \times 10^8 T^{-2} + 1.2 \times 10^{10} T^{-3}$
	$GHSERNI = -5179.159 + 117.854T - 22.096T \ln T - 4.8407 \times 10^3 T^2$

3.2. Numerical Simulation

As a numerical simulation method, the core model of the phase-field method is usually composed of complex equations of partial differential equations. Due to the computational intensity, selecting an appropriate numerical method is crucial for enhancing computational efficiency. Currently, there are three primary methods for solving PDEs: the finite difference method (FDM), spectral methods (SP), and the finite element method (FEM). Among them, the Fourier spectrum method (FSM) is one of the most commonly used spectral methods [30]. By using high-order polynomials or Fourier series for global approximation, FSM can achieve high accuracy with fewer sample points. For time discretization, the semi-implicit method avoids the instability issues of the fully explicit approach while also reducing the computational complexity compared to the fully implicit method. Based on the above analysis, the study employs a semi-implicit Fourier spectral method to solve the Cahn-Hilliard and Allen-Cahn equations with periodic boundary conditions.

The Fourier transform is applied to both sides of the Cahn-Hilliard and Allen-Cahn equations simultaneously. The partial differential equations in the spatial domain can be transformed into ordinary differential equations in the frequency domain. The spatial discretization of these expressions is represented as:

$$\frac{\partial\{c_i\}_k}{\partial t} = -k^2 M_i \left\{ \frac{\partial f_{\mathrm{ch}}}{\partial\{c_i\}_k} \right\} - k^4 M_i \kappa_c \{c_i\}_k + \{\xi_{c_i}(k,t)\}_k \tag{13}$$

$$\frac{\partial\{\eta\}_k}{\partial t} = -M_\eta \frac{\partial f_{ch}}{\partial\{\eta\}_k} - \kappa_\eta M_\eta k^2 \{\eta\}_k + \{\xi_\eta(k,t)\}_k \tag{14}$$

where $\{*\}_k$ is the Fourier transform of the phase field variable in parentheses, and k is the wave vector in Fourier space. In two dimensions, take $\sqrt{k_x^2 + k_y^2}$. A semi-implicit discretization scheme is employed, where the linear terms and the fourth-order differential operator are treated implicitly, while the nonlinear terms are handled explicitly [31]. Thus, the semi-implicit discretized form of the equation can be expressed as:

$$\frac{\{c_i\}_k^{n+1} - \{c_i\}_k^n}{\Delta t} = -k^2 M_i \left(\frac{\partial f_{ch}}{\partial c_i} \right)_k^n - k^4 M_i \kappa_c \{c_i\}_k^{n+1} \tag{15}$$

$$\frac{\{\eta\}_k^{n+1} - \{\eta\}_k^n}{\Delta t} = -M_\eta \left(\frac{\partial f_{ch}}{\partial \eta} \right)_k^n - \kappa_\eta M_\eta k^2 \{\eta\}_k^{n+1} \tag{16}$$

where $\{c_i\}_k^{n+1}$ and $\{c_i\}_k^n$ is the forward Fourier transform of the corresponding parameter when $t = n + 1$ and $t = n$, respectively, and they are expressed as:

$$\{c_i\}_k^{n+1} = \frac{\{c_i\}_k^n - \Delta t k^2 M_i \left(\frac{\partial f_{ch}}{\partial c_i} \right)_k^n}{1 + \Delta t k^4 M_i \kappa_c} \tag{17}$$

$$\{\eta\}_k^{n+1} = \frac{\{\eta\}_k^n - \Delta t M_\eta \left(\frac{\partial f_{ch}}{\partial \eta} \right)_k^n}{1 + \Delta t \kappa_\eta M_\eta k^2} \tag{18}$$

Equation (18) is solved in Fourier space, and the results need to be transformed back into real space using the inverse Fourier transform.

3.3. Result Analysis

A two-dimensional computational domain was used to perform the phase-field simulation for the Ni-Cr-C system. The spatial discretization employed a uniform grid with $Nx = Ny = 128$ and a grid spacing of $\Delta x = \Delta y = 0.5$. To ensure numerical stability and accuracy, the time step Δt was set to 0.01. To obtain the initial microstructure, the noise term was set to 0.001. All parameters used during the simulation are listed in Table 6.

The content and distribution of phases have a decisive impact on the coating's properties. Figure 3a shows the SEM characterization of the Ni-WC coating prepared by plasma-transferred arc welding, where various morphologies of M_7C_3 precipitates were observed [22,32]. The precipitates predominantly exhibit long rod-like structures, with a few appearing in plate-like forms. Figure 3b–d present the simulated results of the growth and coarsening process of M_7C_3 carbides based on the phase-field method. In these figures, the yellow regions are the M_7C_3 precipitates, while the light blue regions are the FCC phase. At the dimensionless time $t^* = 3550$ (Figure 3b), a few M_7C_3 carbide nuclei began to precipitate in the substrate, indicating the onset of M_7C_3 nucleation from the supersaturated austenite. When t^* reached 3590 (Figure 3c), the precipitated structures gradually became apparent, with neighboring nuclei beginning to merge and evolve into short rod-like carbides. As t^* increased to 3630 (Figure 3d), the short rod-like carbides continued to grow and coarsen, eventually transforming into the long rod-like and plate-like carbides observed in Figure 3a. Subsequently, the carbides continued to grow and formed typical eutectic structures. It is noteworthy that during the period $t^* = 3590\sim3630$, the growth rate of the M_7C_3 phase is significantly faster than the nucleation stage, as the growth of the M_7C_3 phase dominated this process. Figure 4a,b are local magnification images of Figure 3a,d, respectively, both of which reveal incompletely grown M_7C_3 carbide nuclei. The form of long rods is mainly adopted by the precipitated M_7C_3 phase, with some being distributed in plate-like morphologies. The simulation results are highly consistent with the experimental observations, verifying the reliability of the model. The morphological evolution mechanism of M_7C_3 carbides can be summarized. Isolated nuclei merge to form rod-like carbides. Multiple nuclei link or aggregate to generate larger carbide structures. Slender rod-like carbides in contact further grow into eutectic carbides. This mechanism is highly similar to the nucleation and growth process of M_7C_3 carbides observed by Bedolla et al. [33] based on SEM analysis.

Table 6. Phase-field simulation parameters.

Parameter	Symbol	Value
Grid number		128×128
Time interval	Δt	0.01
Size of grid	$\Delta x,\ \Delta y$	$\Delta x = \Delta y = 0.5$
Gas constant	R	8.31448 J mol^{-1} K^{-1}
Aging temperature	T	1273 K
Gradient energy coefficient [25]	$\kappa_c,\ \kappa_\eta$	$\kappa_c = 7.05 \times 10^{-9}$ J m^{-1} $\kappa_\eta = 1.4 \times 10^{-10}$ J m^{-1}
Potential height [25]	W	5×10^3 J mol^{-1}
Molar volume	V_m	7×10^{-6} m^3 mol^{-1}
Interfacial thickness	δ	1×10^{-9} m
Kinetic mobility	M_η	0.01 m s^{-1} J^{-1}

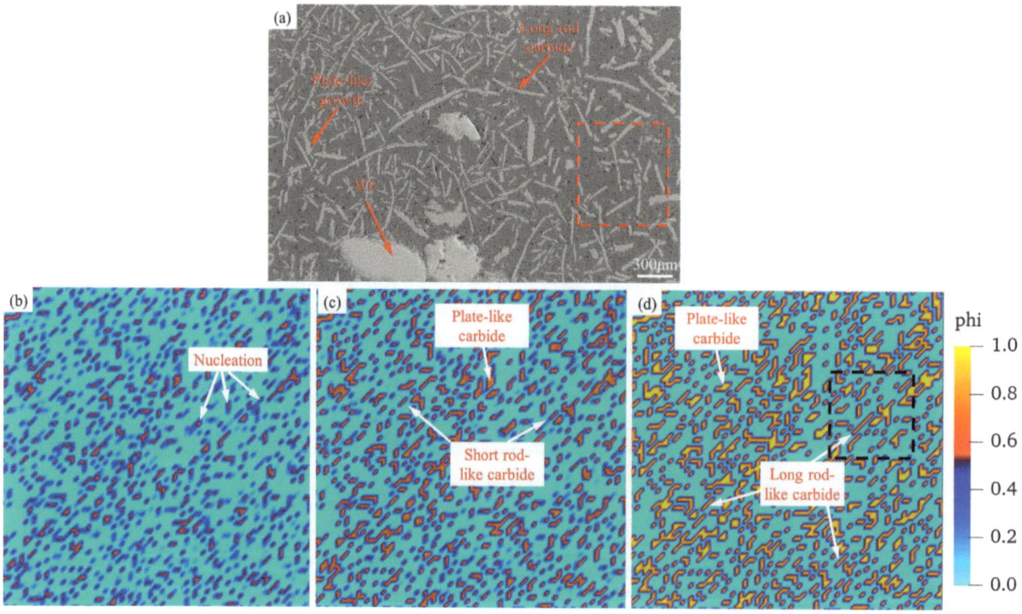

Figure 3. Phase field evolution and experimental comparison of 1273 K. (**a**) Microstructure of Ni-WC coating, (**b–d**) is the dimensionless phase field evolution time (**b**) $t^* = 3550$, (**c**) $t^* = 3590$, (**d**) $t^* = 3630$.

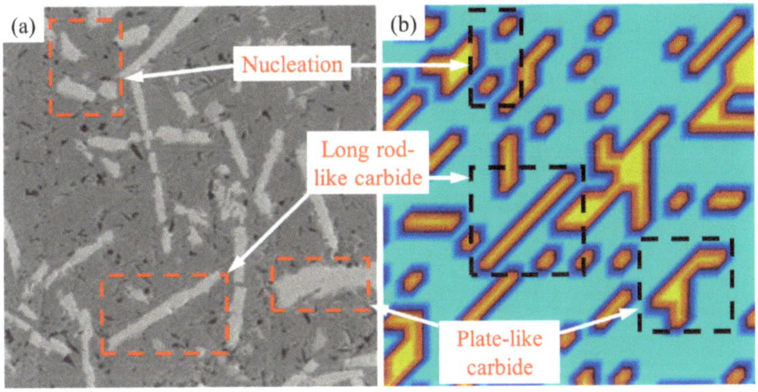

Figure 4. Local magnified image of experimental comparison and phase-field evolution. (**a**) Local magnified image of Figure 3a. (**b**) Local magnified image of Figure 3d.

The evolution of the concentration field for M_7C_3 carbides precipitated in the alloy solution, as shown in Figure 5. During the precipitation process, the spatial distribution of each element is visually represented by the color gradient. The carbide precipitates are indicated by the red regions and the substrate phase is represented by the blue regions. Figure 5a–c and Figure 5d–f show the concentration field distributions of carbon (C) and chromium (Cr) at different times, respectively. During precipitation, Cr enriches in localized regions first, forming supersaturated areas that create thermodynamic conditions favorable for the migration of C atoms. Subsequently, C atoms diffuse towards Cr-rich regions and ultimately form the chemically stable M_7C_3 phase. This reaction mechanism is primarily achieved through the interdiffusion of elements, highlighting the kinetic features of carbide formation. The evolution of the concentration field reveals that, in the initial stages, fine precipitate particles form in localized enriched regions. These particles gradually merge into larger, irregular carbide particles through the proximity of neighboring particles and the adjustment of diffusion interfaces. This process aligns with the Ostwald Ripening mechanism. Smaller particles gradually dissolve, and their material migrates to larger particles, eventually resulting in the coarsening of the carbide phase.

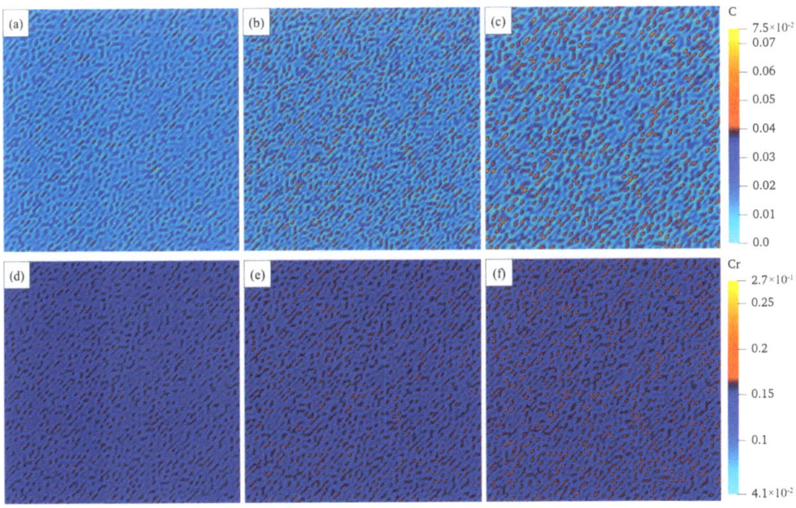

Figure 5. Evolution of concentration field of Ni-Cr-C system at 1273 K temperature. (**a**–**c**) show the evolution of C concentration field. (**d**–**f**) show the evolution of Cr concentration field with dimensionless time (**a**,**d**) $t^* = 3550$; (**b**,**e**) $t^* = 3590$; (**c**,**f**) $t^* = 3630$.

4. Conclusions

In this study, Ni-based WC coatings were prepared on descaling rolls using plasma-transferred arc welding technology. The phase composition and morphology of the coatings were comprehensively analyzed through X-ray diffraction (XRD) and scanning electron microscopy (SEM). The morphological evolution mechanism of M_7C_3 carbides was elucidated based on the phase-field method. The primary conclusions are as follows:

(1) The XRD analysis of the cladding layer indicates that the Ni-WC coatings prepared by plasma-transferred arc welding predominantly consist of phases such as γ-(Fe/Ni), M_7C_3, Fe-Ni, W_2C, FeSi, and Fe_3C. The presence of phases like M_7C3 and W_2C shows that element diffusion and precipitation reactions occurred in the molten pool.

(2) The proposed phase-field method uses thermodynamic data and can effectively reproduce the evolution of M_7C_3 carbides. It models the process from nucleus precipitation to nucleus contact, forming short rod-like carbides, and gradually developing into long rod-like and plate-like carbides. The simulation results are consistent with the experimental results obtained in this study. The validity of the established phase field model in predicting and describing the evolution of M_7C_3 hard phase in coatings is verified. Theoretical support for the design of high-performance welding overlay is provided.

Author Contributions: Conceptualization, L.F. and F.K.; methodology, H.S.; software, F.K.; validation, X.H., C.L. and F.L.; formal analysis, F.K.; investigation, F.K.; resources, C.L.; data curation, L.F.; writing—original draft preparation, F.K.; writing—review and editing, F.K.; visualization, F.K.; supervision, C.L.; project administration, C.L.; funding acquisition, C.L. All authors have read and agreed to the published version of the manuscript.

Funding: This work was supported by "the National Key Research and Development Program of China, Ministry of Science and Technology of the People's Republic of China" (2021YFB3702002), "Applied Basic Research Program of Liaoning Province" (2023JH2/101300226); "Liaoning Province Metallurgical equipment and process control significant laboratory open project fund (2024KFKT-01)"; "Supported by the Fundamental Research Funds for the Liaoning Universities (LJ222410146021)"; "Project for Graduate Education Reform and Technological Innovation and Entrepreneurship of University of Science and Technology Liaoning" (2023YJSCX02).

Institutional Review Board Statement: Not applicable.

Informed Consent Statement: Not applicable.

Data Availability Statement: The data that supports the findings of this study are available within the article.

Conflicts of Interest: The authors declare that they have no known competing financial interests or personal relationships that could have appeared to influence the work reported in this paper.

Abbreviations

List of main abbreviations

PTAW	Plasma-transferred arc welding
SEM	Scanning electron microscopy
XRD	X-ray diffraction
EDM	Electrical discharge machining
PDE	Partial differential equation
FDM	Finite difference method
SP	Spectral methods
FEM	Finite element method
FSM	Fourier spectrum method

References

1. Pohanka, M.; Votavová, H.; Resl, O.; Kotrbáček, P. The Effect of Water Jet Overlaps in a Descaler on the Quality of Surface of the Hot Rolled Steel. *Metals* **2023**, *13*, 1722. [CrossRef]
2. Gongye, F.; Zhou, J.; Peng, J.; Zhang, H.; Peng, S.; Li, S.; Deng, H. Study on the Removal of Oxide Scale Formed on 300 M Steel Special-Shaped Hot Forging Surfaces during Heating at Elevated Temperature by a High-Pressure Water Descaling Process. *Materials* **2023**, *16*, 1745. [CrossRef] [PubMed]
3. Zhou, Y.; Xing, Z.; Zhuang, Q.; Sun, J.; Chu, X. Parameter Optimization of a Surface Mechanical Rolling Treatment Process to Improve the Surface Integrity and Fatigue Property of FV520B Steel by Machine Learning. *Materials* **2024**, *17*, 4505. [CrossRef] [PubMed]
4. Li, N.; Wang, Q.; Dong, F.; Liu, X.; Han, P.; Han, Y. Research progress of coating preparation on light alloys in aviation field: A review. *Materials* **2022**, *15*, 8535. [CrossRef] [PubMed]
5. Deenadayalan, K.; Murali, V. Role of various weight percentages of WC particle on interface thickness and friction-wear property of NiCrBSi-WC composite fabricated using PTAW process. *Mater. Res. Express* **2019**, *6*, 046542. [CrossRef]
6. Qian, S.; Dai, Y.; Guo, Y.; Zhang, Y. Microstructure and wear resistance of multi-layer Ni-based alloy cladding coating on 316L SS under different laser power. *Materials* **2021**, *14*, 781. [CrossRef]
7. Devojno, O.G.; Feldshtein, E.; Kardapolava, M.A.; Lutsko, N.I. On the formation features, microstructure and microhardness of single laser tracks formed by laser cladding of a NiCrBSi self-fluxing alloy. *Opt. Lasers Eng.* **2018**, *106*, 32–38. [CrossRef]
8. Schwanekamp, T.; Marginean, G.; Reuber, M.; Ostendorf, A. Impact of cobalt content and grain growth inhibitors in laser-based powder bed fusion of WC-Co. *Int. J. Refract. Met. Hard Mater.* **2022**, *105*, 105814. [CrossRef]
9. Bartkowski, D.; Bartkowska, A.; Jurči, P. Laser cladding process of Fe/WC metal matrix composite coatings on low carbon steel using Yb: YAG disk laser. *Opt. Laser Technol.* **2021**, *136*, 106784. [CrossRef]
10. Guo, C.; Chen, J.; Zhou, J.; Zhao, J.; Wang, L.; Yu, Y.; Zhou, H. Effects of WC–Ni content on microstructure and wear resistance of laser cladding Ni-based alloys coating. *Surf. Coat. Technol.* **2012**, *206*, 2064–2071. [CrossRef]
11. Wang, T.; Sheng, G.; Liu, Z.K.; Chen, L.-Q. Coarsening kinetics of γ' precipitates in the Ni–Al–Mo system. *Acta Mater.* **2008**, *56*, 5544–5551. [CrossRef]
12. Liang, L.; Qi, Y.; Xue, F.; Bhattacharya, S.; Harris, S.J.; Chen, L.-Q. Nonlinear phase-field model for electrode-electrolyte interface evolution. *Phys. Rev. E—Stat. Nonlinear Soft Matter Phys.* **2012**, *86*, 051609. [CrossRef] [PubMed]
13. Wu, H. A review on the Cahn-Hilliard equation: Classical results and recent advances in dynamic boundary conditions. *Electron. Res. Arch.* **2021**, *30*, 2788–2832. [CrossRef]
14. Ginzburg, V.L. On the theory of superconductivity. *Il Nuovo C. (1955–1965)* **1955**, *2*, 1234–1250. [CrossRef]
15. Ginzburg, V.L.; Ginzburg, V.L.; Landau, L.D. *On the Theory of Superconductivity*; Springer: Berlin/Heidelberg, Germany, 2009.
16. Hillert, M.H. A Theory of Nucleation for Solid Metallic Solutions. Ph.D. Thesis, Massachusetts Institute of Technology, Cambridge, MA, USA, 1956.
17. Cahn, J.W.; Hilliard, J.E. Free energy of a nonuniform system. I. Interfacial free energy. *J. Chem. Phys.* **1958**, *28*, 258–267. [CrossRef]
18. Guozhi, X.; Xiaolong, S.; Dongjie, Z.; Yuping, W.; Pinghua, L. Microstructure and corrosion properties of thick WC composite coating formed by plasma cladding. *Appl. Surf. Sci.* **2010**, *256*, 6354–6358. [CrossRef]
19. Yang, G.; Huang, C.; Song, W.; Li, J.; Lu, J.; Ma, Y.; Hao, Y. Microstructure characteristics of Ni/WC composite cladding coatings. *Int. J. Miner. Metall. Mater.* **2016**, *23*, 184–192. [CrossRef]
20. Ju, J.; Zhou, Y.; Kang, M.; Wang, J. Optimization of Process Parameters, Microstructure, and Properties of Laser Cladding Fe-Based Alloy on 42CrMo Steel Roller. *Materials* **2018**, *11*, 2061. [CrossRef] [PubMed]
21. Buytoz, S.; Kok, M.; Qader, I.N.; Balci, E.; Dagdelen, F. Microstructure of NiCrBSi/WC composite coating deposited on AISI316 stainless steel by TIG coating process. *Surf. Rev. Lett.* **2021**, *28*, 2050050. [CrossRef]
22. Luo, K.; He, J.; Ma, H.; Wang, C.; Lu, J.; Li, C.; Peng, Y.; Wu, N.; Luo, F. Dissolution behaviour of WC particles and evolution of precipitated phases in the plasma transfer arc Ni-based composite coating reinforced by 30 wt% WC particles. *Ceram. Int.* **2024**, *50*, 35065–35078. [CrossRef]
23. Miranville, A. The Cahn–Hilliard equation and some of its variants. *AIMS Math.* **2017**, *2*, 479–544. [CrossRef]
24. Provatas, N.; Elder, K. *Phase-Field Methods in Materials Science and Engineering*; John Wiley & Sons: Hoboken, NJ, USA, 2011.
25. Koyama, T.; Hashimoto, K.; Onodera, H. Phase-field simulation of phase transformation in Fe-Cu-Mn-Ni quaternary alloy. *Mater. Trans.* **2006**, *47*, 2765–2772. [CrossRef]
26. Zhu, J.Z.; Wang, T.; Ardell, A.J.; Zhou, S.H.; Liu, Z.K.; Chen, L.Q. Three-dimensional phase-field simulations of coarsening kinetics of γ' particles in binary Ni-Al alloys. *Acta Mater.* **2004**, *52*, 2837–2845. [CrossRef]
27. Wang, T. An Integrated Approach for Microstructure Simulation: Application to Ni-Al-Mo Alloys. Ph.D. Thesis, Materials Science and Engineering, Penn State University, University Park, PA, USA, 2006.
28. Leonovich, B.I. Thermodynamic analysis and phase equilibria in the Ni-Cr-C system. *Steel Transl.* **2011**, *41*, 371–375. [CrossRef]
29. Lee, B.J. On the stability of Cr carbides. *Calphad* **1992**, *16*, 121–149. [CrossRef]
30. Shen, J.; Tang, T.; Wang, L.L. *Spectral Methods: Algorithms, Analysis and Applications*; Springer Science & Business Media: Berlin/Heidelberg, Germany, 2011.
31. Chen, L.Q.; Shen, J. Applications of semi-implicit Fourier-spectral method to phase field equations. *Comput. Phys. Commun.* **1998**, *108*, 147–158. [CrossRef]

32. Su, Y.H.; Wu, D.G.; Liu, G.C. Morphology and Distribution of M7C3 and TiC In Situ Synthesized in Plasma Surfacing. *Adv. Mater. Res.* **2012**, *557–559*, 1752–1755. [CrossRef]
33. Bedolla-Jacuinde, A.; Arias, L.; Hernández, B. Kinetics of secondary carbides precipitation in a high-chromium white iron. *J. Mater. Eng. Perform.* **2003**, *12*, 371–382. [CrossRef]

MDPI AG
Grosspeteranlage 5
4052 Basel
Switzerland
Tel.: +41 61 683 77 34

Coatings Editorial Office
E-mail: coatings@mdpi.com
www.mdpi.com/journal/coatings

www.ingramcontent.com/pod-product-compliance
Lightning Source LLC
Chambersburg PA
CBHW041424210926
51850CB00013B/3350
9783725834976